BOSCH

Gasoline-engine management

EDITION

Imprint

Published by:
© Robert Bosch GmbH, 1999
Postfach 30 02 20,
D-70442 Stuttgart.
Automotive Equipment Business Sector,
Product-Marketing software products (KH/PDI).

Editor-in-Chief:
Dipl.-Ing. (FH) Horst Bauer.

Editors:
Dipl.-Ing. Karl-Heinz Dietsche,
Dipl.-Ing. (BA) Jürgen Crepin,
Dipl.-Holzw. Folkhart Dinkler.

Layout:
Dipl.-Ing. (FH) Ulrich Adler,
Berthold Gauder, Leinfelden-Echterdingen.

Translation:
Peter Girling.

Technical graphics:
Bauer & Partner, Stuttgart.

Printed in Germany. Imprimé en Allemagne.
1st edition, September 1999.
SAE Society of Automotive Engineers
400 Commonwealth Drive
Warrendale, PA 15096-0001 U.S.A.

(1.0 N)

ISBN 0-7680-0510-8

Authors

Gasoline engine, engine design and operating conditions, gasoline-engine fuels
Dr. rer. nat. H. Schwarz, Dr. rer. nat. B. Blaich.

Emissions control technology, exhaust and evaporative-emissions testing
Dipl.-Ing. (FH) D. Günther, Dr.-Ing. G. König, Dipl.-Ing. E. Schnaibel, Dipl.-Ing. D. Dambach, Dipl.-Ing. (FH) W. Dieter.

A/F mixture formation, air and fuel supply, gasoline-injection systems
Dipl.-Ing. (FH) U. Steinbrenner, Dipl.-Ing. G. Felger, Ing. (grad.) L. Seebald, Dr. rer. nat. W. Huber, Dr.-Ing. W. Richter, Dipl.-Ing. M. Lembke, Dipl.-Ing. H. G. Gerngroß, Dipl.-Ing. A. Kratt. Dr.-Ing. O. Parr, Filterwerk Mann + Hummel, Ludwigsburg; Dipl.-Ing. A. Förster, Aktiengesellschaft Kühnle, Kopp und Kausch, Frankental; Dr.-Ing. H. Hiereth, DaimlerChrysler, Stuttgart.

Ignition, spark plugs
Dipl.-Ing. H. Decker, Dr. rer. nat. A. Niegel.

M-Motronic engine management
Dipl.-Ing. (FH) U. Steinbrenner, Dipl.-Ing. E. Wild, Dipl.-Ing. (FH) H. Barho, Dr.-Ing. K. Böttcher, Dipl.-Ing. (FH) V. Gandert, Dipl.-Ing. W. Gollin, Dipl.-Ing. W. Häming, Dipl.-Ing. (FH) K. Joos, Dipl.-Ing. (FH) M. Mezger, Ing. (grad.) B. Peter.

ME-Motronic engine management
Dipl.-Ing. J. Gerhardt.

MED-Motronic engine management with gasoline direct injection (prospects)
Dr. techn. W. Moser.

Unless otherwise stated, the above are all employees of Robert Bosch GmbH, Stuttgart.

Foreword

All the manuals from the Bosch "Technical Instruction" publication range dealing with gasoline-engine management technology were combined to form this reference book. It is intended to satisfy the thirst for knowledge of a large circle of readers.

In the past years, the rapid developments which took place in engine electronics and systems resulted in important, far-reaching changes in the gasoline engine's equipment and its management.

Modern engine components comply with a very wide range of different requirements, and thanks to their coordinated interaction permit:

- A reduction in fuel consumption,
- The minimization of pollutant emissions,
- The improvement of driving smoothness,
- An improvement of running refinement,
- The optimization of the trouble-free service life of all components attached to the engine.

This reference book contains the subjects which were previously dealt with in the "Automotive Electrics/Electronics" book. It provides comprehensive information ranging from the design and function of various generations of fuel-injection and ignition systems and their components up to present-day gasoline-engine management systems using the M- and ME-Motronic systems. The MED-Motronic for gasoline direct injection is also dealt with briefly.

The reader who is interested in automotive engineering technology is thus provided with a wide range of detailed and easily understood descriptions of the gasoline engine's most important control systems and components.

The editorial staff

Contents

Combustion in the gasoline engine

The spark-ignition or Otto-cycle engine

Operating concept

The spark-ignition or Otto-cycle[1]) powerplant is an internal-combustion (IC) engine that relies on an externally-generated ignition spark to transform the chemical energy contained in fuel into kinetic energy.

Today's standard spark-ignition engines employ manifold injection for mixture formation outside the combustion chamber. The mixture formation system produces an air/fuel mixture (based on gasoline or a gaseous fuel), which is then drawn into the engine by the suction generated as the pistons descend. The future will see increasing application of systems that inject the fuel directly into the combustion chamber as an alternate concept. As the piston rises, it compresses the mixture in preparation for the timed ignition process, in which externally-generated energy initiates combustion via the spark plug. The heat released in the combustion process pressurizes the cylinder, propelling the piston back down, exerting force against the crankshaft and performing work. After each combustion stroke the spent gases are expelled from the cylinder in preparation for ingestion of a fresh charge of air/fuel mixture. The primary design concept used to govern this gas transfer in powerplants for automotive applications is the four-stroke principle, with two crankshaft revolutions being required for each complete cycle.

Fig. 1

Reciprocating piston-engine design concept

OT = TDC (Top Dead Center); UT = BDC (Bottom Dead Center), V_h Swept volume, V_C Compressed volume, s Piston stroke.

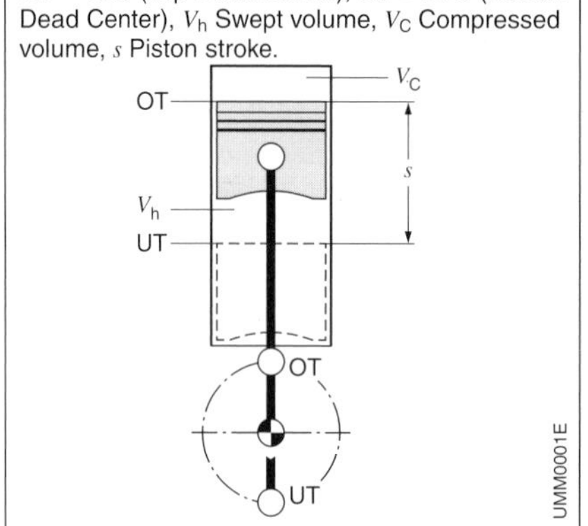

The four-stroke principle

The four-stroke engine employs flow-control valves to govern gas transfer (charge control). These valves open and close the intake and exhaust tracts leading to and from the cylinder:

1st stroke: Induction,
2nd stroke: Compression and ignition,
3rd stroke: Combustion and work,
4th stroke: Exhaust.

Induction stroke
Intake valve: open,
Exhaust valve: closed,
Piston travel: downward,
Combustion: none.

The piston's downward motion increases the cylinder's effective volume to draw fresh air/fuel mixture through the passage exposed by the open intake valve.

Compression stroke
Intake valve: closed,
Exhaust valve: closed,
Piston travel: upward,
Combustion: initial ignition phase.

[1]) After Nikolaus Augst Otto (1832–1891), who unveiled the first four-stroke gas-compression engine at the Paris World Exhibition in 1876.

Otto cycle

As the piston travels upward it reduces the cylinder's effective volume to compress the air/fuel mixture. Just before the piston reaches top dead center (TDC) the spark plug ignites the concentrated air/fuel mixture to initiate combustion.
Stroke volume V_h
and compression volume V_C
provide the basis for calculating the compression ratio
$\varepsilon = (V_h + V_C)/V_C$.
Compression ratios ε range from 7...13, depending upon specific engine design. Raising an IC engine's compression ratio increases its thermal efficiency, allowing more efficient use of the fuel. As an example, increasing the compression ratio from 6:1 to 8:1 enhances thermal efficiency by a factor of 12 %. The latitude for increasing compression ratio is restricted by knock. This term refers to uncontrolled mixture inflammation characterized by radical pressure peaks. Combustion knock leads to engine damage. Suitable fuels and favorable combustion-chamber configurations can be applied to shift the knock threshold into higher compression ranges.

Power stroke
Intake valve: closed,
Exhaust valve: closed,
Piston travel: upward,
Combustion: combustion/post-combustion phase.

The ignition spark at the spark plug ignites the compressed air/fuel mixture, thus initiating combustion and the attendant temperature rise.
This raises pressure levels within the cylinder to propel the piston downward. The piston, in turn, exerts force against the crankshaft to perform work; this process is the source of the engine's power.
Power rises as a function of engine speed and torque ($P = M \cdot \omega$).
A transmission incorporating various conversion ratios is required to adapt the combustion engine's power and torque curves to the demands of automotive operation under real-world conditions.

Exhaust stroke
Intake valve: closed,
Exhaust valve: open,
Piston travel: upward,
Combustion: none.

As the piston travels upward it forces the spent gases (exhaust) out through the passage exposed by the open exhaust valve. The entire cycle then recommences with a new intake stroke. The intake and exhaust valves are open simultaneously during part of the cycle. This overlap exploits gas-flow and resonance patterns to promote cylinder charging and scavenging.

Fig. 2
Operating cycle of the 4-stroke spark-ignition engine

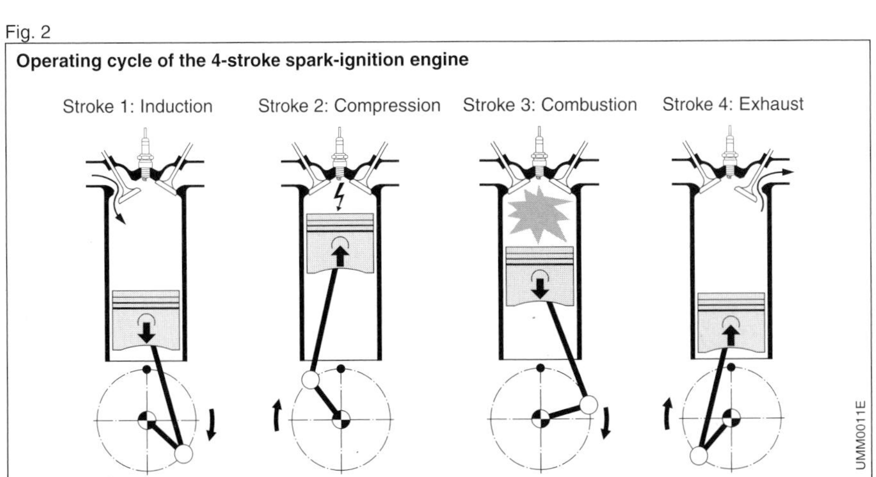

Engine design

While numerous individual design factors affect the levels of noxious emissions generated by an engine, powerplant layout must also reflect a range of other requirements. These include fuel economy, power, torque and preignition tendency along with the desire for smooth and tractable operation. As this implies, every engine design must necessarily be a compromise accomodating a variety of mutually antagonistic objectives.

Compression ratio

Although compression ratio assumes vital significance as a determinant of every engine's thermal efficiency, two factors work to prohibit blanket introduction of ultra-high compression ratios on all vehicles: higher emissions and the greater tendency toward combustion knock.

High compression ratios raise combustion-chamber temperatures. This promotes pre-ignition chemical reactions in the fuel, and can ultimately lead to portions of the air-fuel mixture self-igniting before being reached by the normal flame front. Basically, engines need fuel with higher octane ratings to counter this greater knock tendency, although suitable combustion-chamber configurations can also reduce an engine's preignition tendency to a certain degree.

Yet another factor is the rise in NO_X emissions that results from higher compression ratios and the higher combustion temperatures they produce. This extra heat within the combustion chamber shifts the overall chemical equilibrium of the combustion process toward higher concentrations of NO_X, but even more important is that it accelerates the reactive processes that foster NO_X generation. This consideration has combined with the low octane levels in available unleaded fuels to oblige manufacturers to specify lower compression ratios for countries with extremely stringent emissions laws, such as the USA and Japan, while retaining higher ratios for Europe. One result has been higher fuel-consumption figures for the low-compression versions. Research on vehicles equipped with catalytic converters designed to comply with upcoming European emissions limits is focusing on avoiding the penalties in fuel economy that tend to accompany lower compression ratios. Efforts are concentrating on new design concepts for intake manifolds and combustion chambers along with complex engine-management systems.

Combustion-chamber layout

The shape of the combustion chamber has a considerable influence on generation of unburned hydrocarbons. Because the breeding grounds for hydrocarbon emissions are the pockets and mixture layers found immediately adjacent to the chamber's walls, combustion chambers with multiplanar geometries and large surface areas will tend to produce large quantities of unburned hydrocarbons. Improvements are available from compact combustion chambers featuring limited surface areas and designed to reduce octane requirements by promoting intense charge turbulence. This strategy is suitable for combination with high compression ratios, where it facilitates implementation of lean-burn concepts. The ultimate result is high efficiency and low emissions, as defined charge turbulence immediately around the spark plug tip is important for ensuring reliable ignition of the air/fuel mixture. Low turbulence is characterized by substantial fluctuations from cycle to cycle in the conditions (mixture status, residual gas levels) predominating at the spark plug, so random local variations assume substantial significance when the ignition fires. This leads to variations in the duration of flame-front propagation and from cycle to cycle produces inconsistencies in combustion processes. Induced turbulence within the combustion chamber substantially reduces these fluctuations.

Another decisive factor for both emissions and fuel consumption is the location of the spark plug. Central locations with short flame travel provide fast and relatively complete combustion which result in lower emissions of unburned hydrocarbons (Figure 1). Flame travel can be further reduced by using two spark plugs (twin-spark concept) in each combustion chamber, with benefits for both emissions and fuel economy. Thanks to short flame travel paths, yet another advantage of a compact combustion chamber with either a central spark plug or dual plugs is a lower octane requirement. This asset, in turn, can be exploited with higher compression ratios for improved thermal efficiency.

Four-valve engines featuring two intake and two exhaust valves for each cylinder provide particularly positive effects (Figure 2). Not only does four-valve technology allow compact combustion chambers, with the accompanying short flame paths, it also provides more efficient gas flow.

Fig. 1

Effect of spark-plug position on fuel consumption and HC emissions

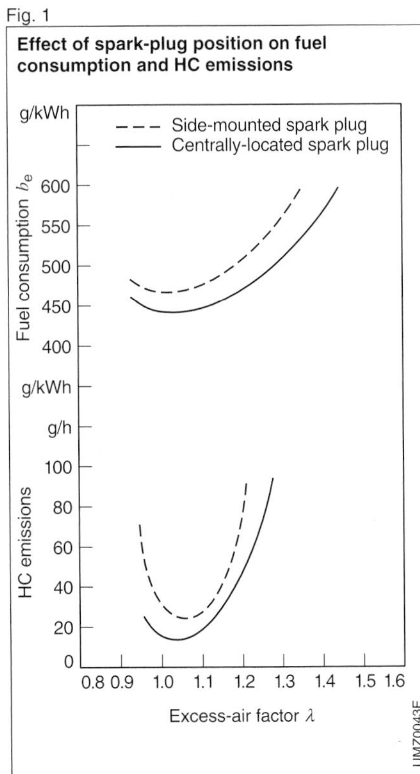

Valve timing

The gas-exchange process in which combusted gases are replaced by fresh mixture within the cylinder is controlled by the intake and exhaust valves. These valves open and close at specified intervals defined as valve timing. Valve timing combines with the valve lift (travel) as regulated by the ramps on the camshaft lobes to define gas flow. The amount of fresh gas entering the cylinder determines the engine's torque and power.

The residual gases are that portion of the combusted mixture that remains within the cylinder instead of being discharged through the open exhaust port. These gases also affect flame propagation and combustion, with collateral effects on emissions of unburned hydrocarbons and nitrous oxides as well as overall thermal efficiency. In the valve-overlap phase the intake and exhaust valves are open simultaneously. During this period fresh

Fig. 2

Reductions in fuel consumption and HC emissions with 4-valves per cylinder

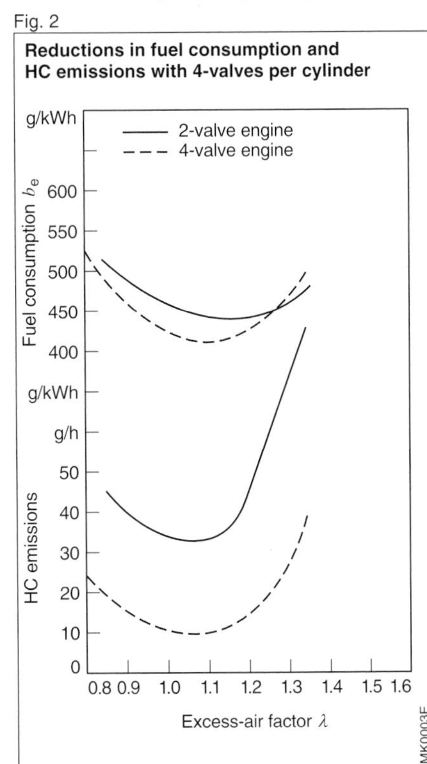

mixture may be discharged through the exhaust valve and/or exhaust gases may flow back into the intake manifold, depending on pressure patterns (Figure 3). This process has a major effect on engine efficiency and levels of unburned hydrocarbons.

Any individual set of valve-timing specifications can only be optimal at a single engine speed. To elucidate the principle: Extending the intake-valve opening period increases the output power at high speeds, but also means increased valve overlap. This overlap leads to increased emissions of unburned hydrocarbons as well as rough running (owing to the larger proportion of residual gases) at the low end of the engine's speed range and at idle. Thus the optimal solution is a variable valve-timing concept capable of adapting to changes in rpm and load factor.

One method is to shift the rotation angle of the intake camshaft on dual-cam engines for increased valve overlap. This strategy provides high top-end performance and handling, while valve overlap at the lower end of the rev range is reduced for low emissions of unburned hydrocarbons.

Intake-manifold geometries

Valve timing is not the only factor that shapes the gas-exchange process: Intake and exhaust-tract configuration are also vital. Periodic pressure waves are generated within the intake manifold during the cylinder's intake stroke. These pressure waves propagate through the intake runners and are reflected at their ends. The idea is to adapt the length and diameter of the runners to the valve timing in such a way that a pressure peak reaches the intake valve just before it closes. This supplementary pressurization effect increases the mass of fresh gas entering the cylinder (Figure 4). Similar principles apply to the exhaust tract. If the exhaust manifold and downstream system are configured to produce a positive pressure differential during valve overlap, the result will be efficient gas flow with benefits in the areas of emissions, power and fuel consumption.

Injection systems that discharge fuel directly in front of the intake valves provide for intake-manifold designs that promote highly efficient gas exchange Because such manifolds only need to distribute air (and not mixture, as for instance when used with a carburetor) their geometry can be optimized for improved fuel economy and reduced emissions.

Intake manifolds that promote swirl have a similar effect to combustion-chamber turbulence and generate a gas-flow

Fig. 3

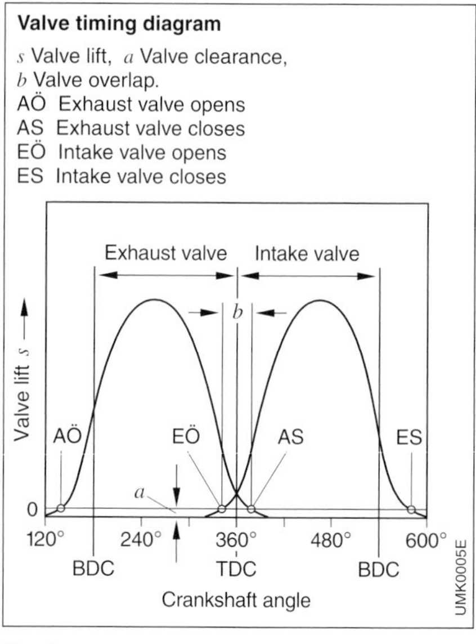

Valve timing diagram

s Valve lift, a Valve clearance,
b Valve overlap.
AÖ Exhaust valve opens
AS Exhaust valve closes
EÖ Intake valve opens
ES Intake valve closes

Fig. 4

Induction boost from manifold geometry (intake-wave ram effect)

V_h Swept volume, V_R Intake-runner volume ($V_R \geq V_h$), l Intake-runner length.

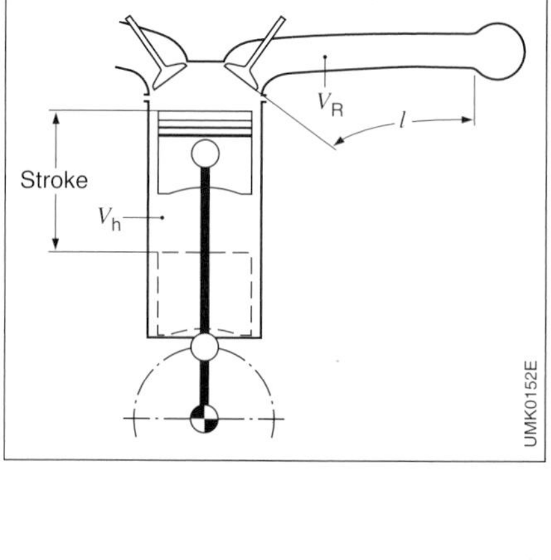

Stroke V_h V_R l UMK0152E

pattern that promotes accelerated conversion of the air-fuel mixture inside the combustion chamber. This improves thermal efficiency while also allowing leaner mixtures. This makes induced intake swirl another option for use in low-emission engine concepts.

Stratified charge

Although spark-ignition engines are generally designed to operate on a homogenous air-fuel mixture, deliberately induced charge stratification can be employed to obtain pronounced changes in the nature of the combustion process.
The object of designing a stratified-charge powerplant is to ensure reliable ignition by placing rich mixture in the immediate vicinity of the spark plug, but then allowing most of the subsequent conversion process to proceed in the lean-mixture region. A very efficient (although relatively complicated) method is to divide the combustion chamber into different sectors, and then use an auxiliary mixture-formation system to supply rich mixture to a spark plug located within a small prechamber. An advantage of this concept is that it ensures reliable ignition despite the lean mixture conditions predominating in the combustion chamber as a whole. As all combustion proceeds either in very rich or extremely lean mixtures, this strategy can be used to achieve substantial reductions in NO_X emissions. At the same time, the stratified-charge engine's greater combustion-chamber surface area leads to far higher emissions of unburned hydrocarbons than in the case of powerplants with an open chamber layout. Another stratification concept is to inject gasoline directly into the combustion chamber. In a process analogous to that used in diesel engines, this creates a rich area next to the spark plug, even though the overall mixture in the cylinder is lean. This kind of direct injection also has distinct disadvantages, such as low specific output, design complexity, etc.
It is also possible to achieve a certain degree of charge stratification by endowing the gas flowing into the combustion chamber with a carefully calculated swirl pattern. The "layer effect" is not very pronounced, and the process is also difficult to control; it fluctuates considerably in response to changes in the engine's instantaneous operating conditions.

Other engine-based strategies

Action directed toward sinking power requirements and thus fuel consumption on the periphery of the engine can also influence exhaust emissions. The possibilities include reducing friction at the pistons and in the drivetrain as well as specifying fans, alternators and other ancillaries with lower power requirements. In this case reductions in fuel consumption are directly reflected in proportional drops in toxic emissions. This is the opposite of the response pattern encountered with most of the concepts based on direct manipulation of the engine's thermodynamic balance.
Under real-world operating conditions, a large proportion of carbon monoxide and unburned-hydrocarbon emissions are generated in the engine's warm-up phase, before it has reached its normal operating temperature. The duration of this phase can be substantially reduced through suitable design of the cooling and lubrication systems. This not only improves fuel economy, but also brings rewards in the form of disproportionately high reductions in emissions of carbon monoxide and unburned hydrocarbons.

Operating conditions

Engine operating range

Engine speed

Higher engine speeds mean increased internal friction losses as well as greater power consumption from ancillary devices. This translates into reduced effective power generation for any given rate of energy supply; efficiency drops. When any specific amount of power is generated at a higher rpm, fuel consumption will be greater than when the same power is produced at a lower engine speed. Naturally, this will also raise emissions.

The effects of this engine-speed factor are reflected by all exhaust components to more or less the same degree.

Engine load factor

Changes in engine load factor affect the individual components in different ways. Higher loads are accompanied by increases in combustion-chamber temperature, reducing the depth of the quench zone adjacent to the chamber's surfaces. The higher exhaust-gas temperatures that accompany high-load operation also promote useful secondary reactions during the expansion and exhaust phases. Higher loads thus reduce emissions of unburned hydrocarbons relative to power output.

CO emissions present a similar picture, with the higher process temperatures promoting secondary reactions to produce CO_2 during the expansion (combustion) phase.

NO_X emissions display a completely opposite response pattern. The higher combustion-chamber temperatures characteristic of high load factors promote the formation of NO_X to produce a disproportionate rise in NO_X emissions as loads increase.

Vehicle speed

The extra power required to maintain higher road speeds also increases fuel consumption.

The processes described in the preceding section compensate for the effects that higher fuel consumption might otherwise be expected to have on emissions of hydrocarbons and carbon monoxide; emissions of these exhaust components thus remain relatively insensitive to vehicle speed. At the same time, the curve for NO_X emissions directly mirrors vehicle speed.

Dynamic operation

In dynamic operation a spark-ignition engine develops substantially higher emissions than during static operation. This is due to the imperfect adaptation of mixture formation during the transitional phases associated with dynamic operation. When the throttle valve opens abruptly, a portion of the fuel supplied by a throttle-body injection unit or carburetor condenses and remains inside the intake manifold. These systems rely on acceleration-enrichment strategies to compensate. With carburetors, in particular, it is not possible to control this enrichment with sufficient accuracy to ensure that all cylinders receive an optimal air-fuel mixture during transitional phases, and higher emissions of unburned hydrocarbons and carbon monoxide are the result.

On the other hand, multipoint injection systems designed to discharge fuel directly in front of the cylinders' intake valves have a distinct advantage in this area, to the extent that acceleration enrichment is usually completely unnecessary once the engine is warm. This better injection-system performance is available under all dynamic operating conditions because there is no need to alternately recharge and evacuate an extra fuel storage device – which is what the intake manifold acts as in systems with single-point injection. This pattern also affects fuel consumption: the superiority of fuel injection over carburetors in the realm of fuel economy is proportional to the how dynamically the vehicle is operated.

Air-fuel mixture

Air/fuel ratio

Because the A/F ratio is a crucial factor in defining an engine's emissions (Figure 2), the engine-management system is of immense importance in determining exhaust content.

CO emissions

In the rich range (air deficiency), CO emissions and air-fuel mixture feature virtually parallel progression curves. Within the lean range (with excess air) CO emissions stay at very low levels while remaining largely insensitive to changes in A/F ratio. In the sector around the stoichiometric $\lambda = 1$ mixture, the primary factor determining levels of CO emissions is how uniformly the fuel is distributed to the individual cylinders. A combination of some rich cylinders and some lean ones operating together will produce higher mean overall emissions than would emerge with all cylinders running at a uniform excess-air factor λ.

HC emissions

Although HC emissions mimic CO emissions by falling in response to higher excess-air factors during operation in the rich range, the hydrocarbons start to rise again in the lean range. Minimum HC emissions occur at roughly $\lambda = 1.1...1.2$. This increase in HC emissions within the lean range is caused by the deeper quench zone that arises from lower combustion-chamber temperatures. With extremely lean mixtures this effect is amplified by slower combustion, which can culminate in ignition miss with an attendant drastic rise in HC emissions. Excess-air factors in this range mark the powerplant's lean-burn limit.

NO_X emissions

The relationship between NO_X and the excess-air factor λ is the reverse of the pattern described above: in the rich range, emissions respond to higher excess-air factors and their higher oxygen concentrations by climbing. In the lean range, NO_X emissions react to increases in excess-air factor by falling, as the progressive reduction in mixture densities results in lower combustion-chamber temperatures. Maximum NO_X emissions are encountered with moderate excess air in the $\lambda = 1.05...1.1$ range.

Fig. 1
Only a thoroughly atomized injection spray can provide the homogenous mixture needed for efficient combustion and low emissions of unburned hydrocarbons

Air-fuel mixture formation

Optimal combustion sequences within the spark-ignition engine are obtained from homogenous mixtures, formed when efficient atomization produces fuel droplets that are as minute as possible (Figure 1).
Because inefficient mixture formation induces inconsistent flame-front propagation, it also leads to substantially higher emissions of unburned hydrocarbons (HC components).
Mixture formation and mixture distribution are closely related processes. Poor mixture generation of the kind encountered when carburetors operate at the top end of the load range lead to large fuel droplets being deposited in the bends inside the intake manifold. At this point the amount of fuel delivered to each individual cylinder is largely determined by purely random factors. In addition, uneven distribution has a negative influence on toxic emissions. HC and CO emissions both rise, as does fuel consumption, while power generation drops.
Injection systems designed to spray fuel into the area immediately in front of the intake valves provide extremely uniform fuel distribution. The intake manifold transports only an extremely even flow of air, while the injection system meters uniform quantities of fuel directly into all cylinders.

Fig. 2
Effect of excess-air factor λ and ignition timing α_z on exhaust emissions and fuel consumption

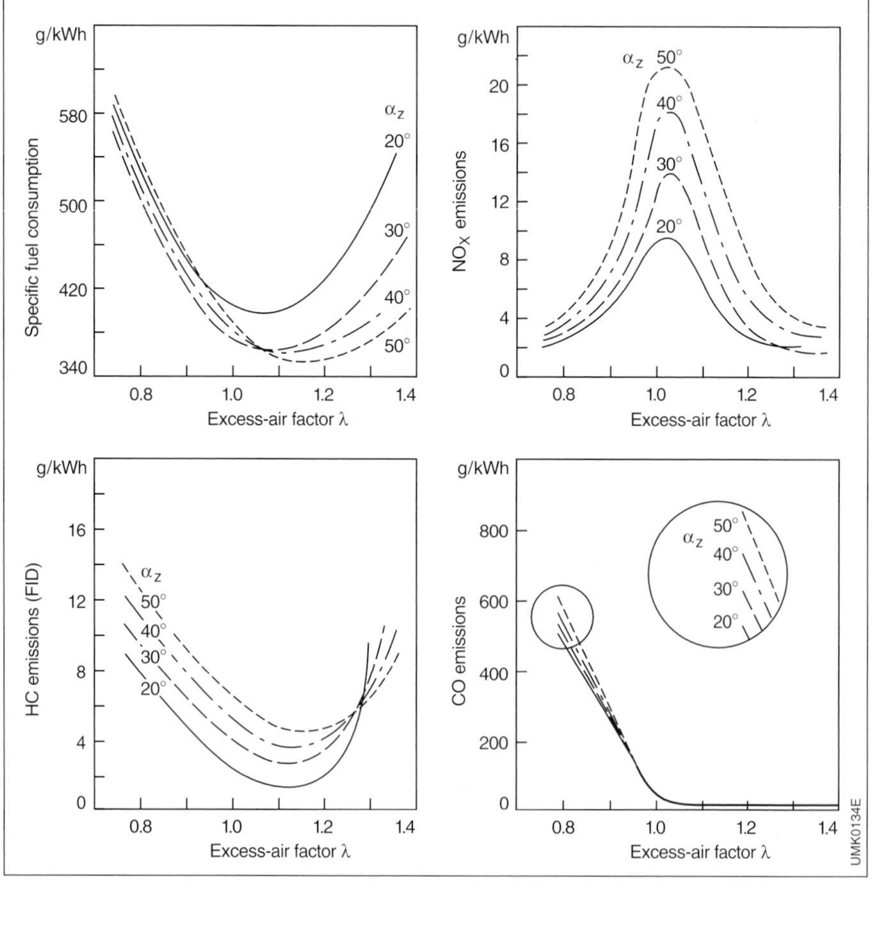

Ignition

Ignition of the air-fuel mixture – defined here as the time that elapses between arc formation and the development of a stable flame front – has a decisive influence on the combustion process. The quality of ignition is determined by the timing of the arcing process and the available ignition energy.
High excess energy provides consistent ignition with positive effects on the stability of the combustion process from cycle to cycle. Low levels of cyclic variation lead to smoother engine operation as well as reductions in emissions of unburned hydrocarbons. These facts result in the following priorities for spark plugs:

- Wide electrode gap to maximize activated gas volume,
- Exposed and unobscured arcing path to ensure optimal access to the air-fuel mixture,
- Thin electrodes and projecting spark position to minimize heat dissipation through electrodes and cylinder wall.

Under critical ignition conditions such as idle, both smoother operation and considerable reductions in HC emissions can be obtained by increasing the spark-plug electrode gap. Higher ignition energy furnishes similar benefits. Ignition systems with extended arcing durations, providing correspondingly higher levels of energy transfer to the mixture, are superior for igniting lean mixtures.
A/F ratio is joined by ignition timing among the factors with the greatest effect on emissions (Figure 2).

HC emissions
Advancing the ignition timing raises levels of unburned hydrocarbons, as the lower exhaust-gas temperatures inhibit progress of secondary reactions in the combustion and exhaust phases. This trend does eventually reverse itself, but only at extremely lean mixture ratios. Combustion in lean mixtures takes place so slowly that it is still in progress when the exhaust valve opens. As a result, if timing advance is limited to only minimal levels, the engine will reach its lean-burn limit sooner when it runs on low λ excess-air factors.

NO_X emissions
Throughout the entire A/F ratio range, NO_X emissions increase along with increasing ignition advance. This is due to the high combustion-chamber temperatures that result from advanced ignition timing; this extra heat not only shifts the combustion process' chemical equilibrium toward greater NO_X formation, but – even more significant – also greatly accelerates the rate at which this NO_X generation takes place.

CO emissions
CO emissions are essentially insensitive to changes in ignition timing, as they are almost entirely a function of the A/F ratio.

Fuel consumption
The effects of timing advance on fuel consumption are the exact opposite of its influence on emissions. If the combustion sequence is to remain optimal at higher excess-air factors λ, then ignition timing must be advanced to compensate for the slower combustion rate. Thus advanced ignition timing means lower fuel consumption and more torque.
Complex ignition-timing control mechanisms, capable of independent optimization of firing points in all engine operating ranges, are vital for homing in on the best compromise between the conflicting demands of fuel economy and exhaust emissions.

Fuels for gasoline engines

Minimum requirements for these fuels are contained in various national standards. European Standard EN 228 defines the unleaded fuel on the market in Europe (Euro-Super).
DIN 51 607 defines the German specifications for unleaded fuels; DIN 51 600 the specifications for premium leaded gasoline.

Components

Fuels for spark-ignition engines are basically hydrocarbon compounds, but can also contain oxygenous organic compounds or other additives for improved performance. The basic classes are regular and premium fuel, with the latter having enhanced knock resistance for use in high-compression engines.

Unleaded gasoline (DIN 51 607)
Unleaded gasoline is indispensable for vehicles that rely on catalytic converters for exhaust-gas treatment, as lead would damage the layers of noble metals in the converter and render it inoperative.
Unleaded fuels are a mixture composed of special high-grade, high-octane components, in which resistance to preignition can be further enhanced through the addition of nonmetallic additives. Maximum lead content is limited to 13 mg/l.

Leaded gasoline (DIN 51 600)
Environmental considerations dictate that leaded fuels be used exclusively in those engines with exhaust valves that require the combustion products of lead-alkyl compounds for lubrication. This basically applies only to a small number of older vehicles, and sales of leaded gasoline are decreasing steadily. Currently available "Super Plus" provides the same anti-knock protection as leaded gasoline. In most European countries maximum lead content is restricted to 0.15 g/l.

Specifications

Density (DIN 51 757)
European Standard EN 228 limits the fuel density range to 725...780 kg/m^3. Because premium fuels generally include a higher proportion of aromatic compounds, they are denser than regular gasoline and also have a slightly higher calorific value.

Knock protection (octane rating)
The octane rating defines resistance to preignition in fuels for spark-ignition engines. Higher octane ratings indicate a greater resistance to knock. Two different procedures are in international use for defining octane ratings; these are the Research Method and the Motor Method (DIN 51 756; ASTM D 2699 and ASTM D 2700).

RON, MON
The number determined in testing using the Research Method is the Research Octane Number, or RON. It can be considered as the essential index of acceleration knock.
The Motor Octane Number, or MON, is derived from testing according to the Motor Method. The MON basically provides an indication of the tendency to knock at high speeds. MON figures are lower than those for RON.
Octane numbers up to 100 specify the percentage by volume of iso-octane contained in a mixture with n-heptane at the point where the mixture's knock resistance in a test engine is identical to that of the fuel being tested.
Iso-octane, with its extreme resistance to knock, is assigned the RON/MON octane number 100, and n-heptane, with its low resistance to knock, the number 0.

Enhancing knock resistance
Normal (untreated) straight-run gasoline has only modest resistance to knock. Various refinery components must be added to obtain a fuel with an adequate octane rating. The highest-possible octane level must be maintained throughout the fuel's entire boiling range.

Knock inhibitors

The most effective knock inhibitors are organic lead compounds. These can raise the octane number by several points, with the exact amount depending on the specific hydrocarbon structure. Both DIN 51 600 and most European national standards limit lead content to a maximum of 150 mg per litre of fuel. Environmental considerations have combined with increasingly widespread use of catalytic converters to produce a steady reduction in the use of lead alkyls.

Volatility

Gasoline must satisfy stringent volatility requirements to ensure satisfactory engine operation. The fuel must contain a large enough proportion of highly-volatile components to ensure good cold starting, but volatility should not be so high that it causes hot-starting and handling problems (vapor lock) when the fuel is hot.

In addition, environmental considerations demand that evaporative losses be maintained at minimal levels. Volatility is defined in various ways.

Boiling curve

Three ranges on the boiling curve are significant for their effect on performance. These ranges can be defined based on fuel evaporation rates at three different temperatures.

Vapor pressure

DIN 51 600 and DIN 51 606 limit the fuel vapor pressure at 38 °C to 0.7 bar for summer gasoline and 0.9 for winter gasoline. The actual curves for vapor pressure over temperature are very sensitive to variations in the composition of the fuel.

Vapor/liquid ratio

This specification serves an an index of a fuel's tendency to form vapor bubbles. It is the volume of vapor generated by a specific quantity of fuel under a defined pressure at a set temperature.

Additives

Along with the structure of the hydrocarbons (refinery components), it is the additives that determine the ultimate quality of any given fuel. Additives are generally combined in packages containing individual components with various attributes. Extreme care and precision are vital both in testing additives and in determining their optimal concentrations. It is essential to avoid any undesirable side-effects. Both the definition of additive component levels and their physical mixing with the gasoline should be performed by the fuel manufacturer.

Anti-aging additives

These agents are added to fuels to improve their stability during storage, and are particularly important in fuels containing cracked components. They inhibit oxidation with atmospheric oxygen and prevent catalytic reactions with metal ions.

Protection against corrosion

The entrainment of water into the fuel system can lead to corrosion. This can be effectively counteracted by the use of anti-corrosive additives which form a protective layer beneath the film of water.

Intake-tract contamination inhibitors

Detergent additives ensure that the intake system (throttle valve, injectors, intake valves) remains free of contamination and deposits; this satisfies a prerequisite for trouble-free operation and minimal exhaust emissions.

Anti-icing additives

These additives are intended to prevent water vapor in the intake air from freezing on the throttle valve. Alcohols, for instance dissolve ice crystals, while other additives inhibit the formation of ice deposits on the throttle valve.

System development

Gasoline fuel-injection systems and ignition systems

The gasoline fuel injection and the ignition are what get the engine running. The gasoline is injected into the intake manifold onto the engine's intake valves. The resulting air/fuel mixture is drawn into the combustion chambers when the pistons move downwards. When they move back up again, the A/F mixture is compressed and at the ignition point ignited by a spark (thus spark-ignition engine) generated by the spark plug. The resulting combustion energy forces the piston downwards and via the conrod the piston's linear motion is converted to crankshaft rotation.
Originally, the two subsystems "gasoline injection" and "ignition" controlled the individual parameters such as injected fuel quantity and ignition point completely independently of each other. Exchange of information between the two subsystems was either totally impossible or only possible to a very limited extent. This meant that the in part mutually contradictive requirements from each of these systems could only be implemented inside the respective system, but not in "system-overlapping" form. This ceased to be a problem when Bosch combined gasoline injection and ignition in a single system.
The combination of gasoline injection and ignition in the "Motronic" engine-management system enabled the optimisation of the control parameters for injection and ignition while taking into account the various demands made on the combustion process. Fig. 1 shows the development history of the Bosch gasoline-injection and ignition systems.

Gasoline-injection systems

Gasoline injection permits fuel to be precisely metered to the engine as a function of the operating and load status, and taking into account environmental influences. The A/F mixture is controlled so that the toxic content of the exhaust gas is kept to a low level.

Multipoint fuel injection systems using the continuous-injection principle

The K-Jetronic mechanical-hydraulic gasoline injection system was installed in series-production vehicles from 1973 until 1995. K-Jetronic continually meters the fuel to the engine as a function of the intake air quantity. It was possible to extend the K-Jetronic by means of a Lambda closed-loop control in order to obtain low exhaust-gas values.
More extensive demands, which were not least of all concerned with better exhaust-gas quality, led to the K-Jetronic being expanded by the addition of an ECU, a primary-pressure regulator, and a pres-

Fig. 1

Bosch gasoline-injection systems and ignition systems: History of developments

Gasoline-injection systems:	
D-Jetronic	1967–1979
K-Jetronic	1973–1995
L-Jetronic	1973–1986
LH-Jetronic	1981–1998
KE-Jetronic	1982–1996
Mono-Jetronic	1987–1997
Ignition systems:	
Coil ignition (CI)	1934–1986
Transistorized ignition (TI)	1965–1993
Semiconductor ignition	1983–1998
Combined ignition and gasoline-injection systems:	
M-Motronic	Since 1979
KE-Motronic	1987–1996
Mono-Motronic	Since 1989

sure actuator for control of the A/F mixture composition. The KE-Jetronic was installed between 1982 and 1996.

Injection systems with intermittent fuel injection

The L-Jetronic, an electronic fuel-injection system with analog technology (1973 till 1986), intermittently injects the fuel as a function of the quantity of air drawn into the engine, the engine speed, and a number of other actuating variables. The L3 Jetronic is a system which uses digital technology. This means that it can take over additional control functions which would otherwise have been impossible with analog technology, the overall result being that the injected fuel quantity is better adapted to the engine's various operating requirements.
On the LH-Jetronic (1981–1998), instead of the air quantity drawn into the engine being measured, a hot-wire air-mass meter registers and measures the air mass. This enables correct A/F mixture formation independent of the environmental conditions.

Single-point injection system with intermittent injection

In the electronic injection system Mono-Jetronic (1987–1997) fitted in small to medium-sized automobiles, a single-injector is used which is located in the throttle body at a central point directly above the throttle valve. This system is also referred to as throttle-body injection or TBI. Engine speed and throttle-valve setting are the controlled variables for fuel metering.

Ignition systems

The ignition is responsible for igniting the compressed A/F mixture at precisely the correct ignition point, and thus initiating the mixture's combustion. In the spark-ignition (SI) engine, ignition is by an electric spark in the form of a brief arc discharge across the spark-plug electrodes.
Correctly operating ignition is an absolute must if the catalytic converter is to operate efficiently. Misfiring results in damage to the catalyst (or its destruction) due to the afterburning of the mixture which was incompletely combusted as a result of misfiring. In the course of time, electronic components gradually replaced the ignition system's mechanical parts.
The ignition point is calculated from the engine's speed and load status (by measuring the air pressure in the intake manifold). To calculate the ignition point, the conventional coil ignition (1934–1986), and the transistorized coil ignition (1965–1993) used mechanical principles, whereas the semiconductor ignition and distributorless semiconductor ignition systems (1983–1998) resort to electronic calculation via characteristic maps.

Subsystem combination

The above-named gasoline-injection systems and ignition systems are not invariably assigned to one another. A variety of different ignition systems can be combined with the different gasoline injection systems.

Motronic engine management

Motronic combines the gasoline-injection and ignition systems to form a common engine-management system. The basic fuel-injection systems together with an electronic ignition form the basis for both the ignition and fuel-injection systems in the Motronic. The KE-Motronic is based on the continuous-injection system KE-Jetronic, the Mono-Motronic on the intermittent-injection Mono-Jetronic, and the M-Motronic on the intermittent-injection manifold injection system L-Jetronic.
The integration of the electronic throttle control (ETC) in the M-Motronic resulted in the ME-Motronic. In the future MED-Jetronic system, gasoline direct injection, electronic ignition, and ETC are all combined to form a single system.

Emissions-control technology

Exhaust-gas constituents

The quality of the air we breathe is affected by a range of factors. Emissions from industry, private households and power plants all join traffic as significant sources of emissions (Figure 1).

The following basic principle holds true for all internal-combustion engines: Complete combustion within the engine's cylinders is a physical impossibility, even when more than enough oxygen is available. The levels of toxic emissions in the exhaust gas directly mirror the efficiency of the combustion process, with less-complete combustion leading to more emissions. The object is to modify the composition of the spark-ignition engine's exhaust-gases with expedients such as the catalytic converter (Figure 3).

The ultimate goal behind all strategies for reducing the entire spectrum of legally regulated pollutants is to achieve maximum fuel economy, good performance and tractability while simultaneously generating only minimal toxic emissions. In addition to a high proportion of harmless components, the spark-ignition engine's exhaust gases also contain combustion byproducts (Figure 2); which, in high concentrations, represent a potential hazard for the environment. These pollutants make up about 1% of the exhaust gas, with this 1% consisting almost entirely of carbon monoxide (CO), oxides of nitrogen (NO_X) and hydrocarbons (usually designated HC). The effect of the air-fuel mixture on relative concentrations of these substances is of particular interest: the response pattern for NO_X is exactly the opposite to that for CO and HC.

Fig. 1

Total emissions in Germany, 1996

In percent by weight.
Not including emissions from natural sources.
Total emissions: 935 Mt (metric megatons).

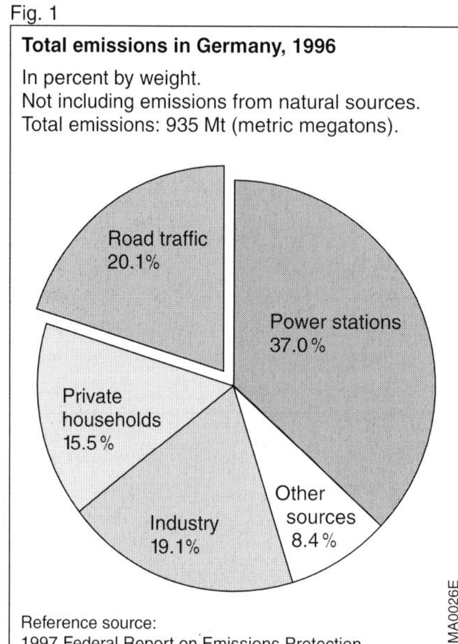

Reference source:
1997 Federal Report on Emissions Protection

Fig. 2

Components of road-traffic emissions

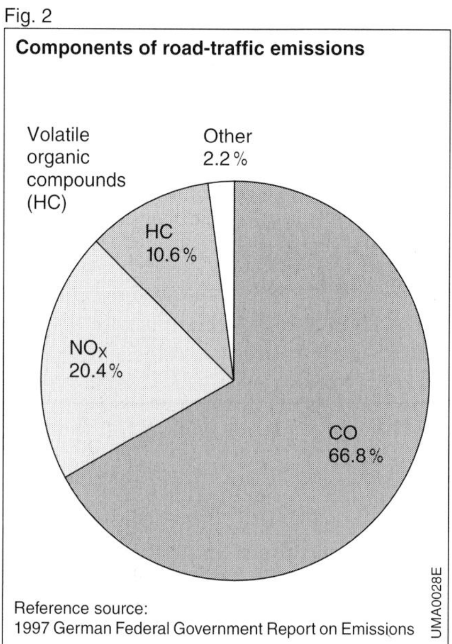

Reference source:
1997 German Federal Government Report on Emissions

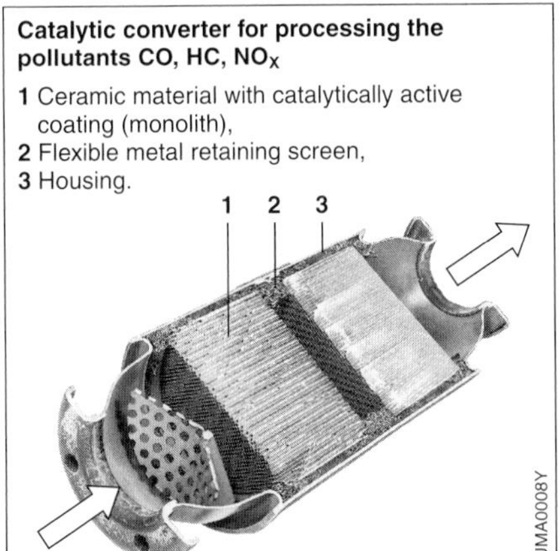

Catalytic converter for processing the pollutants CO, HC, NO_X

1 Ceramic material with catalytically active coating (monolith),
2 Flexible metal retaining screen,
3 Housing.

Fig. 3

Primary components

The primary components in exhaust gas are nitrogen (N_2), carbon dioxide (CO_2) and water vapor (H_2O). These are not toxic substances.
Nitrogen is the most abundant element in the atmosphere. Although not directly involved in the combustion process, at roughly 71 % it is the main component in exhaust gas. Small amounts of nitrogen do, though, react with oxygen to form nitrous oxides.
Complete combustion converts the hydrocarbons contained in the fuel's chemical bonds into carbon dioxide, which makes up about 14% of the exhaust gas. Reduction of CO_2 is becoming increasingly significant, as it is a suspected contributor to the "greenhouse effect." Because CO_2 is one of the products of complete combustion (which may proceed within the exhaust gas), the only way to reduce CO_2 emissions is via reductions in fuel consumption.
The hydrogen chemically bonded within the fuel burns to produce water vapor, most of which condenses as it cools. This is the vapor cloud that can be seen emerging from exhaust pipes in cool weather.

Combustion byproducts

The most important byproducts of the combustion process are carbon monoxide (CO), hydrocarbons (HC) and nitrous oxides (NO_X).

Carbon monoxide (CO) is a colorless and odorless gas produced by incomplete combustion. It can cause asphyxiation by impairing the blood's ability to absorb oxygen. This is why an engine should never be allowed to run in an enclosed area unless an exhaust-gas extraction system is in operation.

Hydrocarbons (HC) in exhaust gas stem from either hydrocarbon compounds newly formed during combustion, or from residual unburned hydrocarbons. Aliphatic hydrocarbons are odorless and have a low boiling point. Closed-chain aromatic hydrocarbons (benzol, toluol, polycyclic hydrocarbons) emit a distinct odor, and it is suspected that long-term exposure may also be carcinogenic. Partially-oxidated hydrocarbons (aldehydes, cetones, etc.) emit a disagreeable odor. They respond to sunlight by reacting to form substances which are considered to be carcinogenic in case of extended exposure to higher concentrations.

Nitrous oxides (NO_X) are the result of secondary reactions occurring in all combustion processes that use air. The main forms are the NO and NO_2 produced when oxygen combines with atmospheric nitrogen during high-temperature combustion. Colorless and odorless NO gradually converts to NO_2 in the atmosphere. Pure NO_2 is a reddish-brown gas with a pungent odor. At the levels that occur in exhaust gases and in highly polluted air, NO_2 can irritate the mucous membranes in the respiratory system

Sulfur dioxide (SO_2) is produced when sulfur contained in the fuel is combusted. A relatively small proportion of this pollutant stems from motorized traffic. The catalytic converter cannot treat the SO_2 in the exhaust gas, and its effectiveness in treating other exhaust components also suffers when SO_2 is present. As a result, efforts are being directed toward reducing levels of sulfur in gasoline and diesel fuel.

Exhaust-gas treatment

Lambda closed-loop control

Of currently available methods, lambda closed-loop control systems incorporating a catalytic converter are the most effective when it comes to cleaning the exhaust gases from spark-ignition engines. None of the available alternatives are capable of reaching anywhere near the same low emissions levels.

Currently available ignition and fuel-injection systems can achieve extremely low levels of emissions, and catalytic converters allow further reductions in the critical hydrocarbon (HC), carbon monoxide (CO) and nitrous oxide (NO_X) components within the exhaust gas.
The 3-way or selective catalytic converter performs particularly well. It is able to reduce the emissions of hydrocarbons, carbon monoxide and nitrous oxides by more than 98 %, provided that the engine operates within a very narrow scatter range (<1%) centered around the stoichiometric A/F ratio ($\lambda = 1.0$). While consistent maintenance of this restricted tolerance range is necessary under all operating conditions, not even modern injection systems can comply without asssistance. The answer is to employ the lambda closed-loop control system which relies on a closed-loop control circuit to consistently maintain the air-fuel mixture entering the engine within the optimal range known as the "catalyst window" (Figure 1).

Implementing this concept entails monitoring exhaust-gas composition as the basis for making instantaneous corrections in the mixture's fuel content. The monitoring device is the oxygen or lambda sensor. Because this probe displays a voltage jump when the mixture is precisely stoichiometric ($\lambda = 1$), the signal it generates indicates whether the mixture is richer or leaner than $\lambda = 1$.

Lambda oxygen sensor

The oxygen sensor is installed in the exhaust tract, where it monitors the flow of exhaust gases from all cylinders. Conceptually, it is a galvanic oxygen concentration cell with a solid-state electrolyte.

Design

The solid-state electrolyte is an impermeable zirconium dioxide ceramic unit stabilized with yttrium oxide, open on one side and closed on the other. Gas-permeable platinum electrodes are mounted on both the inner and outside surfaces.
The outside platinum electrode acts as a miniature catalyst to support reactions in the incoming exhaust gases and bring them into a state of stoichiometric balance. The side exposed to the exhaust gases also has a porous ceramic layer (Spinell coating) to protect it against contamination. A metal tube with numerous slots guards the ceramic body against im-

[1]) The stoichiometric air-fuel ratio is the mass ratio of 14.7 kg air to 1 kg gasoline theoretically necessary for complete combustion. The excess-air factor or air ratio λ (lambda) indicates the deviation of the actual air-fuel ratio from the theoretically required ratio:

$$\lambda = \frac{\text{actual inducted air mass}}{\text{theoretical air requirement}}$$

Fig. 1

Lambda oxygen sensor: Control range and reductions in harmful exhaust emissions

1 Without catalytic exhaust treatment,
2 With catalytic exhaust treatment,
3 Lambda oxygen sensor voltage curve.

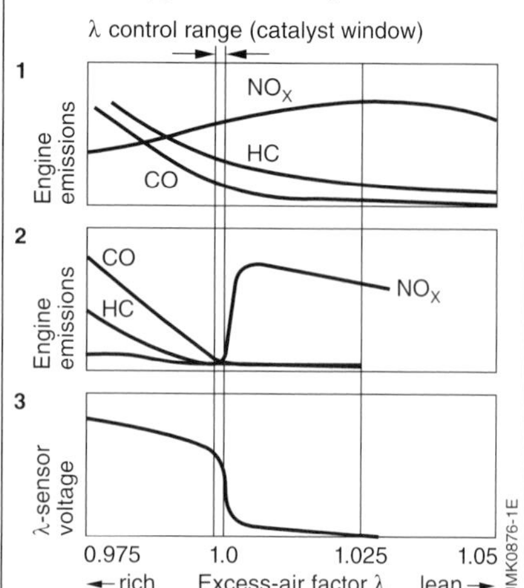

pacts and thermal shocks. The inner cavity is open to the atmosphere, which serves as the unit's reference gas (Figure 2).

Operating concept (two-state sensor)

Two-state sensor operation is based on the Nernst Principle. The sensor's ceramic material conducts oxygen ions at temperatures of roughly 350 °C and above. Disparities in oxygen levels on the respective sides of the sensor will result in generation of electrical voltage between the two surfaces, and it is this voltage that serves as the index of how much oxygen levels vary on the two sides of the sensor. The amount of residual oxygen contained within an internal combustion engine's exhaust gas fluctuates sharply in response to variations in the induction mixture's A/F ratio. Even operation with excess fuel in the mixture produces exhaust gases with residual oxygen. To cite an example, volumetric concentrations of oxygen in the exhaust remain as high as 0.2...0.3 even when the engine is operated at $\lambda = 0.95$. Because the exhaust gas' oxygen content varies to reflect differences in A/F ratio, the former can be used to monitor the latter. Oxygen-sensitive voltage generation in the oxygen sensor ranges from 800...1000 mV for rich mixtures ($\lambda < 1$) to levels as low as approximately 100 mV for lean mixtures ($\lambda > 1$). The transition from rich to lean corresponds to 450...500 mV.

Fig. 2

Lambda oxygen sensor: Position in exhaust pipe (schematic)

1 Ceramic coating, **2** Electrodes, **3** Contacts, **4** Housing contacts, **5** Exhaust pipe, **6** Ceramic support shield (porous), **7** Exhaust gas, **8** Ambient air, *U* Voltage.

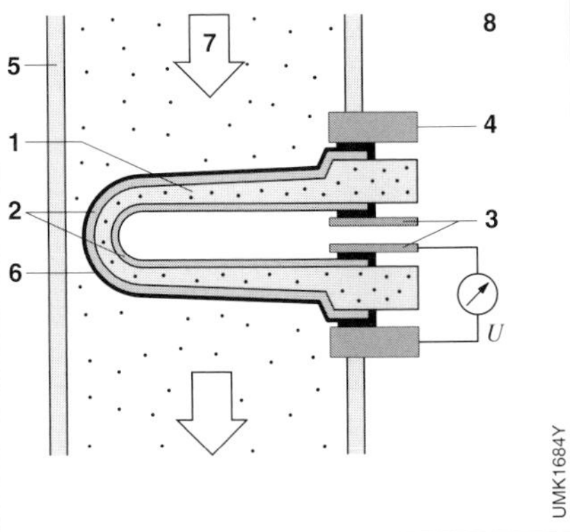

The temperature of the ceramic body joins the oxygen content of the exhaust gas as a decisive factor, as conductivity for oxygen ions varies according to how hot the ceramic is. As a result, the curve for sensor voltage over excess-air factor λ (static characteristic curve) is extremely sensitive to temperature variations, and the basic operating data are understood as applicable at a working temperature of roughly 600 °C. Yet another parameter characterized by extreme sensitivity to thermal variations is the response time that elapses before mixture changes are registered as voltage shifts. Although lag is measured in seconds with the sensor cooler than 350 °C, the sensor responds within less than 50 ms once it has heated to its normal operating temperature of 600 °C. This is reflected in the control pattern employed immediately after engine starts; the lambda closed-loop control remains inactive and the engine relies on open-loop mixture regulation until the sensor heats to its minimum operating temperature of approximately 350 °C.

Installation location

Excessive temperatures reduce service life. Thus installation positions where the sensor would be exposed to temperatures in excess of 850 °C during extended WOT operation are ruled out, although brief temperature peaks extending to 930 °C may be tolerated.

Unheated Lambda oxygen sensor

A ceramic support tube and a spring washer retain and seal the active finger-type ceramic elements within the sensor casing (structure similar to that of the heated oxygen sensor in Figure 3, but without heater element). A contact element extending between the support tube and the active ceramic elements furnishes electrical continuity between the interior electrode and the outside connection wire.

The metallic seal connects the external electrode with the sensor casing. A metallic shield that simultaneously serves as the support surface for the spring washer retains the sensor's internal components while also protecting the inside of the unit against contamination. For connection to the outside, the connecting wire is crimped to the sensor's contact element and protected against moisture and physical damage by a heat-resistant cap. A guard tube featuring a special geometry is installed in the exhaust-side of the casing to protect the ceramic element against combustion residue; this tube includes a number of slots specially arranged to protect the sensor against thermal and chemical stress factors.

Heated Lambda oxygen sensor

This sensor relies on an electric heater element to warm the ceramic material when the engine is operating at low load factors (and thus producing low-temperature exhaust gas). At higher load factors the sensor's temperature is determined by the exhaust gas. Because heated O_2 sensors (Figure 3) function even when mounted at substantial distances from the engine, extended WOT operation is not a problem. At the same time, the internal heater quickly warms the unit to its working temperature. This allows the sensor – and thus the closed-loop mixture control system – to assume operation within 20...30 seconds after the engine has started. The heated O_2 sensor helps ensure low and stable emissions thanks to consistent maintenance of optimal operating temperatures.

Planar Lambda oxygen sensor

In its basic operating concept the planar sensor (Figure 4) corresponds to the heated finger-type element sensor by generating a response curve with a characteristic jump at $\lambda = 1$. At the design level the planar sensor is distinguished from the finger-type unit by the following features:

- The solid-body electrolyte consists of ceramic layers,
- A solid ceramic sealant retains the sensor element within the sensor casing,
- A dual-wall guard tube effectively protects the sensor element against excessive thermal and physical stresses.

The individual active layers (Figure 5) are manufactured using silk-screening techniques. Stacking laminated layers with various configurations makes it possible to integrate a heater within the sensor element.

Wide-band Lambda oxygen sensor

The wide-band sensor expands on the principle of the Nernst unit (two-state sensor function) by incorporating a second chamber, the pump cell. It is through a small slot in this pump cell that the exhaust gas enters the actual monitoring chamber (diffusion gap) in the Nernst cell. Figure 6 provides a schematic of this sensor's design. This configuration contrasts with the layout used in the two-state sensor by

Fig. 3

Heated O_2 sensor

1 Sensor housing, **2** Ceramic support tube, **3** Connection wire, **4** Guard tube with slots, **5** Active ceramic sensor layer, **6** Contact, **7** Protective cap, **8** Heater element, **9** Crimped connections for heater element, **10** Spring washer.

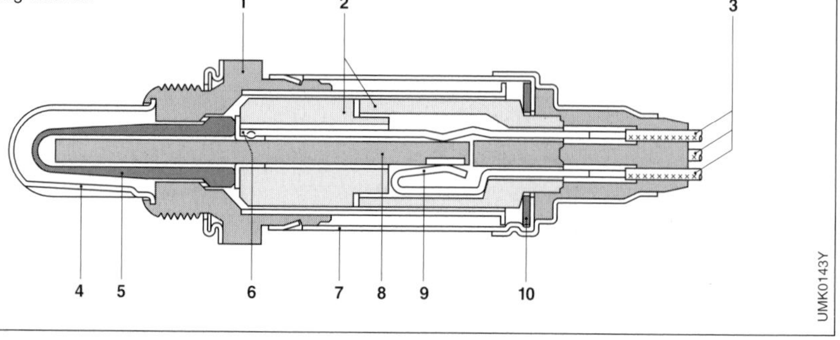

maintaining a consistently stoichiometric A/F ratio in the chamber. Electronic circuitry modulates the voltage supply to maintain the composition of the gas in the monitoring chamber at a consistent $\lambda = 1$. The pump cell responds to lean exhaust by discharging oxygen from the diffusion gap to the outside, but reacts to rich exhaust gas by pumping oxygen from the surrounding exhaust gas into the diffusion gap, reversing the direction of the current. Because the pumping current is also proportional to the oxygen concentration and/or oxygen deficiency, it serves as an index of the excess-air factor of the exhaust gas. An integral heater unit ensures an operating temperature of at least 600 °C.

While the two-state unit uses the voltage at the Nernst cell as a direct measurement signal, the wide-band sensor employs special processing and control circuitry to set the pumping current, which is then monitored and measured as an index of the exhaust gas' excess-air factor. Because sensor operation is no longer dependent on the step-function response of the Nernst cell, air factors ranging from 0.7 to 4 can be monitored as a continuous progression, and lambda control of the engine can proceed based on a reference spectrum, instead of depending solely upon a single point.

Operation of lambda closed-loop control

The oxygen sensor relays a voltage signal to the electronic engine-management

Operational layers in planar O_2 sensor

1 Porous protective layer, **2** External electrode, **3** Sensor laminate, **4** Internal electrode, **5** Reference air laminate, **6** Insulation layer, **7** Heater, **8** Heater laminate, **9** Connection contacts.

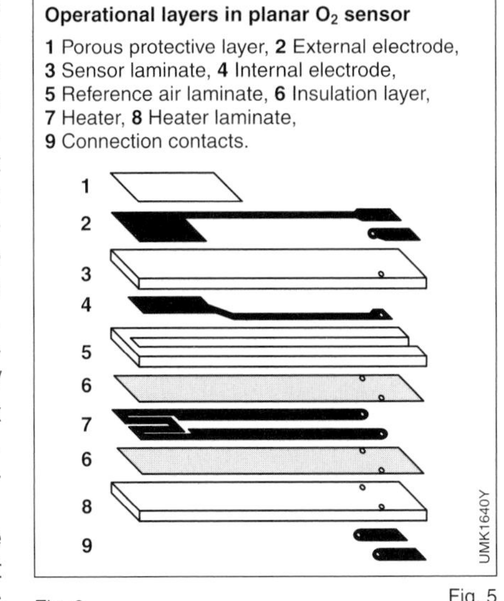

Fig. 5

Fig. 6

Principal design of a continuous-action wide-band Lamba oxygen sensor showing the sensor's installation in the exhaust pipe

1 Nernst cell, **2** Reference cell, **3** Heater, **4** Diffusion gap, **5** Pump cell, **6** Exhaust pipe.

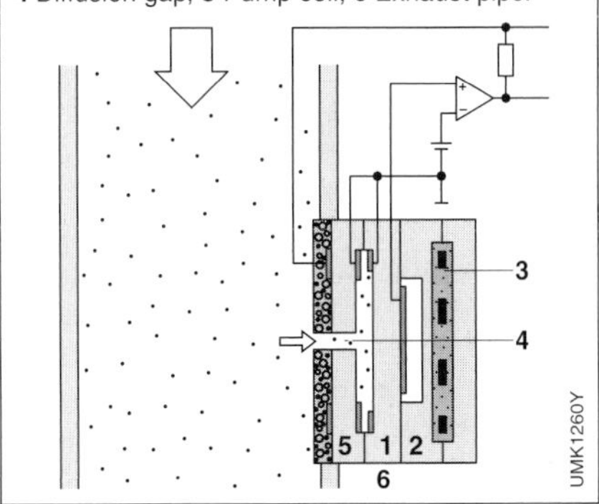

Fig. 4

Planar O_2 sensor

1 Guard tube, **2** Ceramic seal assembly, **3** Sensor housing, **4** Ceramic support tube, **5** Planar sensor element, **6** Protective cap, **7** Connection wire.

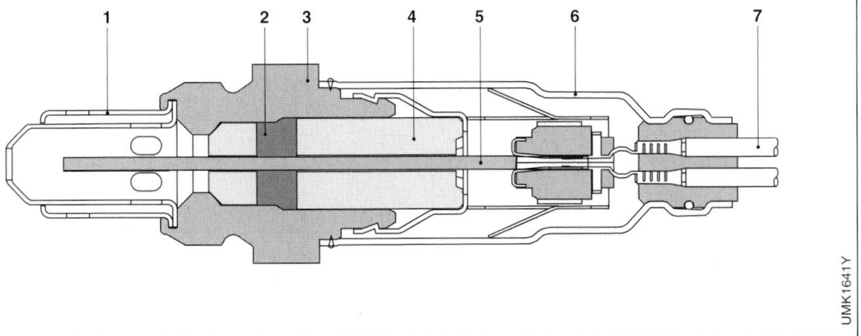

1 2 3 4 5 6 7

UMK1641Y

unit, which then issues a command to the mixture-formation unit (injection system or electronically-controlled carburetor) to enrichen or lean out the mixture, as indicated by the oxygen sensor's signal voltage (Figure 7). The system thus counters lean mixtures by increasing the injected fuel quantity and rich mixtures by reducing it.

Two-state control

The engine-management ECU in a two-state control system converts the signal from the oxygen sensor into a two-state signal.

Each jump in the oxygen sensor's voltage provokes a reaction by shifting the lambda closed-loop control parameters in the opposite direction (Figure 8), with the system responding to lean readings with enrichment and vice versa. Typical spikes in the control parameter's step function are in the 3 % range. This means that instantaneous fuel-discharge quantities are multiplied by factor:

- 1.00 under standard conditions,
- 1.03 with lean mixtures, and
- 0.97 with rich mixtures.

Following the jump in the control parameter the "control factor" undergoes a ramp conversion to return the system to operation at a mean value and compensate for interference factors in the pilot control. The control frequency is basically defined by the time that elapses between formation of fresh mixture and registration of the resulting exhaust gas at the lambda oxygen sensor (transport delay, response lag).

This transport delay is the period that always elapses before the oxygen sensor can react to rich or lean induction mixtures with a corresponding voltage shift. This voltage shift, in turn, is the precondition for all mixture adjustments. Yet another transport-delay period must then elapse before the new mixture arrives at the oxygen sensor. Thus the minimum time for one control cycle corresponds to no less than twice the transport delay. Because this transport delay is extremely sensitive to variations in the engine's speed and load factor, ramp rates for implementation in response to jumps in the control parameter also vary to compensate for these two factors and maintain an essentially consistent control oscillation.

Up to now the assumption has been that the system always responds to jumps in the O_2 sensor's voltage by dialing in optimal exhaust-gas compositions. However, the intensity of the voltage jump varies according to the composition and temperature of the gas, so the voltage change displays a slight stoichiometric offset. A controlled rich and lean offset is thus employed to compensate for all of the factors that can distort the sensor's response curve. This strategy maintains the control parameter for a regulated dwell period t_v even in the face of a new sensor jump. This dwell period is stored in a program map with definitions for various engine speeds and load factors.

Fig. 7

Schematic diagram of lambda closed-loop mixture control

1 Mass airflow sensor, **2** Engine, **3a** Oxygen sensor 1, **3b** Oxygen sensor 2 (only if required), **4** Catalytic converter, **5** Injectors, **6** ECU.
U_S Sensor voltage, U_V Valve control voltage, V_E Injection quantity.

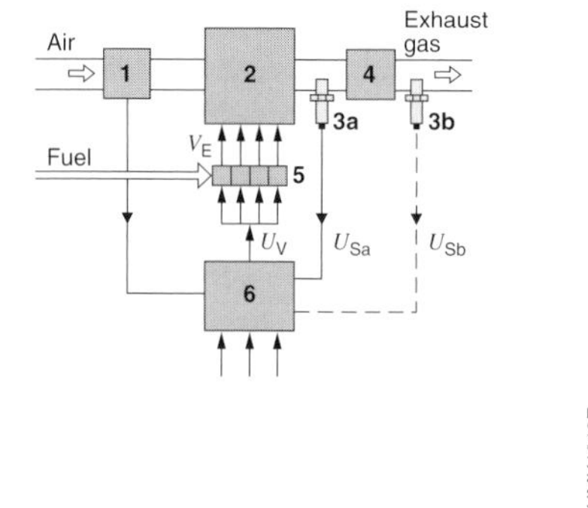

Dual-sensor control

In systems designed to comply with the most stringent emissions regulations, the primary oxygen sensor (the "cat-forward" unit located on the catalyst's engine side) is supplemented by a second sensor located behind the converter. Because it monitors the exhaust gas after it has passed through the catalyst and attained stoichiometric balance, it helps provide

more precise operation, and is suitable for use in correcting the closed-loop control data provided by the cat-forward sensor.
Lag arising from gas transit periods renders it impossible to implement lambda control strategies based exclusively on a cat-back sensor. Instead, the cat-back catalytic converter manipulates dwell period t_V in order to slowly correct the cat-forward catalytic converter.
The cat-back unit also can be used to compensate for shifts in the characteristic response curve of the cat-forward oxygen sensor. Systems featuring two oxygen sensors are therefore characterised by substantially improved long-term emission-control stability.

Fig. 8
Control parameter curves with regulated lambda shift (dual-threshold control)
t_V Residence time after sensor jump.

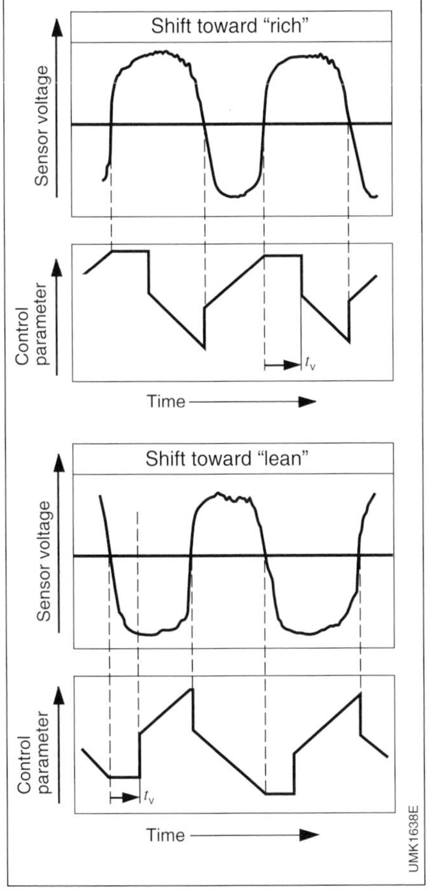

Continuous lambda control
The planar wide-band oxygen sensor is an advanced version of the sensors described above. It consists of a combination of two cells incorporating special electronic control circuitry.
While the two-state sensor can only indicate two states – rich and lean – with a corresponding voltage jump, the wide-band sensor monitors deviations from $\lambda = 1$ by transmitting a continuous signal. This wide-band sensor thus makes it possible to implement lambda control strategies based on continuous instead of two-state information.
The advantages are:
- A substantial improvement in dynamic response, with quantified data on deviations from the specified gas composition, and
- The option of adjusting to any desired mixture strength, i.e., including air factors other than $\lambda = 1$.

The second option is especially significant for strategies seeking to exploit the fuel-saving potential of lean operation (lean-burn concepts). It will be noted that this entails using catalysts capable of converting the nitrous oxides in the exhaust gas during lean operation.

Catalytic exhaust-gas treatment

Catalytic-converter systems
Four different kinds of catalytic-converter system are available to suit different emissions concepts and applications.

Oxidation catalytic converter
The oxidation (or single-bed) catalytic converter operates with excess air, and employs oxidation, i.e. combustion, to convert hydrocarbons and carbon monoxide into water vapor and carbon dioxide. Oxidation catalytic converters provide virtually no reduction in nitrous oxides. With fuel-injection engines, the oxygen required for oxidation is usually obtained from lean induction mixtures ($\lambda > 1$). Carburetor engines rely on engine-driven centrifugal pumps or self-priming air

valves to inject "secondary air" into the exhaust stream before it reaches the catalytic converter (Figure 9a).
Oxidation catalytic converters were originally introduced in 1975 to comply with then-current US emissions regulations, but are now virtually extinct.

Dual-bed catalytic converter

The dual-bed catalytic converter consists of two catalytic elements installed in series (hence the name "dual-bed"). This strategy works only when the engine is operated on a rich mixture ($\lambda < 1$), viz., with air deficiency. The exhaust gas flows through a reduction catalytic converter before proceeding through an oxidation catalytic converter, with air being injected between these two elements. The first catalyst converts nitrous oxides, while the second transforms hydrocarbons and carbon monoxide. Because it depends on rich induction mixtures to work, the dual-bed concept is the least attractive strategy from the fuel-economy standpoint. An advantage is that it is suitable for use in combination with simple mixture-formation systems, without electronic control. A further disadvantage is that ammonia (NH_3) is produced during reduction of nitrous oxides in lean mixtures. A portion of this ammonia then reoxidizes back into nitrous oxides during subsequent air injection.
With this design, conversion of NO_X is significantly less effective than with a single-bed 3-way catalytic converter operating with lambda closed-loop control.
The dual-bed catalytic converter was once popular among US vehicle manufacturers, but is now rare. In the US dual-bed concepts were often used together with lambda mixture control, but this strategy is not only very complex, it also suffers from the problems with nitrous oxide emissions described above (Figure 9b).

3-way catalytic converter

The prime asset of the three-way (or single-bed) catalytic converter is its ability to remove large proportions of all three pollutants (thus 3-way).
The primary requirement is that the engine's induction mixture – and with it the exhaust gas – maintain a consistently stoichiometric A/F ratio (refer to section on "Lambda closed-loop control"). The 3-way catalytic converter combines with lambda closed-loop control to form the most effective pollutant-reduction system currently available. This is why it is used to comply with the most stringent emissions limits (Figure 9c).
Three-way catalytic converters for after-market installation are also available in kit form. Although these kits obviously cannot achieve the high levels of conversion achieved using lambda closed-loop control, they are able to reduce pollutants by roughly 50 %.

NO_X storage catalytic converter

Exhaust gases from engines that operate on air/fuel mixtures with only limited oxygen ($\lambda > 1$ with lean-burn concepts, direct-injection systems on part-throttle, etc.) display substantially higher concentrations of NO_X than the exhaust generated by conventional powerplants.
The NO_X storage catalyst displays the greatest potential for reducing concentrations of NO_X in the exhaust gas. It uses the oxygen available in lean exhaust gases to store nitrous oxides as nitrates on its active surfaces. However, the storage catalyst must be regenerated when its capacity is exhausted. The regeneration strategy entails a temporary switch to engine operation on a homogenous rich mixture to promote reduction of nitrates to nitrogen in a process largely supported by CO. The engine-management ECU relies on stored data describing the converter's absorption and desorption properties as the basis for regulating the storage and regeneration phases. oxygen sensors located in both cat-forward and cat-back positions monitor the emissions values.
The cycles in which the engine operates on the homogenous rich mixture last only a few seconds. A vital consideration is transition tuning, to avoid undesirable changes in engine response in the form of sudden torque jumps.

Substrate systems

The catalytic converter (or, more precisely, the "catalytic exhaust-gas converter") consists of a metal housing, a substrate and the actual active catalytic layer (the catalyst).

There are three different substrate systems:

- Pellets (obsolete),
- Ceramic monoliths, and
- Metallic monoliths.

Ceramic monoliths

These are ceramic bodies perforated by several thousands of channels serving as exhaust channels. The ceramic material is a magnesium-aluminum silicate designed to withstand extreme heat. The monolith, which is extremely sensitive to mechanical tension, is mounted within a metal housing. Between the housing walls and the substrate is a flexible metal screen made of high-alloy steel wire featuring a diameter of approximately 0.25 mm. The screen must be flexible enough to compensate for such factors as production tolerances, the different expansion coefficients of housing and substrate material, mechanical stresses associated with vehicle operation, and the gas forces exerted against the ceramic body (Figure 10). Ceramic mono-

Fig. 9

Catalytic systems

a Single-bed oxidation catalytic converter, **b** Dual-bed catalytic converter, **c** Single-bed catalytic converter.
1 Mixture formation/injection system, **2** Secondary-air injection, **3** Oxidation catalytic converter for HC, CO, **4** NO_X reduction catalytic converter, **5** Electronic control unit, **6** Lambda O_2 sensor, **7** 3-way catalytic converter for NO_X, HC, CO. U_S Sensor voltage, U_V valve control voltage.

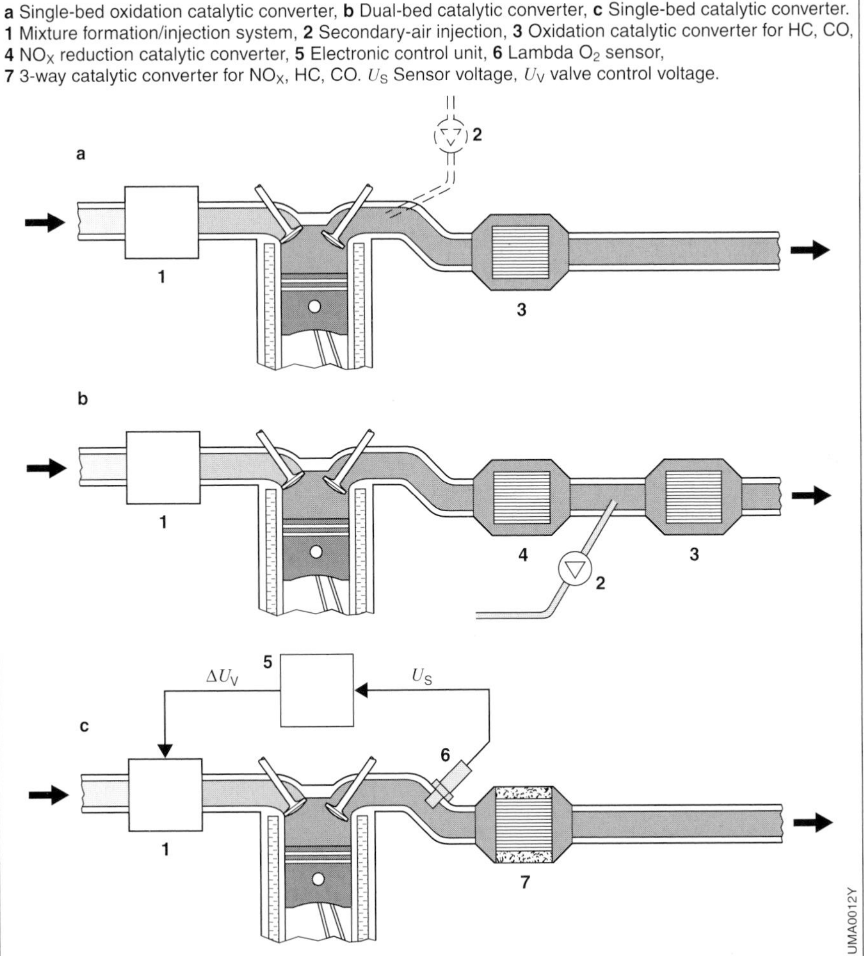

liths are currently the most frequent substrate concept for catalytic converters. This configuration is used by all European manufacturers and has largely superseded the earlier pellet technology in the US and Japan.

Metallic monoliths

Metallic monoliths have seen only limited use up to now. They are primarily installed as pre-catalysts (start-up catalysts) mounted in the immediate vicinity of the engine, where they provide more rapid catalytic conversion following cold starts. The major impediment to application as primary catalysts is their high cost compared to ceramic monoliths.

Coatings

While the catalytically-active substances can be applied directly to pellets, ceramic and metallic monoliths are dependent on a “wash coat” of aluminum oxide. This substrate coating increases the effective surface area of the catalytic converter by a factor of roughly 7,000. In oxidation-type catalytic converters the actual catalytic layer applied to the wash coat consists of the noble metals platinum and palladium, while platinum and rhodium are used in 3-way catalytic converters (Figure 10). Platinum accelerates the oxidation of hydrocarbons and carbon monoxide, while rhodium promotes reduction of nitrous oxides. The noble metals in each catalytic converter usually amount to between 2 and 3 grams.

Operating conditions

In the catalytic converter, as with the O_2 sensor, operating temperature is a vital factor. The unit must warm beyond roughly 250 °C before it can assume genuinely effective pollutant conversion, while ideal conditions for high conversion rates and long service life prevail in a temperature range of approximately 400...800 °C. Heat in the 800 °C to 1,000 °C temperature range sinters the noble metals to the Al_2O_3 substrate layer, reducing the effective catalytic surface and promoting thermal aging. The amount of time spent operating in this range is thus a consideration of vast importance. The radical rise in thermal fatigue that sets in above 1,000 °C leads to rapid degeneration in the catalytic converter, which soon becomes virtually useless. These thermal considerations effectively limit the range of potential installation positions, and the ultimate choice must necessarily assume the form of a compromise. Coatings with improved thermal stability (with the critical threshold being raised to approximately 950 °C) are expected to ease the situation in the future. When operated under favorable conditions a catalytic converter

Fig. 10

3-way catalytic converter with O_2 sensor

1 Lambda O_2 sensor, **2** Ceramic monolith, **3** Flexible metal screen, **4** Heat-insulated dual shell, **5** Platinum, rhodium coating, **6** Ceramic or metallic substrate.

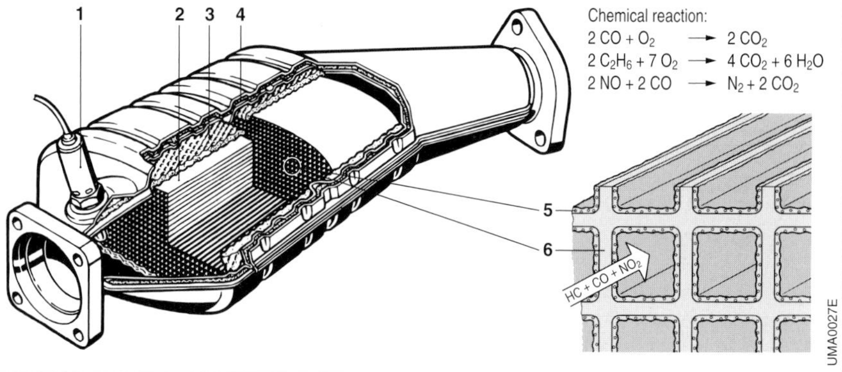

Chemical reaction:
$2\,CO + O_2 \rightarrow 2\,CO_2$
$2\,C_2H_6 + 7\,O_2 \rightarrow 4\,CO_2 + 6\,H_2O$
$2\,NO + 2\,CO \rightarrow N_2 + 2\,CO_2$

$HC + CO + NO_2$

UMA0027E

can last for up to 100,000 kilometers (62,000 miles). On the other hand, engine malfunctions such as misfiring can raise catalyst temperatures beyond 1,400 °C, melting the substrate materials and leading to the converter's complete destruction. A major element in preventing these kinds of developments is an extremely reliable, maintenance-free ignition system; electronic ignition systems make an important contribution by satisfying these criteria. Yet another condition for reliable long-term operation is the exclusive use of unleaded fuel for engine operation. Lead deposits form in and on the pores of the active catalytic surfaces to reduce their number. Residue from engine oil can also "poison" the catalytic converter.

Other options

Lean-burn concepts

Pollutant reduction based on the catalytic converter is an "external process" without any direct influence on the engine's internal combustion process. Yet another strategy relies on modifying "internal processes" by focusing on such factors as combustion-chamber design, valve timing, exhaust-gas recirculation, compression ratio, ignition timing and A/F ratios. By directly affecting the combustion process, these strategies can exert a considerable influence on exhaust emissions, even if the ultimate effects are not as pronounced as those achieved with catalytic exhaust-gas treatment. Modifications to these "internal processes" are applied in lean-burn concepts.
The lambda excess-air factor – the air/fuel mixture ratio used to operate the engine – has a dramatic effect on concentrations of hydrocarbons (HC), carbon monoxide (CO) and nitrous oxides (NO_X) while also serving as a prime determinant of fuel economy. HC and CO emissions rise in the rich range and sink to minimum levels under lean operation. This pattern is reflected by fuel consumption. Nitrous oxides present a contrasting picture by peaking under slightly lean mixtures ($\lambda = 1.05$).

Prior to 1970, engines were designed to run on rich mixtures to ensure high performance and good handling. Then increasingly severe emissions legislation forced designers to raise A/F ratios, and engines had to start operating on excess air. The primary benefits of the new leaner mixtures were reductions in emissions of HC and CO, and substantial improvements in fuel economy, but these though were all at the cost of higher nitrous-oxide emission levels. Specifically for lean-burn operation therefore, in order not to impair handling and driveability, the designers were forced to continually improve both the engines themselves and their mixture-formation systems. More precise definition and control of ignition timing also became imperative. These developments have lead to increasing use of engine-management systems featuring electronic ignition as a means of ensuring optimal spark advance to maximize fuel economy and minimize emissions.

Lean-burn engine

Consistent optimisation of combustion-chamber design combined with flanking measures outside the chamber (for instance, promoting intake swirl) led to the design of a lean-burn engine capable of operating at excess-air factors in the $\lambda \approx 1.4$ range. Although the lean-burn engine combines lower emissions with improved fuel economy, it still depends on catalytic exhaust treatment to bring HC and CO levels into compliance with "severe" emissions standards. Because it has not yet proven possible to meet the strict US emissions regulations with a lean-burn engine, this concept has remained in an outsider's role, despite its attractive fuel-economy figures.

Thermal afterburners

In the days before today's catalytic treatment of exhaust emissions became standard, early attempts at emissions reduction focused on thermal afterburning. This method retains the exhaust gases in a high-temperature atmosphere for a specified period to burn exhaust components that failed to combust in the engine's cylinders. Supplementary air injection is required to support this combustion process during operation in the rich range (λ = 0.9...1.0), but during lean operation (λ = 1.1...1.2) the exhaust gas contains enough oxygen to support the process unassisted.
Today thermal afterburners are totally insignificant, due to their having no potential for meeting low NO_X limits. However, the concept can be employed to reduce emissions of HC and CO in the warm-up phase, before the catalytic converter reaches its normal operating temperature. Thus thermal aftertreatment with secondary-air injection represents an option for compliance with tomorrow's more stringent limits by reducing the emissions produced by the engine in its warm-up phase.

Secondary-air injection

Supplementary air can be injected immediately downstream from the combustion chamber to promote secondary combustion in the hot exhaust gases. This "exothermic" reaction not only reduces levels of hydrocarbons (HC) and carbon monoxide (CO), it also heats the catalytic converter.
This process substantially enhances the catalytic converter's conversion rate in the warm-up phase. The primary components of the secondary-air injection system (Figure 11) are the:
– Electric secondary-air pump (6),
– Secondary-air valve (5) and the
– Non-return valve (4).

Thermal reactors

When the engine is run with rich air-fuel mixtures, it generates exhaust gases with very high concentrations of HC and CO. This mixture of exhaust gases and air is retained in the thermal reactor for as long a period as possible and ignited there at high temperatures in order to oxidise the pollutants. Although thermal reactors can reduce HC emissions by roughly 50 %, the concept also leads to increases of up to 15 % in fuel consumption. This is why thermal reactors were used for only a brief period prior to the advent of catalytic-converter technology.

Overrun fuel cutoff

Yet another strategy for reducing emissions of HC and CO relies on switching off the fuel supply during closed-throttle operation (overrun). Overrun generates high levels of vacuum within the engine's intake tract and therefore in the combustion chambers. The mixture's low oxygen content makes it difficult to ignite during this type of operation, and combustion remains incomplete, leading to higher emissions of hydrocarbons and carbon monoxide. Complete interruption of the fuel supply during overrun operation prevents production of uncombusted pollutants.
In systems such as KE-Jetronic, continuous injection ensures smooth and seamless transitions between the function's active and passive states. The overrun cutoff responds to coolant temperature. To inhibit continuous "hunting"

Fig. 11

Secondary air injection

1 Intake air, **2** Engine, **3** Secondary air, **4** 1-way check valve, **5** Secondary-air valve, **6** Electric secondary-air pump, **7** Lambda O_2 sensor, **8** Catalytic converter, **9** Exhaust gas.

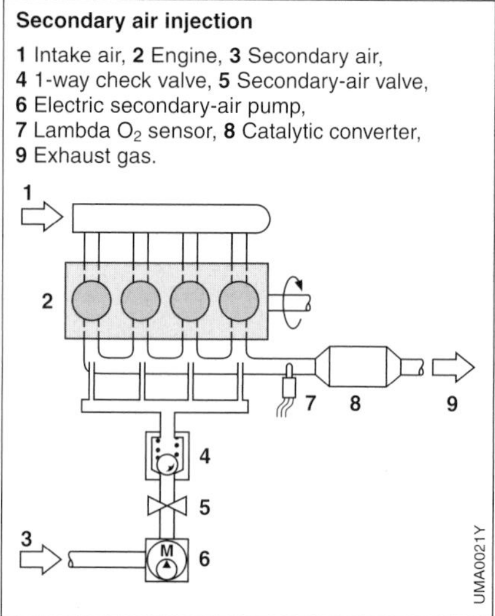

during steady-state operation, the system varies the activation point according to the direction in which engine speed changes (hysteresis). The activation thresholds used with the engine warmed to its normal operating temperature are defined as low as possible to maximize fuel savings.
Pollutant emissions from the spark-ignition engine can be curtailed by choosing from a wide and variegated range of available options. The ultimate selection of technical solutions therefore, is affected by numerous considerations, among which official legislation on emissions is far from being the least important.

Testing exhaust and evaporative emissions

Test technology

Test cycles

To precisely measure a passenger car's emission levels, the vehicle must be tested in an emissions test cell under standardized conditions designed to accurately reflect real-world driving conditions. Compared with highway driving, a major advantage of operation in a test cell is that it permits precisely defined speed curves to be closely adhered to, and there is no need to adapt and react to traffic conditions. This is essential for conducting emissions tests that will provide mutually comparable results.
The test vehicle is parked with its wheels resting on special rollers. The rollers' resistance to motion can be adjusted to simulate friction losses and aerodynamic drag, and inertial mass can be added to simulate the vehicle's weight. The required cooling is furnished by a fan mounted a short distance from the vehicle.
Emissions are measured based on a precisely-defined, simulated driving cycle. The exhaust gases generated during this cycle are collected for subsequent analysis of pollutant mass.
Although officially prescribed procedures for collecting exhaust gases and determining emissions levels have been standardized at the international level, this does not apply to the actual driving cycles. In some countries regulations governing exhaust emissions are supplemented by limits on evaporative emissions from the fuel system.

Chassis dynamometer

To ensure comparable emissions data, the temporal progression curves for the speeds and forces acting on the vehicle during the simulated cycle on the chassis dynamometer must precisely coincide with those for highway operation. Eddy-current brakes and DC motors produce loads to simulate vehicular inertia, rolling resistance and aerodynamic drag. These speed-sensitive loads are applied to the vehicle through the rollers, and represent the resistance forces that the vehicle must overcome during the cycle. Rapid couplings are employed to connect the rollers to various inertial masses as a means of simulating the vehicle's mass. The progression curves for braking force must be maintained in a precise relationship to vehicle speed and inertial mass, as any deviations will lead to inaccurate test results. Test results are also influenced by such factors as atmospheric humidity, ambient temperature and barometric pressure.

Driving cycles

To ensure that test results remain mutually comparable, the speeds used on the dynamometer must accurately reflect actual highway operation. Testing is based on a standardized driving cycle in which shift points, braking manœuvres, idling and stationary phases have all been selected to reflect the traffic conditions typically encountered in large urban areas. Seven different test cycles are in use at the international level. Within Europe, the EU Stage III sequence (valid from January 2000) will feature a

shorter driving cycle (the 40-second preliminary phase is deleted), but the US is adding the SFTP test containing a special assessment of vehicles with air conditioning as well as several new operating modes.
Usually a driver sits in the vehicle, maintaining the speed sequence indicated on a display screen.

Test samples and dilution procedures (CVS method)

With European adoption of the constant-volume sampling (CVS) method in 1982, there is now basically one single internationally recognized procedure for collecting exhaust gases.

Test sampling and emissions analysis

The dilution process employs the following principle:
The exhaust gases emitted by the test vehicle are diluted with fresh air at a ratio of 10:1 before being extracted using a special system of pumps. These pumps are arranged to maintain a precise, constant ratio between the flow volumes for exhaust gas and fresh air, i.e., the air feed is adjusted to reflect the vehicle's instantaneous exhaust volume. Throughout the test a constant proportion of the diluted exhaust gas is extracted for collection in one or several sample bags. Upon completion of the test cycle, the pollutant concentration in these bags precisely mirrors mean concentration levels in the overall mixture of fresh air and exhaust gases. As it is possible to monitor the total volume of fresh air and exhaust mixture, pollutant concentration levels can be employed to calculate the masses of these substances emitted during the test cycle.
Advantages of this procedure: The condensation of the water vapor contained in the exhaust gas is avoided, otherwise this would lead to a sharp reduction in the NO_X losses in the bag. In addition, dilution greatly inhibits the tendency of the exhaust components (especially the hydrocarbons) to engage in mutual reactions.

Fig. 1

Test layouts

a For US Federal Test (shown here with venturi system), **b** For European test (shown here with rotary-piston compressor).
1 Brake, **2** Rotating mass, **3** Exhaust gas, **4** Air filter, **5** Dilution air, **6** Cooler, **7** Test-sample venturi nozzle, **8** Gas temperature, **9** Pressure, **10** Venturi nozzle, **11** Fan, **12** Sample bag, **13** Rotary-piston blower, **14** To discharge.
ct Exhaust gases in transition phase, s Exhaust gases in stabilized phase, ht Exhaust gases from hot test.

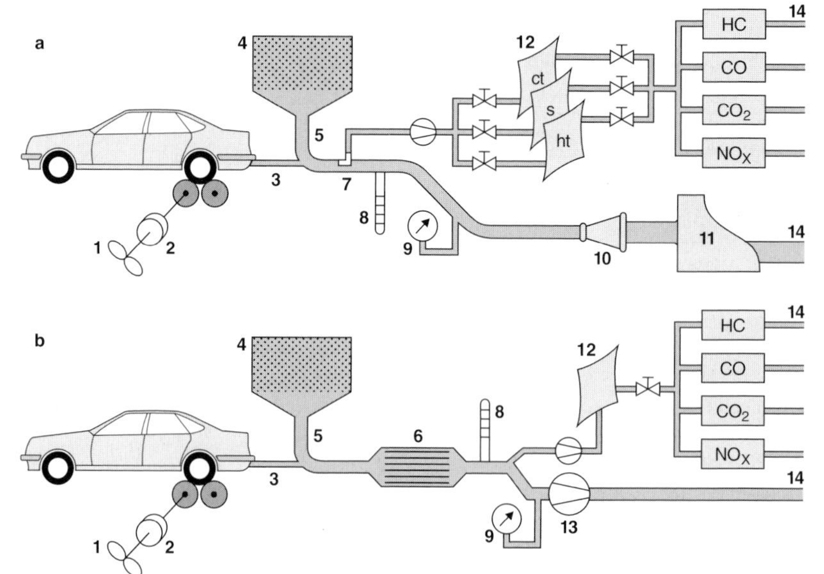

However, dilution does mean that pollutant concentrations decrease proportionally to mean dilution ratio, so high-precision analyzers become essential. Standardized devices are available for analyzing the concentrations of individual substances in the test bags.

Dilution equipment

Either one of two different but equally acceptable pump arrangements is generally employed to maintain the constant flow volume required for testing. In the first, a standard blower extracts the mixture of fresh air and exhaust gas through a venturi tube; the second concept relies on a special vane pump (Roots blower). Both methods are capable of metering flow volume with an acceptable degree of accuracy.

Quantifying fuel-system evaporative losses (evaporation tests)

In addition to and separate from the emissions generated during the engine's combustion process, motor vehicles also emit hydrocarbons (HC) in the form of evaporative emissions escaping from the gas tank and fuel system. Actual levels vary according to fuel-system design and fuel temperature. Some countries (including the USA and a number of countries in Europe) already limit maximum acceptable levels of evaporative emissions.

SHED test

The SHED test is the most common procedure for determining evaporative emissions. It comprises two test phases – distinguished by different conditioning procedures – conducted in a gas-tight enclosure (SHED tent).

The first phase of the test proceeds with the fuel tank filled to approximately 40 % of its overall capacity. The test fuel is warmed from its initial temperature of 10...14.5 °C, with actual monitoring of HC concentrations within the enclosure starting once it reaches 15.5 °C. The fuel temperature is increased by 14 °C in the following hour, after which testing is concluded with a final sampling of the HC concentration. Evaporative emissions levels are determined by comparing the initial and final measurements. The vehicle's windows and trunk lid must remain open for the duration of the test.

The vehicle is prepared for the second phase of testing by being taken for a "warm-up drive" through the applicable official urban test cycle. The vehicle is then parked in the test chamber for one hour; the test monitors the increases in HC concentrations produced by the vehicle as it cools.

The sum of the results from both tests must be less than the current limit of 2 g hydrocarbon vapor. A more stringent SHED test procedure has now been mandated for the US.

ECE/EU test cycle and limits

The ECE/EU test cycle relies on a hypothetical driving curve (Figure 2) designed to serve as a reasonably accurate reflection of operating conditions in an urban center. In 1993 this test cycle was supplemented by a rural phase including speeds of up to 120 km/h. This new ECE/EU test cycle is currently mandatory in the following countries: Austria, Belgium, Denmark, Finland, France, Germany, Great Britain, Greece, Ireland,

Fig. 2

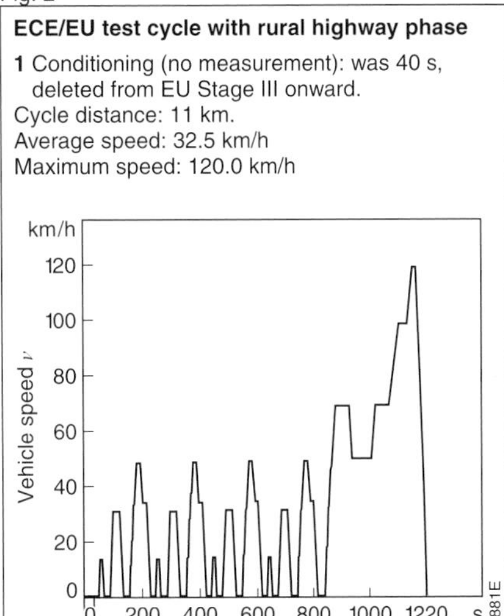

Testing exhaust and evaporative emissions

Italy, Luxembourg, Netherlands, Portugal, Spain, Sweden.

Testing proceeds as follows:
After initial conditioning (vehicle parked at a room temperature of 20...30 °C for at least 6 hours), the actual test cycle commences following a cold start and a 40-second warm-up period (this preliminary phase has been deleted from EU Stage III and future tests). During the test, the CVS method is used to collect exhaust gases in a sample bag. The European test mirrors standard practice by converting the mass pollutant levels of the gas contained in the bags for quantification relative to test distance. Hydrocarbons and nitrous oxides are currently subject to a cumulative limit (HC + NO_X), but EU Stage III will see the introduction of separate and distinct limits for these two substances.
More stringent limits applicable to all vehicles regardless of engine piston displacement have been in effect since 1992. The data from the corresponding directive, EEC 91/441 (EU Stage I), are provided in Table 1 in the section on emissions limits. This standard also prescribes limits for evaporative emissions. The directive EEC/94/12 (EU Stage II) brought further reductions in emissions limits for 1996/97.
The European limits are slated for further tightening (Stage III and IV, 2000 and 2005):
- Cold start at –7 °C (starting in 2002),
- EOBD (European On-Board Diagnosis) for emissions-relevant components,
- Stricter evaporative emissions test,
- Long-term reliability (80,000; 100,000 km) and arrangements to monitor performance in the field,
- Exhaust sampling begins immediately after the vehicle starts.

US test cycles

FTP-75 test cycle

The FTP (Federal Test Procedure) 75 test cycle comprises three phases. The sequences and speed curves are designed to reflect the conditions measured in actual morning commuter traffic on the streets of Los Angeles (Figure 3a):
The test vehicle is first conditioned by being left parked for 12 hours at an ambient temperature of 20...30 °C. It is then started and driven through the prescribed test cycle:
Phase ct: Diluted exhaust gases are collected in Bag 1 during the cold transition phase.
Phase s: Exhaust samples are diverted to Bag 2 at the beginning of the stabilized phase (after 505 s) without any interruption in the program sequence. The engine is switched off for a 10-minute pause immediately following completion of the stabilized phase (after 1,372 seconds).
Phase ht: The engine is restarted for the hot test (lasting 505 seconds). The speed sequence used in this phase is identical to the one employed for the cold transition test. The exhaust gases generated in this phase are collected in a third bag. As the probes should not remain in the bags for more than 20 minutes, the samples from the previous phases are analyzed before the hot test.
The exhaust-gas sample from the third bag is then analyzed upon completion of this final driving sequence. The weighted sum of the masses of all pollutants (HC, CO and NO_X; ct 0.43, s1, ht 0.57) is then calculated relative to the distance covered and expressed as emissions per mile.The maximum permitted emission quantities differ in the various countries. This test procedure is used throughout the US including California ("Emissions limits," Table 2) and in several other countries (Table 4).

SFTP cycles

Testing according to the SFTP standard is slated for introduction between 2001 and 2004. The process combines three driving cycles, FTP 75, SC03 and US06, and expands upon earlier procedures to embrace the following supplementary operating conditions (Figure 3b, c):
- Aggressive driving,
- Abrupt changes in vehicle speed,

- Engine start and initial acceleration from standing start,
- Operation with frequent speed changes of minimal amplitude,
- Parked periods, and
- Operation with air conditioner.

Following initial conditioning, the SC03 and US06 cycles proceed through the ct phase of FTP 75 without exhaust gases being collected, although other preconditioning options are available.

The SC03 cycle proceeds at 30 °C with a relative humidity of 40 % (vehicles with air conditioning only). The individual driving cycles are weighted as follows:

- Vehicles with air-conditioning:
  35 % FTP 75 + 37 % SC03 + 28 % US06
- Vehicles without air-conditioning:
  72 % FTP 75 + 28 % US06.

Vehicles must absolve the SFTP and FTP 75 test cycles separately (Tables 2 and 3 in "Emissions limits").

Test cycles for determining average fleet fuel consumption

Each vehicle manufacturer must determine average fuel consumption for its vehicle fleet as a whole. Penalties are imposed upon manufacturers who fail to meet specified limits, while a bonus is available when test results fall below a prescribed threshold. Fuel consumption is determined using the exhaust gases generated in two test cycles: the FTP 75 test cycle (55 %) and the highway test (45 %). After prior conditioning (parked at 20...30 °C for 12 hours) the vehicle

Testing exhaust and evaporative emissions

Fig. 3

US test cycles

	a	b	c	d
Test cycle	**FTP75**	**SC03**	**US06**	**Highway**
Cycle distance:	17.87 km	5.76 km	12.87 km	16.44 km
Cycle duration:	1877 s + 600 s Pause	594 s	600 s	765 s
Average cycle speed:	34.1 km/h	34.9 km/h	77.3 km/h	77.4 km/h
Maximum cycle speed:	91.2 km/h	88.2 km/h	129.2 km/h	96.4 km/h

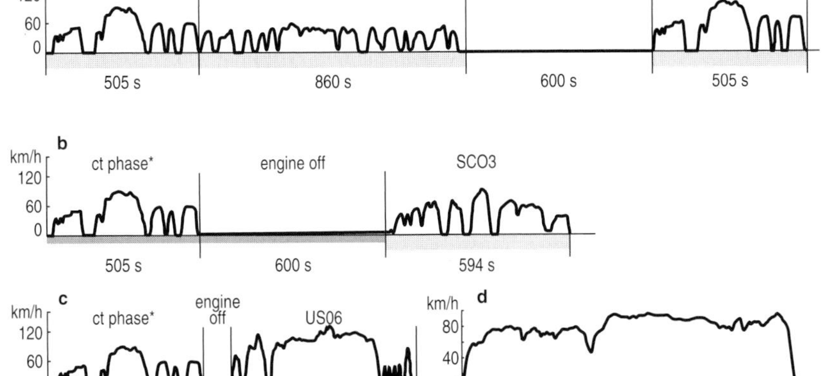

* ct Transition phase; s Stabilized phase, ht Hot test
Phases with exhaust-gas collection
Conditioning (instead, other driving cycles can also be used here)

UWT0003-1E

makes one unmonitored run through the highway test cycle; this is followed by a second cycle, this time with collection of the exhaust gases. The emissions are then used as the basis for calculating fuel consumption (Figure 3d).

Every newly licensed vehicle must continue to comply with these limits (regardless of vehicle weight or displacement) over a minimum distance of 50,000 miles. The US grants waivers for vehicles from specific model years under certain conditions, and two qualification mileages are available: 50,000 and 100,000 miles. The approved limits for 100,000 miles are higher (wear factor).
Legislation has been passed ("Clean Air Act") which, while containing numerous initiatives for protection of the environment, also prescibes stricter emissions limits for vehicles manufactured from the 1994 model year onward (Table 2). California enacted stricter emissions standards as early as 1993 and is also planning further drastic action.
The cold-start enrichment required when vehicles are started at cold temperatures produces brief peaks in emissions that are not covered by the current test procedure (at an ambient temperature of 20...30 °C). The Clean Air Act seeks to limit this cold-start pollution with the introduction of an emissions test conducted at –6.7 °C, although this test only allocates a limit to carbon monoxide.

Fig. 4

Japanese test cycles

a 11 mode cycle (cold test)

Cycle distance:	1.021 km
Cycle no./test:	4
Average speed:	30.6 km/h
Maximum speed:	60 km/h

b 10·15 mode cycle (hot test)

Cycle distance:	4.16 km
Cycle no./test:	1
Average speed:	22.7 km/h
Maximum speed:	70 km/h

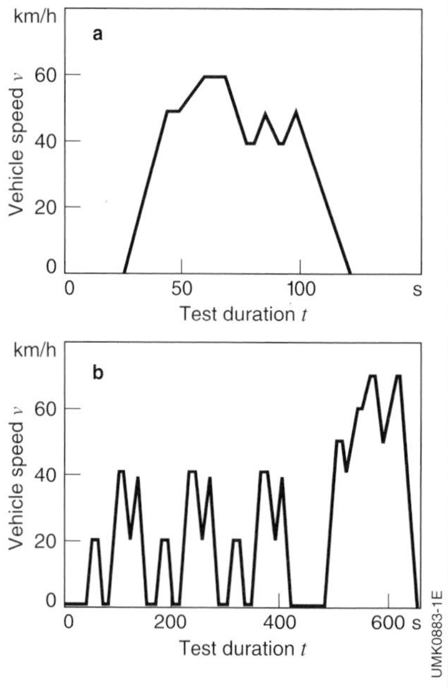

UMK0883-1E

Japanese test cycle

Two test cycles with different hypothetical driving curves are included in this composite test:
Following a cold start, the vehicle proceeds to absolve the 11-mode cycle four times in succession, and all four cycles are evaluated. The vehicle runs through the 10·15 mode test as a hot test (Figure 4).
Preconditioning for the hot start includes the prescribed idling-emissions test, and proceeds as follows:
First, the vehicle is driven through a 15-minute warm-up phase at 60 km/h. Concentrations of HC, CO and CO_2 are then measured in the exhaust pipe. The 10·15 hot test starts after a further 5-minute warm-up period at 60 km/h. The CVS method forms the basis for exhaust-gas analysis in both the 11 mode and the 10·15 mode tests. The diluted gas is collected in separate bags. Emission limits for the cold test are specified in g/test, while the limits for hot testing are expressed relative to distance, viz., converted to grams per kilometer (Table 5 in "Emissions limits" section).
Japanese regulations include limits on evaporative emissions as determined in testing using the SHED method.

Exhaust-gas analyzers

Infrared test chamber (schematic)

1 Receiver chamber with compensation volumes V_1 and V_2, **2** Flow sensor, **3** Test cell,
4 Rotating "chopper disc" with motor,
5 Infrared projector.

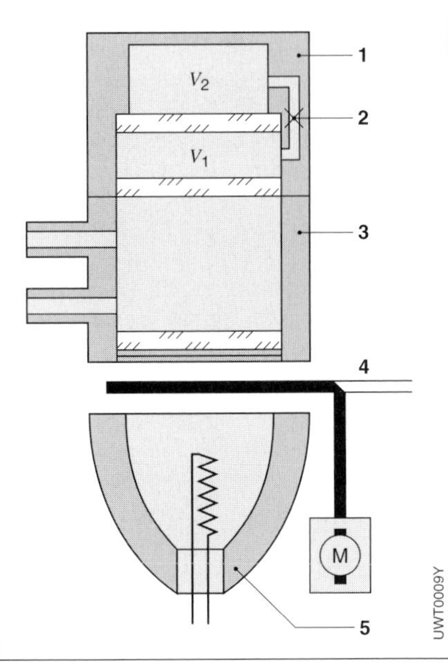

Fig. 5

Legislation reflects governmental efforts to reduce the quantities of toxic substances in exhaust gases by mandating regular periodic emissions testing for vehicles already in service. In Germany compliance with specified CO limits is verified at prescribed intervals in an emissions inspection (AU, or Abgasuntersuchung) as defined in § 29 of the StVZO (FMVSS/CUR). Exhaust-gas analyzers are also indispensable tools for general automotive service, useful for correct mixture adjustment as well as efficient fault diagnosis on the engine.

Test procedures

It is necessary to carry out precise measurements of the individual exhaust-gas components. While test laboratories rely on complex procedures, automotive service facilities have adopted the infrared method on a widespread basis. The concept is based on the fact that individual exhaust-gas components absorb infrared light at different specific rates according to their characteristic wavelengths.

Available units include single-component analyzers (e.g., for CO) as well as devices for measuring several substances (for CO/HC, CO/CO_2, $CO/HC/CO_2$, etc.).

<u>Test chamber (Fig. 5)</u>

Infrared radiation is transmitted from an emitter (5) heated to approximately 700 °C. The infrared beam passes through a measuring cell (3) before entering the receiver chamber (1). If CO content is being measured, then the sealed receiver chamber contains a gaseous atmosphere with a defined CO content. This gas absorbs a portion of the CO-specific radiation. This absorption process is accompanied by an increase in the temperature of the gas, which then generates a gas current flowing from volume V_1 and through a flow sensor on its way to compensating volume V_2. A rotating "chopper disc" (4) induces a rhythmic interruption in the beam to produce an alternating flow between volumes V_1 and V_2. The flow sensor converts this motion into an alternating electrical signal. When a test gas with a variable CO content flows through the measuring cell it absorbs radiant energy in quantities proportional to its CO content; this energy is then no longer available in the receiver chamber.

As this leads to a reduction in the base flow to the receiver chamber, the deviation from the alternating base signal serves as an index of the CO content in the test gas.

<u>Gas path (Fig. 6, next page)</u>

A probe (1) is employed to extract the test gas from the vehicle's exhaust system. The tester's integral diaphragm pump (6) extracts the gas, drawing it through the loose-mesh filter screen (2) and into the water trap (3) to remove condensation and larger particulates prior to subsequent cleansing in the fine-mesh filter

(4). The solenoid valve (5) located upstream from the diaphragm pump switches the entry to the test chamber (9) from exhaust gas to air and the system automatically recalibrates to zero. Back-up filters in the supply orifices for both exhaust gas and air ensure that particulates do not enter the test chamber, which is also sealed against condensation of the kind that could enter the system if the external water trap were allowed to overflow. The restriction in the tank (10) pressurizes the safety reservoir (8) to induce a flow through the bypass circuit and to the test chamber. Gravity pulls any moisture ingested into the system back into the tank, whence it escapes back into the atmosphere. The pressure switch (7) monitors gas flow to ensure that adequate amounts of gas are drawn into the system. The restrictor in the safety reservoir raises pressure levels at the pump discharge to activate the pressure switch, which will consequently be released if the gas flow is interrupted, simultaneously triggering a warning display to alert the operator.

Testing the catalytic converter

On vehicles with closed-loop mixture control a representative component can be used for indirect assessment of the catalytic converter's operation. The best proxy is CO, which should not exceed 0.2% by volume in the gas emerging from the catalytic converter when lambda is maintained at a precise level of 1.00 (+0.01).

Lambda, in turn, is determined based on the composition of the exhaust gases emerging from the catalytic converter. The exhaust-gas analyzer determines lambda with the required accuracy using measurements based on the CO, HC, CO_2 and O_2 in the exhaust gas along with defined constants for NO and fuel composition.

O_2 concentrations are monitored with an electrochemical probe.

Fig. 6

Gas flow path in CO tester

1 Probe, **2** Filter screen, **3** Water separator, **4** Fine-mesh filter, **5** Solenoid valve, **6** Diaphragm pump, **7** Pressure switch, **8** Safety reservoir, **9** Test chamber, **10** Tank.

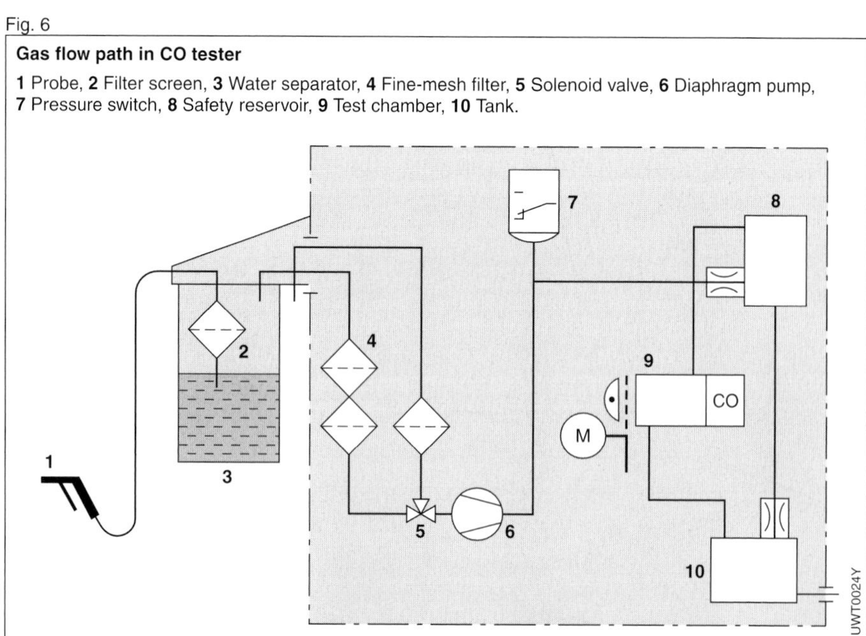

Current (1998) emissions limits for gasoline engines

Table 1
EU emissions limits as measured in ECE/EU test cycle

Standards	Introduction	CO g/km	HC g/km	NO_X g/km	HC+NO_X g/km
EU Stage I	07.92	2.72	–	–	0.97
EU Stage II	01.96	2.2	–	–	0.5
EU Stage III	01.00	2.3	0.2	0.15	–
EU Stage IV	01.05	1.0	0.1	0.08	–

Table 2
US Federal (49 state) and California emissions limits. FTP 75 test cycle

	Model year	Standards	CO g/mile	HC g/mile	NO_X g/mile
US Federal	1994	Level 1	3.4	0.25	0.4
	2004 [1])	Level 2	1.7	0.125 [2])	0.2
California	[3])	TLEV [4])	3.4	0.125 [2])	0.4
	[3])	LEV [5])	3.4	0.075 [2])	0.2
	[3])	ULEV [6])	1.7	0.04 [2])	0.2

[1]) Proposal. [2]) NMOG = Non Methanic Organic Gases. [3]) Introduction varies according to manufacturer's NMOG fleet average (both vehicle and total fleet are certified).
[4]) Transitional Low Emission Vehicles. [5]) Low Emission Vehicles. [6]) Ultra Low Emission Vehicles.

Table 3
US emissions limits. STFP test cycle

	NMHC [1])+NO_X	$CO_{Composite}$ [2]) g/mile	CO_{SC03} [2]) g/mile	CO_{US06} [2]) g/mile
up to 50,000 miles	0.65	3.4	3.0	9.0
50,000 to 100,000 miles	0.91	4.2	3.7	11.1

[1]) Non Methane HC. [2]) The manufacturer has the option of selecting $CO_{Composite}$ or CO_{SC03} and CO_{US06} limits.

Table 4
Emissions limits for Argentina, Australia, Brazil, Canada, Mexico, Norway, Switzerland, and South Korea measured with FTP 75 test cycle

Country	Introduction	CO g/km	HC g/km	NO_X g/km	Evap. emissions (HC) g/test
Argentina	01.97	2.0	0.3	0.6	6.0
Australia	01.97	1.9	0.24	0.57	1.9
Brazil	01.97	2.0	0.3	0.6	6.0
Canada	01.98	2.1	THC [2]) 0,25; NMHC [3]) 0,16	0.24	2.0
Mexico	01.95	2.1	0.25	0.62	2.0
Norway	01.89	2.1	0.25	0.62	2.0
Switzerland [1])	10.87	2.1	0.25	0.62	2.0
South Korea	01.91 01.00	2.1	0.25 0.16	0.62 0.25	2.0

[1]) EU/ECE regulations recognized since 10/95. [2]) THC =Total HC. [3]) NMHC = Non Methane HC.

Table 5
Japanese emissions limits measured in Japanese test cycle

Test procedure	CO	HC	NO_X	Evap. emissions
10·15-mode (g/km)	2.1...2.7 (0,67)	0.25...0.39 (0,08)	0.25...0.48 (0,08)	–
11-mode (g/test)	60.0...85.0 (19,0)	7.0...9.5 (2,2)	4.4...6.0 (1,4)	–
SHED (g/Test)	–	–	–	2.0

( ) planned figures

Gasoline-engine management

Technical requirements

Spark-ignition (SI) engine torque

The power P furnished by the spark-ignition engine is determined by the available net flywheel torque and the engine speed.
The net flywheel torque consists of the force generated in the combustion process minus frictional losses (internal friction within the engine), the gas-exchange losses and the torque required to drive the engine ancillaries (Figure 1).
The combustion force is generated during the power stroke and is defined by the following factors:
- The mass of the air available for combustion once the intake valves have closed,
- The mass of the simultaneously available fuel, and
- The point at which the ignition spark initiates combustion of the air/fuel mixture.

Primary engine-management functions

The engine-management system's first and foremost task is to regulate the engine's torque generation by controlling all of those functions and factors in the various engine-management subsystems that determine how much torque is generated.

Cylinder-charge control

In Bosch engine-management systems featuring electronic throttle control (ETC), the "cylinder-charge control" subsystem determines the required induction-air mass and adjusts the throttle-valve opening accordingly. The driver exercises direct control over throttle-valve opening on conventional injection systems via the physical link with the accelerator pedal.

Mixture formation

The "mixture formation" subsystem calculates the instantaneous mass fuel requirement as the basis for determining the correct injection duration and optimal injection timing.

Fig. 1

Driveline torque factors

1 Ancillary equipment (alternator, a/c compressor, etc.),
2 Engine,
3 Clutch,
4 Transmission.

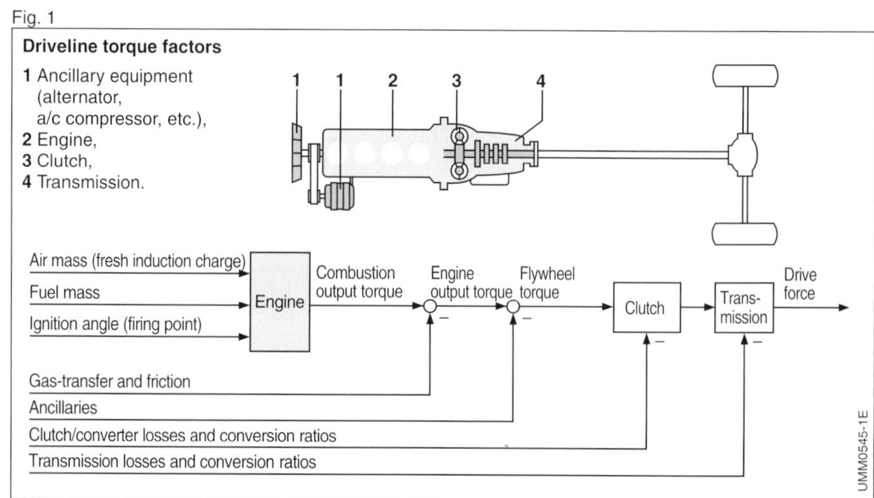

Ignition
Finally, the "ignition" subsystem determines the crankshaft angle that corresponds to precisely the ideal instant for the spark to ignite the mixture.

The purpose of this closed-loop control system is to provide the torque demanded by the driver while at the same time satisfying strict criteria in the areas of
- Exhaust emissions,
- Fuel consumption,
- Power,
- Comfort and convenience, and
- Safety.

Cylinder charge

Elements

The gas mixture found in the cylinder once the intake valve closes is referred to as the cylinder charge, and consists of the inducted fresh air-fuel mixture along with residual gases.

Fresh gas
The fresh mixture drawn into the cylinder is a combination of fresh air and the fuel entrained with it. While most of the fresh air enters through the throttle valve, supplementary fresh gas can also be drawn in through the evaporative-emissions control system (Figure 2). The air entering through the throttle-valve and remaining in the cylinder after intake-valve closure is the decisive factor defining the amount of work transferred through the piston during combustion, and thus the prime determinant for the amount of torque generated by the engine. In consequence, modifications to enhance maximum engine power and torque almost always entail increasing the maximum possible cylinder charge. The theoretical maximum charge is defined by the volumetric capacity.

Residual gases
The portion of the charge consisting of residual gases is composed of
- The exhaust-gas mass that is not discharged while the exhaust valve is open and thus remains in the cylinder, and
- The mass of recirculated exhaust gas (on systems with exhaust-gas recirculation, Figure 2).

The proportion of residual gas is determined by the gas-exchange process.
Although the residual gas does not participate directly in combustion, it does influence ignition patterns and the actual combustion sequence. The effects of this residual-gas component may be thoroughly desirable under part-throttle operation.
Larger throttle-valve openings to compensate for reductions in fresh-gas filling

Fig. 2

Cylinder charge in the spark-ignition engine

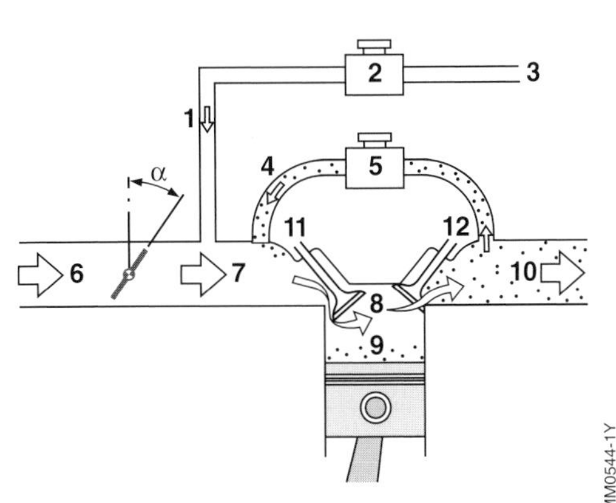

1 Air and fuel vapor,
2 Purge valve with variable aperture,
3 Link to evaporative-emissions control system,
4 Exhaust gas,
5 EGR valve with variable aperture,
6 Mass airflow (barometric pressure p_U),
7 Mass airflow (intake-manifold pressure p_s),
8 Fresh air charge (combustion-chamber pressure p_B),
9 Residual gas charge (combustion-chamber pressure p_B),
10 Exhaust gas (back-pressure p_A),
11 Intake valve,
12 Exhaust valve,
α Throttle-valve angle.

are needed to meet higher torque demand. These higher angles reduce the engine's pumping losses, leading to lower fuel consumption. Precisely regulated injection of residual gases can also modify the combustion process to reduce emissions of nitrous oxides (NO_x) and unburned hydrocarbons (HC).

Control elements

Throttle valve

The power produced by the spark-ignition engine is directly proportional to the mass airflow entering it. Control of engine output and the corresponding torque at each engine speed is regulated by governing the amount of air being inducted via the throttle valve. Leaving the throttle valve partially closed restricts the amount of air being drawn into the engine and reduces torque generation. The extent of this throttling effect depends on the throttle valve's position and the size of the resulting aperture.
The engine produces maximum power when the throttle valve is fully open (WOT, or wide open throttle).
Figure 3 illustrates the conceptual correlation between fresh-air charge density and engine speed as a function of throttle-valve aperture.

Gas exchange

The intake and exhaust valves open and close at specific points to control the transfer of fresh and residual gases. The ramps on the camshaft lobes determine both the points and the rates at which the valves open and close (valve timing) to define the gas-exchange process, and with it the amount of fresh gas available for combustion.
Valve overlap defines the phase in which the intake and exhaust valves are open simultaneously, and is the prime factor in determining the amount of residual gas remaining in the cylinder. This process is known as "internal" exhaust-gas recirculation. The mass of residual gas can also be increased using "external" exhaust-gas recirculation, which relies on a supplementary EGR valve linking the intake and exhaust manifolds. The engine ingests a mixture of fresh air and exhaust gas when this valve is open.

Pressure charging

Because maximum possible torque is proportional to fresh-air charge density, it is possible to raise power output by compressing the air before it enters the cylinder.

Dynamic pressure charging

A supercharging (or boost) effect can be obtained by exploiting dynamics within the intake manifold. The actual degree of boost will depend upon the manifold's configuration as well as the engine's instantaneous operating point (essentially a function of the engine's speed, but also affected by load factor). The option of varying intake-manifold geometry while the vehicle is actually being driven, makes it possible to employ dynamic precharging to increase the maximum available charge mass through a wide operational range.

Mechanical supercharging

Further increases in air mass are available through the agency of

Fig. 3

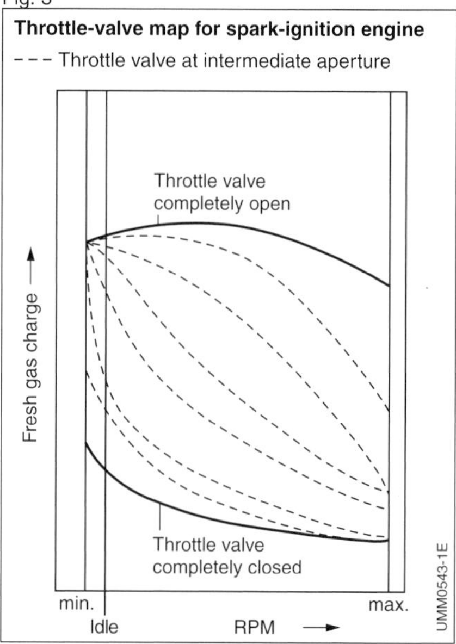

mechanically driven compressors powered by the engine's crankshaft, with the two elements usually rotating at an invariable relative ratio. Clutches are often used to control compressor activation.

Exhaust-gas turbochargers

Here the energy employed to power the compressor is extracted from the exhaust gas. This process uses the energy that naturally-aspirated engines cannot exploit directly owing to the inherent restrictions imposed by the gas expansion characteristics resulting from the crankshaft concept. One disadvantage is the higher back-pressure in the exhaust gas exiting the engine. This back-pressure stems from the force needed to maintain compressor output.
The exhaust turbine converts the exhaust-gas energy into mechanical energy, making it possible to employ an impeller to precompress the incoming fresh air. The turbocharger is thus a combination of the turbine in the exhaust-fas flow and the impeller that compresses the intake air.
Figure 4 illustrates the differences in the torque curves of a naturally-aspirated engine and a turbocharged engine.

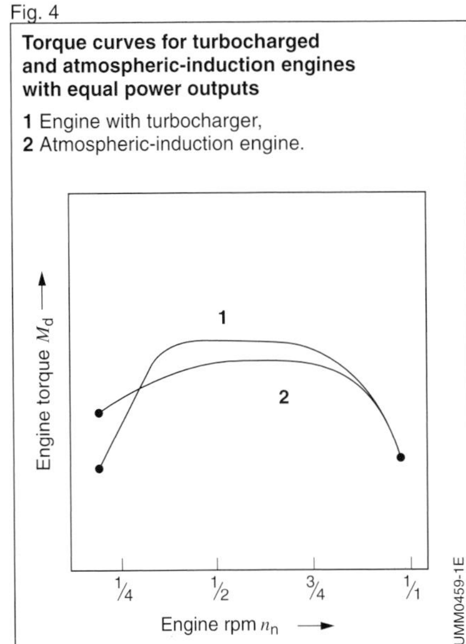

Fig. 4

Torque curves for turbocharged and atmospheric-induction engines with equal power outputs

1 Engine with turbocharger,
2 Atmospheric-induction engine.

Mixture formation

Parameters

Air-fuel mixture

Operation of the spark-ignition engine is contingent upon availability of a mixture with a specific air/fuel (A/F) ratio. The theoretical ideal for complete combustion is a mass ratio of 14.7:1, referred to as the stoichiometric ratio. In concrete terms this translates into a mass relationship of 14.7 kg of air to burn 1 kg of fuel, while the corresponding volumetric ratio is roughly 9,500 litres of air for complete combustion of 1 litre of fuel.

The air-fuel mixture is a major factor in determining the spark-ignition engine's rate of specific fuel consumption. Genuine complete combustion and absolutely minimal fuel consumption would be possible only with excess air, but here limits are imposed by such considerations as mixture flammability and the time available for combustion.

The air-fuel mixture is also vital in determining the efficiency of exhaust-gas treatment system. The current state-of-the-art features a 3-way catalytic converter, a device which relies on a stoichiometric A/F ratio to operate at maximum efficiency and reduce undesirable exhaust-gas components by more than 98 %.

Current engines therefore operate with a stoichiometric A/F ratio as soon as the engine's operating status permits

Certain engine operating conditions make mixture adjustments to non-stoichiometric ratios essential. With a cold engine for instance, where specific adjustments to the A/F ratio are required. As this implies, the mixture-formation system must be capable of responding to a range of variable requirements.

Excess-air factor

The designation l (lambda) has been selected to identify the excess-air factor (or air ratio) used to quantify the spread between the actual current mass A/F ratio and the theoretical optimum (14.7:1):

λ = Ratio of induction air mass to air requirement for stoichiometric combustion.

$\lambda = 1$: The inducted air mass corresponds to the theoretical requirement.

$\lambda < 1$: Indicates an air deficiency, producing a corresponding rich mixture. Maximum power is derived from λ = 0.85...0.95.

$\lambda > 1$: This range is characterized by excess air and lean mixture, leading to lower fuel consumption and reduced power. The potential maximum value for λ – called the "lean-burn limit (LML)" – is essentially defined by the design of the engine and of its mixture formation/induction system. Beyond the lean-burn limit the mixture ceases to be ignitable and combustion miss sets in, accompanied by substantial degeneration of operating smoothness.

In engines featuring systems to inject fuel directly into the chamber, these operate with substantially higher excess-air factors (extending to $\lambda = 4$) since combustion proceeds according to different laws.

Spark-ignition engines with manifold injection produce maximum power at air deficiencies of 5...15 % (λ = 0.95...0.85), but maximum fuel economy comes in at 10...20 % excess air (λ = 1.1...1.2).

Figures 1 and 2 illustrate the effect of the excess-air factor on power, specific fuel consumption and generation of toxic emissions. As can be seen, there is no single excess-air factor which can simultaneously generate the most favorable levels for all three factors. Air factors of λ = 0.9...1.1 produce "conditionally optimal" fuel economy with "conditionally optimal" power generation in actual practice.

Once the engine warms to its normal operating temperature, precise and consistent maintenance of $\lambda = 1$ is vital for the 3-way catalytic treatment of exhaust gases. Satisfying this requirement entails exact monitoring of induction-air mass and precise metering of fuel mass.

Optimal combustion from current engines equipped with manifold injection relies on formation of a homogenous mixture as well as precise metering of the injected fuel quantity. This makes effective atomization essential. Failure to satisfy this requirement will foster the formation of large droplets of condensed fuel on the walls of the intake tract and in the combustion chamber. These droplets will fail to combust completely and the ultimate result will be higher HC emissions.

Fig. 1

Effects of excess-air factor λ on power P and specific fuel consumption b_e.

a Rich mixture (air deficiency),
b Lean mixture (excess air).

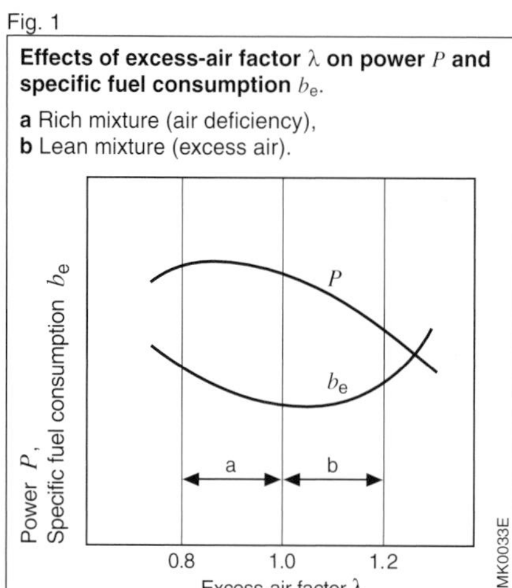

Fig. 2

Effect of excess-air factor λ on untreated exhaust emissions

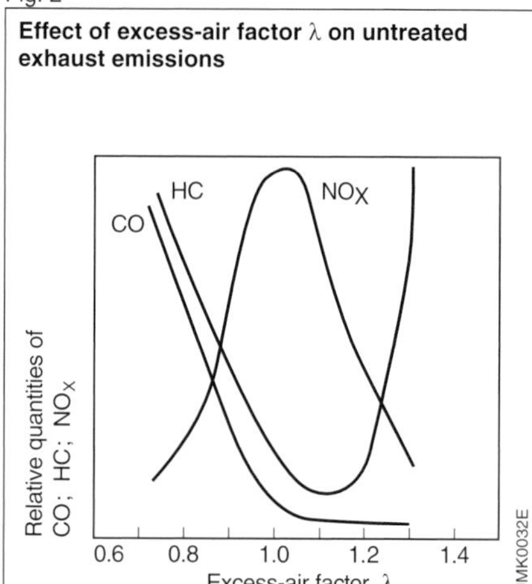

Adapting to specific operating conditions

Certain operating states cause fuel requirements to deviate substantially from the steady-state requirements of an engine warmed to its normal temperature, thus necessitating corrective adaptations in the mixture-formation apparatus. The following descriptions apply to the conditions found in engines with manifold injection.

Cold starting

During cold starts the relative quantity of fuel in the inducted mixture decreases: the mixture "goes lean." This lean-mixture phenomenon stems from inadequate blending of air and fuel, low rates of fuel vaporization, and condensation on the walls of the inlet tract, all of which are promoted by low temperatures. To compensate for these negative factors, and to facilitate cold starting, supplementary fuel must be injected into the engine.

Post-start phase

Following low-temperature starts, supplementary fuel is required for a brief period, until the combustion chamber heats up and improves the internal mixture formation. This richer mixture also increases torque to furnish a smoother transition to the desired idle speed.

Warm-up phase

The warm-up phase follows on the heels of the starting and immediate post-start phases. At this point the engine still requires an enriched mixture to offset the fuel condensation on the intake-manifold walls. Lower temperatures are synonymous with less efficient fuel processing (owing to factors such as poor mixing of air and fuel and reduced fuel vaporization). This promotes fuel precipitation within the intake manifold, with the formation of condensate fuel that will only vaporize later, once temperatures have increased. These factors make it necessary to provide progressive mixture enrichment in response to decreasing temperatures.

Idle and part-load

Idle is defined as the operating status in which the torque generated by the engine is just sufficient to compensate for friction losses. The engine does not provide power to the flywheel at idle. Part-load (or part-throttle) operation refers to the range of running conditions between idle and generation of maximum possible torque. Today's standard concepts rely exclusively on stoichiometric mixtures for the operation of engines running at idle and part-throttle once they have warmed to their normal operating temperatures.

Full load (WOT)

At WOT (wide-open throttle) supplementary enrichment may be required. As Figure 1 indicates, this enrichment furnishes maximum torque and/or power.

Acceleration and deceleration

The fuel's vaporization potential is strongly affected by pressure levels inside the intake manifold. Sudden variations in manifold pressure of the kind encountered in response to rapid changes in throttle-valve aperture cause fluctuations in the fuel layer on the walls of the intake tract. Spirited acceleration leads to higher manifold pressures. The fuel responds with lower vaporization rates and the fuel layer within the manifold runners expands. A portion of the injected fuel is thus lost in wall condensation, and the engine goes lean for a brief period, until the fuel layer restabilizes. In an analogous, but inverted, response pattern, sudden deceleration leads to rich mixtures. A temperature-sensitive correction function (transition compensation) adapts the mixture to maintain optimal operational response and ensure that the engine receives the consistent air/fuel mixture needed for efficient catalytic-converter performance.

Trailing throttle (overrun)

Fuel metering is interrupted during trailing throttle. Although this expedient saves fuel on downhill stretches, its primary purpose is to guard the catalytic converter against overheating stemming from poor and incomplete combustion (misfiring).

Air supply

Air filters

By preventing air-borne dust being drawn into the engine with the intake air, the air filter helps inhibit internal engine wear.
On paved roads, the air's average dust content is about 1 mg/m^3; on unpaved roads however, and during construction work, it can range as high as 40 mg/m^3. This means that depending on roads and operating conditions, a medium-sized engine can draw in up to 50 mg every 1,000 km (600–700 miles).

Passenger-car air filters

Paper elements, contained in centrally located or fender-mounted housings serve as the air filters on passenger cars (Figs. 1 and 2). In addition to filtering the intake air, these units preheat and regulate its temperature, as well as attenuating the intake noise. Intake-air temperature regulation is important for smooth response, and also for exhaust-gas emissions. The temperatures for part and full-throttle operation can differ.
The hot air is taken off adjacent to the exhaust system and directed by means of a flap mechanism into the air filter entrance so that it can mix with the cold intake air. Regulation is usually automatic, using either a pneumatic vacuum unit connected to the intake manifold or expansion elements. Since it improves A/F mixture formation and distribution, the controlled (and thus constant) intake-air temperature has positive effects upon engine power, fuel consumption, and exhaust emissions.
In addition, and this applies particularly at very low outside temperatures, the heating-up of the intake air reduces the duration of the warm-up phase after the engine has started.

Fig. 1
Central air filter for passenger cars
1 Fresh-air intake, **2** Warm-air intake, **3** Outlet for warm/fresh air mixture, **4** Vacuum unit.

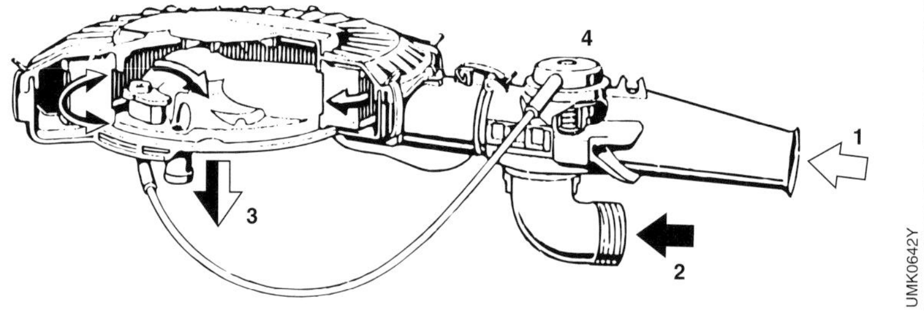

Fig. 2
Fender-mounted passenger-car air filter
1 Fresh-air intake, **2** Warm-air intake,
3 Outlet for warm/fresh air mixture.

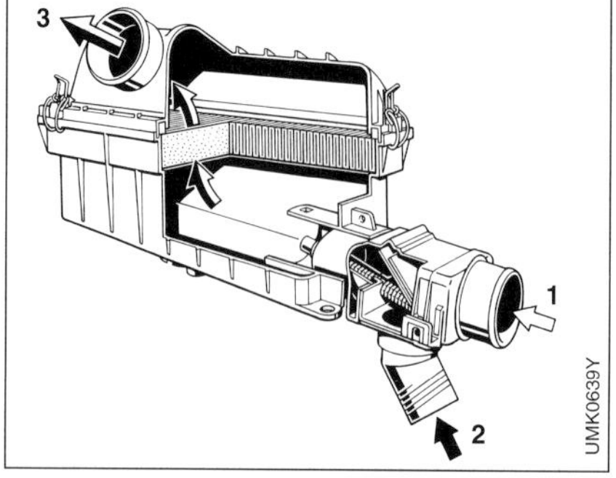

Fig. 3
Paper air filter with cyclone for commercial vehicles
1 Air intake,
2 Air outlet,
3 Cyclone air vanes,
4 Filter element,
5 Dust bowl.

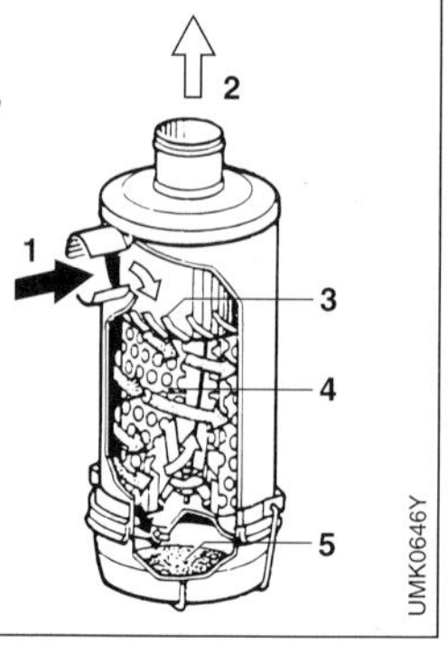

Passenger-car air filters are in the form of centrally located or fender-mounted filters with paper elements. Characteristic for these filters is their high filtering efficiency which is independent of the loading. It is a simple matter to change the paper cartridges at the intervals prescribed by the vehicle manufacturer. Passenger-car air filters must be carefully aligned to the particular engine in order to optimise power output, fuel consumption, intake-air temperature, and damping.

Commercial-vehicle air filters

Most of the air filters used in commercial vehicles are of the paper-element type, although oil-bath filters are found in isolated cases. Characteristic for paper filters is their high filtering efficiency in all load ranges and their higher level of flow resistance as levels of retained contaminants increase. The paper air filter can be supplemented by a cyclone prefilter directly incorporated in the housing to save space (Fig. 3). This is the preferred combination and is at present in widespread use. Service entails replacing the element and/or emptying the dust cup.
For paper air filters, the service interval is often indicated by a special display.
Regarding servicing, the vehicle manufacturer's instructions are to be complied with. For service simplification, specially aligned dust-removal valves can be incorporated depending upon the magnitude of the engine's air pulsation.
Cyclone's serve to extend the filter's useful life and, therefore, also the service intervals. Guide vanes in the cyclone cause the air to go into rotation, so that a large proportion of the dust is removed before the air reaches the downstream air filter. Cyclones are suitable for installation upstream of the paper air filter and/or the oil-bath air filter. Due to their inadequate filtration efficiency, cyclones are unsuitable for use as the only engine-air filter. Pertinent standard: DIN 71 459.

Intake-noise damping

In order for legal regulations concerning the vehicle's overall noise level to be complied with, it is necessary that the intake-air noise caused by passenger-car and commercial-vehicle air filters be damped. This damping is implemented almost exclusively by designing the air filter to act as a reflection sound absorber having the special shape of a Helmholtz resonator.
Acting as a suction resonator, the Helmholtz resonator has a damping effect in the area of its resonant frequency. When acting as a throughflow resonator it amplifies at resonant frequency, above which it features a wide damping range.
Assuming that the air filter is of sufficient size (a good empirical value for 4-stroke engines is 15 to 20 times the piston displacement of 1 cylinder), intake noise can generally be damped by between 10 and 20 dB(A). In case noise is excessive at specific frequencies, special supplementary dampers must be used (Fig. 4).

Fig. 4

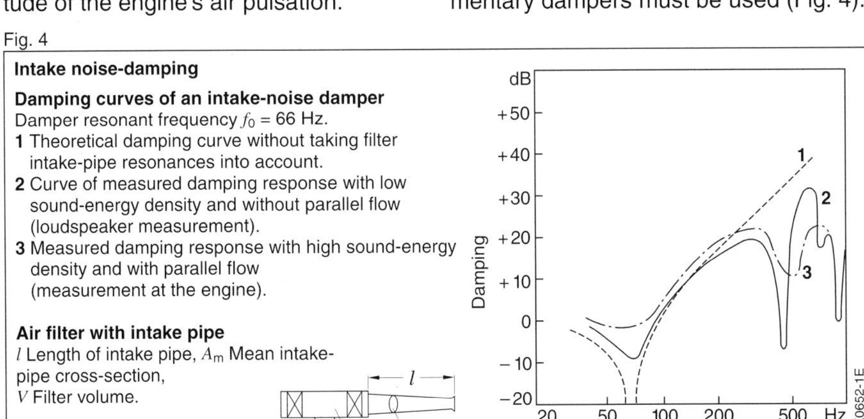

Intake noise-damping

Damping curves of an intake-noise damper
Damper resonant frequency f_0 = 66 Hz.
1 Theoretical damping curve without taking filter intake-pipe resonances into account.
2 Curve of measured damping response with low sound-energy density and without parallel flow (loudspeaker measurement).
3 Measured damping response with high sound-energy density and with parallel flow (measurement at the engine).

Air filter with intake pipe
l Length of intake pipe, A_m Mean intake-pipe cross-section,
V Filter volume.

Fuel supply

Fuel-supply system

The system for the supply of fuel to the engine comprises the following main components:

- Fuel tank
- Fuel lines
- Electric fuel pump
- Fuel filter
- Fuel rail (only for multipoint fuel injection)
- Fuel-pressure regulator.

Together, it is the task of these components to provide the engine with sufficient fuel no matter what operating conditions are concerned (cold-start, hot-fuel delivery, idle, and full-load).
An electrically driven fuel pump pumps the fuel from the fuel tank and through the fuel filter to the injection valve (injector). The electromagnetically controlled injector injects a precisely metered quantity of fuel into the engine's intake manifold. Surplus fuel flows back to the fuel tank via a fuel-pressure regulator which provides for constant fuel pressure in the system (Fig. 1).
On single-point fuel-injection (SPI) systems, fuel is injected through a single injector situated directly above the throttle valve.
In the case of multipoint fuel-injection (MPI) systems, each cylinder is allocated its own injector which is situated in the intake manifold directly before the intake valve concerned. Fuel is supplied to the individual injectors through a socalled fuel rail.

Fuel tank

According to the German equivalent of the FMVSS/CUR, the fuel tank must be corrosion-resistant and not leak even at a pressure defined as double the normal operating pressure, but at least at 0.3 bar overpressure. Suitable openings, safety valves etc. must be provided to permit excess pressure to escape. Fuel must not escape past the filler cap, nor through the pressuree-qualization devices. This also applies in the case of road shocks, or in curves, or when the vehicle is tilted. The fuel tank must be remote from the engine so that the ignition of the fuel is not to be expected even in the event of an accident. Further, more stringent regulations apply in the case of vehicles with open cabs, and for tractors and buses

Fuel lines

Fuel lines must be installed so that they cannot be adversely effected by torsional motion, engine movement, or similar, phenomena.
They can be manufactured from seamless flexible metal tubing, or flame- and fuel-resistant synthetic hose. They must be protected against mechanical damage.
All fuel-conducting components must be protected against heat which could otherwise impair correct operation. They must be positioned so that the possibility of dripping or evaporating fuel accumulating on hot components, or being ignited by electrical devices, is ruled out. Fuel lines on buses are not to pass through the passenger or driver compartment, and gravity-feed systems are forbidden.

Electric fuel pump

The use of electric pumps in passenger cars places particularly stringent demands on the pumps regarding their function, noise, size, and service life. Having a variety of different pump sizes available means that under every possible operating condition (cold start, hotfuel delivery, idle, and full load), these demands can be fulfilled for a very wide range of different injection systems and engines. Electric fuel pumps are available as in-line (Fig. 1 a), or in-tank versions (Fig. 1 b).
The in-line pump is situated outside the fuel tank in the line between tank and fuel filter, and is attached to the vehicle's body platform.

Fig. 1

Fuel-supply system (Example using multipoint fuel injection)

With electric fuel pump: **a** In-line, **b** In-tank.
1 Fuel tank, **2** Electric fuel pump, **3** Fuel filter, **4** Fuel rail, **5** Fuel injector,
6 Fuel-pressure regulator.

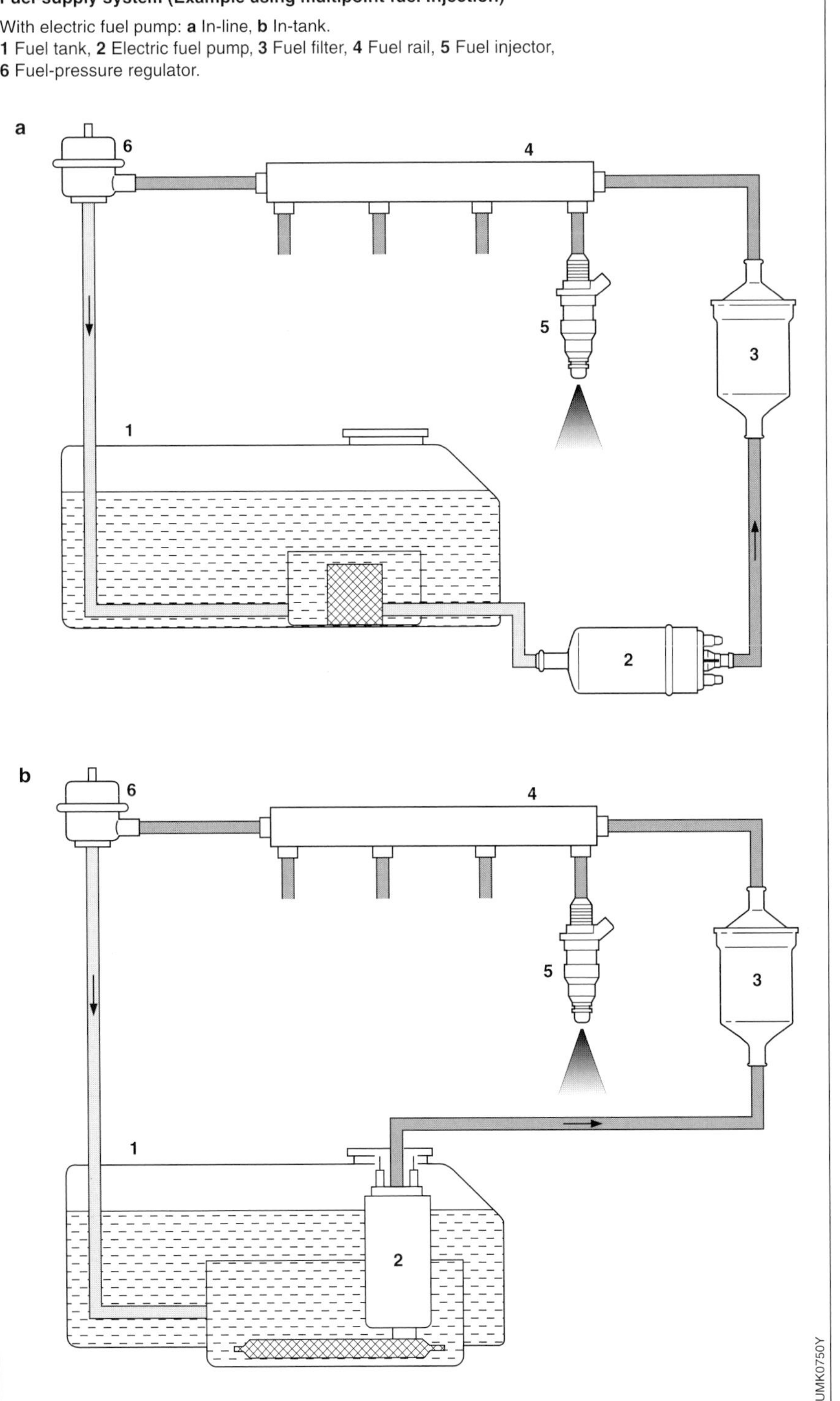

As the name implies, in-tank pumps are fitted inside the fuel tank itself in a special holder which is usually provided with an intake-side fuel strainer, a fuel-level indicator, a fuel-swirl pot which serves as a fuel reservoir, and electrical and hydraulic connections to the outside

Assignment

Commencing with the start of engine cranking, the fuel pump runs non-stop and delivers fuel continuously from the fuel tank and through the fuel filter to the engine. A fuel-pressure regulator ensures constant pressure in the injection system (constant difference between the fuel pressure and the intakemanifold pressure). Excess fuel flows back to the tank. In order to maintain the necessary fuel pressure under all operating conditions the delivery quantity always exceeds the maximum fuel quantity required by the engine.

The fuel pump is switched on by the engine-management system. A safety circuit prevents fuel delivery when the ignition is switched on but the engine is not turning.

Design

The electric fuel pump's main functional components are (Fig. 2):

- Pump cover,
- Electric motor, and
- Pump stage.

Pump cover

The electrical connections, the check valve, and the pressure-side hydraulic connection are incorporated in the pump cover.

The check valve serves to maintain the system pressure at a given level after pump switch-off in order to prevent the formation of vapor bubbles in the fuel system due to high fuel temperatures.

Interference-suppression devices can also be integrated in the pump cover.

Electric motor

The electric motor comprises a permanent-magnet system and an armature, the configuration of which is determined by the specified system pressure. The electric motor and the pump stage are contained in a common housing and permanently surrounded by fuel. This measure serves to keep them cool and permits high motor outputs without complex sealing being needed between pump stage and electric motor.

Pump stage

Since the principle used depends upon the area of application of the pump in question, there are several pump-stage versions available.

Fig. 2

Electric fuel pump (Example)

1 Pump stage, **2** Electric motor, **3** Pump cover.

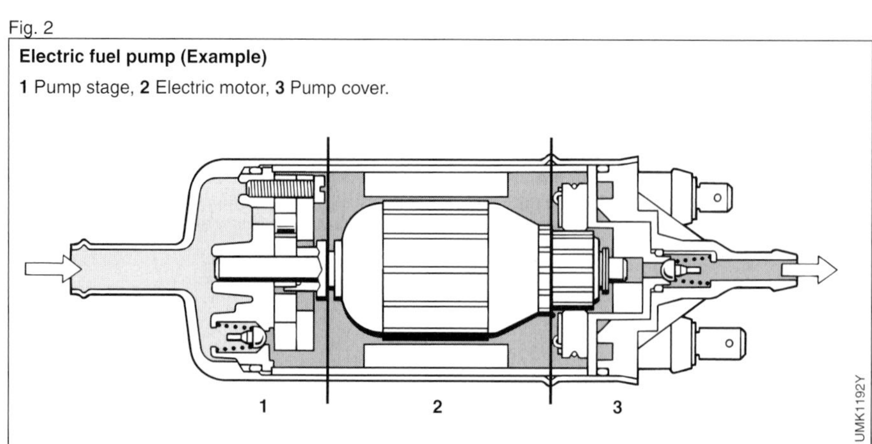

Types (Fig. 3)

Positive-displacement pumps
Roller-cell pumps for pressures up to 650 kPA and inner-gear pumps for pressures up to 400 kPa, are based on the positive-displacement principle.
Positive-displacement pumps predominate in multipoint injection systems.

Roller-cell pump: This comprises an eccentric chamber in which a slotted rotor rotates. Each slot houses a loose roller. When the rotor turns, centrifugal force assisted by the build up of fuel pressure forces these rollers outwards against the outer roller track and the leading edges of the roller grooves. The rollers now act as circulating seals, whereby a chamber is formed between adjacent rollers, the slotted roller, and the outer roller track. The pumping action results from the continuous reduction of chamber volume following the closing of the kidney-shaped intake port. When the discharge port opens, the fuel flows through the electric motor and exits from the roller-cell pump via the pressure-side pump cover.

Inner-gear pump. Comprises an internal driving wheel and an eccentrically arranged rotor which rotates along with the driving wheel. The driving wheel has one less tooth than the rotor, and this leads to a changing chamber volume when rotation takes place. In the period of increasing chamber volume, the chamber is connected to the intake port and fuel is drawn in. As soon as the rotor turns past the point of maximum chamber volume, the chamber is closed with respect to the intake area. Chamber volume now decreases, the outlet port opens, and fuel is forced out via the motor stage and through the outlet.

Flow-type pumps:
Peripheral pumps for pressures up to 400 kPa, and side-channel pumps used as the booster pumps for positive-displacement pumps for pressures up to 30 kPa, are all based upon the hydrokinetic principle. The booster pump is used to

Fig. 3
Types of electric fuel pump
a Roller-cell pump,
b Inner-gear pump,
c Peripheral pump,
d Side-channel pump.

a

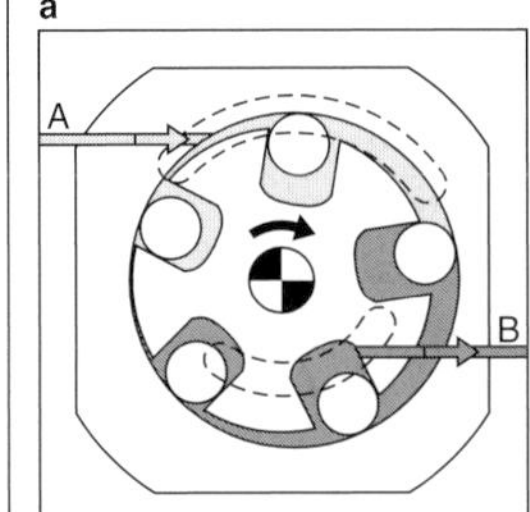

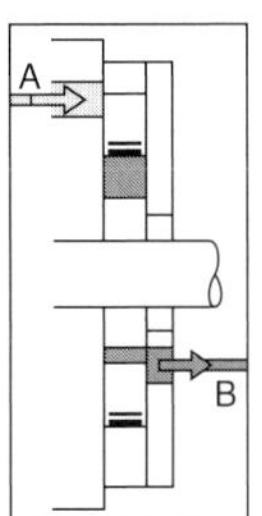

b

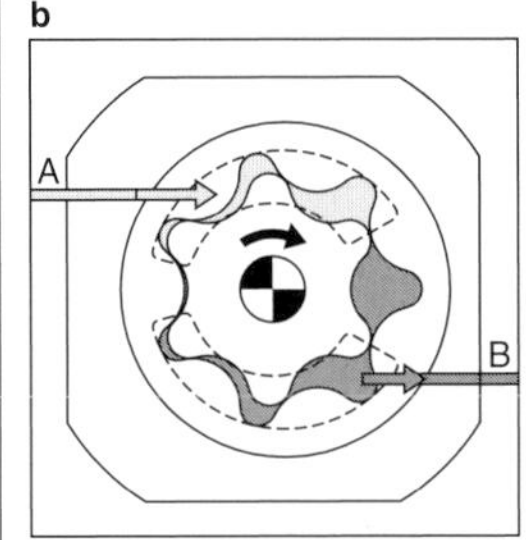

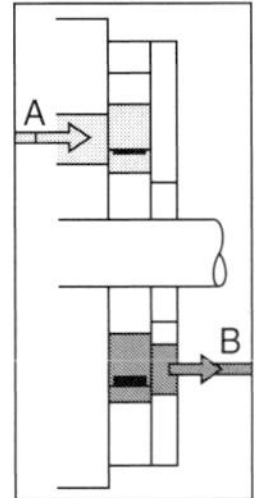

c

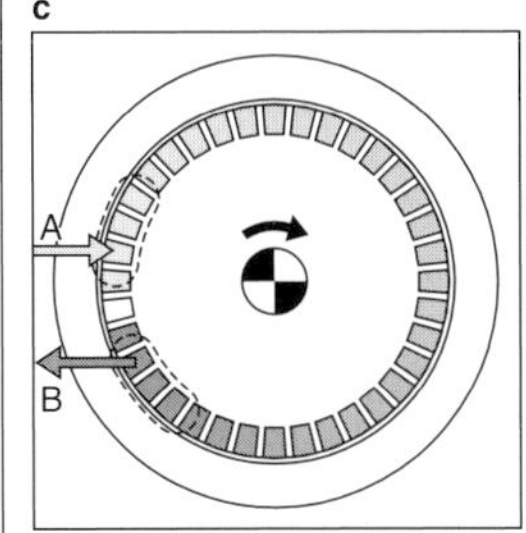

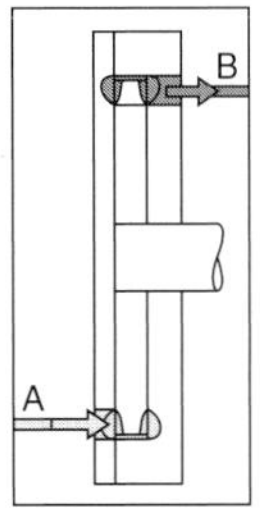

d

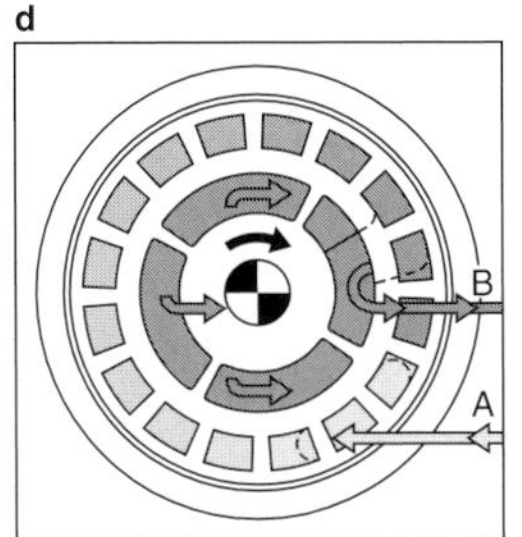

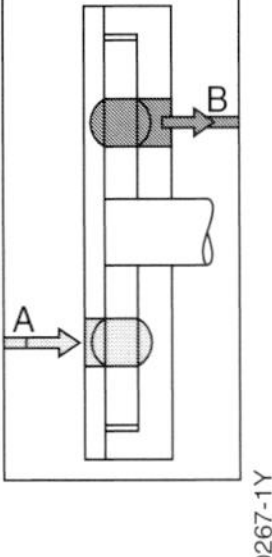

UMK0267-1Y

improve the hot-fuel handling characteristics, and removes vapor bubbles which may form, or even prevents them forming at all.
Flow-type pumps are especially suitable for applications in which noise must be avoided as far as possible.

Peripheral pumps:
The peripheral pump comprises an impeller wheel with numerous blades around its circumference. The pump housing is divided into two sections, and a channel surrounds the impeller blades around the housing's complete periphery (= peripheral). The blades in the rotating impeller ring fling the pumped liquid outwards and into the channel in which almost continuous pressure buildup takes place as a result of pulse exchange. Fuel flow is practically pulsation-free.

Side-channel pump:
The side-channel pump functions similarly to the peripheral pump, fuel also being delivered as a result of centrifugal force. The main difference is in the design of the impeller which has less blades, and in the shape and configuration of the flow channels which are only at the sides adjacent to the blades (= side-channel). The pressure of between 20 and 30 kPa generated by this configuration is lower than that with the peripheral pump. The side-channel pump is used preferably as a pre-stage (booster) pump.

Fuel filters

Since it operates with an extremely high level of precision, the spark-ignition engine's fuel-injection system needs efficiently cleaned fuel. The particles in the fuel which cause wear, and the water which leads to corrosion and swelling, are removed by special filter installations, or simple filters fitted in the fuel circuit.
The solid particles which would lead to wear are removed by a variety of effects which include, straining, diffusion, impact, and blockage. The filtration, or retention, efficiency of these individual processes depends upon particle size and throughflow velocity. The filter thickness (and with it the particle residence time in the filter material) is matched to these factors.
When a contaminated liquid flows through a filter, contaminant particles are deposited on the filter surface which accumulate to form a finely structured "filter cake" which gets thicker in the course of time. The filter cake has retention (or filtering) qualities which are identical to those of the actual filter material. In the case of fuel filters therefore, maximum efficiency is only attained after the filter cake has formed.
Paper has firmly established itself as the best filter material in comparison to felt and edge (or disk) filters. A paper fleece (Fig. 4) is built up from paper fibers and impregnated with a resinous substance. This fleece is integrated in the fuel circuit in such a manner that as far as possible, fuel flows through each filter surface at identical speed. Regular replacement of the fuel filter ensures that the fuel-injection system is efficiently protected against contamination and wear.

Fig. 4
Fuel-filter paper seen under the scanning electron microscope (SEM).

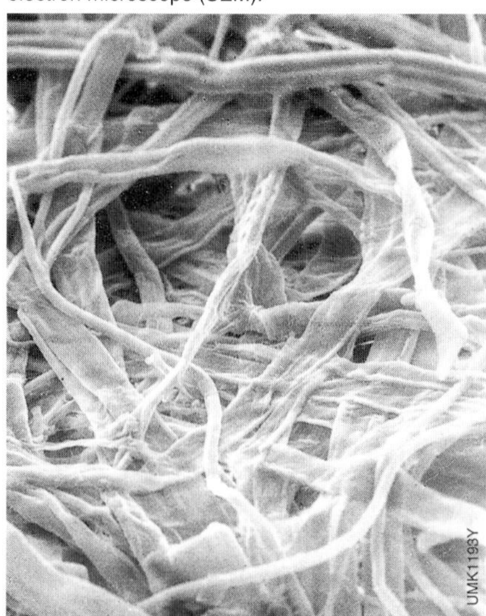

Fuel rail (fuel distributor)

In the case of multipoint injection, the fuel flows through the fuel rail where it is evenly distributed to all injectors. In addition to the injectors, the fuel rail usually incorporates the fuel-pressure regulator and possibly also a pressure attenuator. The fuel rail's dimensions are carefully selected to inhibit local fuel-pressure fluctuations which could otherwise be triggered due to resonances occurring during the opening and closing of the injectors. This prevents the injection quantities from reacting to changes in load and engine speed. Depending upon the particular vehicle type and its special requirements, the fuel rail can be made of steel, aluminum, or plastic. It may also include an integral test valve, which can be used to bleed off pressure for servicing as well as for test purposes.

Fuel-pressure regulator

The injected fuel quantity should be determined exclusively by injection duration. Thus the difference between the fuel pressure in the fuel rail and the pressure in the intake tract must remain constant. A means is thus required for adjusting the fuel pressure to reflect variations in the load-sensitive manifold pressure. The fuel-pressure regulator therefore regulates the amount of fuel returning to the tank so that a constant pressure drop is maintained across the injectors. With multipoint injection, the pressure regulator is generally positioned at the far end of the fuel rail to avoid impairing the flow within the rail. However, it can also be mounted in the fuel return line.
With single-point fuel injection, the fuel-pressure regulator is integrated in the central injection unit.
The fuel-pressure regulator is designed as a diaphragm-controlled overflow pressure regulator. A rubber-fiber diaphragm divides the pressure regulator into two sections: fuel chamber and spring chamber. The spring presses against a valve holder integrated within the diaphragm. This force causes a flexibly mounted valve plate to push against a valve seat. When the pressure exerted against the diaphragm by the fuel exceeds that of the spring, the valve opens and allows fuel to flow directly back to the tank until the diaphragm assembly returns to a state of equilibrium, with equal pressure exerted on both of its sides.
In the case of multipoint injection, a pneumatic line is provided between the spring chamber and the intake manifold downstream from the throttle valve, allowing the chamber to respond to changes in manifold vacuum. Thus the pressures at the diaphragm correspond to those at the injectors. As a result, the pressure drop at the injectors remains constant, as it is determined solely by the spring force and surface area of the diaphragm (Fig. 5).
With single-point injection, vent openings ensure that the same ambient pressure is present in the pressure regulator's spring chamber as at the injector's injection point.

Fig. 5

Fuel-pressure regulator

1 Intake-manifold connection, **2** Spring, **3** Valve holder, **4** Diaphragm, **5** Valve, **6** Fuel inlet, **7** Fuel return.

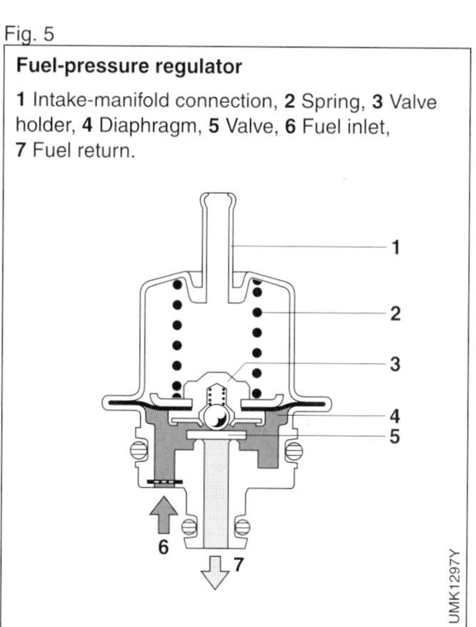

Ignition

Function

The function of the ignition system is to initiate combustion in the compressed air/fuel mixture by igniting it at precisely the right instant. In the spark-ignition engine, this function is assumed by an electric spark in the form of a short-duration discharge arc between the spark plug's electrodes.
Consistently reliable ignition is vital for efficient catalytic-converter operation. Ignition miss allows uncombusted gases to enter the catalytic converter, leading to its damage or destruction from overheating when these gases burn inside it.

Technical requirements

An electrical arc with an energy content of approximately 0.2 mJ is required for each sustainable ignition of a stoichiometric mixture, while up to 3 mJ may be needed for richer or leaner mixtures. This energy is only a fraction of the total (ignition) energy contained in the ignition spark. If the available ignition energy is inadequate, the mixture cannot ignite since ignition fails to take place, and the result is that the engine starts to misfire. This is why the system must supply levels of ignition energy that are high enough to always ensure reliable inflammation of the air/fuel mixture, even under the most severe conditions. A small ignitable mixture cloud passing by the arc is enough to initiate the process. The mixture cloud ignites and propagates combustion through the remaining mixture in the cylinder. Efficient mixture formation and easy access of the mixture cloud to the spark will improve ignition response, as will extended spark durations and larger electrode gaps (longer arcs). The location and length of the spark are determined by the spark plug's design dimensions. Spark duration is governed by the design and configuration of the ignition system along with the instantaneous ignition conditions.

Ignition timing

Ignition timing and its adjustment

Approximately two milliseconds elapse between the instant when the mixture ignites and its complete combustion. Assuming consistent mixture strength, this period will remain invariable. This means that the ignition spark must arc early enough to support generation of optimal combustion pressure under all operating conditions.
Standard practice defines ignition timing relative to top dead center, or TDC on the crankshaft. Advance angles are then quantified in degrees before TDC, with the corresponding figure being known as the ignition (timing advance) angle. Moving the ignition point back toward TDC is referred to as "retarding" the timing and displacing it forward toward an earlier ignition (firing) point is "advancing" it (Figure 1).
Ignition timing must be selected so that the following criteria are complied with:
- Maximum engine power,
- Maximum fuel economy,
- Prevention of engine knock, and
- "Clean" exhaust gas.

Fig. 1

Position of crankshaft and piston at the ignition (firing) point with advanced ignition

TDC Top Dead Center, BDC Bottom Dead Center, Z Ignition point.

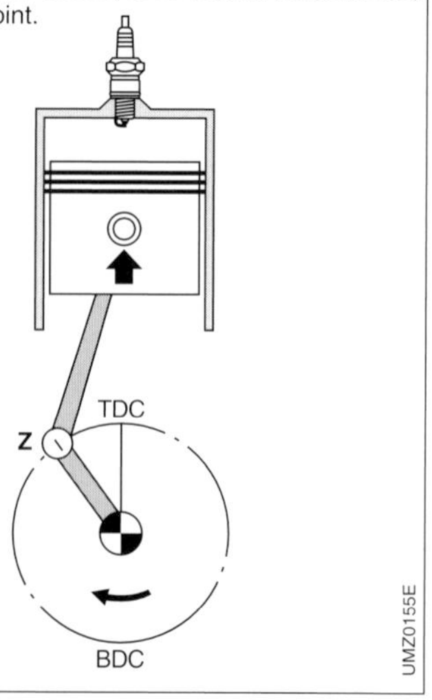

It is impossible to fulfill all the above demands simultaneously, and a compromise must be reachd from case to case. The most favorable firing point at a given torque depends upon a variety of different factors. These are in particular, engine speed, engine load, engine design, fuel, and the particular operating conditions (e.g. starting, idle, WOT, overrun).
Engine knock is due to the abrupt combustion of portions of the air-fuel mixture which have not yet been reached by the advancing flame front triggered by the ignition spark. In this case, the firing point is too far advanced. Combustion knock not only leads to increases in combustion-chamber temperature, which in turn can cause pre-ignition, but also to marked increases in pressure. Such abrupt ignition events generate pressure oscillations which are superimposed on the normal pressure characteristic (Fig. 2).
Today, the high compressions employed in spark-ignition engines involve a far greater risk of combustion knock than was the case with the compression ratios which were common in the past. One differentiates between two different forms of "knock":

- Acceleration knock at low engine speeds and high load (clearly audible as pinging), and
- High-speed knock at high engine speeds and high load.

For the engine, high-speed knock is a particularly critical factor, since the other engine noises generated at such speeds make it inaudible. This is why audible knock is not a faithful index of preignition tendency. At the same time, electronic means are available for precise detection. Consistent knock causes severe engine damage (destruction of cylinder-head gaskets, bearing damage, "holed" piston crowns) as well as spark-plug damage.
Preignition tendency depends upon such factors as engine design (for instance: combustion chamber layout, homogenous air-fuel mixture, efficient induction flow passages) and fuel quality.

Ignition timing and emissions

The effects of the excess-air factor λ and ignition timing on specific fuel consumption and exhaust emissions are demonstrated in Figures 3 and 4. Specific fuel consumption responds to leaner mixtures with an initial dip before rising

Fig. 2

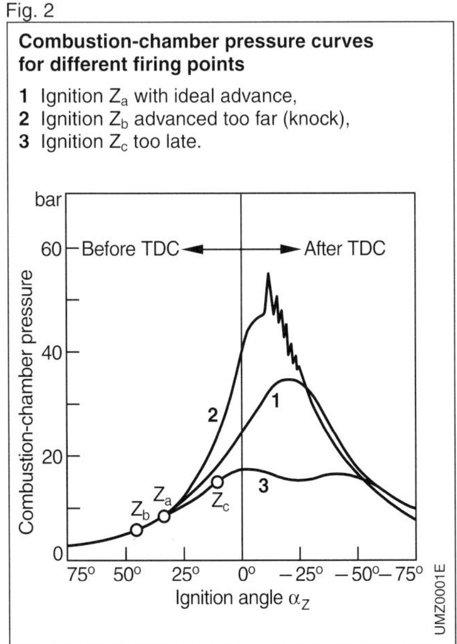

Combustion-chamber pressure curves for different firing points

1 Ignition Z_a with ideal advance,
2 Ignition Z_b advanced too far (knock),
3 Ignition Z_c too late.

Fig. 3

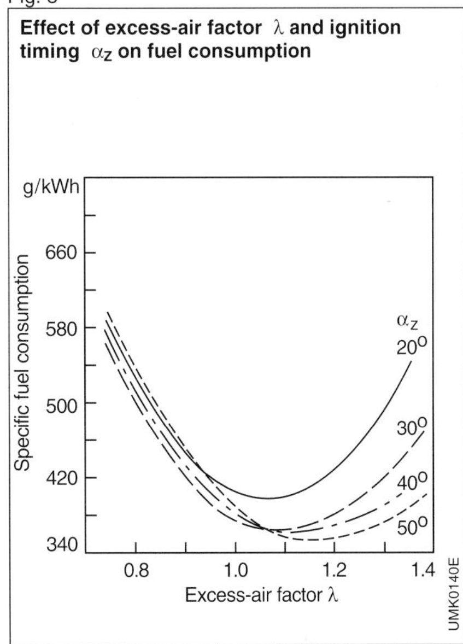

Effect of excess-air factor λ and ignition timing α_Z on fuel consumption

from $\lambda = 1.1...1.2$. Increases in the excess-air factor are accompanied by a corresponding increase in the optimal ignition advance angle, which is defined here as the timing that will minimize specific fuel consumption. The relationship between specific fuel consumption and excess-air factor (assuming optimal ignition timing) can be explained as follows: The air deficiency encountered in the "fuel-rich" range leads to incomplete combustion, while substantial shifts toward the lean misfire limit (LML) will start to cause delayed combustion and misfiring, ultimately leading to higher levels of specific fuel consumption. The optimal ignition advance angle increases at higher excess-air ratios owing to the slower rate of flame-front propagation enountered in lean mixtures; the ignition timing must be advanced to compensate for these delays.

HC emissions, which bottom out at $\lambda = 1.1$, display a similar response pattern. The initial rise within the lean range can be attributed to the flame being extinguished due to the cooling on the walls of the combustion chamber. Extremely lean mixtures produce delayed combustion and failure to ignite, phenomena which occur with increasing frequency as the lean misfire limit is approached. Below $\lambda = 1.2$, further ignition advance will lead to higher HC emissions, but it will also shift the lean misfire limit to accomodate mixtures with even less fuel. This is why an increase in ignition advance lowers the levels of HC emissions in the lean range beyond $\lambda = 1.25$.

Emissions of nitrous oxides (NO_x) display a completely different pattern by rising in response to higher oxygen (O_2) concentrations and maximum peak combustion temperatures. The result is the characteristic bell-shaped curve for NO_x emissions. These rise up to $\lambda \approx 1.05$ in response to the accompanying increases in O_2 concentrations and peak combustion temperatures. Then, beyond $\lambda = 1.05$, NO_x generation displays a sharp drop as the mixture continues further into the lean range, owing to the

Fig. 4

Excess-air factor λ and ignition timing α_z on exhaust emissions

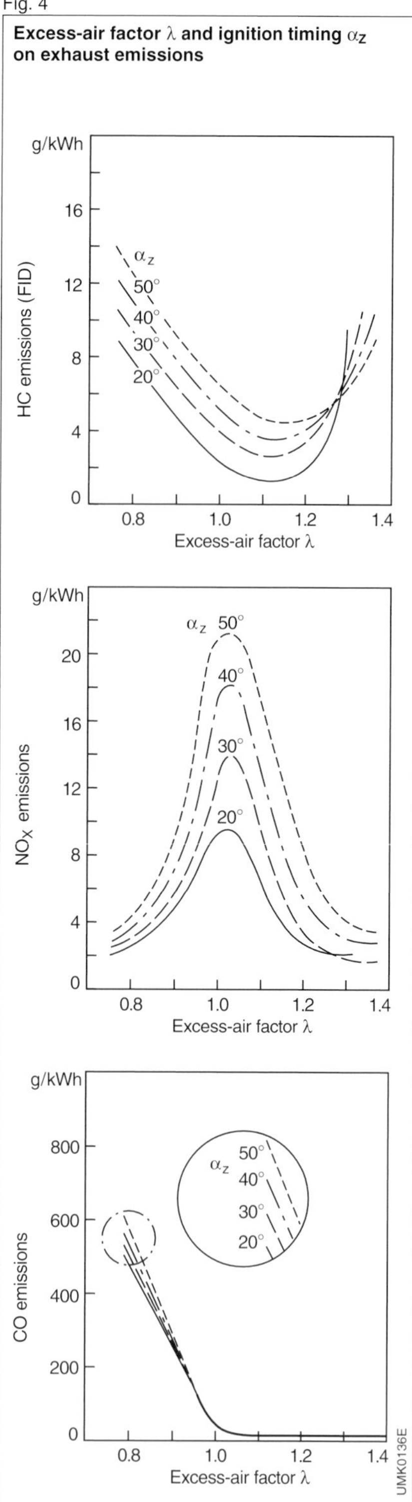

rapid reduction in peak temperatures that accompanies higher levels of mixture dilution. This response pattern also accounts for the extreme sensitivity with which NO_x emissions respond to changes in ignition timing, escalating sharply as advance is increased.
Because a mixture of $\lambda = 1$ is needed to implement emissions-control concepts relying on the 3-way catalytic converter, adjusting the ignition advance angle is the only remaining option for optimizing emissions.

Inductive ignition systems

The spark-ignition engine's inductive (coil) ignition system generates the high-tension voltage to provide the energy then employed to create an arc at the spark plug. While inductive ignition systems rely on coils to store ignition energy, an available alternative is storage in a condenser (→ so-called high-voltage capacitor-discharge ignition/CDI). The inductive ignition circuit's components are the driver (output amplifier) stage, the coil and the spark plug

Ignition coil

Function

The ignition coil stores the required ignition energy and generates the high voltages required to produce an arc at the firing point.

Design and function

Ignition-coil operation is based on an inductive concept. The coil consists of two magnetically coupled copper coils (primary and secondary windings). The energy stored in the primary winding's magnetic field is transmitted to the secondary side. Current and voltage are transformed in accordance with the turns ratio of the primary and secondary windings (Fig. 1).

Modern ignition coils feature an iron core, composed of individual metal plates inside a synthetic casing. Within this casing the primary winding is wound around a bobbin mounted directly on the core. These elements are concentrically enclosed by the secondary winding, which is designed as a disc or chamber winding for improved insulation resistance. For effective insulation of core and windings, these elements are all enclosed in epoxy resin inside the casing. Specific design configurations are selected to reflect individual operational requirements.

Ignition driver stage

Assignment and function

Ignition driver stages featuring multi-stage power transistors switch the flow of primary current through the coil, replacing the contact-breaker points employed in earlier systems.
In addition, this ignition driver stage is also responsible for limiting primary current and primary voltage. The primary voltage is limited to prevent excessively steep increases of secondary voltage,

Fig. 1

Ignition coils (schematic)

Rotating distribution: **a** Single-spark coil.
Static distribution: **b** Single-spark coil,
c Dual-spark coil.

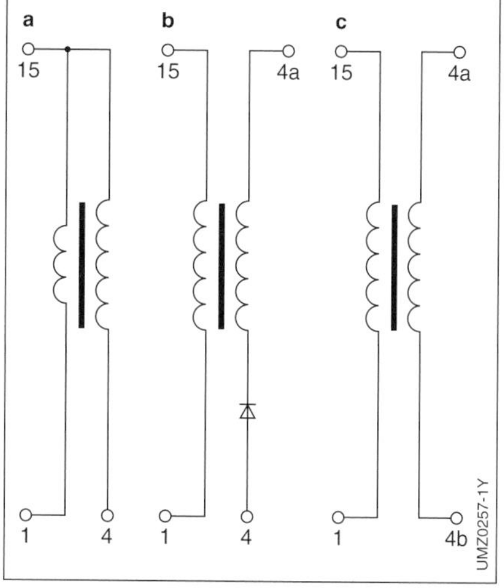

UMZ0257-1Y

which could damage components within the high-tension circuit. Restrictions on primary current hold the ignition system's energy output to the specified level.
The ignition system's driver stage may be internal (integrated within the ignition ECU) or external (mounted locally).

High-voltage generation

The ignition ECU switches on the ignition driver stage for the calculated dwell period. It is within this period that the primary current within the coil climbs to its specified intensity.
The energy for the ignition system is stored in the coil's magnetic field and defined by the levels of the coil's primary current and primary inductance.
At the firing point the ignition driver stage interrupts the current flow through the primary winding to induce flux in the magnetic field and generate secondary voltage in the coil's secondary winding.
The ultimate level of secondary voltage (secondary voltage supply) depends upon a number of factors. These include the amount of energy stored in the ignition system, the capacity of the windings and the coil's transformation ratio as well as the secondary load factor and the restrictions on primary voltage imposed by the ignition system's driver stage.
The secondary voltage must always exceed the level required to produce an arc at the spark plug (ignition-voltage requirement), and the spark energy must always be high enough to reliably initiate combustion in the mixture, even in the face of secondary arcing.
When primary current is switched on, this induces an undesired voltage (switch-on voltage) of roughly 1...2 kV in the secondary winding whose polarity opposes that of the high voltage. It is essential that this is prevented from generating an arc (switch-on arc) at the spark-plug.
In systems with conventional rotating voltage distribution, this switch-on spark is effectively suppressed by the distributor's spark gap. On distributorless ignition systems with non-rotating (static) voltage distribution featuring dedicated ignition coils, a diode in the high-voltage circuit performs this function.
With distributorless (static) spark distribution and dual-spark coils, the high arcing voltage associated with two spark plugs connected in series effectively suppresses the switch-on spark without any need for supplementary counter-measures.

Voltage distribution

High-tension voltage must be on hand at the spark plug at the moment of ignition (firing point). This function is the responsibility of the high-voltage distribution system.

Rotating voltage distribution

Systems using a rotating voltage-distribution concept rely on a mechanical ignition distributor to relay the high voltage from a single ignition coil to the individual cylinders. This type of voltage-distribution has ceased to be relevant in the current generation of engine-management systems.

Static voltage distribution

Distributorless ignition (otherwise known as static or electronic ignition) is available in two different versions:

System equipped with single-spark ignition coils

Each cylinder is equipped with its own ignition coil and driver stage, which the engine-management ECU triggers sequentially in the defined firing order. Because internal voltage loss within a distributor is no longer a consideration, the coils can be extremely compact. The preferred installation location is directly above the spark plug. Static distribution with single-spark ignition coils is universally suited for use with any number of cylinders. While there are no inherent restrictions on adjusting ignition advance (timing), these units do require a supplementary synchronization arrangement furnished by a camshaft sensor.

System equipped with dual-spark ignition coils

Each set of two cylinders is supplied by a single ignition driver stage and one coil, with each end of the latter's secondary winding being connected to a different spark plug. The cylinders are paired so that the compression stroke on one will coincide with the exhaust stroke on the other.

When the ignition fires an arc is generated at both spark plugs simultaneously. Because it is important to ensure that the spark produced during the exhaust stroke will ignite neither residual nor fresh incoming gases, this system is characterized by restrictions on adjusting ignition advance (timing). This system does not require a synchronization sensor at the camshaft.

Connectors and interference suppressors

High-voltage cables

The high voltage from the ignition coil must be able to reach the spark plugs. On coils not mounted in direct electrical contact with the spark plugs this function is performed by special high-voltage cables featuring outstanding high-voltage strength and synthetic insulation. Fitted with the appropriate terminals, these cables provide the electrical connections between the high-voltage components.

Because every high-voltage lead represents a capacitive load for the ignition system and reduces the available supply of secondary voltage accordingly, cables should always be as short as possible.

Interference resistors, interference suppression

The pulse-shaped, high-tension discharge that characterizes every arc at the spark plug also represents a source of radio interference. The current peaks associated with discharge are limited by suppression resistors in the high-voltage circuit. To hold radiation of interference emanating from this circuit to a minimum, the suppression resistors should be installed as close as possible to the actual interference source.

Resistors (capacitors) for interference suppression are generally installed in the spark-plug cable terminals, while rotating distributors also include rotor-mounted resistors. Spark plugs with integral suppression resistors are also available.

It is important to remember that higher levels of resistance in the secondary circuit are synonymous with corresponding energy loss in the ignition circuit, and result in a reduction in the energy available for firing the spark plug.

Partial or comprehensive encapsulation of the ignition system can be implemented to obtain further reductions in interference radiation.

Spark plug

The spark plug creates the electrical arc that ignites the air-fuel mixture within the combustion chamber.

The spark plug is a ceramic-insulated, high-voltage conductor leading into the combustion chamber. Once arcing voltage is reached, electrical energy flows between the center and ground electrodes to convert the remainder of the ignition-coil energy into a spark.

The level of the voltage required for ignition depends upon a variety of factors including electrode gap, electrode geometry, combustion-chamber pressure, and the instantaneous A/F ratio at the firing point.

Spark-plug electrodes are subject to wear in the course of normal engine operation, and this wear leads to progressively higher voltage requirements. The ignition system must be capable of providing enough secondary voltage to ensure that adequate ignition voltage always remains available, regardless of the operating conditions encountered in the intervals between spark-plug replacements.

Gasoline-injection systems

Carburetors and gasoline-injection systems are designed for a single purpose: To supply the engine with the optimal air-fuel mixture for any given operating conditions. Gasoline injection systems, and electronic systems in particular, are better at maintaining air-fuel mixtures within precisely defined limits, which translates into superior performance in the areas of fuel economy, comfort and convenience, and power. Increasingly stringent mandates governing exhaust emissions have led to a total eclipse of the carburetor in favor of fuel injection.
Although current systems rely almost exclusively on mixture formation outside the combustion chamber, concepts based on internal mixture formation – with fuel being injected directly into the combustion chamber – were actually the foundation for the first gasoline-injection systems. As these systems are superb instruments for achieving further reductions in fuel consumption, they are now becoming an increasingly significant factor.

Overview

Systems with external mixture formation

The salient characteristic of this type of system is the fact that it forms the air-fuel mixture outside the combustion chamber, inside the intake manifold.

Multipoint fuel injection

Multipoint fuel injection forms the ideal basis for complying with the mixture-formation criteria described above. In this type of system each cylinder has its own injector discharging fuel into the area directly in front of the intake valve. Representative examples are the various versions of the KE and L-Jetronic systems (Figure 1).

Mechanical injection systems
The K-Jetronic system operates by injecting continually, without an external drive being necessary. Instead of being determined by the injection valve, fuel mass is regulated by the fuel distributor.

Combined mechanical-electronic fuel injection
Although the K-Jetronic layout served as the mechanical basis for the KE-Jetronic system, the latter employs expanded data-monitoring functions for more precise adaptation of injected fuel quantity to specific engine operating conditions.

Electronic injection systems
Injection systems featuring electronic control rely on solenoid-operated injection

Fig. 1

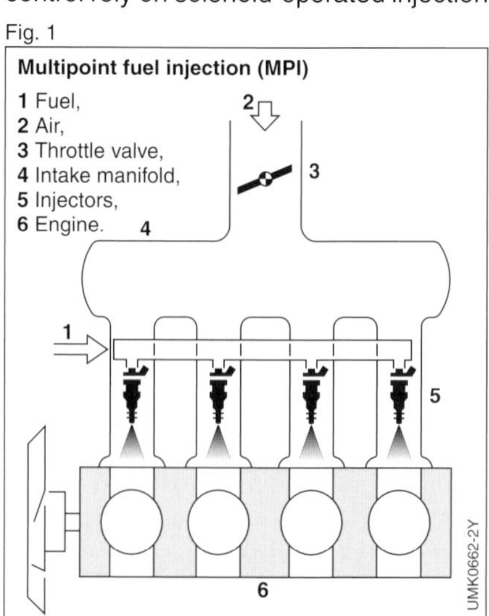

valves for intermittent fuel discharge. The actual injected fuel quantity is regulated by controlling the injector's opening time (with the pressure-loss gradient through the valve being taken into account in calculations as a known quantity).
Examples: L-Jetronic, LH-Jetronic, and Motronic as an integrated engine-management system.

Single-point fuel injection

Single-point (throttle-body injection (TBI)) fuel injection is the concept behind this electronically-controlled injection system in which a centrally located solenoid-operated injection valve mounted upstream from the throttle valve sprays fuel intermittently into the manifold. Mono-Jetronic and Mono-Motronic are the Bosch systems in this category (Figure 2).

Systems for internal mixture formation

Direct-injection (DI) systems rely on solenoid-operated injection valves to spray fuel directly into the combustion chamber; the actual mixture-formation process takes place within the cylinders, each of which has its own injector (Figure 3). Perfect atomization of the fuel emerging from the injectors is vital for efficient combustion.
Under normal operating conditions, DI engines draw in only air instead of the combination of air and fuel common to conventional injection systems. This is one of the new system's prime advantages: It banishes all potential for fuel condensation within the runners of the intake manifold. External mixture formation usually provides a homogenous, stoichiometric air-fuel mixture throughout the entire combustion chamber. In contrast, shifting the mixture-preparation process into the combustion chamber provides for two distinctive operating modes:
With stratified-charge operation, only the mixture directly adjacent to the spark plug needs to be ignitable. The remainder of the air-fuel charge in the combustion chamber can consist solely of fresh and residual gases, without unburned fuel. This strategy furnishes an extremely lean overall mixture for idling and part-throttle operation, with commensurate reductions in fuel consumption.
Homogenous operation reflects the conditions encountered in external mixture formation by employing uniform consistency for the entire air-fuel charge throughout the combustion chamber. Under these conditions all of the fresh air within the chamber participates in the combustion process. This operational mode is employed for WOT operation.
MED-Motronic is used for closed-loop control of DI gasoline engines.

Fig. 2

Throttle-body fuel injection (TBI)

1 Fuel,
2 Air,
3 Throttle valve,
4 Intake manifold,
5 Injector,
6 Engine.

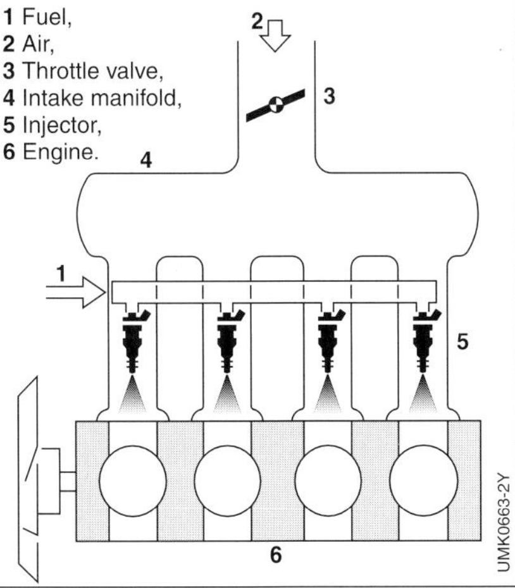

1
2
3
4
5
6
UMK0663-2Y

Fig. 3

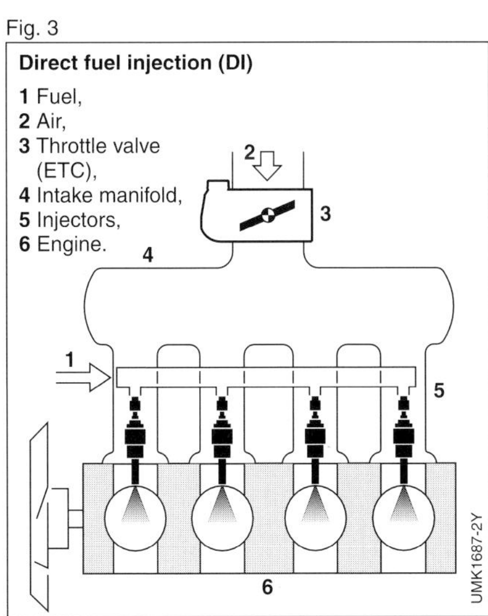

Direct fuel injection (DI)

1 Fuel,
2 Air,
3 Throttle valve (ETC),
4 Intake manifold,
5 Injectors,
6 Engine.

The story of fuel injection

The story of fuel injection extends back to cover a period of almost one hundred years.

The Gasmotorenfabik Deutz was manufacturing plunger pumps for injecting fuel in a limited production series as early as 1898.

A short time later the uses of the venturi-effect for carburetor design were discovered, and fuel-injection systems based on the technology of the time ceased to be competitive.

Bosch started research on gasoline-injection pumps in 1912. The first aircraft engine featuring Bosch fuel injection, a 1,200-hp unit, entered series production in 1937; problems with carburetor icing and fire hazards had lent special impetus to fuel-injection development work for the aeronautics field. This development marks the beginning of the era of fuel injection at Bosch, but there was still a long path to travel on the way to fuel injection for passenger cars.

1951 saw a Bosch direct-injection unit being featured as standard equipment on a small car for the first time. Several years later a unit was installed in the 300 SL, the legendary production sports car from Daimler-Benz.

In the years that followed, development on mechanical injection pumps continued, and ...

In 1967 fuel injection took another giant step forward: The first electronic injection system: the intake-pressure-controlled D-Jetronic!

In 1973 the air-flow-controlled L-Jetronic appeared on the market, at the same time as the K-Jetronic, which featured mechanical-hydraulic control and was also an air-flow-controlled system.

In 1976, the K-Jetronic was the first automotive system to incorporate a Lambda closed-loop control.

1979 marked the introduction of a new system: Motronic, featuring digital processing for numerous engine functions. This system combined L-Jetronic with electronic program-map control for the ignition. The first automotive microprocessor!

In 1982, the K-Jetronic model became available in an expanded configuration, the KE-Jetronic, including an electronic closed-loop control circuit and a Lambda oxygen sensor.

These were joined by Bosch Mono-Jetronic in 1987: This particularly cost-efficient single-point injection unit made it feasible to equip small vehicles with Jetronic, and once and for all made the carburetor absolutely superfluous.

By the end of 1997, around 64 million Bosch engine-management systems had been installed in countless types of vehicles since the introduction of the D-Jetronic in 1967. In 1997 alone, the figure was 4.2 million, comprised of 1 million throttle-body injection (TBI) systems and 3.2 million multipoint fuel-injection (MPI) systems.

Bosch gasoline fuel injection from the year 1954

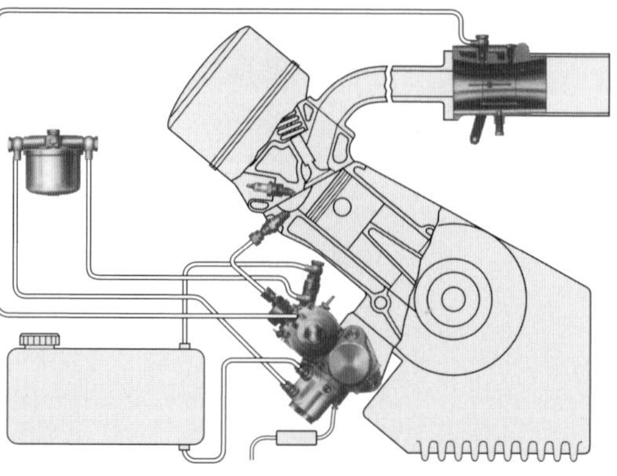

K-Jetronic

System overview

The K-Jetronic is a mechanically and hydraulically controlled fuel-injection system which needs no form of drive and which meters the fuel as a function of the intake air quantity and injects it continuously onto the engine intake valves.
Specific operating conditions of the engine require corrective intervention in mixture formation and this is carried out by the K-Jetronic in order to optimize starting and driving performance, power output and exhaust composition. Owing to the direct air-flow sensing, the K-Jetronic system also allows for engine variations and permits the use of facilities for exhaust-gas aftertreatment for which precise metering of the intake air quantity is a prerequisite.
The K-Jetronic was originally designed as a purely mechanical injection system. Today, using auxiliary electronic equipment, the system also permits the use of lambda closed-loop control.
The K-Jetronic fuel-injection system covers the following functional areas:
- Fuel supply,
- Air-flow measurement and
- Fuel metering.

Fuel supply

An electrically driven fuel pump delivers the fuel to the fuel distributor via a fuel accumulator and a filter. The fuel distributor allocates this fuel to the injection valves of the individual cylinders.

Air-flow measurement

The amount of air drawn in by the engine is controlled by a throttle valve and measured by an air-flow sensor.

Fuel metering

The amount of air, corresponding to the position of the throttle plate, drawn in by the engine serves as the criterion for metering of the fuel to the individual cylinders. The amount of air drawn in by the engine is measured by the air-flow sensor which, in turn, controls the fuel distributor. The air-flow sensor and the fuel distributor are assemblies which form part of the mixture control unit. Injection occurs continuously, i.e. without regard to the position of the intake valve. During the intake-valve closed phase, the fuel is "stored". Mixture enrichment is controlled in order to adapt to various operating conditions such as start, warm-up, idle and full load. In addition, supplementary functions such as overrun fuel cutoff, engine-speed limiting and closed-loop lambda control are possible.

Fig. 1
Functional schematic of the K-Jetronic

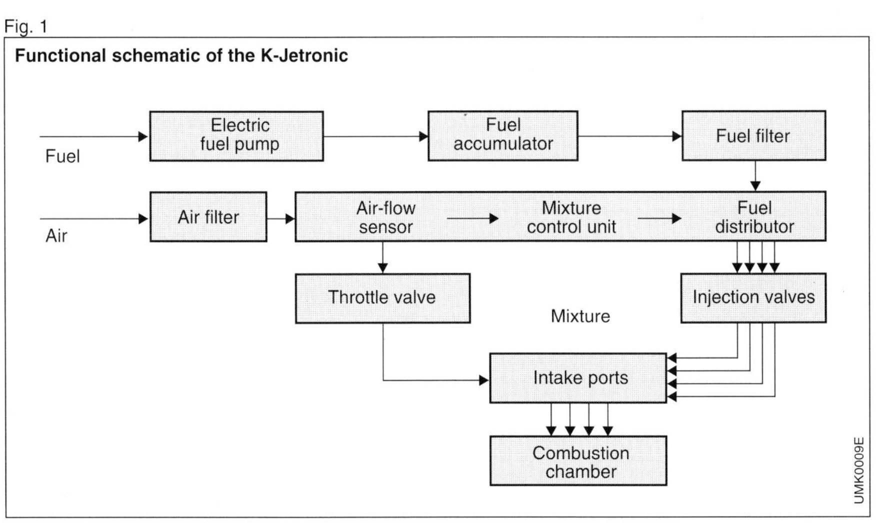

Fuel supply

The fuel supply system comprises
- Electric fuel pump,
- Fuel accumulator,
- Fine filter,
- Primary-pressure regulator and
- Injection valves.

An electrically driven roller-cell pump pumps the fuel from the fuel tank at a pressure of over 5 bar to a fuel accumulator and through a filter to the fuel distributor. From the fuel distributor the fuel flows to the injection valves. The injection valves inject the fuel continuously into the intake ports of the engine. Thus the system designation K (taken from the German for continuous). When the intake valves open, the mixture is drawn into the cylinder.
The fuel primary-pressure regulator maintains the supply pressure in the system constant and reroutes the excess fuel back to the fuel tank.
Owing to continual scavenging of the fuel supply system, there is always cool fuel available. This avoids the formation of fuel-vapor bubbles and achieves good hot starting behavior.

Electric fuel pump

The electric fuel pump is a roller-cell pump driven by a permanent-magnet electric motor.
The rotor plate which is eccentrically mounted in the pump housing is fitted with metal rollers in notches around its circumference which are pressed against the pump housing by centrifugal force and act as rolling seals. The fuel is carried in the cavities which form between the rollers. The pumping action takes place when the rollers, after having closed the inlet bore, force the trapped fuel in front of them until it can escape from the pump through the outlet bore (Figure 4). The fuel flows directly around the electric motor. There is no danger of explosion, however, because there is never an ignitable mixture in the pump housing.

Fig. 2

Schematic diagram of the K-Jetronic system with closed-loop lambda control

1 Fuel tank, **2** Electric fuel pump, **3** Fuel accumulator, **4** Fuel filter, **5** Warm-up regulator, **6** Injection valve, **7** Intake manifold, **8** Cold-start valve, **9** Fuel distributor, **10** Air-flow sensor, **11** Timing valve, **12** Lambda sensor, **13** Thermo-time switch, **14** Ignition distributor, **15** Auxiliary-air device, **16** Throttle-valve switch, **17** ECU, **18** Ignition and starting switch, **19** Battery.

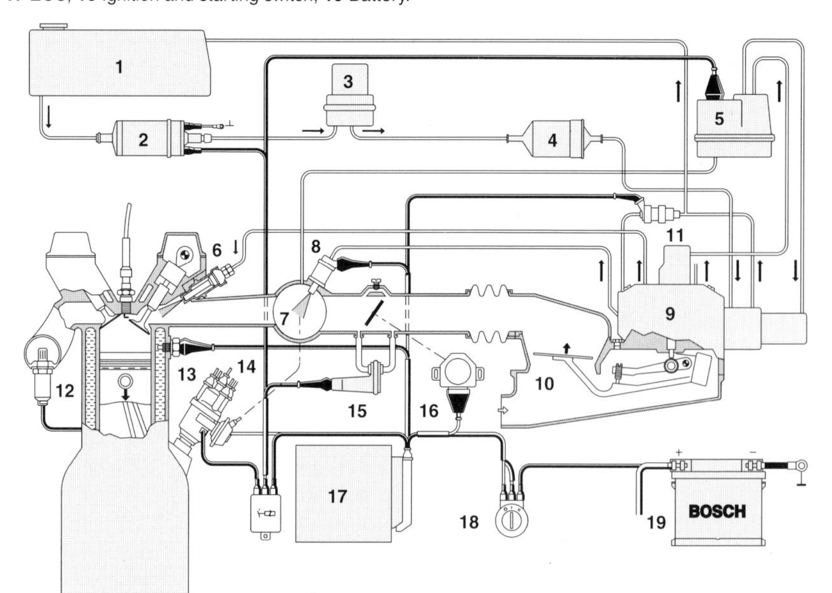

The electric fuel pump delivers more fuel than the maximum requirement of the engine so that compression in the fuel system can be maintained under all operating conditions. A check valve in the pump decouples the fuel system from the fuel tank by preventing reverse flow of fuel to the fuel tank.

The electric fuel pump starts to operate immediately when the ignition and starting switches are operated and remains switched on continuously after the engine has started. A safety circuit is incorporated to stop the pump running and, thus, to prevent fuel being delivered if the ignition is switched on but the engine has stopped turning (for instance in the case of an accident).

The fuel pump is located in the immediate vicinity of the fuel tank and requires no maintenance.

Fuel accumulator

The fuel accumulator maintains the pressure in the fuel system for a certain time after the engine has been switched off in order to facilitate restarting, particularly when the engine is hot. The special design of the accumulator housing (Figure 5) deadens the sound of the fuel pump when the engine is running.

The interior of the fuel accumulator is divided into two chambers by means of a diaphragm. One chamber serves as the accumulator for the fuel whilst the other represents the compensation volume and is connected to the atmosphere or to the fuel tank by means of a vent fitting. During operation, the accumulator chamber is filled with fuel and the diaphragm is caused to bend back against the force of the spring until it is halted by the stops in the spring chamber. The diaphragm remains in this position, which corresponds to the maximum accumulator volume, as long as the engine is running.

Electric fuel pump

1 Suction side, **2** Pressure limiter, **3** Roller-cell pump, **4** Motor armature, **5** Check valve, **6** Pressure side.

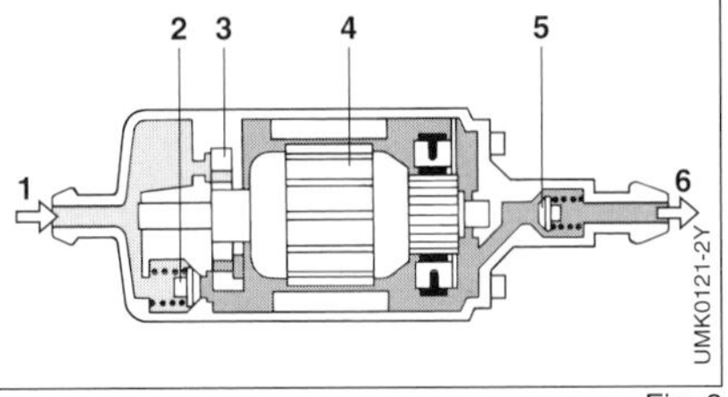

Fig. 3

Fig. 4

Operation of roller-cell pump

1 Suction side, **2** Rotor plate, **3** Roller, **4** Roller race plate, **5** Pressure side.

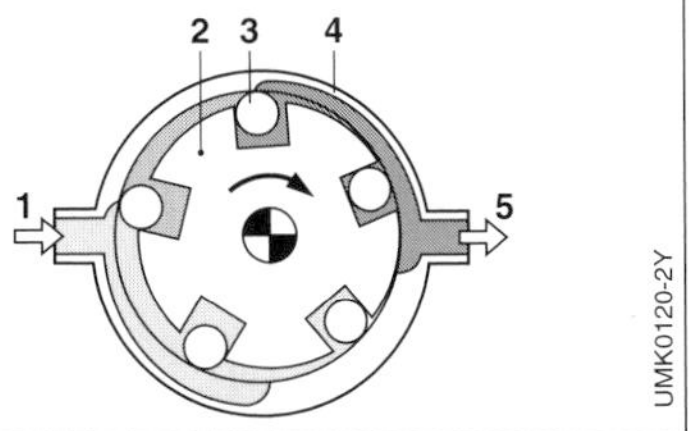

Fig. 5

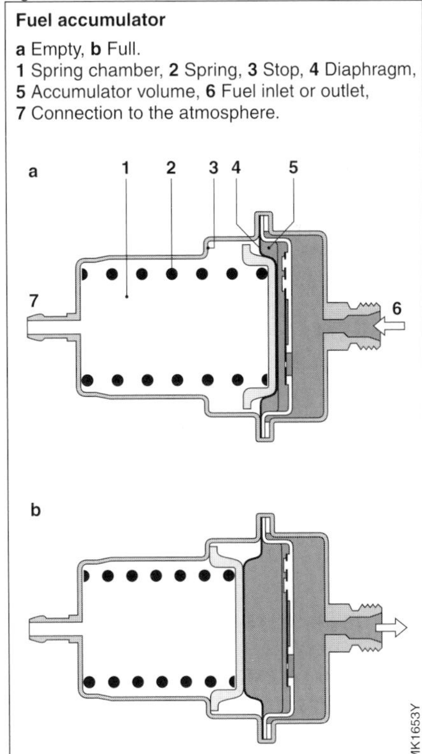

Fuel accumulator

a Empty, **b** Full.
1 Spring chamber, **2** Spring, **3** Stop, **4** Diaphragm, **5** Accumulator volume, **6** Fuel inlet or outlet, **7** Connection to the atmosphere.

Fuel filter

The fuel filter retains particles of dirt which are present in the fuel and which would otherwise have an adverse effect on the functioning of the injection system. The fuel filter contains a paper element with a mean pore size of 10 µm backed up by a fluff trap. This combination ensures a high degree of cleaning.

The filter is held in place in the housing by means of a support plate. It is fitted in the fuel line downstream from the fuel accumulator and its service life depends upon the amount of dirt in the fuel. It is imperative that the arrow on the filter housing showing the direction of fuel flow through the filter is observed when the filter is replaced.

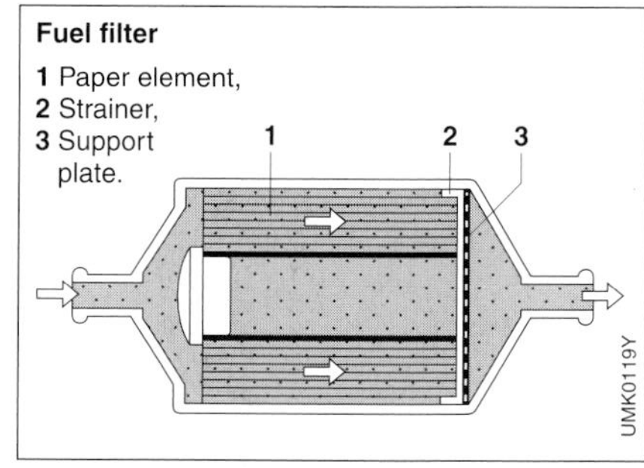

Fuel filter
1 Paper element, **2** Strainer, **3** Support plate.

Fig. 6

Primary-pressure regulator

The primary-pressure regulator maintains the pressure in the fuel system constant.

It is incorporated in the fuel distributor and holds the delivery pressure (system pressure) at about 5 bar. The fuel pump always delivers more fuel than is required by the vehicle engine, and this causes a plunger to shift in the pressure regulator and open a port through which excess fuel can return to the tank.

The pressure in the fuel system and the force exerted by the spring on the pressure-regulator plunger balance each other out. If, for instance, fuel-pump delivery drops slightly, the plunger is shifted by the spring to a corresponding new position and in doing so closes off the port slightly through which the excess fuel returns to the tank. This means that less fuel is diverted off at this point and the system pressure is controlled to its specified level.

When the engine is switched off, the fuel pump also switches off and the primary pressure drops below the opening pressure of the injection valves. The pressure regulator then closes the return-flow port and thus prevents the pressure in the fuel system from sinking any further (Fig. 8).

Fuel-injection valves

The injection valves open at a given pressure and atomize the fuel through oscillation of the valve needle. The injection valves inject the fuel metered to them into the intake passages and onto the intake valves. They are secured in special

Fig. 7

Primary-pressure regulator fitted to fuel distributor
a In rest position, **b** In actuated position.
1 System-pressure entry, **2** Seal, **3** Return to fuel tank, **4** Plunger, **5** Spring.

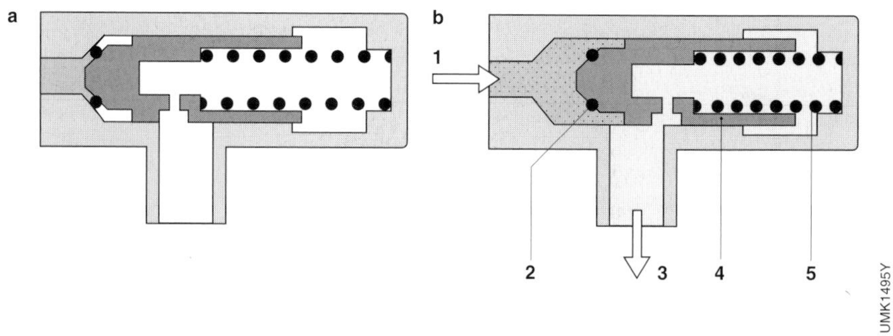

holders to insulate them against the heat radiated from the engine. The injection valves have no metering function themselves, and open of their own accord when the opening pressure of e.g. 3.5 bar is exceeded. They are fitted with a valve needle (Fig. 9) which oscillates ("chatters") audibly at high frequency when fuel is injected. This results in excellent atomization of the fuel even with the smallest of injection quantities. When the engine is switched off, the injection valves close tightly when the pressure in the fuel-supply system drops below their opening pressure. This means that no more fuel can enter the intake passages once the engine has stopped.

Air-shrouded fuel-injection valves

Air-shrouded injection valves improve the mixture formation particularly at idle. Using the pressure drop across the throttle valve, a portion of the air inducted by the engine is drawn into the cylinder through the injection valve (Fig. 20): The result is excellent atomization of the fuel at the point of exit (Fig. 10). Air-shrouded injection valves reduce fuel consumption and toxic emission constituents.

Pressure curve after engine switchoff

Firstly pressure falls from the normal system pressure (**1**) to the pressure-regulator closing pressure (**2**). The fuel accumulator then causes it to increase to the level (**3**) which is below the opening pressure (**4**) of the injection valves.

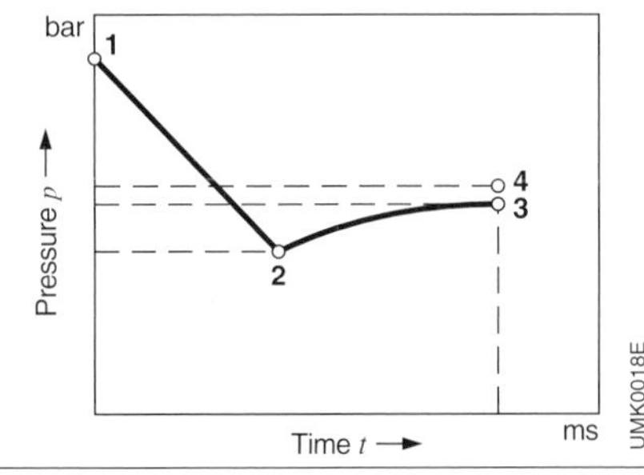

Fig. 8

Fig. 9

Fuel-injection valve

a In rest position,
b In actuated position.
1 Valve housing,
2 Filter,
3 Valve needle,
4 Valve seat.

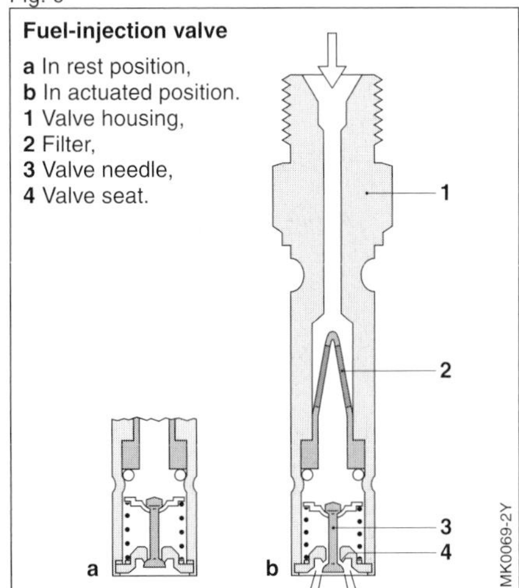

Fig. 10
Spray pattern of an injection valve without air-shrouding (left) and with air-shrouding (right).

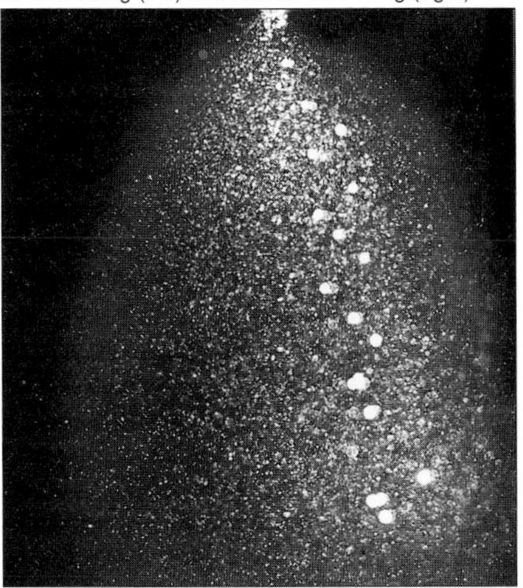

Fuel metering

The task of the fuel-management system is to meter a quantity of fuel corresponding to the intake air quantity.
Basically, fuel metering is carried out by the mixture control unit. This comprises the air-flow sensor and the fuel distributor.
In a number of operating modes however, the amount of fuel required deviates greatly from the "standard" quantity and it becomes necessary to intervene in the mixture formation system (see section "Adaptation to operating conditions").

Air-flow sensor

The quantity of air drawn in by the engine is a precise measure of its operating load. The air-flow sensor operates according to the suspended-body principle, and measures the amount of air drawn in by the engine.
The intake air quantity serves as the main actuating variable for determining the basic injection quantity. It is the appropriate physical quantity for deriving the fuel requirement, and changes in the induction characteristics of the engine have no effect upon the formation of the air-fuel mixture. Since the air drawn in by the engine must pass through the air-flow sensor before it reaches the engine, this means that it has been measured and the control signal generated before it actually enters the engine cylinders. The result is that, in addition to other measures described below, the correct mixture adaptation takes place at all times.

Principle of the air-flow sensor

a Small amount of air drawn in: sensor plate only lifted slightly, **b** Large amount of air drawn in: sensor plate is lifted considerably further.

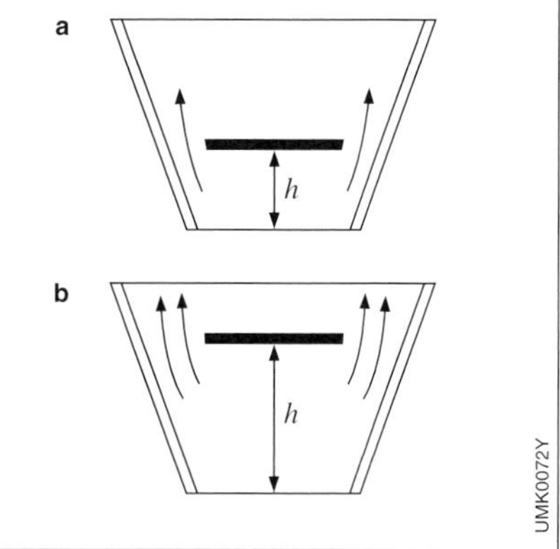

Fig. 11

Fig. 12

Updraft air-flow sensor

a Sensor plate in its zero position,
b Sensor plate in its operating position.

1 Air funnel,
2 Sensor plate,
3 Relief cross-section,
4 Idle-mixture adjusting screw,
5 Pivot,
6 Lever,
7 Leaf spring.

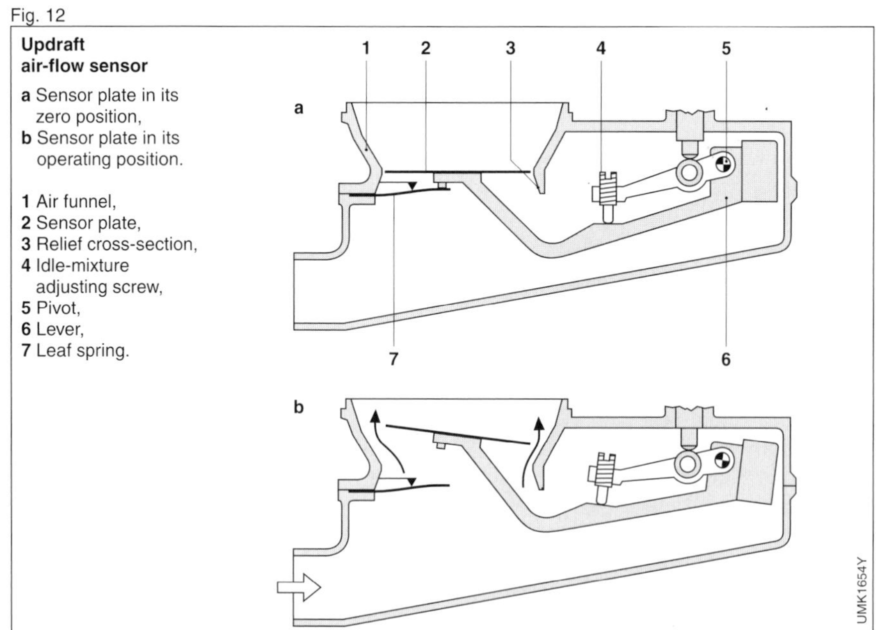

The air-flow sensor is located upstream of the throttle valve so that it measures all the air which enters the engine cylinders. It comprises an air funnel in which the sensor plate (suspended body) is free to pivot. The air flowing through the funnel deflects the sensor plate by a given amount out of its zero position, and this movement is transmitted by a lever system to a control plunger which determines the basic injection quantity required for the basic functions. Considerable pressure shocks can occur in the intake system if backfiring takes place in the intake manifold. For this reason, the air-flow sensor is so designed that the sensor plate can swing back in the opposite direction in the event of misfire, and past its zero position to open a relief cross-section in the funnel. A rubber buffer limits the downward stroke (the upwards stroke on the downdraft air-flow sensor). A counterweight compensates for the weight of the sensor plate and lever system (this is carried out by an extension spring on the downdraft air-flow sensor). A leaf spring ensures the correct zero position in the switched-off phase.

Barrel with metering slits

1 Intake air, **2** Control pressure, **3** Fuel inlet, **4** Metered quantity of fuel, **5** Control plunger, **6** Barrel with metering slits, **7** Fuel distributor.

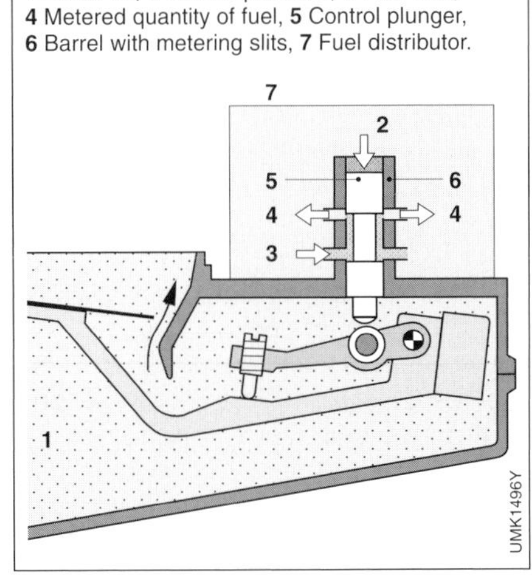

Fig. 13

Fuel distributor

Depending upon the position of the plate in the air-flow sensor, the fuel distributor meters the basic injection quantity to the individual engine cylinders. The position of the sensor plate is a measure of the amount of air drawn in by the engine. The position of the plate is transmitted to the control plunger by a lever.

Fig. 14

Barrel with metering slits and control plunger

a Zero (inoperated position), **b** Part load, **c** Full load.
1 Control pressure, **2** Control plunger, **3** Metering slit in the barrel, **4** Control edge, **5** Fuel inlet, **6** Barrel with metering slits.

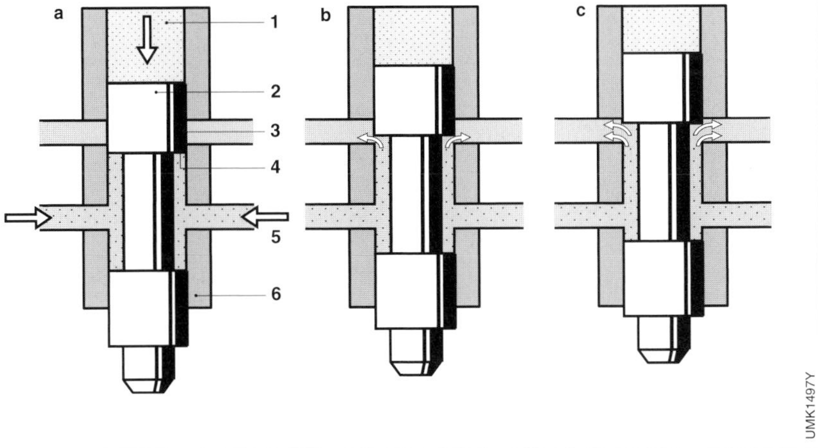

Depending upon its position in the barrel with metering slits, the control plunger opens or closes the slits to a greater or lesser extent. The fuel flows through the open section of the slits to the differential pressure valves and then to the fuel injection valves. If sensor-plate travel is only small, then the control plunger is lifted only slightly and, as a result, only a small section of the slit is opened for the passage of fuel. With larger plunger travel, the plunger opens a larger section of the slits and more fuel can flow. There is a linear relationship between sensor-plate travel and the slit section in the barrel which is opened for fuel flow.
A hydraulic force generated by the so-called control pressure is applied to the control plunger. It opposes the movement resulting from sensor-plate deflection. One of its functions is to ensure that the control plunger follows the sensor-plate movement immediately and does not, for instance, stick in the upper end position when the sensor plate moves down again. Further functions of the control pressure are discussed in the sections "Warm-up enrichment" and "Full-load enrichment".

Control pressure
The control pressure is tapped from the primary pressure through a restriction bore (Figure 16). This restriction bore serves to decouple the control-pressure circuit and the primary-pressure circuit from one another. A connection line joins the fuel distributor and the warm-up regulator (control-pressure regulator).
When starting the cold engine, the control pressure is about 0.5 bar. As the engine warms up, the warm-up regulator increases the control pressure to about 3.7 bar (Figure 26).
The control pressure acts through a damping restriction on the control plunger and thereby develops the force which opposes the force of the air in the air-flow sensor. In doing so, the restriction dampens a possible oscillation of the sensor plate which could result due to pulsating air-intake flow.
The control pressure influences the fuel distribution. If the control pressure is low, the air drawn in by the engine can deflect the sensor plate further. This results in the control plunger opening the metering slits further and the engine being allocated more fuel. On the other hand, if the control pressure is high, the air drawn in by the engine cannot deflect the sensor plate so far and, as a result, the engine receives less fuel. In order to fully seal off the control-pressure circuit with absolute certainty when the engine has been switched off, and at the same time to maintain the pressure in the fuel circuit, the return line of the warm-up regulator is fitted with a check valve. This (push-up) valve is attached to the primary-pressure regulator and is held open during operation by the pressure-regulator plunger. When the engine is switched off and the plunger of the primary-pressure regulator returns to its zero position, the check valve is closed by a spring (Figure 17).

Barrel with metering slits
The slits are shown enlarged (the actual slit is about 0.2 mm wide).

Fig. 15

Differential-pressure valves
The differential-pressure valves in the fuel distributor result in a specific pressure drop at the metering slits.
The air-flow sensor has a linear characteristic. This means that if double the quantity of air is drawn in, the sensor-

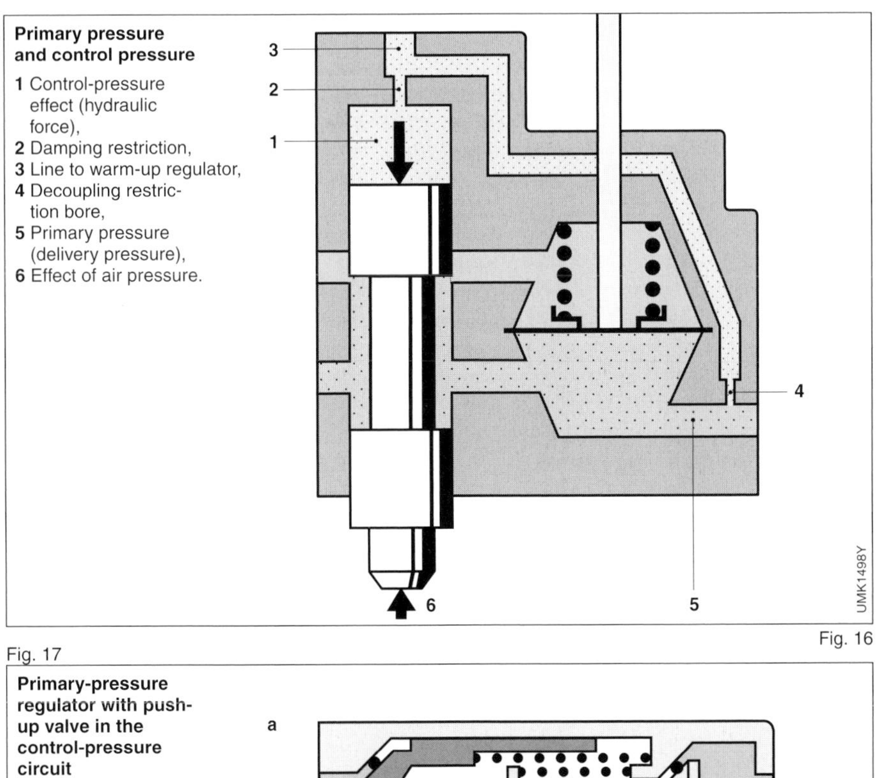

Primary pressure and control pressure

1 Control-pressure effect (hydraulic force),
2 Damping restriction,
3 Line to warm-up regulator,
4 Decoupling restriction bore,
5 Primary pressure (delivery pressure),
6 Effect of air pressure.

Fig. 16

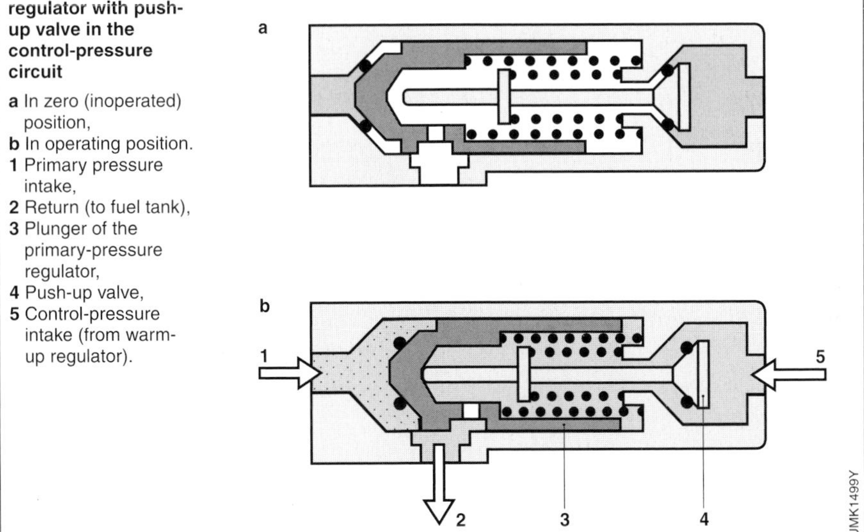

Fig. 17

Primary-pressure regulator with push-up valve in the control-pressure circuit

a In zero (inoperated) position,
b In operating position.
1 Primary pressure intake,
2 Return (to fuel tank),
3 Plunger of the primary-pressure regulator,
4 Push-up valve,
5 Control-pressure intake (from warm-up regulator).

plate travel is also doubled. If this travel is to result in a change of delivered fuel in the same relationship, in this case double the travel equals double the quantity, then a constant drop in pressure must be guaranteed at the metering slits (Figure 14), regardless of the amount of fuel flowing through them.

The differential-pressure valves maintain the differential pressure between the upper and lower chamber constant regardless of fuel throughflow. The differential pressure is 0.1 bar.

The differential-pressure valves achieve a high metering accuracy and are of the flat-seat type. They are fitted in the fuel

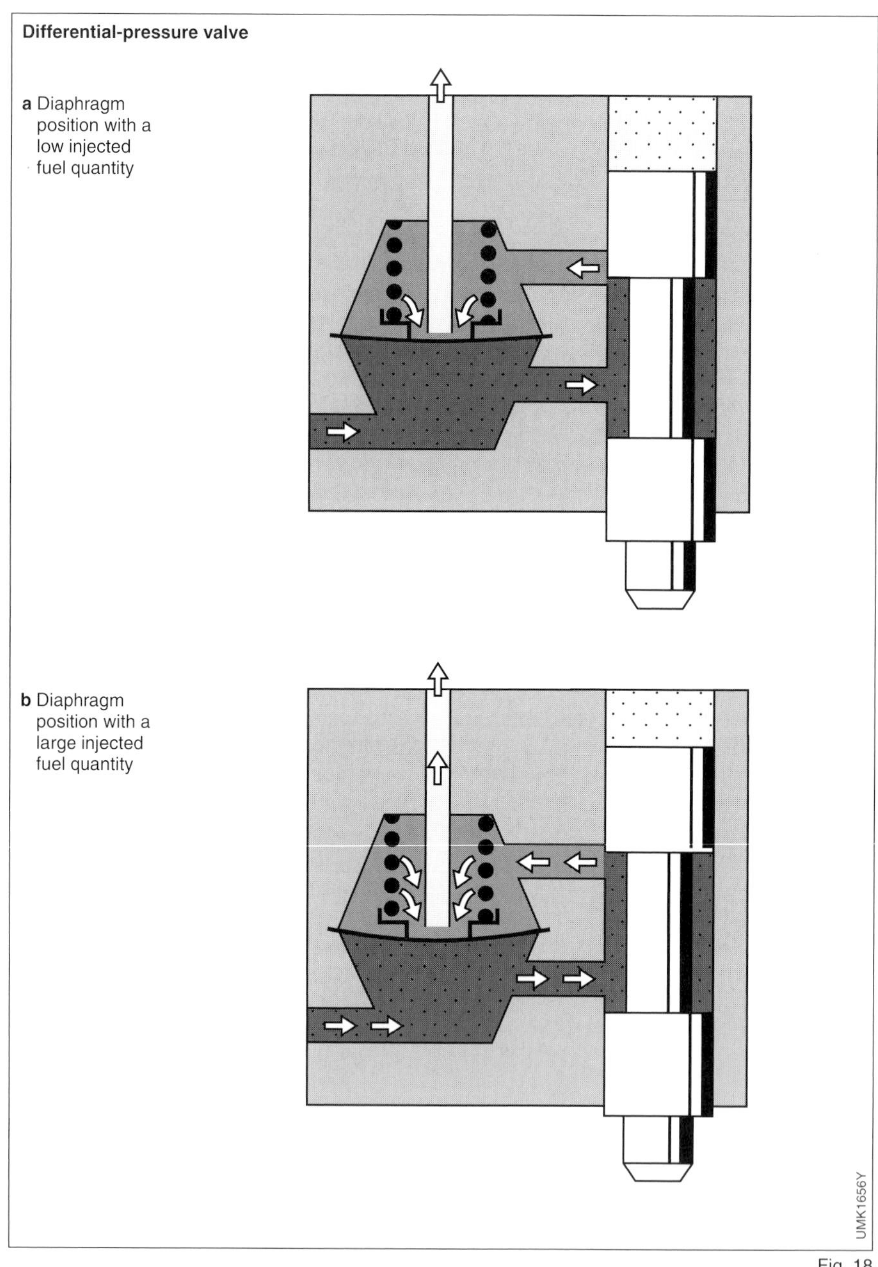

Fig. 18

distributor and one such valve is allocated to each metering slit. A diaphragm separates the upper and lower chambers of the valve (Figures 18 and 19). The lower chambers of all the valves are connected with one another by a ring main and are subjected to the primary pressure (delivery pressure). The valve seat is located in the upper chamber. Each upper chamber is connected to a metering slit and its corresponding connection to the fuel-injection line. The upper chambers are completely sealed off from each other. The diaphragms are spring-loaded and it is this helical spring that produces the pressure differential.

Fuel distributor with differential-pressure valves

1 Fuel intake (primary pressure), **2** Upper chamber of the differential-pressure valve, **3** Line to the fuel-injection valve (injection pressure), **4** Control plunger, **5** Control edge and metering slit, **6** Valve spring, **7** Valve diaphragm, **8** Lower chamber of the differential-pressure valve.

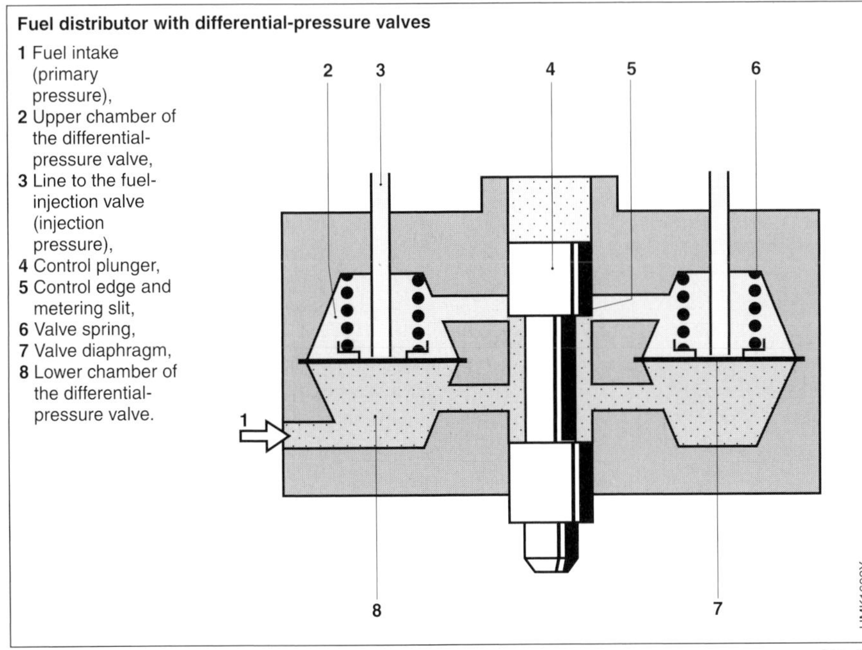

Fig. 19

If a large basic fuel quantity flows into the upper chamber through the metering slit, the diaphragm is bent downwards and enlarges the valve cross-section at the outlet leading to the injection valve until the set differential pressure once again prevails.

If the fuel quantity drops, the valve cross-section is reduced owing to the equilibrium of forces at the diaphragm until the differential pressure of 0.1 bar is again present.

This causes an equilibrium of forces to prevail at the diaphragm which can be maintained for every basic fuel quantity by controlling the valve cross-section.

Mixture formation

The formation of the air-fuel mixture takes place in the intake ports and cylinders of the engine.

The continually injected fuel coming from the injection valves is "stored" in front of the intake valves. When the intake valve is opened, the air drawn in by the engine carries the waiting "cloud" of fuel with it into the cylinder. An ignitable air-fuel mixture is formed during the induction stroke due to the swirl effect.

Fig. 20

Mixture formation with air-shrouded fuel-injection valve

1 Fuel-injection valve, **2** Air-supply line, **3** Intake manifold, **4** Throttle valve.

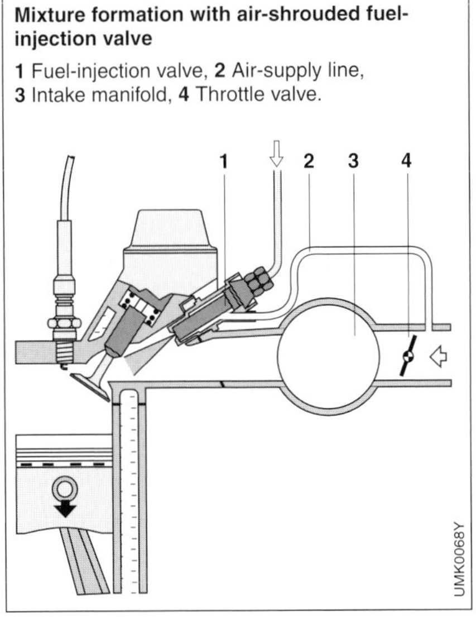

Air-shrouded fuel-injection valves favor mixture formation since they atomize the fuel very well at the outlet point (Figures 10, 20).

Adaptation to operating conditions

In addition to the basic functions described up to now, the mixture has to be adapted during particular operating conditions. These adaptations (corrections) are necessary in order to optimize the power delivered, to improve the exhaust-gas composition and to improve the starting behavior and driveability.

Basic mixture adaptation

The basic adaptation of the air-fuel mixture to the operating modes of idle, part load and full load is by appropriately shaping the air funnel in the air-flow sensor (Figures 21 and 22).

If the funnel had a purely conical shape, the result would be a mixture with a constant air-fuel ratio throughout the whole of the sensor plate range of travel (metering range). However, it is necessary to meter to the engine an air-fuel mixture which is optimal for particular operating modes such as idle, part load and full load. In practice, this means a richer mixture at idle and full load, and a leaner mixture in the part-load range. This adaptation is achieved by designing the air funnel so that it becomes wider in stages.

If the cone shape of the funnel is flatter than the basic cone shape (which was specified for a particular mixture, e.g. for $\lambda = 1$), this results in a leaner mixture. If the funnel walls are steeper than in the basic model, the sensor plate is lifted further for the same air throughput, more fuel is therefore metered by the control plunger and the mixture is richer. Consequently, this means that the air funnel can be shaped so that it is poss-ible to meter mixtures to the engine which have different air-fuel ratios depending upon the sensor-plate position in the funnel (which in turn corresponds to the particular engine operating mode i.e. idle, part load and full load). This results in a richer mixture for idle and full load (idle and full-load enrichment) and, by contrast, a leaner mixture for part load.

Influence of funnel-wall angle upon the sensor-plate deflection for identical air throughput

a The basic funnel shape results in stroke "h",
b Steep funnel walls result in increased stroke "h" for identical air throughput,
c Flatter funnel shape results in reduced deflection "h" for identical air throughput.
A Annular area opened by the sensor plate (identical in a, b and c).

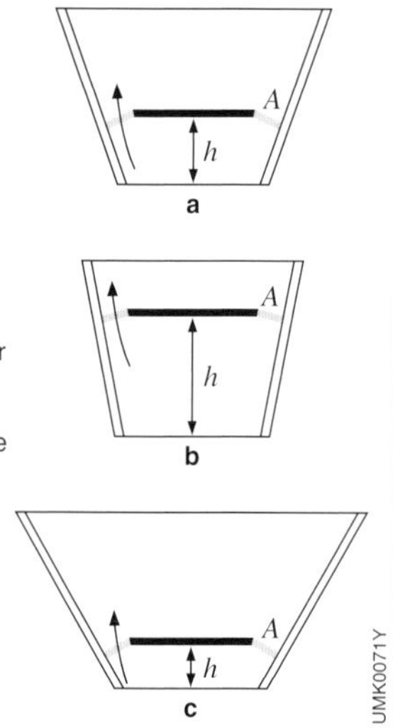

Fig. 21

Fig. 22

Adaptation of the air-funnel shape

1 For maximum power, **2** For part load, **3** For idle.

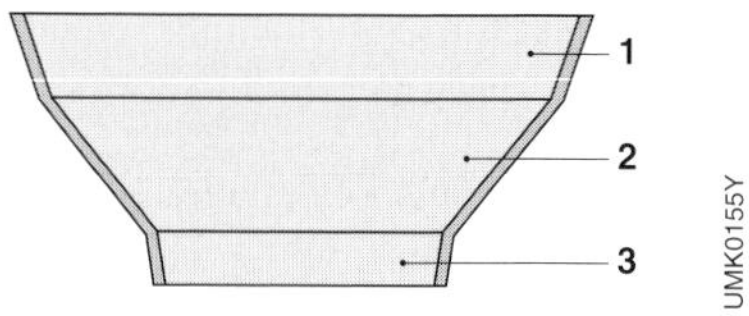

Cold-start enrichment

Depending upon the engine temperature, the cold-start valve injects extra fuel into the intake manifold for a limited period during the starting process.

In order to compensate for the condensation losses due to condensation on the cold cylinder walls, and in order to facilitate starting the cold engine during cold starting, extra fuel must be injected at the instant of start-up. This extra fuel is injected by the cold-start valve into the intake manifold. The injection period of the cold-start valve is limited by a thermo-time switch depending upon the engine temperature.

This process is known as cold-start enrichment and results in a "richer" air-fuel

Cold-start valve in operated state

1 Electrical connection, **2** Fuel supply with strainer, **3** Valve (electromagnet armature), **4** Solenoid winding, **5** Swirl nozzle, **6** Valve seat.

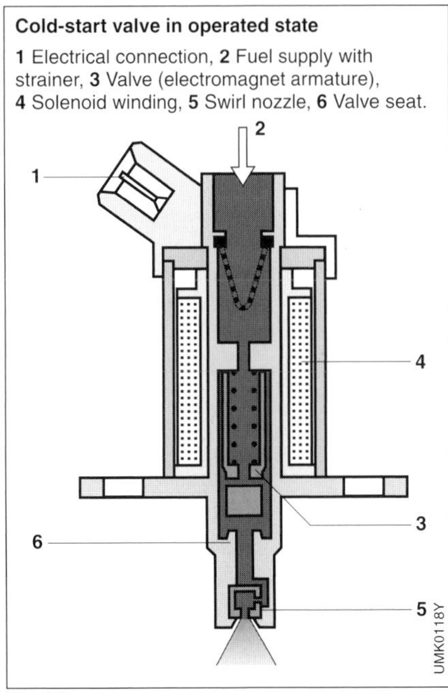

Fig. 23

mixture, i.e. the excess-air factor λ is temporarily less than 1.

Cold-start valve

The cold-start valve (Figure 23) is a solenoid-operated valve. An electromagnetic winding is fitted inside the valve. When unoperated, the movable electromagnet armature is forced against a seal by means of a spring and thus closes the valve. When the electromagnet is energized, the armature which consequently has lifted from the valve seat opens the passage for the flow of fuel through the valve. From here, the fuel enters a special nozzle at a tangent and is caused to rotate or swirl.

The result is that the fuel is atomized very finely and enriches the mixture in the manifold downstream of the throttle valve. The cold-start valve is so positioned in the intake manifold that good distribution of the mixture to all cylinders is ensured.

Thermo-time switch

The thermo-time switch limits the duration of cold-start valve operation, depending upon temperature.

Thermo-time switch

1 Electrical connection, **2** Housing, **3** Bimetal, **4** Heating filament, **5** Electrical contact.

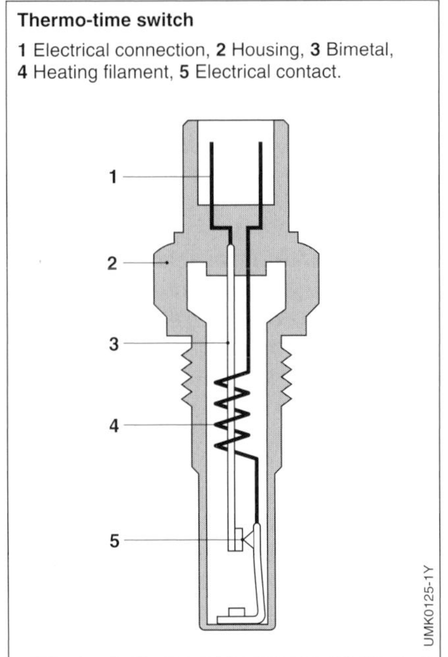

Fig. 24

The thermo-time switch (Figure 24) consists of an electrically heated bimetal strip which, depending upon its temperature opens or closes a contact. It is brought into operation by the ignition/starter switch, and is mounted at a position which is representative of engine temperature. During a cold start, it limits the "on" period of the cold-start valve. In case of repeated start attempts, or when starting takes too long, the cold-start valve ceases to inject.

Its "on" period is determined by the thermo-time switch which is heated by engine heat as well as by its own built-in heater. Both these heating effects are necessary in order to ensure that the "on" period of the cold-start valve is limited under all conditions, and engine flooding prevented. During an actual cold start, the heat generated by the built-in heater is mainly responsible for the "on" period (switch off, for instance, at –20 °C after 7.5 seconds). With a warm engine, the thermo-time switch has already been heated up so far by engine heat that it remains open and prevents the cold-start valve from going into action.

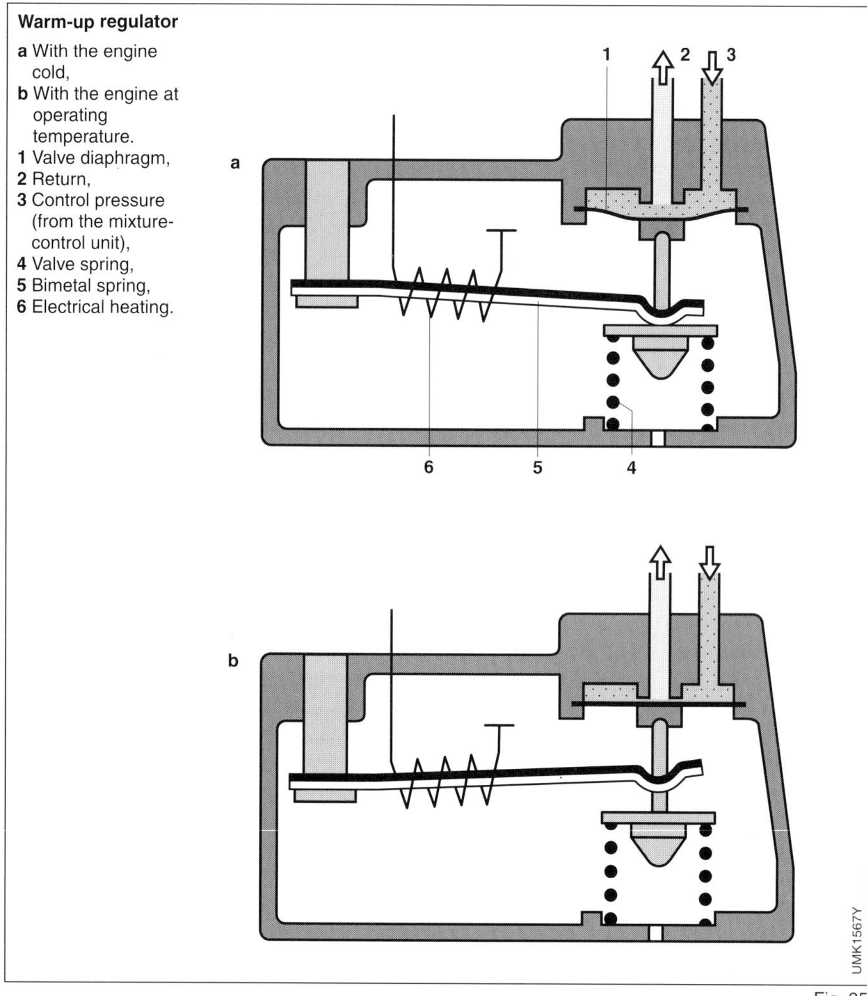

Warm-up regulator
a With the engine cold,
b With the engine at operating temperature.
1 Valve diaphragm,
2 Return,
3 Control pressure (from the mixture-control unit),
4 Valve spring,
5 Bimetal spring,
6 Electrical heating.

Fig. 25

Warm-up enrichment

Warm-up enrichment is controlled by the warm-up regulator. When the engine is cold, the warm-up regulator reduces the control pressure to a degree dependent upon engine temperature and thus causes the metering slits to open further (Figure 25).

At the beginning of the warm-up period which directly follows the cold start, some of the injected fuel still condenses on the cylinder walls and in the intake ports. This can cause combustion misses to occur. For this reason, the air-fuel mixture must be enriched during the warm-up ($\lambda < 1.0$). This enrichment must be continuously reduced along with the rise in engine temperature in order to prevent the mixture being over-rich when higher engine temperatures have been reached. The warm-up regulator (control-pressure regulator) is the component which carries out this type of mixture control for the warm-up period by changing the control pressure.

Warm-up regulator

The change of the control pressure is effected by the warm-up regulator which is fitted to the engine in such a way that it ultimately adopts the engine temperature. An additional electrical heating system enables the regulator to be matched precisely to the engine characteristic.

The warm-up regulator comprises a spring-controlled flat seat (diaphragm-type) valve and an electrically heated bimetal spring (Figure 25).
In cold condition, the bimetal spring exerts an opposing force to that of the valve spring and, as a result, reduces the effective pressure applied to the underside of the valve diaphragm. This means that the valve outlet cross-section is slightly increased at this point and more fuel is diverted out of the control-pressure circuit in order to achieve a low control pressure. Both the electrical heating system and the engine heat the bimetal spring as soon as the engine is cranked. The spring bends, and in doing so reduces the force opposing the valve spring which, as a result, pushes up the diaphragm of the flat-seat valve. The valve outlet cross-section is reduced and the pressure in the control-pressure circuit rises.
Warm-up enrichment is completed when the bimetal spring has lifted fully from the valve spring. The control pressure is now solely controlled by the valve spring and maintained at its normal level. The control pressure is about 0.5 bar at cold start and about 3.7 bar with the engine at operating temperature (Figure 26).

Idle stabilization
In order to overcome the increased friction in cold condition and to guarantee smooth idling, the engine receives more air-fuel mixture during the warm-up phase due to the action of the auxiliary air device.
When the engine is cold, the frictional resistances are higher than when it is at operating temperature and this friction must be overcome by the engine during idling. For this reason, the engine is allowed to draw in more air by means of the auxiliary-air device which bypasses the throttle valve. Due to the fact that this auxiliary air is measured by the air-flow sensor and taken into account for fuel metering, the engine is provided with more air-fuel mixture. This results in idle stabilization when the engine is cold.

Auxiliary-air device
In the auxiliary-air device, a perforated plate is pivoted by means of a bimetal spring and changes the open cross-section of a bypass line. This perforated plate thus opens a correspondingly large cross-section of the bypass line, as a function of the temperature, and this cross-section is reduced with increasing engine temperature and is ultimately closed. The bimetal spring also has an electrical heating system which permits the opening time to be restricted dependent upon the engine type. The in-

Fig. 26

Warm-up regulator characteristics at various operating temperatures

Enrichment factor 1.0 corresponds to fuel metering with the engine at operating temperature.

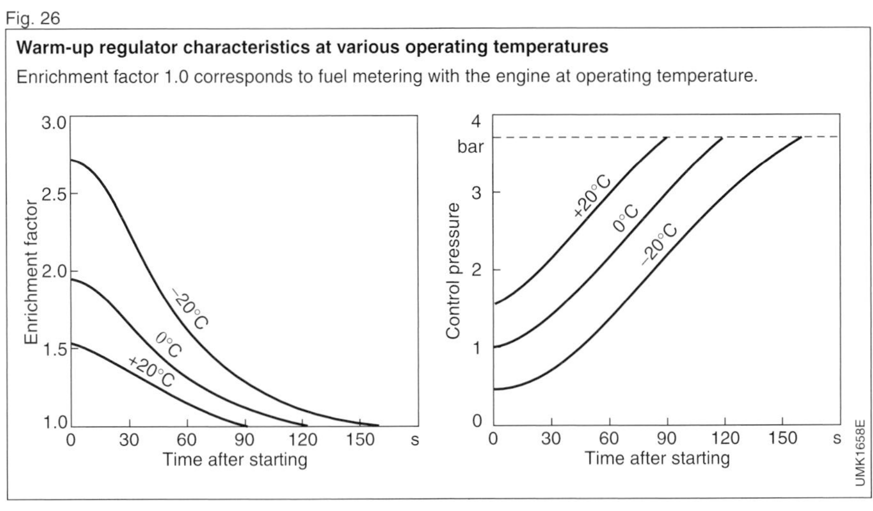

stallation location of the auxiliary-air device is selected such that it assumes the engine temperature. This guarantees that the auxiliary-air device only functions when the engine is cold (Figure 27).

Full-load enrichment

Engines operated in the part-load range with a very lean mixture require an enrichment during full-load operation, in addition to the mixture adaptation resulting from the shape of the air funnel.
This extra enrichment is carried out by a specially designed warm-up regulator. This regulates the control pressure depending upon the manifold pressure (Figures 28 and 30).
This model of the warm-up regulator uses two valve springs instead of one. The outer of the two springs is supported on the housing as in the case with the normal-model warm-up regulator. The inner spring however is supported on a diaphragm which divides the regulator into an upper and a lower chamber. The manifold pressure which is tapped via a hose connection from the intake manifold downstream of the throttle valve acts in the upper chamber. Depending upon the model, the lower chamber is subjected to atmospheric pressure either directly or by means of a second hose leading to the air filter.
Due to the low manifold pressure in the idle and part-load ranges, which is also present in the upper chamber, the diaphragm lifts to its upper stop. The inner spring is then at maximum pretension. The pretension of both springs, as a result, determines the particular control pressure for these two ranges. When the throttle valve is opened further at full load, the pressure in the intake manifold increases, the diaphragm leaves the upper stops and is pressed against the lower stops.
The inner spring is relieved of tension and the control pressure reduced by the specified amount as a result. This results in mixture enrichment.

Auxiliary-air device

1 Electrical connection, **2** Electrical heating, **3** Bimetal spring, **4** Perforated plate.

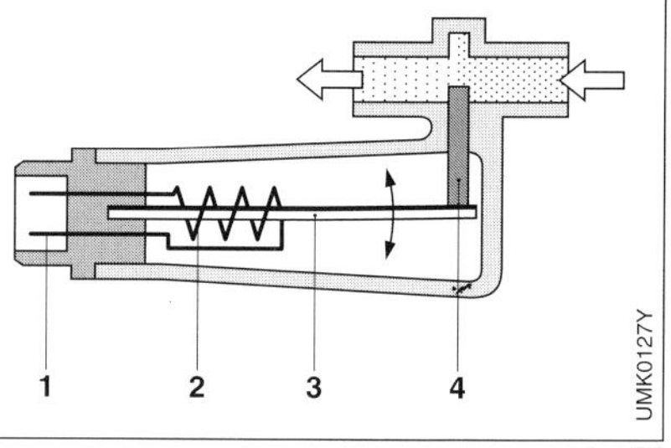

1
2
3
4
UMK0127Y

Fig. 29

Fig. 28

Dependence of the control pressure on engine load

Control pressure
Idle and part load
Full load
Engine load
UMK0019E

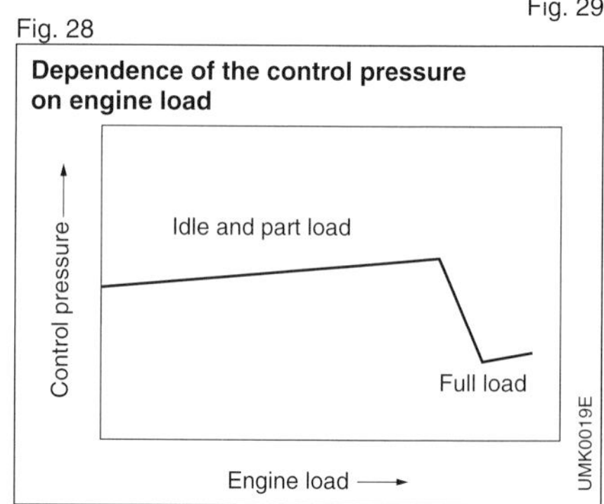

Fig. 29

Acceleration response

Behavior of the K-Jetronic when the throttle valve is suddenly opened.

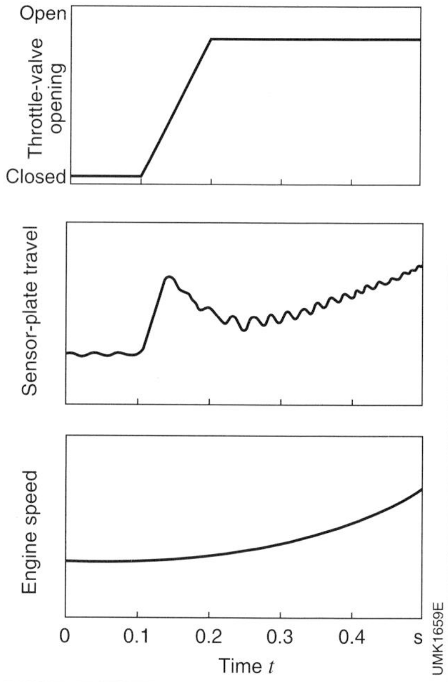

Open
Throttle-valve opening
Closed
Sensor-plate travel
Engine speed
0
0.1
0.2
0.3
0.4
s
Time t
UMK1659E

Acceleration response

The good acceleration response is a result of “overswing” of the air-flow sensor plate (Figure 29).

Transitions from one operating condition to another produce changes in the mixture ratio which are utilized to improve driveability.

If, at constant engine speed, the throttle valve is suddenly opened, the amount of air which enters the combustion chamber, plus the amount of air which is needed to bring the manifold pressure up to the new level, flow through the airflow sensor. This causes the sensor plate to briefly “overswing” past the fully opened throttle point. This “overswing” results in more fuel being metered to the engine (acceleration enrichment) and ensures good acceleration response.

Fig. 30

Warm-up regulator with full-load diaphragm

a During idle and part load,
b During full load.

1 Electrical heating,
2 Bimetal spring,
3 Vacuum connection (from intake manifold),
4 Valve diaphragm,
5 Return to fuel tank,
6 Control pressure (from fuel distributor),
7 Valve springs,
8 Upper stop,
9 To atmospheric pressure,
10 Diaphragm,
11 Lower stop.

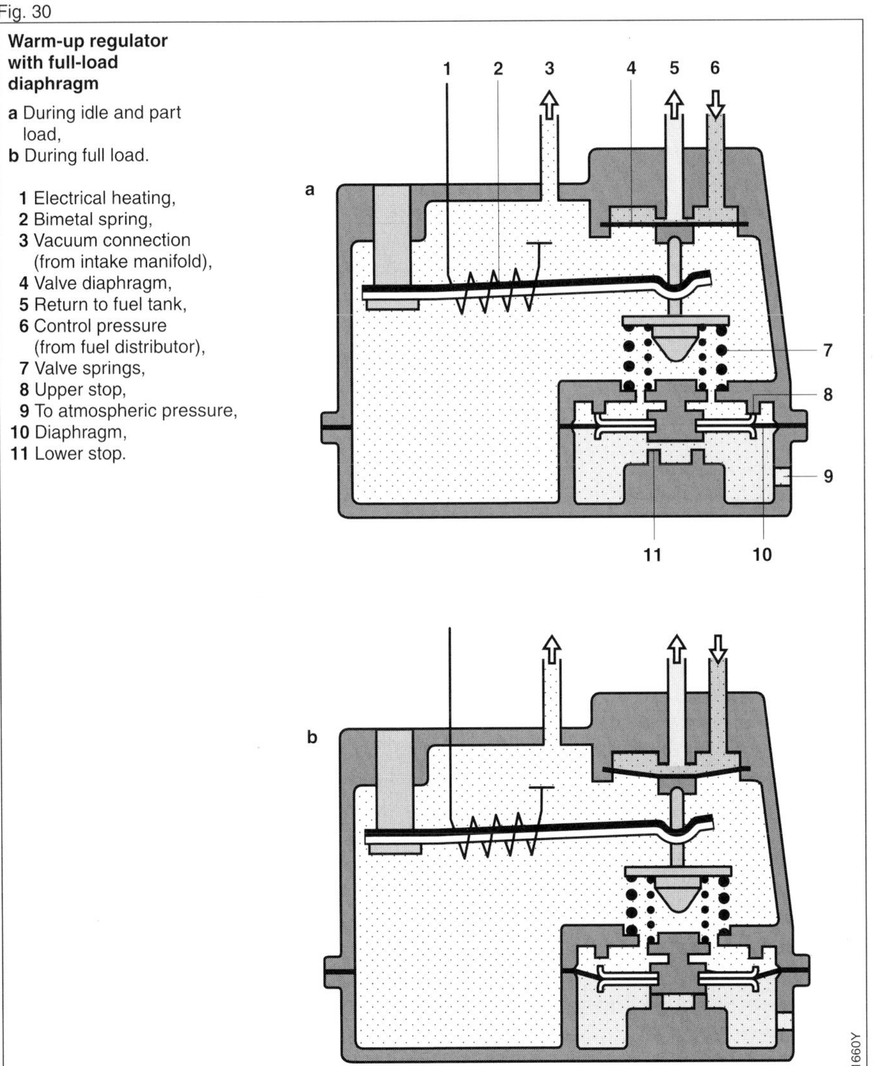

Supplementary functions

Overrun fuel cutoff

Smooth fuel cutoff effective during overrun responds as a function of the engine speed. The engine-speed information is provided by the ignition system. Intervention is via an air bypass around the sensor plate. A solenoid valve controlled by an electronic speed switch opens the bypass at a specific engine speed. The sensor plate then reverts to zero position and interrupts fuel metering. Cutoff of the fuel supply during overrun operation permits the fuel consumption to be reduced considerably not only when driving downhill but also in town traffic.

Engine speed limiting

The fuel supply can be cut off to limit the maximum permissible engine speed.

Lambda closed-loop control

Open-loop control of the air-fuel ratio is not adequate for observing extremely low exhaust-gas limit values. The lambda closed-loop control system required for operation of a three-way catalytic converter necessitates the use of an electronic control unit on the K-Jetronic. The important input variable for this control unit is the signal supplied by the lambda sensor.

In order to adapt the injected fuel quantity to the required air-fuel ratio with $\lambda = 1$, the

Fig. 31

Additional components for lambda closed-loop control

1 Lambda sensor,
2 Lambda closed-loop controller,
3 Frequency valve (variable restrictor),
4 Fuel distributor,
5 Lower chambers of the differential-pressure valves,
6 Metering slits,
7 Decoupling restrictor (fixed restrictor),
8 Fuel inlet,
9 Fuel return line.

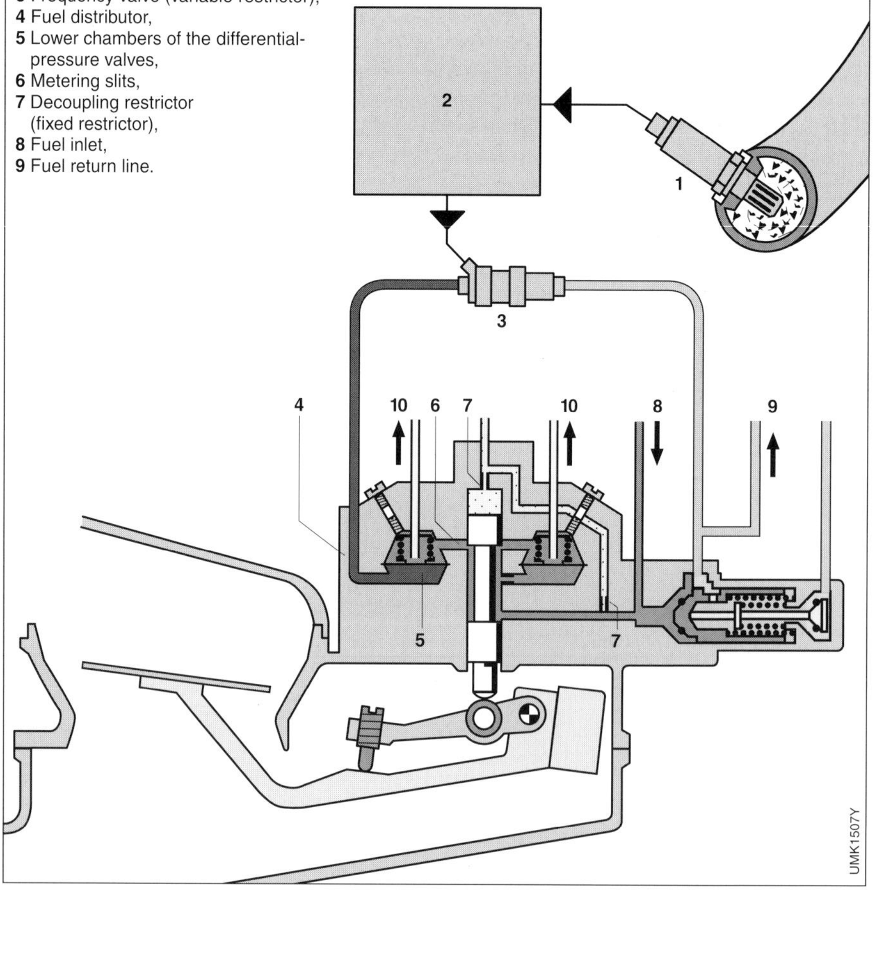

pressure in the lower chambers of the fuel distributor is varied. If, for instance, the pressure in the lower chambers is reduced, the differential pressure at the metering slits increases, whereby the injected fuel quantity is increased. In order to permit the pressure in the lower chambers to be varied, these chambers are decoupled from the primary pressure via a fixed restrictor, by comparison with the standard K-Jetronic fuel distributor. A further restrictor connects the lower chambers and the fuel return line.
This restrictor is variable: if it is open, the pressure in the lower chambers can drop. If it is closed, the primary pressure builds up in the lower chambers. If this restrictor is opened and closed in a fast rhythmic succession, the pressure in the lower chambers can be varied dependent upon the ratio of closing time to opening time. An electromagnetic valve, the frequency valve, is used as the variable restrictor. It is controlled by electrical pulses from the lambda closed-loop controller.

Fig. 32
Components of the K-Jetronic system

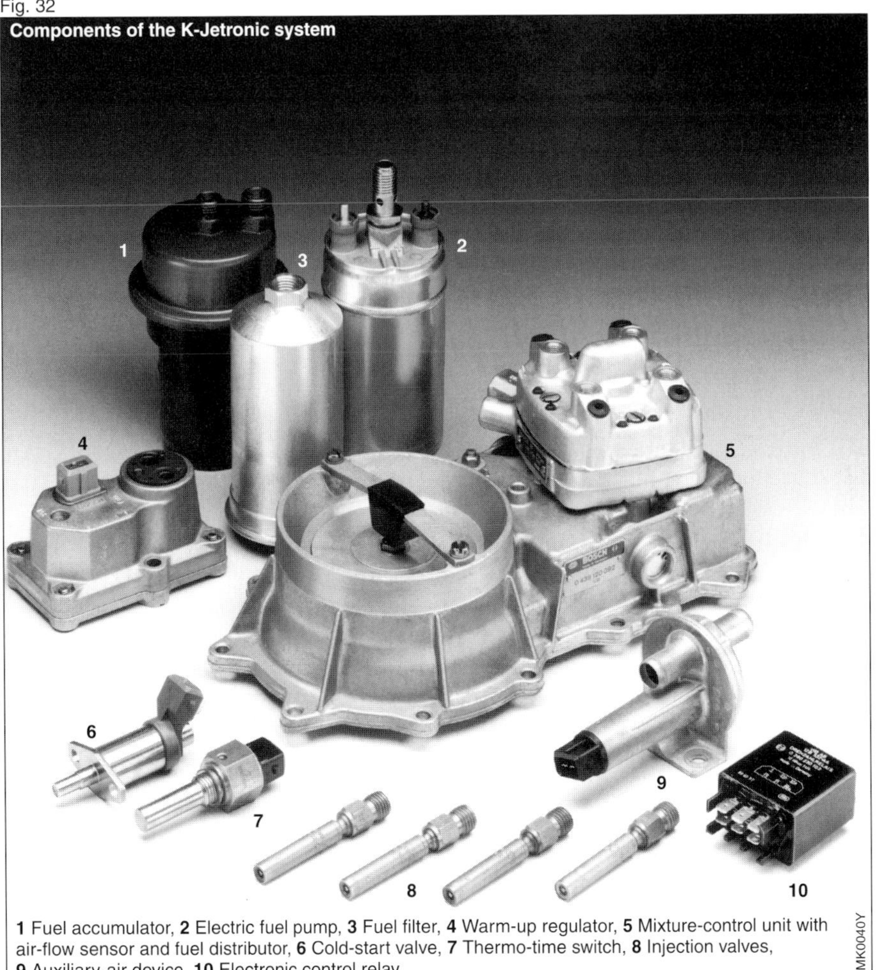

1 Fuel accumulator, **2** Electric fuel pump, **3** Fuel filter, **4** Warm-up regulator, **5** Mixture-control unit with air-flow sensor and fuel distributor, **6** Cold-start valve, **7** Thermo-time switch, **8** Injection valves, **9** Auxiliary-air device, **10** Electronic control relay.

Exhaust-gas treatment

Lambda sensor

The Lambda sensor inputs a voltage signal to the ECU which represents theinstantaneous composition of the air-fuel mixture.

The Lambda sensor is installed in the engine exhaust manifold at a point which maintains the necessary temperature for the correct functioning of the sensor over the complete operating range of the engine.

Operation

The sensor protrudes into the exhaust-gas stream and is designed so that the outer electrode is surrounded by exhaust gas, and the inner electrode is connected to the atmospheric air.

Basically, the sensor is constructed from an element of special ceramic, the surface of which is coated with microporous platinum electrodes. The operation of the sensor is based upon the fact that ceramic material is porous and permits diffusion of the oxygen present in the air (solid electrolyte). At higher temperatures, it becomes conductive, and if the oxygen concentration on one side of the electrode is different to that on the other, then a voltage is generated between the electrodes. In the area of stoichiometric airfuel mixture (λ = 1.00), a jump takes place in the sensor voltage output curve. This voltage represents the measured signal.

Construction

The ceramic sensor body is held in a threaded mounting and provided with a protective tube and electrical connections. The surface of the sensor ceramic body has a microporous platinum layer which on the one side decisively influences the sensor characteristic while on the other serving as an electrical contact. A highly adhesive and highly porous ceramic coating has been applied over the platinum layer at the end of the ceramic body that is exposed to the exhaust gas. This protective layer prevents the solid particles in the exhaust gas from eroding the platinum layer. A protective metal sleeve is fitted over the sensor on the electrical connection end and crimped to the sensor housing. This sleeve is provided with a bore to ensure pressure compensation in the sensor interior, and also serves as the support for the disc spring. The connection lead is crimped to the contact element and is led through an insulating sleeve to the outside of the sensor. In order to keep combustin deposits in the exhaust gas away from the ceramic body, the end of the exhaust sensor which protrudes into the exhaust-gas flow is protected by a special tube having slots so designed that the exhaust gas and the solid particles entrained in it do not come into direct contact with the ceramic body.

In addition to the mechanical protection thus provided, the changes in sensor temperature during transition from one operating mode to the other are effectively reduced.

The voltage output of the λ sensor, and its internal resistance, are dependent upon temperature. Reliable functioning of the sensor is only possible with exhaust-gas temperatures above 360 °C (unheated version), and above 200 °C (heated version).

Fig. 33

Control range of the lambda sensor and reduction of pollutant concentrations in exhaust

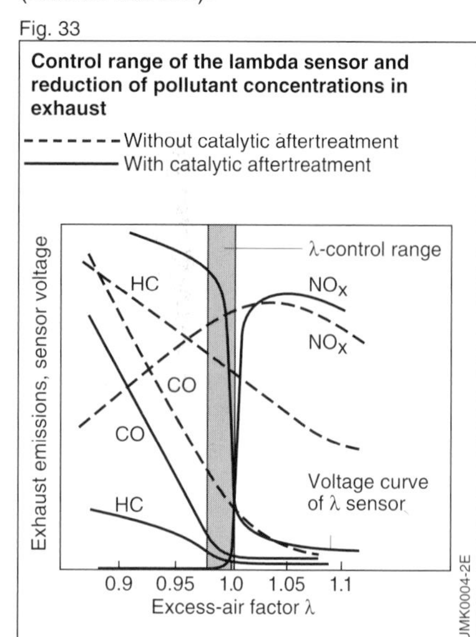

Heated Lambda oxygen sensor

To a large extent, the design principle of the heated Lambda sensor is identical to that of the unheated sensor.

The active sensor ceramic is heated internally by a ceramic heating element with the result that the temperature of the ceramic body always remains above the function limit of 350 °C.

The heated sensor is equipped with a protective tube having a smaller opening. Amongst other things, this prevents the sensor ceramic from cooling down when the exhaust gas is cold. Among the advantages of the heated Lambda sensor are the reliable and efficient control at low exhaust-gas temperatures (e.g. at idle), the minimum effect of exhaust-gas temperature variations, the rapid coming into effect of the Lambda control following engine start, short sensor-reaction time which avoids extreme deviations from the ideal exhaust-gas composition, versatility regarding installation because the sensor is now independent of heating from its surroundings.

Lambda closed-loop control circuit

By means of the Lambda closed-loop control, the air-fuel ratio can be maintained precisely at λ= 1.00.

The Lambda closed-loop control is an add-on function which, in principle, can supplement every controllable fuel-management system. It is particularly suitable for use with Jetronic gasoline-injection systems or Motronic. Using the closed-loop control circuit formed with the aid of the Lambda sensor, deviations from a specified air-fuel ratio can be detected and corrected. This control principle is based upon the measurement of the exhaust-gas oxygen by the Lambda sensor. The exhaust-gas oxygen is a measure for the composition of the air-fuel mixture supplied to the engine. The Lambda sensor acts as a probe in the exhaust pipe and delivers the information as to whether the mixture is richer or leaner than λ = 1.00.

In case of a deviation from this λ = 1.00 figure, the voltage of the sensor output signal changes abruptly. This pronounced change is evaluated by the ECU which is provided with a closed-loop control circuit for this purpose. The injection of fuel to the engine is controlled by the fuel-management system in accordance with the information on the composition of the air-fuel mixture received from the Lambda sensor. This control is such that an airfuel ratio of λ = 1 is achieved. The sensor voltage is a measure for the correction of the fuel quantity in the air-fuel mixture.

Fig. 34

Positioning of the lambda sensor in a dual exhaust system

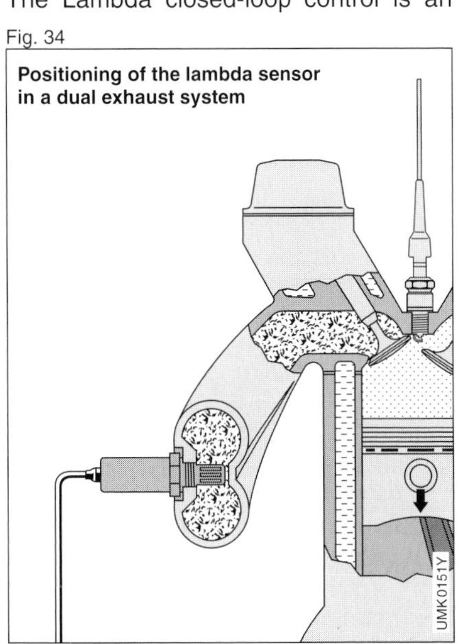

Fig. 35

Location of the lambda sensor in the exhaust pipe (schematic)

1 Sensor ceramic, **2** Electrodes, **3** Contact, **4** Electrical contacting to the housing, **5** Exhaust pipe, **6** Protective ceramic coating (porous), **7** Exhaust gas, **8** Air. *U* voltage.

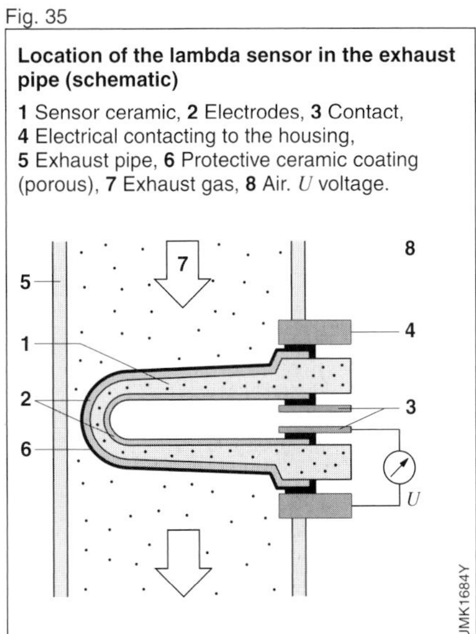

The signal which is processed in the closed-loop control circuit is used to control the actuators of the Jetronic installation. In the fuel-management system of the K-Jetronic (or carburetor system), the closed-loop control of the mixture takes place by means of an additional control unit and an electromechanical actuator (frequency valve). In this manner, the fuel can be metered so precisely that depending upon load and engine speed, the air-fuel ratio is an optimum in all operating modes. Tolerances and the ageing of the engine have no effect whatsoever. At values above λ = 1.00, more fuel is metered to the engine, and at values below λ = 1.00, less. This continuous, almost lag-free adjustment of the air-fuel mixture to λ = 1.00, is one of the prerequisites for the efficient after-treatment of the exhaust gases by the downstream catalytic converter.

Control functions at various operating modes

Start

The Lambda sensor must have reached a temperature of above 350 °C before it outputs a reliable signal. Until this temperature has been reached, the closed-loop mode is suppressed and the air-fuel mixture is maintained at a mean level by means of an open-loop control. Starting enrichment is by means of appropriate components similar to the Jetronic installations not equipped with Lambda control.

Acceleration and full load (WOT)

The enrichment during acceleration can take place by way of the closed-loop control unit. At full load, it may be necessary for temperature and power reasons to operate the engine with an air-fuel ratio which deviates from the λ = 1 figure. Similar to the acceleration range, a sensor signals the full-load operating mode to the closed-loop control unit which then switches the fuel-injection to the open-loop mode and injects the corresponding amount of fuel.

Deviations in air-fuel mixture

The Lambda closed-loop control operates in a range between λ = 0.8...1.2 in which normal disturbances (such as the effects of altitude) are compensated for by controlling λ to 1.00 with an accuracy of ±1 %. The control unit incorporates a circuit which monitors the Lambda sensor and prevents prolonged marginal operation of the closed-loop control. In such cases, open-loop control is selected and the engine is operated at a mean λ-value.

Fig. 36

Heated lambda sensor

1 Sensor housing, **2** Protective ceramic tube, **3** Connection cable, **4** Protective tube with slots, **5** Active sensor ceramic, **6** Contact element, **7** Protective sleeve, **8** Heater, **9** Clamp terminals for heater.

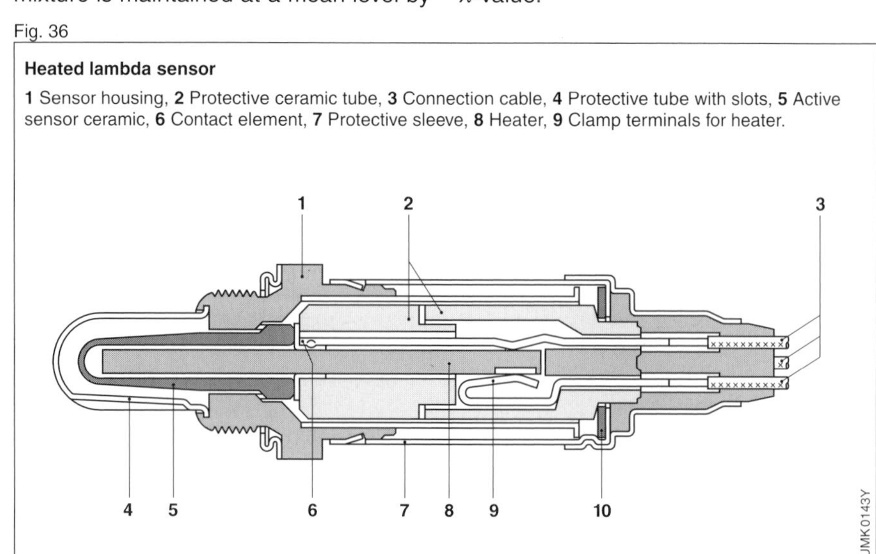

Lambda closed control-loop

The Lambda closed control-loop is superimposed upon the air-fuel mixture control. The fuel quantity to be injected, as determined by the air-fuel mixture control, is modified by the Lambda closed-loop control in order to provide optimum combustion.

U_λ Lambda-sensor signal

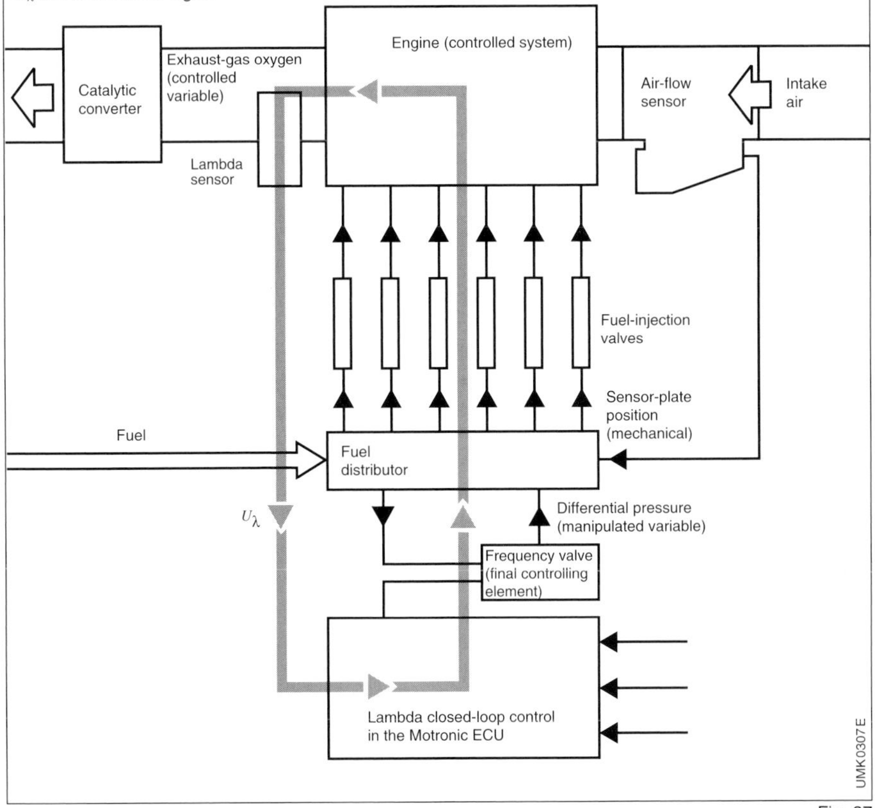

Engine (controlled system)
Catalytic converter
Exhaust-gas oxygen (controlled variable)
Lambda sensor
Air-flow sensor
Intake air
Fuel-injection valves
Sensor-plate position (mechanical)
Fuel
Fuel distributor
U_λ
Differential pressure (manipulated variable)
Frequency valve (final controlling element)
Lambda closed-loop control in the Motronic ECU

UMK0307E

Fig. 37

Fig. 38

View of the unheated (front) and heated lambda sensors

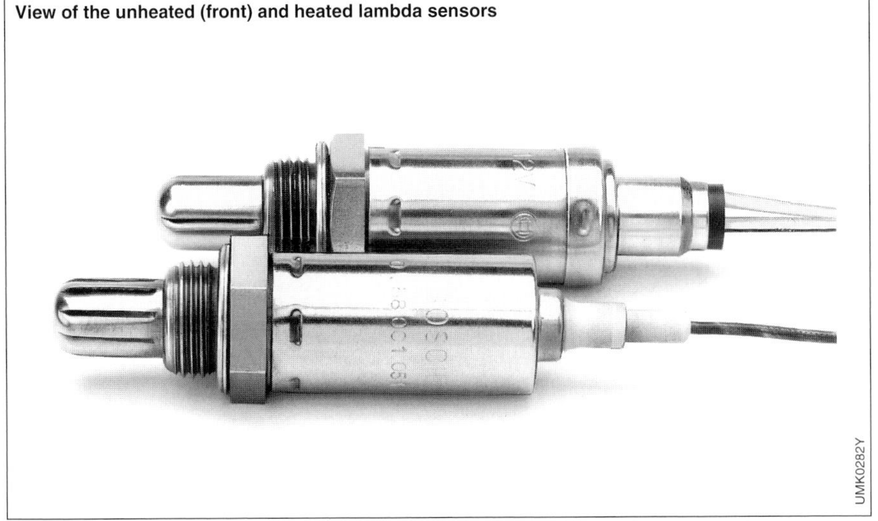

UMK0282Y

Electrical circuitry

If the engine stops but the ignition remains switched on, the electric fuel pump is switched off.
The K-Jetronic system is equipped with a number of electrical components, such as electric fuel pump, warm-up regulator, auxiliary-air device, cold-start valve and thermo-time switch. The electrical supply to all of these components is controlled by the control relay which, itself, is switched by the ignition and starting switch.
Apart from its switching functions, the control relay also has a safety function. A commonly used circuit is described below.

Function

When cold-starting the engine, voltage is applied to the cold-start valve and the thermo-time switch through terminal 50 of the ignition and starting switch. If the cranking process takes longer than between 8 and 15 seconds, the thermo-time switch switches off the cold-start valve in order that the engine does not "flood". In this case, the thermo-time switch performs a time-switch function.

If the temperature of the engine is above approximately +35 °C when the starting process is commenced, the thermo-time switch will have already open-circuited the connection to the start valve which,

Fig. 39

Circuit without voltage applied

1 Ignition and starting switch,
2 Cold-start valve,
3 Thermo-time switch,
4 Control relay,
5 Electric fuel pump,
6 Warm-up regulator,
7 Auxiliary-air device.

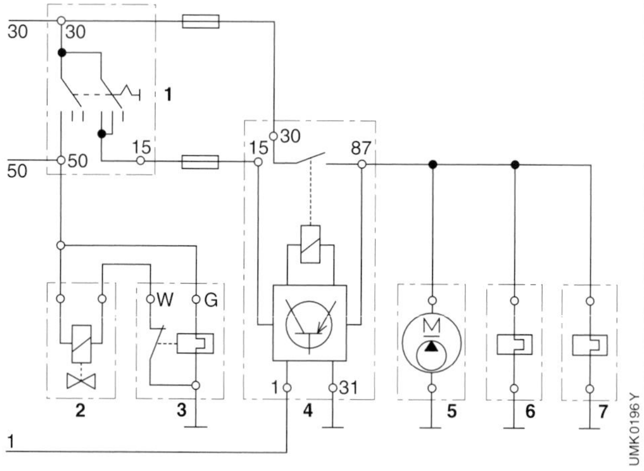

Fig. 40

Starting (with the engine cold)

Cold-start valve and thermo-time switch are switched on. The engine turns (pulses are taken from terminal 1 of the ignition coil). The control relay, electric fuel pump, auxiliary-air device and warm-up regulator are switched on.

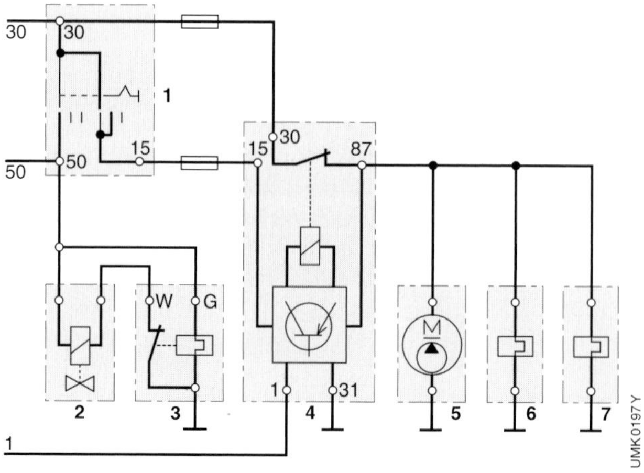

consequently, does not inject extra fuel. In this case, the thermo-time switch functions as a temperature switch.
Voltage from the ignition and starting switch is still present at the control relay which switches on as soon as the engine runs. The engine speed reached when the starting motor cranks the engine is high enough to generate the "engine running" signal which is taken from the ignition pulses coming from terminal 1 of the ignition coil. An electronic circuit in the control relay evaluates these pulses. After the first pulse, the control relay is switched on and applies voltage to the electric fuel pump, the auxiliary-air device and the warm-up regulator. The control relay remains switched on as long as the ignition is switched on and the ignition is running. If the pulses from terminal 1 of the ignition coil stop because the engine has stopped turning, for instance in the case of an accident, the control relay switches off approximately 1 second after the last pulse is received.

This safety circuit prevents the fuel pump from pumping fuel when the ignition is switched on but the engine is not turning.

Fig. 41

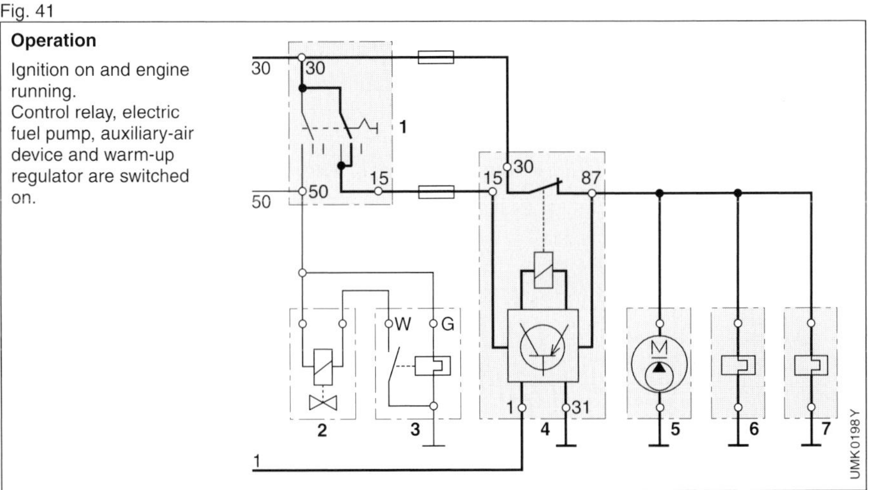

Operation

Ignition on and engine running.
Control relay, electric fuel pump, auxiliary-air device and warm-up regulator are switched on.

Fig. 42

Ignition on but engine stopped

No pulses can be taken from terminal 1 of the ignition coil. The control relay, electric fuel pump, auxiliary-air device and warm-up regulator are switched off.

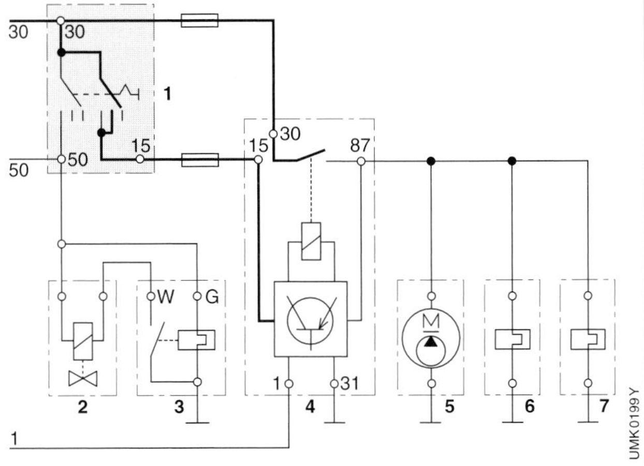

30 30 1 15 50 50 15 30 87 W G M 1 31 1 2 3 4 5 6 7 UMK0199Y

Workshop testing techniques

Bosch customer service

Customer service quality is also a measure for product quality. The car driver has more than 10,000 Bosch Service Agents at his disposal in 125 countries all over the world. These workshops are neutral and not tied to any particular make of vehicle. Even in sparsely populated and remote areas of Africa and South America the driver can rely on getting help very quickly. Help which is based upon the same quality standards as in Germany, and which is backed of course by the identical guarantees which apply to customer-service work all over the world. The data and performance specs for the Bosch systems and assemblies of equipment are precisely matched to the engine and the vehicle. In order that these can be checked in the workshop, Bosch developed the appropriate measurement techniques, test equipment, and special tools and equipped all its Service Agents accordingly.

Testing techniques for K-Jetronic

Apart from the regular replacement of the fuel filter as stipulated by the particular vehicle's manufacturer, the K-Jetronic gasoline-injection system requires no special maintenance work.

In case of malfunctions, the workshop expert has the following test equipment, together with the appropriate test specs, at his disposal:

- Injector tester
- Injected-quantity comparison tester
- Pressure-measuring device, and
- Lambda closed-loop control tester (only needed if Lambda control is fitted).

Together with the relevant Test Instructions and Test Specifications in a variety of different languages, this uniform testing technology is available throughout the world at the Bosch Service Agent workshops and at the majority of the workshops belonging to the vehicle manufacturers. Purposeful trouble-shooting and technically correct repairs cannot be performed at a reasonabe price without this equipment. It is therefore inadvisable for the vehicle owner to attempt to carry out his own repairs.

Fig. 43

Injector tester

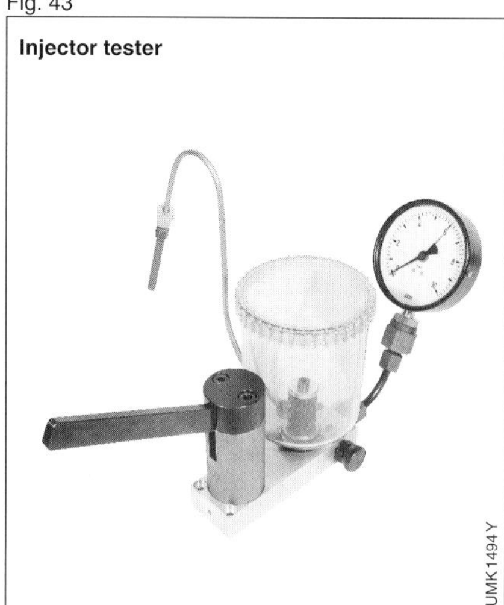

Injector tester

The injector tester (Fig. 43) was developed specifically for testing the K- and KE-Jetronic injectors when removed from the engine. The tester checks all the functions of the injector which are essential for correct engine running:

- Opening pressure,
- Leakage integrity,
- Spray shape,
- Chatter.

Those injectors whose opening pressure is outside tolerance are replaced. For the leak test, the pressure is slowly increased up to 0.5 bar below the opening pressure and held at this point. Within 60 secs, no droplet of fuel is to form at the injector. During the chatter test, the injector must generate a "chattering" noise without a fuel droplet being formed. Serviceable injectors generate a fully atomized spray pattern. "Pencil" jets and "bundled" jets are not to form.

Injected-quantity comparison tester

Without removing the fuel distributor from the vehicle, a comparitive measurement is made to determine the differences in the delivered quantities from the various fuel-distributor outlets (this applies to all engines of up to maximum eight cylinders.

Fig. 44). And since the test is performed using the original injectors it is possible to ascertain at the same time whether any scatter in the figures results from the fuel distributor itself or from the injectors.
The tester's small measuring tubes serve for idle measurement and its larger measuring tubes for part-load or full-load measurement.
Connection to the fuel distributor is by means of eight hoses. The injectors are pulled out of their mountings on the engine and inserted in the automatic couplings at the ends of the hoses. Each automatic coupling incorporates a push-up valve which prevents fuel escaping on hoses which are not connected (e.g. on 6-cylinder systems. Fig. 44). A further hose returns the fuel to the tank.

Pressure-measuring device

This is used to measure all the pressures which are important for correct K-Jetronic operation:

- Primary (system) pressure: Provides information on the performance of the fuel-supply pump, on fuel-filter flow resistance, and on the condition of the primary-pressure regulator.
- Control pressure: Important for assessment of all operating conditions (for instance: Cold/warm engine; part load/full load; fuel-enrichment functions, occasionally pressure at high altitudes).
- Leakage integrity of the complete system. This is particularly important with regard to the cold-start and hot-start behavior. Automatic couplings in the hoses prevent the escape of fuel.

Lambda closed-loop-control tester

On K-Jetronic systems with Lambda closed-loop control, this tester serves to check the duty factor of the Lambda-sensor signal (using simulation of the "rich"/ "lean" signal), and the "open-loop/closed-loop control function". Special adapter lines are available for connection to the Lambda-sensor cable of the various vehicle models. Measured values are shown on an analog display.

Fig. 44

Injected-quantity comparison tester (connected to a 6-cylinder installation)

1 Fuel-distributor injection lines,
2 Injectors,
3 Automatic couplings,
4 Comparison-tester hoses,
5 Small measuring tube,
6 Large measuring tube,
7 Return line to fuel tank.

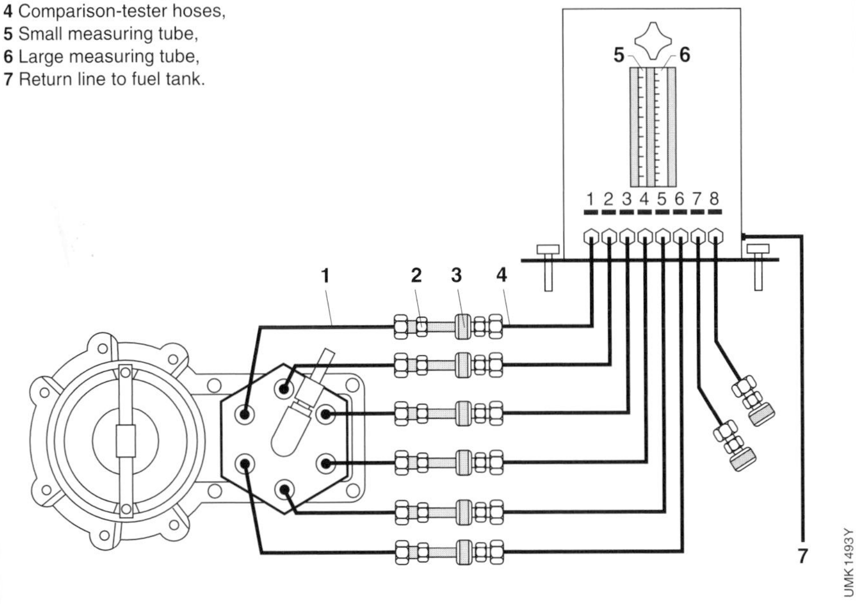

KE-Jetronic

Outline of system

A mechanical hydraulic injection system provides the basis for the KE-Jetronic, in the same way as on the K-Jetronic. This basic system is supplemented by an electronic control unit (ECU) in order to increase flexibility and to enable further functions.

Further components are:
- The sensor for the air quantity inducted by the engine,
- The pressure actuator which intervenes in the mixture composition, and
- The pressure regulator which maintains the primary pressure constant and exerts a specific closing function when the engine is switched off.

Function

A sensor plate deflected by the air stream controls the fuel-metering plunger and thus opens the metering slits to a greater or lesser extent. In the basic function, the KE-Jetronic meters the fuel as a function of the air quantity inducted by the engine, this being the main control variable.
However, by contrast with the K-Jetronic system, the KE-Jetronic injection system also processes other operating data of the engine via sensors whose output signals are processed by an electronic control unit. This electronic control unit controls an electro-hydraulic pressure actuator which adapts the injected-fuel quantity as required to the various operating conditions. If a malfunction occurs, the KE-Jetronic reverts to its basic function. The driver then still has an injection system which functions well when the engine is hot.

Advantages of the KE-Jetronic system

Low fuel consumption

With conventional fuel-management systems, the differences in the length of the intake passages result in different air-fuel mixtures at the individual cylinders. With the KE-Jetronic injection system, each cylinder is allocated its own fuel-injection valve which injects continually onto the intake valve. The injected fuel vaporizes and mixes intensively with the air drawn in by the piston. This means that, in addition to the precise metering, the fuel is evenly distributed between the individual engine cylinders. Since the intake manifolds serve only to carry the intake air to the cylinders, the condensation of the fuel on the manifold walls, a phenomenon in conventional systems which increases the fuel consumption, is practically ruled out.
The KE-Jetronic system ensures that considerably less fuel is used particularly during the warm-up phase, during acceleration enrichment, at full-load and during overrun when it switches off the fuel supply.

Adaptation to operating conditions

The fuel requirement differs greatly from the normal figures during the post-start period, during warm-up and during acceleration and full-load.
By means of commands received from its ECU (electronic control unit), the KE-Jetronic intervenes in the preparation of the air-fuel mixture and increases or decreases the injected fuel quantity accordingly.
The KE-Jetronic system incorporates additional sensors for detecting the engine temperature, the throttle-valve position (load signal) and the sensor-plate deflection (corresponds approximately to the change in engine power as a function of time). With the aid of these sensors, the ECU commands the hydraulic pressure actuator to either "lean-off" the mixture or "enrich" it as appropriate.
The KE-Jetronic responds rapidly to the variations in engine operating conditions, and improves the torque characteristics as well as the engine flexibility. This results in distinct advantages when driving at energy-efficient low engine speeds and in as high a gear as possible, and also an improvement of driveability. Reliable starting is another of the outstanding features of the KE-Jetronic

system. The smooth overrun fuel cutoff responds to engine speed and temperature and cuts off the supply of fuel during deceleration. There are no unpleasant jerks when the fuel supply cuts back in again. This system results in a reduction of the fuel consumption and, since combustion ceases during fuel cutoff, there is no emission of toxic exhaust gases.

Cleaner exhaust gases

The prerequisite for minimum pollutants in the exhaust gas is as complete combustion of the fuel as possible. The KE-Jetronic supplies each cylinder with precisely the amount of fuel appropriate for the particular engine operating condition and for changes in loading. For instance, the required air-fuel mixture is precisely maintained at the level necessary for minimum toxic emissions by reducing the post-start enrichment as soon as possible, or by rapid response of the acceleration enrichment function. Further improvements in the exhaust gas can be achieved by employing the lambda closed-loop control together with exhaust-gas aftertreatment using a catalytic converter (Fig. 1).

Higher power output per liter

The efficiently designed air-intake system of the KE-Jetronic permits an increase in power due to improved cylinder charge. The fuel-injection paths are shorter, with the result that there are no acceleration flat spots.

In a similar way to all the other Jetronic systems, the KE-Jetronic achieves a marked increase in engine power for the same piston displacement but not at the cost of increased fuel consumption. It allows economical engines to be designed, featuring high power-output per liter, whilst at the same time displaying good flexibility coupled with excellent driveability.

Fig. 1

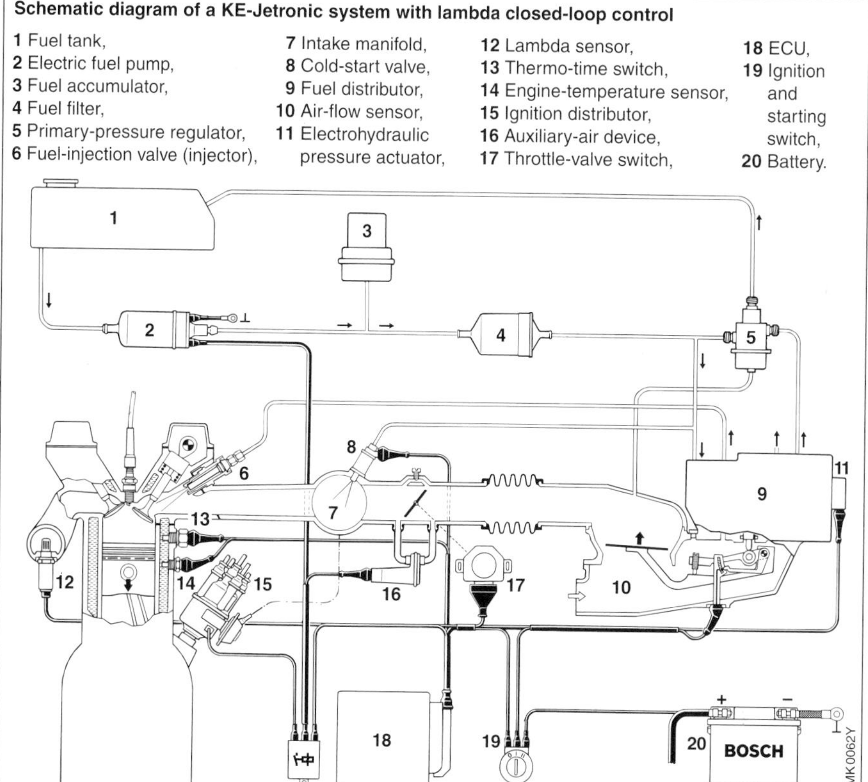

Schematic diagram of a KE-Jetronic system with lambda closed-loop control

1 Fuel tank, **2** Electric fuel pump, **3** Fuel accumulator, **4** Fuel filter, **5** Primary-pressure regulator, **6** Fuel-injection valve (injector), **7** Intake manifold, **8** Cold-start valve, **9** Fuel distributor, **10** Air-flow sensor, **11** Electrohydraulic pressure actuator, **12** Lambda sensor, **13** Thermo-time switch, **14** Engine-temperature sensor, **15** Ignition distributor, **16** Auxiliary-air device, **17** Throttle-valve switch, **18** ECU, **19** Ignition and starting switch, **20** Battery.

Fuel supply

The fuel supply system comprises

- Electric fuel pump (Fig. 2),
- Fuel accumulator,
- Fuel filter (Fig. 4),
- Primary-pressure regulator, and
- Fuel-injection valves.

With regard to the components used, the KE-Jetronic fuel system differs only slightly from that of the K-Jetronic system. An electric roller-cell pump feeds fuel from the tank to the pressure accumulator at a pressure of over 5 bar, and, from there, through the fuel filter to the fuel distributor. From the fuel distributor, the fuel flows to the fuel-injection valves. The fuel-injection valves inject the fuel continuously into the intake ports of the engine. This is why the system is designated KE (taken from the German for continuous and electronic). When the intake valves open, the mixture is drawn into the cylinders.

The primary-pressure regulator maintains the supply pressure in the system constant, and returns the surplus fuel to the tank. Due to the constant flow of fuel through the fuel-supply system, cool fuel is always available. This prevents the formation of vapor bubbles and ensures good hot-starting characteristics.

Electric fuel pump
The electric fuel pump is a roller-cell pump driven by a permanent-magnet electric motor.
The rotor plate which is eccentrically mounted in the pump housing is fitted with metal rollers in notches around its circumference which are pressed against the pump housing by centrifugal force and act as rolling seals. The fuel is carried in the cavities which form between the rollers. The pumping action takes place when the rollers, after having closed the inlet bore, force the trapped fuel around in front of them until it can escape from the pump through the outlet bore (Figure 3). The fuel flows directly around the electric motor. There is no danger of explosion, however, because there is never an ignitable mixture in the pump housing.

The electric fuel pump always delivers more fuel than the engine needs, so that there is always sufficient pressure in the fuel system under all conditions. A check valve in the pump decouples the fuel system from the fuel tank by preventing return flow of fuel to the fuel tank.
The electric fuel pump starts to run immediately when the ignition and starting switch is operated and remains switched on continuously after the en-

Electric fuel pump

1 Suction side, **2** Pressure limiter, **3** Roller-cell pump, **4** Motor armature, **5** Check valve, **6** Pressure side.

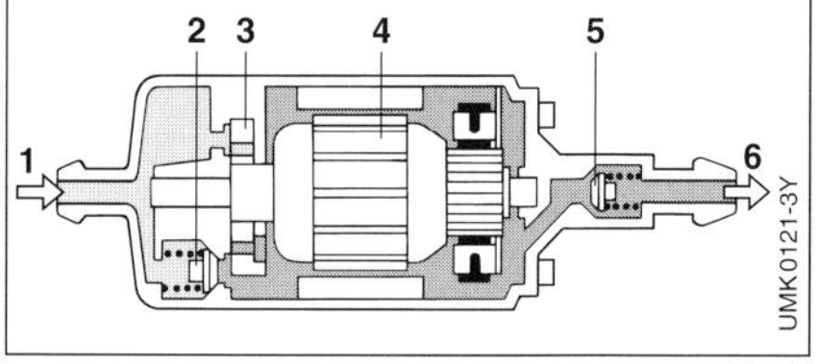

Fig. 2

Fig. 3

Operation of roller-cell pump

1 Suction side, **2** Rotor plate, **3** Roller, **4** Roller race plate, **5** Pressure side.

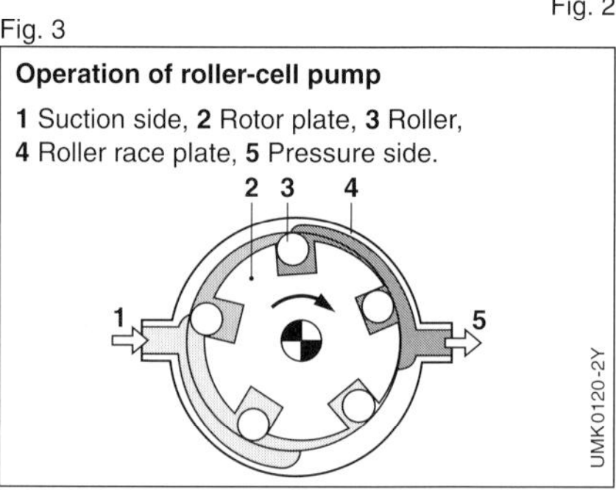

Fig. 4

Fuel filter

1 Paper filter, **2** Strainer, **3** Support plate.

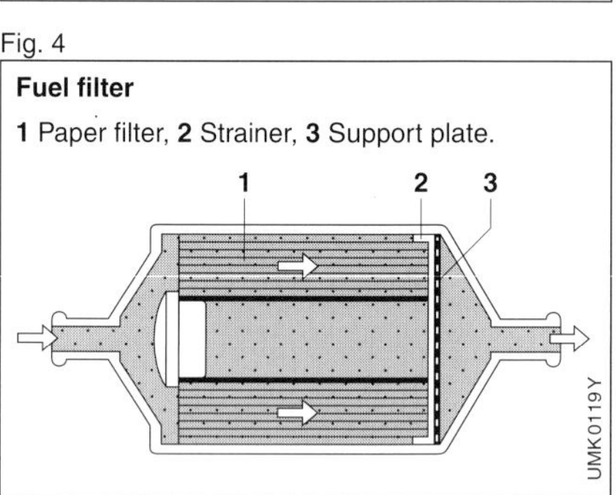

gine has started. A safety circuit is incorporated to stop the pump running and fuel being delivered if the ignition is switched on but the engine has stopped turning (for instance as might occur in the case of an accident). The fuel pump is located in the direct vicinity of the fuel tank and requires no maintenance.

Fuel accumulator

The fuel accumulator maintains the pressure in the fuel system for a certain time after the engine has been switched off in order to facilitate restarting, particularly when the engine is hot. The special design (Figure 5) of the accumulator housing is such that it deadens the sound of the electric fuel pump.

The interior of the fuel accumulator is divided into two chambers by means of a diaphragm. One chamber serves as the accumulator for the fuel, whilst the other represents the compensation volume and is connected to the atmosphere by means of a vent fitting, either directly or through the fuel-tank ventilation system. During operation, the accumulator chamber is filled with fuel and the diaphragm is caused to bend back against the force of the spring until it is halted by the stops in the spring chamber. The diaphragm remains in this position, which corresponds to the maximum accumulator volume, as long as the engine is running.

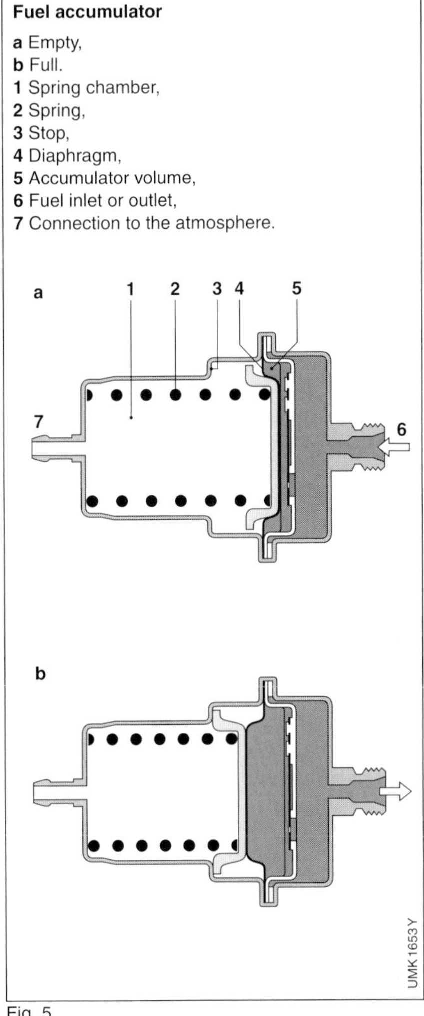

Fuel accumulator

a Empty,
b Full.
1 Spring chamber,
2 Spring,
3 Stop,
4 Diaphragm,
5 Accumulator volume,
6 Fuel inlet or outlet,
7 Connection to the atmosphere.

Fig. 5

Fuel filter

The fuel filter retains particles of dirt which are present in the fuel and which would otherwise adversely effect the functioning of the injection system. The filter contains a paper element with a medium pore size of 10 µm, which is backed up by a fluff strainer (Figure 4). This combination of filter element and fluff strainer ensures a high degree of filtration.

A support plate secures the filter in its metal housing. The filter life depends on the fuel's purity. The filter is installed in the fuel line downstream of the fuel accumulator. When the filter is changed, it is imperative to observe the through-flow direction as indicated by the arrow on the housing.

Primary-pressure regulator

The primary-pressure regulator keeps the supply pressure constant.

In contrast to the K-Jetronic, in which a warm-up regulator regulates the control pressure, the hydraulic counterpressure acting upon the control plunger in the KE-Jetronic is identical to the primary pressure. The control pressure must be held constant since any variation of the control pressure has a direct effect upon the air-fuel ratio. This also applies particularly even if fuel delivery from the supply pump, and injected fuel quantity, vary considerably.

Figure 6 shows a section through the primary-pressure regulator. The fuel enters on the left. The return fuel connection from the fuel distributor is located at the right. The return line to the tank is connected at the top. As soon as the fuel pump starts and generates pressure, the control diaphragm of the pressure accumulator moves downwards. The pressure of the counterspring forces the valve body to follow the diaphragm until, after a very short distance, it encounters a stop and the pressure-controlled function starts. The fuel returning from the fuel distributor, comprising the fuel flowing through the pressure actuator plus the control-plunger leakage can now flow back through the open valve seat to the fuel tank together with the excess fuel. When the engine is switched off, the electric fuel pump also stops. If the system pressure then drops, the valve plate moves back up again and subsequently pushes the valve body upwards against the force of the counterspring until the seal closes the return line to the tank.

Fig. 6

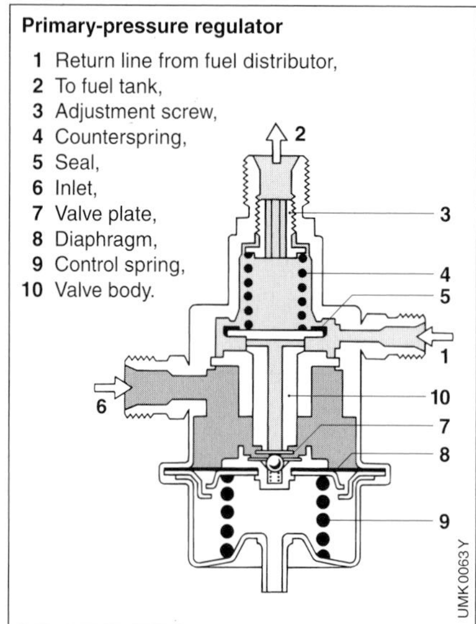

Primary-pressure regulator

1 Return line from fuel distributor,
2 To fuel tank,
3 Adjustment screw,
4 Counterspring,
5 Seal,
6 Inlet,
7 Valve plate,
8 Diaphragm,
9 Control spring,
10 Valve body.

The pressure in the fuel supply system then sinks rapidly to below the injection-valve opening pressure so that the injection valves then close tightly due to the force excerted by the valve-needle spring. The system pressure then increases again to the value determined by the fuel accumulator (Fig. 7).

Fuel-injection valves

At a certain pressure, the injection valves open against the pressure from the valve-needle spring and inject fuel into the intake ports. The fuel is atomized by the operation of the valve needle. They inject the fuel allocated by the fuel distributor into the intake port directly onto the cylinder intake valves. The injection valves are secured in a special holder in order to insulate them from the engine heat. The insulation prevents vapor bubbles forming in the fuel-injection lines which would otherwise lead to poor starting behavior when the engine is hot.

The injection valves have no metering function. They open of their own accord when the opening pressure of, for instance, 3.5 bar is exceeded. They are fitted with a needle valve (Figure 9) whose needle vibrates ("chatters") audibly at high frequency when fuel is

Fig. 7

Pressure curve after engine switch-off

First of all, the pressure drops from the normal primary pressure (**1**) to the closing pressure of the pressure regulator (**2**). It then rises again, due to the effect of the fuel accumulator, to the value (**3**) which is still below the injection-valve opening pressure (**4**).

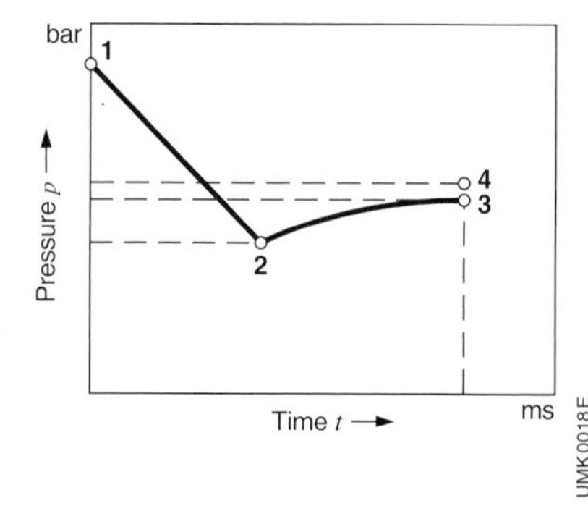

being injected. This results in excellent fuel atomization, even with the smallest of injected quantities.

When the engine is switched off, the fuel-injection valves close tightly as soon as the fuel-system pressure drops below their opening pressure. This means that no more fuel can enter the intake ports and thus reach the intake valves of the engine, once the engine has been switched off.

Air-shrouded fuel-injection valves

Air-shrouded fuel-injection valves improve fuel induction particularly at idle. The air-shrouding principle is based upon the fact that a portion of the air drawn in by the engine enters through the fuel-injection valves (Figure 10) with the result that fuel is particularly well atomized at the point of exit. Air-shrouded valves reduce fuel consumption and lower the level of toxic emissions.

Fig. 8

KE-Jetronic fuel-injection valve (injector): Spray patterns with and without air-shrouding
With air shrouding (shown on the right) the air has a permanent effect and atomizes the fuel even better than in the conventional injector without air-shrouding (shown on the left).

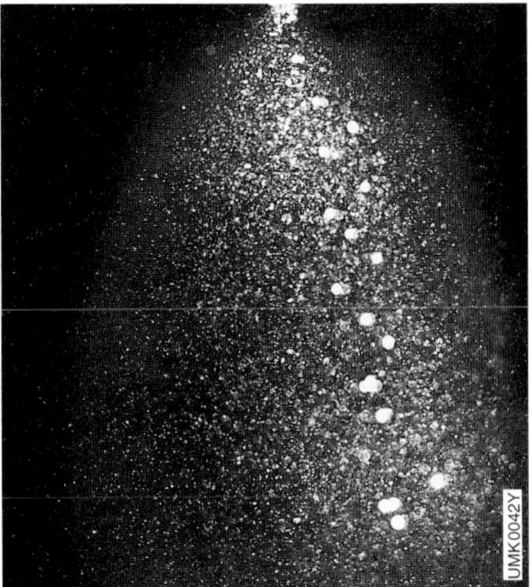

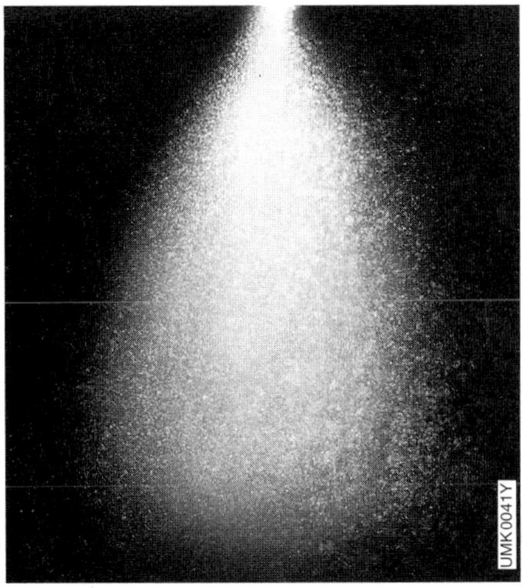

UMK0041Y

Fig. 9

Fuel-injection valve (injector)

a Closed,
b Open,
1 Valve housing,
2 Filter,
3 Valve needle,
4 Valve seat.

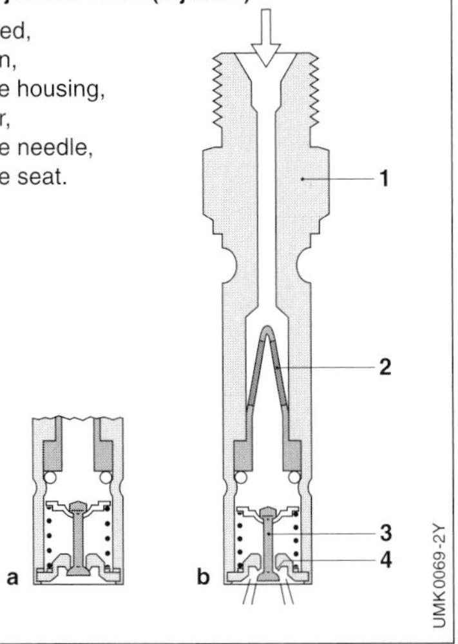

1
2
3
4
a
b
UMK0069-2Y

Fig. 10

Air-shrouded fuel-injection valve (injector)

1 Injector,
2 Air-supply line,
3 Intake manifold,
4 Throttle valve.

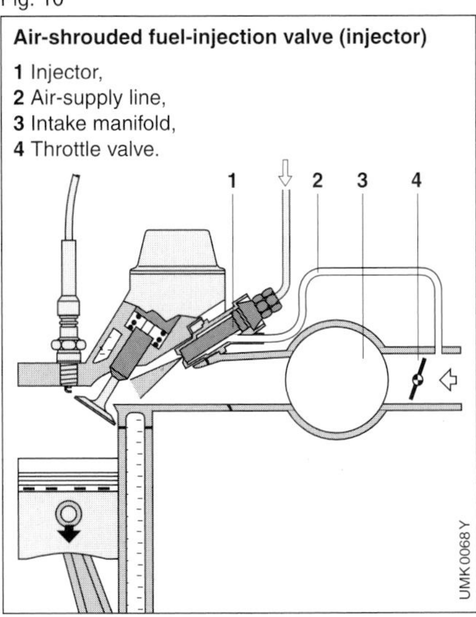

Fuel metering

The task of fuel induction is to meter to the engine a quantity of fuel corresponding to the inducted air quantity.
Basically, fuel metering takes place through the air-flow sensor and the fuel distributor. In a number of operating conditions, however, the amount of fuel required deviates greatly from the "standard" quantity and it becomes necessary to intervene in the mixture-formation system.

Air-flow sensor

The quantity of air drawn in by the engine is a precise measure of its operating load. The air-flow sensor operates according to the suspended-body principle, and measures the amount of air drawn in by the engine (Figure 12).
The intake air quantity serves as the main actuating variable for determining the basic injection quantity. It is the appropriate physical quantity for deriving the fuel requirement, and changes in the induction characteristics of the engine have no effect upon the formation of the air-fuel mixture. Since the air drawn in by the engine must pass through the air-flow sensor before it reaches the engine, this means that it has been measured and the control signal generated before it actually enters the engine cylinders. The result is that, in addition to other measures described below, the correct mixture adaptation takes place at all times.

Principle of the air-flow sensor

a Small amount of air drawn in: Sensor plate only lifted slightly. **b** Large amount of air drawn in: Sensor plate is lifted considerably further.

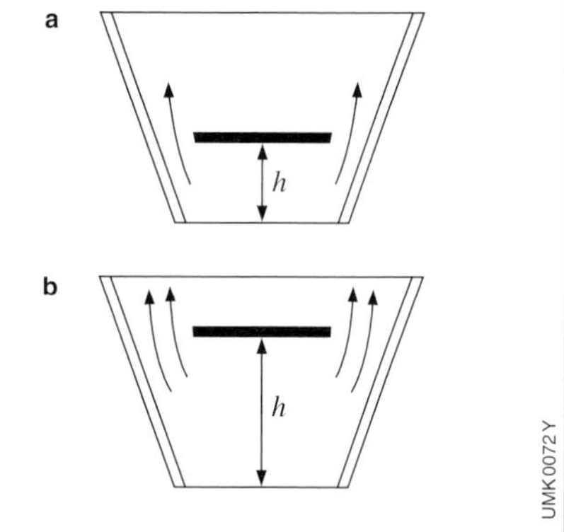

Fig. 12

Fig. 11

Updraft air-flow sensor

a Sensor plate in its zero (inoperative) position,
b Sensor plate in its operating position.
1 Air funnel,
2 Sensor plate,
3 Relief cross-section,
4 Idle-mixture adjusting screw,
5 Pivot,
6 Lever,
7 Leaf spring.

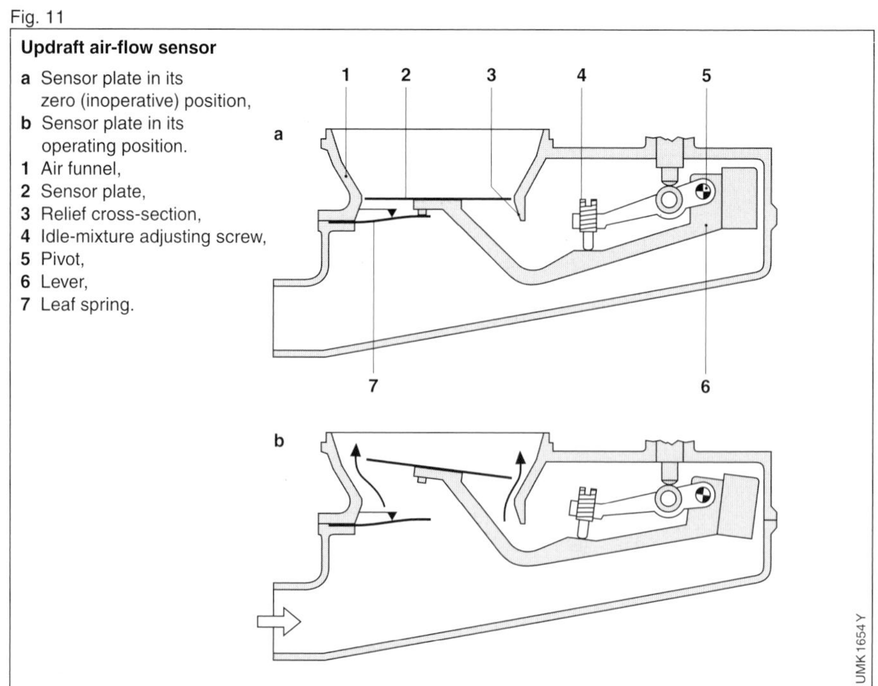

Barrel with metering slits and control plunger

a Zero (inoperative position), **b** Part load, **c** Full load.
1 Fuel inlet, **2** Control plunger, **3** Metering slit in the barrel, **4** Control edge, **5** Barrel,
6 Axial seal, **7** Throttling restriction.

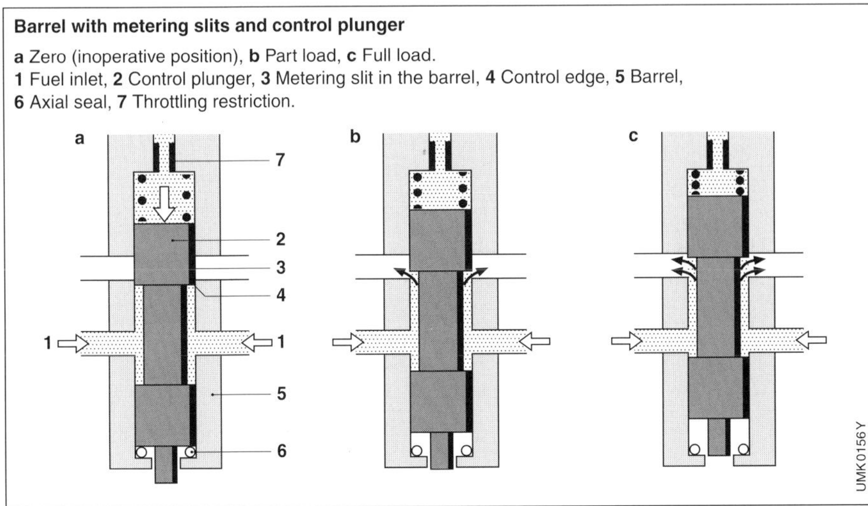

Fig. 13

The air-flow sensor is located upstream of the throttle valve so that it measures all the air which enters the engine cylinders.

The air-flow sensor comprises an air funnel in which the sensor flap (suspended body) is free to pivot. The air flowing through the funnel deflects the sensor plate by a given amount out of its zero position (Figure 11).

The movements of the sensor plate are transmitted by a lever system to a control plunger which determines the basic injection quantity required for the basic functions. Considerable pressure shocks can occur in the intake system if backfiring takes place in the intake manifold. For this reason, the air-flow sensor is so designed that the sensor plate can swing back in the opposite direction in event of misfiring, and past its zero position to open a relief cross-section in the funnel.

A rubber buffer limits the downward stroke (the upward stroke on the downdraft air-flow sensor).

A counterweight compensates for the weight of the sensor plate and lever system (this is performed by means of an extension spring on the downdraft air-flow sensor).

An adjustable leaf spring is fitted to ensure the correct zero position in the switched-off phase.

Fuel distributor

Depending upon the position of the plate in the air-flow sensor, the fuel distributor meters the basic injection quantity to the individual engine cylinders.

The position of the sensor plate is a measure of the amount of air drawn in by the engine. The position of the plate is transmitted to the control plunger by a lever. Depending upon its position in the barrel with metering slits, the control plunger opens or closes the slits to a greater or lesser extent. The fuel flows through the open section of these slits to the differential-pressure valves and then to the fuel-injection valves (Figure 13).

If sensor-plate travel is only very small, the control plunger is lifted only slightly and, as a result, only a small section of the slit is opened for the passage of fuel. On the other hand, with a larger plunger travel, the plunger opens a larger section of the slits and more fuel can flow. There is a linear relationship between sensor-plate travel and the slit section in the barrel which is opened for fuel flow.

A hydraulic force is applied to the control plunger, and acts in opposition to the movement resulting from the sensor-plate deflection. A constant air-pressure drop at the sensor plate is the result, and this ensures that the control plunger always follows the movement of the sensor-plate lever. In some versions, a pressure spring is used to assist this hydraulic force (Figure 14). It prevents the control plunger from being drawn up due to vacuum effects when the system cools down.

It is imperative that the primary pressure be accurately controlled, otherwise variations would have a direct effect upon the air-fuel ratio (or λ value). A damping throttle (Figure 14) serves to damp oscillations that could be caused by sensor-plate forces. When the engine is switched off, the control plunger sinks until it comes to rest against an axial seal ring (Figures 13 to 15). This is secured by an adjustable screw and can be set to the correct height to ensure that the metering slits are closed correctly by the plunger when it is in its zero position.

Whereas with the K-Jetronic, the zero position of the plunger is determined by its abutting against the sensor-plate lever, with the KE-Jetronic, the plunger rests upon the axial seal ring due to the force applied to it by the residual primary pressure. This measure serves to prevent pressure loss due to leakage past the control plunger, and thus prevents the fuel accumulator from emptying through the control-plunger gap. The fuel accumulator must remain full, because it has the job of maintaining the primary pressure above the fuel-vapor pressure which is applicable for the particular fuel temperature prevailing when the engine is switched off.

Differential-pressure valves

The differential-pressure valves in the fuel distributor serve to generate a given pressure drop at the metering slits.

The air-flow sensor has a linear characteristic. This means that if double the quantity of air is drawn in, the sensor-plate travel is also doubled. If this (linear) travel is to result in a change of the basic injection quantity in the same relationship, then a constant pressure drop must be guaranteed at the metering slits regardless of the amount of fuel flowing through them (Figure 13).

The differential-pressure valves maintain the drop in pressure between the upper and lower chambers constant, regardless of fuel throughflow. The differential pressure is usually 0.2 bar.

Fig. 14

Fuel distributor with differential-pressure valves

1 Fuel inlet (primary pressure),
2 Upper chamber of differential-pressure valve,
3 Line to the fuel-injection valve (injector),
4 Control plunger,
5 Control edge and metering slit,
6 Valve spring,
7 Valve diaphragm,
8 Lower chamber of differential-pressure valve,
9 Axial seal ring,
10 Pressure spring,
11 Fuel from the electro-hydraulic pressure actuator,
12 Throttling restriction,
13 Return line.

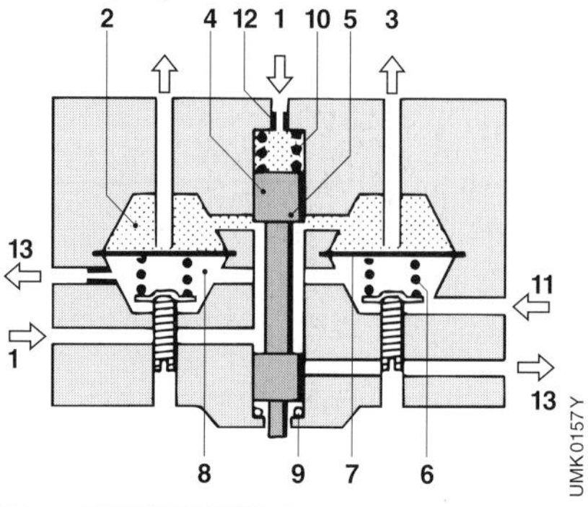

Fig. 15

Differential pressure valve

a Operating position with small injected fuel quantity,
b Operating position with large injected fuel quantity.

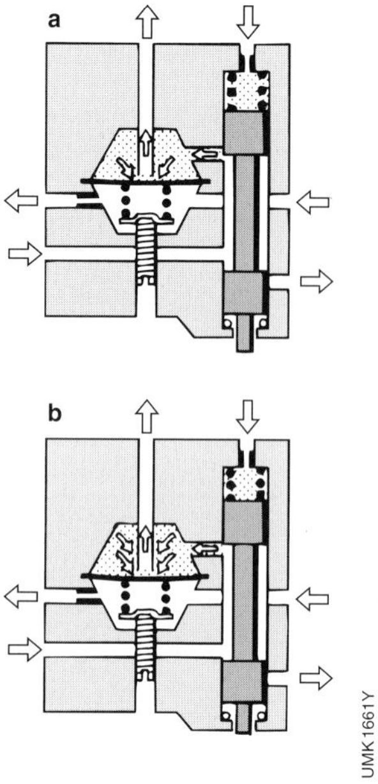

This ensures a high degree of metering accuracy.
The differential-pressure valves are of the flat-seat type and are located in the fuel distributor. They are each allocated to one control slit. A diaphragm separates the upper chamber from the lower chamber of the valve (Figures 14 to 16).
The lower chambers of all valves are interconnected by means of a ring main, as well as to the electro-hydraulic pressure actuator. The valve seat is located in the upper chamber. Each upper chamber is connected to a metering slit and its corresponding fuel-injection line.
The upper chambers are completely sealed off from each other. The pressure differential at the metering slit is determined by the force of the helical spring in the lower chamber, together with the effective diaphragm diameter and the electro-hydraulic pressure actuator.
If a large basic injection quantity flows into the upper chamber, the diaphragm bends downwards and opens the outlet cross-section of the valve until the set differential pressure is reached again. If throughflow quantity drops, the valve cross-section is reduced due to the equilibrium of forces at the diaphragm until a pressure differential of 0.2 bar prevails again. This means that an equilibrium of forces exists at the diaphragm which can be maintained for every basic injection quantity by controlling the valve cross-section (Figure 15).
An additional fine filter with a separator for ferromagnetic contamination is fitted in the fuel line to the electro-hydraulic pressure actuator.

Mixture formation

The formation of the air-fuel mixture takes place in the intake ports and cylinders of the engine.
The continually injected fuel coming from the injection valves is "stored" in front of the intake valves. When the intake valve is opened, the air drawn in by the engine carries the waiting "cloud" of fuel with it into the cylinder. An ignitable air-fuel mixture is formed during the induction stroke due to the swirl effect.

Fig. 16

Air-flow sensor with a section through the fuel distributor

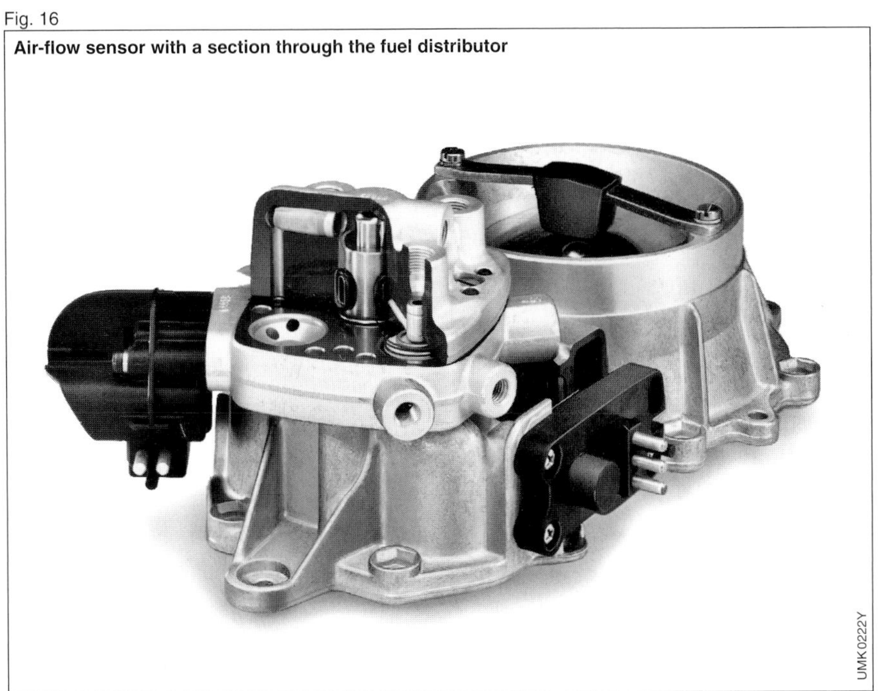

Adaptation to operating conditions

In addition to the basic functions described up to now, the mixture has to be adapted during particular operating conditions. These adaptations (corrections) are necessary in order to optimize the power delivered, to improve the exhaust-gas composition and to improve the starting behavior and driveability. Thanks to additional sensors for the engine temperature and the throttle valve position (load signal), the control unit of the KE-Jetronic can perform these matching tasks better than a mechanical system.

Basic adaptation

The basic adaptation of the air-fuel mixture to the operating modes of idle, part load and full load is by appropriately shaping the air funnel in the air-flow sensor (Fig. 17).

If the funnel had a purely conical shape, the result would be a mixture with a constant air-fuel ratio throughout the whole of the sensor plate range of travel (metering). However, it is necessary to meter to the engine an air-fuel mixture which is optimal for particular operating modes such as idle, part load and full load. In practice, this means a richer mixture at idle and full load and a leaner mixture in the part-load range. This adaptation is achieved by designing the air funnel so that it becomes wider in stages (Fig. 18).

If the funnel is flatter than the basic form (which was specified for a given mixture, e.g. at $\lambda = 1$), then the mixture is leaner. If, on other hand, the funnel walls are steeper than in the basic form, the mixture is richer because the sensor plate deflects further for the same air throughflow and the control plunger meters more fuel. Consequently, this means that the air funnel can be shaped so that it is possible to meter mixtures to the engine which have different air-fuel ratios depending upon the sensor-plate position in the funnel (which in turn corresponds to the particular engine operating mode i.e. idle, part load and full load). In the case of the KE-Jetronic, the air funnel is preferably so shaped that an air-fuel mixture with $\lambda = 1$ results across the whole operating range.

Electronic control unit (ECU)

The electronic control unit evaluates the data delivered by the various sensors concerning the engine operating condition, and, from the status, generates a control signal for the electro-hydraulic pressure actuator (Fig. 19).

Registration of operating data

Additional criteria, above and beyond the information coming from the intake-air quantity, are required in order to determine the optimum fuel quantity

Fig. 17

Influence of funnel-wall angle upon the sensor-plate deflection for identical air throughput

a The basic funnel shape results in stroke "h",
b Steep funnel walls result in increased stroke "h" for identical air throughput,
c Flatter funnel shape results in reduced deflection "h" for identical air throughput.

A Annular section exposed by sensor-plate deflection (identical for a, b, and c).

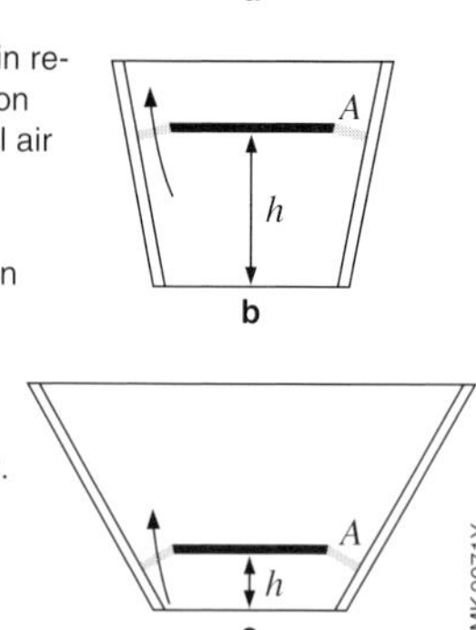

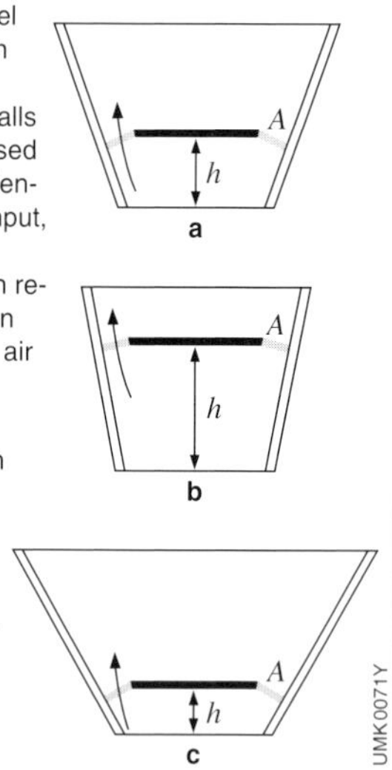

Fig. 18

Adaptation of the air-funnel shape

1 For maximum power, **2** For part load, **3** For idle.

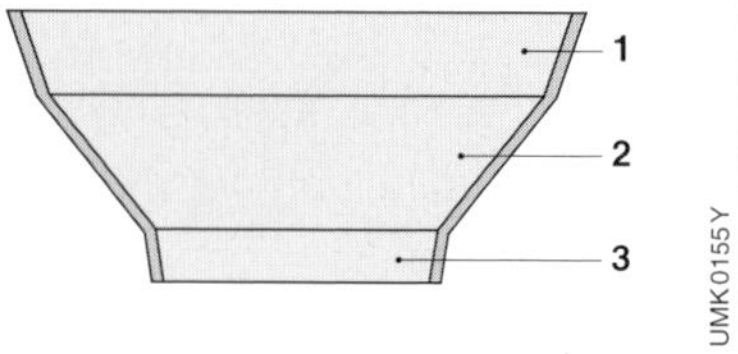

required by the engine. These must be registered by sensors and signalled to the electronic control unit.

Table 1. Adaptations

Performance characteristics	Sensor
Full load Idle	Throttle-valve switch
Engine speed	Ignition-triggering system (usually in the ignition distributor)
Start	Ignition and starting switch
Engine temperature	Engine-temperature sensor
Air pressure	Aneroid-box sensor
Air-fuel mixture	Lambda sensor

The sensors are described in conjunction with the relevant adaptation function.

Design and function

Depending upon the functional scope, the electronic circuitry uses either analog techniques or mixed analog/digital techniques. Starting with the "Europe" unit, the module for idle-mixture control and for lambda closed-loop control can be added. ECUs with a more extensive range of functions are designed using digital techniques. The electronic components are installed on a PC board and include ICs (e.g. operational amplifiers, comparators and voltage stabilizers), transistors, diodes, resistors and capacitors. The PC boards are inserted in the ECU housing which can be equipped with a pressure-equalization element. The ECU is connected to the battery, to the sensors and to the actuator by a 25-pole plug.

The ECU processes the incoming signals from the different sensors and, on the basis of this, calculates the control current for the electro-hydraulic pressure actuator.

Voltage stabilization

The ECU must be powered by a stable voltage which remains constant regardless of the voltage of the vehicle electrical system. The current applied to the pressure actuator, which depends upon the incoming sensor signals carrying the data on the engine operating conditions, is generated from this stabilized voltage, the stabilization of which takes place in a special IC.

Input filters

Input filters filter out any interference which may be present in the incoming signals from the sensors.

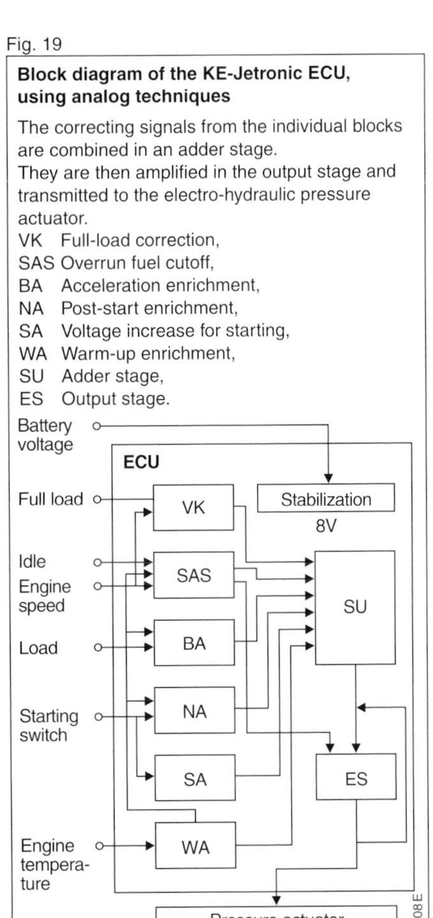

Fig. 19

Block diagram of the KE-Jetronic ECU, using analog techniques

The correcting signals from the individual blocks are combined in an adder stage.
They are then amplified in the output stage and transmitted to the electro-hydraulic pressure actuator.

VK Full-load correction,
SAS Overrun fuel cutoff,
BA Acceleration enrichment,
NA Post-start enrichment,
SA Voltage increase for starting,
WA Warm-up enrichment,
SU Adder stage,
ES Output stage.

Adder

Here, the evaluated sensor signals are combined. The electrically processed corrective signals are added in an operational circuit and then transmitted to the current regulator.

Output stage

The output stage generates the control signal for the pressure actuator, whereby it is possible to input opposing currents into the pressure actuator in order to increase or decrease the pressure drop. The magnitude of the current in the pressure actuator can be adjusted at will in the positive direction by means of a permanently triggered transistor. The current is reversed during "overrun" (overrun fuel cutoff), and influences the differential pressure at the differential-pressure valves so that the flow of fuel to the injection valves is interrupted.

Additional output stages

If necessary, additional output stages can be included. These can trigger the valves for EGR, and control the bypass cross-section around the throttle valve as required for idle-mixture control, to mention but two applications.

Electro-hydraulic pressure actuator

Depending upon the operating mode of the engine and the resulting current signal received from the ECU, the electro-hydraulic pressure actuator varies the pressure in the lower chambers of the differential-pressure valves. This changes the amount of fuel delivered to the injection valves.

Design

The electro-hydraulic pressure actuator (Figure 20) is mounted on the fuel distributor. The actuator is a differential-pressure controller which functions according to the nozzle/baffle-plate principle, and its pressure drop is controlled by the current input from the ECU. In a housing of non-magnetic material, an armature is suspended on a frictionless taut-band suspension element, between two double magnetic poles. The armature is in the form of a diaphragm plate made from resilient material.

Fig. 20

Electro-hydraulic pressure actuator fitted to the fuel distributor

The control signal from the ECU intervenes in the position of the baffle plate (11). This, in turn, varies the fuel pressure in the upper chamber of the differential-pressure valves and, as a result, the quantity of fuel delivered to the injection valves (injectors). Using this principle, adaptation and correction functions can be incorporated.

1 Sensor plate,
2 Fuel distributor,
3 Fuel inlet (primary pressure),
4 Fuel to the injection valves,
5 Fuel return to the pressure regulator,
6 Fixed restriction,
7 Upper chamber,
8 Lower chamber,
9 Diaphragm,
10 Pressure actuator,
11 Baffle plate,
12 Nozzle,
13 Magnetic pole,
14 Air gap.

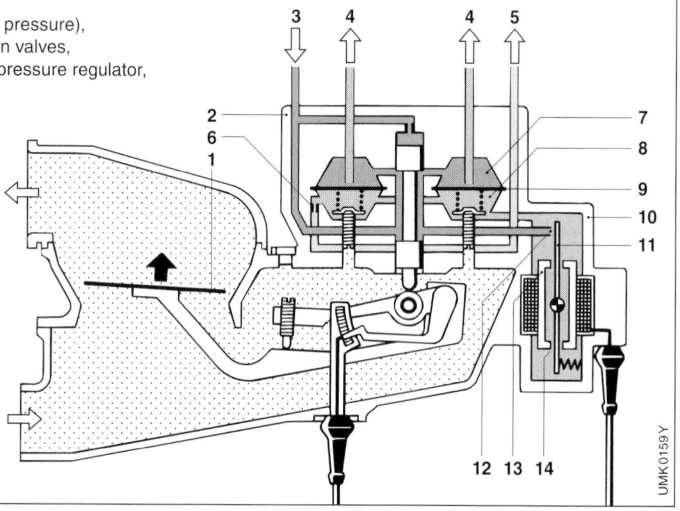

UMK0159Y

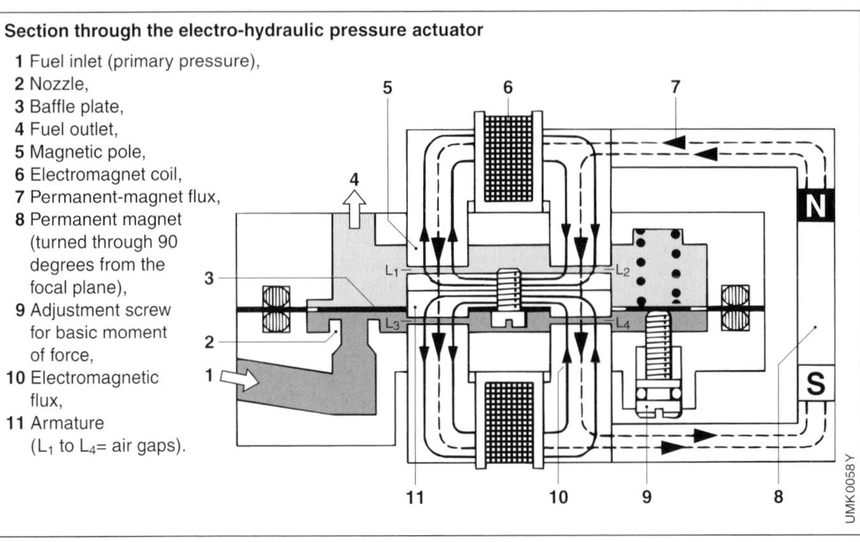

Fig. 21

Function

The magnetic flux of a permanent magnet (broken lines in Figure 21) and that of an electromagnet (unbroken lines) are superimposed upon each other in the magnetic poles and their air gaps. The permanent magnet is actually turned through 90 degrees referred to the focal plane. The paths taken by the magnetic fluxes through the two pairs of poles are symmetrical and of equal length, and flow from the poles, across the air gaps to the armature, and then through the armature.

In the two diagonally opposed air gaps (Figure 21 L_2, L_3), the permanent-magnet flux, and the electro-magnet flux resulting from the incoming ECU control signal are added, whereas in the other two air gaps (Figure 21 L_1, L_4) the fluxes are subtracted from each other. This means that, in each air gap, the armature, which moves the baffle plate, is subjected to a force of attraction proportional to the square of the magnetic flux.

Since the permanent-magnet flux remains constant, and is proportional to the control current from the ECU flowing in the electromagnet coil, the resulting torque is proportional to this control current. The basic moment of force applied to the armature has been selected so that, when no current is applied from the ECU, there results a basic differential pressure which corresponds preferably to $\lambda = 1$. This also means that, in the case of control current failure, limp-home facilities are available without any further correction measures being necessary.

The jet of fuel which enters through the nozzle attempts to bend the baffle plate away against the prevailing mechanical and magnetic forces. Taking a fuel throughflow which is determined by a fixed restriction located in series with the pressure actuator, the difference in pressure between the inlet and outlet is proportional to the control current applied from the ECU. This means that the variable pressure drop at the nozzle is also proportional to the ECU control current, and results in a variable lower-chamber pressure. At the same time, the pressure in the upper chambers changes by the same amount. This, in turn, results in a change in the difference at the metering slits between the upper-chamber pressure and the primary pressure and this is applied as a means for varying the fuel quantity delivered to the injection valves.

As a result of the small electromagnetic time constants, and the small masses which must be moved, the pressure actuator reacts extremely quickly to

variations in the control current from the ECU.
If the direction of the control current is reversed, the armature pulls the baffle plate away from the nozzle and a pressure drop of only a few hundredths of a bar occurs at the pressure actuator. This can be used for auxiliary functions such as overrun fuel cutoff and engine-speed limitation. The latter function is performed by interrupting the flow of fuel to the injection valves.

Cold-start enrichment

Depending upon the engine temperature, the cold-start valve injects an additional quantity of fuel for a limited period of time during starting.
This is carried out in order to compensate for the losses resulting from condensation on the cylinder walls and in order to facilitate starting the cold engine. For this purpose, due to the fact that during starting the pronounced variations in the engine speed would result in a false air-flow signal, it is necessary for the ECU to provide a fixed load signal during cranking which is weighted with an engine-temperature factor.
This additional quantity of fuel is injected by the cold-start valve into the intake manifold. The "on" period of the cold-start valve is limited by a thermo-time switch as a function of the engine temperature.
This process is known as cold-start enrichment and results in a "richer" air-fuel mixture, i.e. the excess-air factor λ is temporarily less than 1.

Cold-start valve

The cold-start valve (Figure 22) is a solenoid-operated valve. The solenoid coil is located inside the valve, and when the valve is closed, it is sealed off by its movable armature being pressed against the seal by a spring.
When the solenoid is energized, the armature is lifted from its seat and permits fuel to flow into a so-called swirl nozzle which causes it to rotate. The result is that the fuel is atomized extremely finely and enriches the mixture in the intake manifold downstream of the throttle valve. The cold-start valve is so

Fig. 22

Cold-start valve in operated state

1 Electrical connection, **2** Fuel supply with strainer, **3** Valve (electromagnet armature), **4** Solenoid winding, **5** Swirl nozzle, **6** Valve seat.

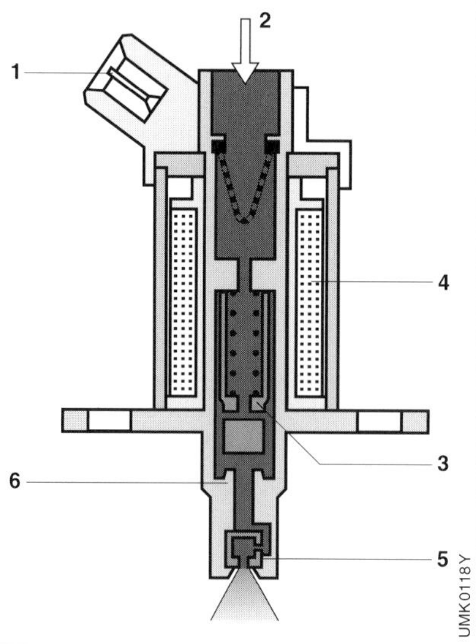

Fig. 23

Thermo-time switch

1 Electrical connection, **2** Housing, **3** Bimetal strip, **4** Heating filaments, **5** Electrical contact.

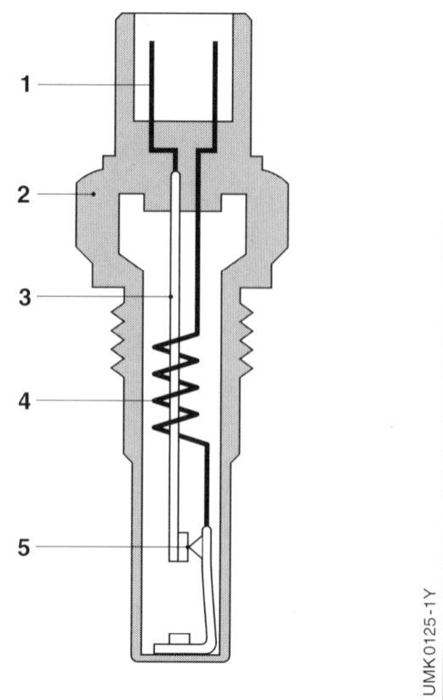

positioned in the inlet manifold that a favorable distribution of the air-fuel mixture to all cylinders is ensured.

Thermo-time switch

The thermo-time switch limits the duration of cold-start valve operation, depending upon temperature.
The thermo-time switch (Figure 23) consists of an electrically heated bimetal strip which, depending upon its temperature, opens or closes a contact. It is brought into operation by the ignition/starter switch, and is mounted at a spot which is representative of engine temperature.
During a cold start, it limits the "on" period of the cold-start valve. In case of repeated start attempts, or when starting takes too long, the cold-start valve ceases to inject. Its "on" period is determined by the thermo-time switch which is heated by engine heat as well as by its own built-in heater. Both these heating effects are necessary in order to ensure that the "on" period of the cold-start valve is limited under all conditions, and engine flooding is prevented.

Fig. 24

Engine-temperature sensor

1 Electrical connection,
2 Housing,
3 NTC resistor.

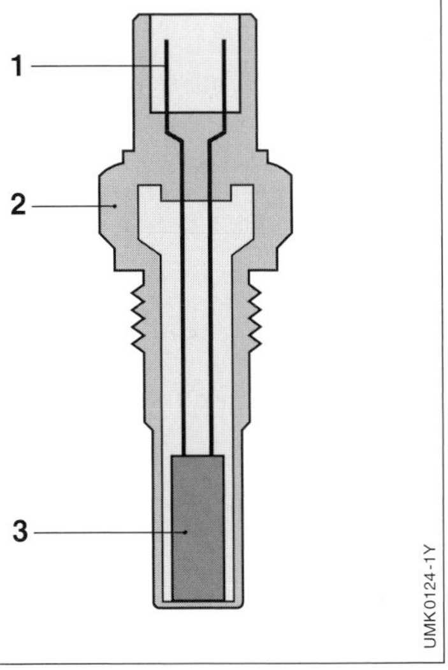

During an actual cold start, the heat generated by the built-in heater is mainly responsible for the "on" period (switch off, for instance, at –20 °C after 7.5 seconds). With a warm engine, the thermo-time switch has already been heated up so far by engine heat that it remains open and prevents the cold-start valve from going into action.

Post-start enrichment

Enrichment with additional fuel improves the post-start performance at low temperatures.
This function is calibrated to give satisfactory throttle response at all temperatures together with minimum fuel consumption.
Post-start enrichment is dependent upon temperature and time and, starting from a temperature-dependent initial value, is decreased practically as a linear function of time. This means that the enrichment duration is a function of the initial temperature.
The ECU maintains the temperature-dependent mixture enrichment at its maximum level for about 4.5 seconds and then reduces to zero. For instance, following a start at 20 °C, the reduction to zero takes 20 seconds.

Engine-temperature sensor

The engine temperature is measured by the engine-temperature sensor which provides the ECU with a corresponding electric signal.
The engine-temperature sensor (Figure 24) is mounted in the engine block on air-cooled engines. With water-cooled engines, it projects into the coolant.
The sensor "reports" the particular engine temperature to the ECU in the form of a resistance value. The ECU then controls the electro-hydraulic pressure actuator which carries out the appropriate adaptation of the injected fuel quantity during the post-start and warm-up periods. The temperature sensor consists of an NTC resistor embedded in a threaded sleeve.

NTC stands for Negative Temperature Coefficient, the decisive characteristic of this resistor. When the temperature increases, the electrical resistance of the semiconductor resistor decreases.

Warm-up enrichment
During warm-up, the engine receives extra fuel depending upon the temperature, the load and the engine speed.
The engine-temperature sensor registers the coolant temperature and reports this to the ECU which then converts this data into a control signal for the electro-hydraulic pressure actuator. The mixture adaptation through the pressure actuator is arranged such that perfect combustion is achieved at all temperatures while, at the same time, keeping the fuel enrichment as low as possible.

Acceleration enrichment
During acceleration, the KE-Jetronic meters additional fuel to the engine as long as it is still cold.
If the throttle is opened abruptly, the air-fuel mixture is momentarily leaned-off, and a short period of mixture enrichment is needed to ensure good transitional response.
As a result of the change in the load signal (referred to time), the ECU recognizes when acceleration is taking place and, as a result, triggers the acceleration enrichment. This prevents the familiar "flat spot". When the engine is cold, it requires additional enrichment due to the less than optimum air-fuel mixing and due to the possibility of fuel being deposited on the intake-manifold walls.
The maximum value for acceleration enrichment is a function of the temperature. The acceleration enrichment is triggered at ≤ 80 °C by a needle-shaped enrichment pulse with a duration of 1 second. The enrichment quantity is higher the colder the engine, and is also dependent upon changes in load.
The speed with which the pedal is depressed when accelerating is determined from the deflection of the air-flow sensor. This has only a very slight lag referred to the throttle-plate movement. This signal, which corresponds to the change in the intake-air quantity and therefore approximately to the engine power, is registered by the potentiometer in the air-flow sensor and passed to the ECU which controls the pressure actuator accordingly. The potentiometer curve is non-linear, so that the acceleration signal is at maximum when accelerating from idle. It decreases along with the increase in engine power. The result is a reduction in the ECU circuit complexity.

Sensor-plate potentiometer
The potentiometer in the air-flow sensor (Figure 25) is manufactured using film techniques on a ceramic base.
A brush-type wiper moves across the potentiometer track. The brushes consist of a number of fine wires which are welded to a lever. The individual wires apply only a very low pressure to the potentiometer track with the result that wear is extremely low. Due to the large number of wires in the brush, excellent electrical contact is guaranteed even on a rough track surface and also when the wiper moves quickly.
The potentiometer lever is attached to the sensor-plate shaft but is electrically insulated from it. The wiper voltage is tapped off by a second wiper brush which is electrically connected to the main wiper.
The wiper is designed to travel past the ends of the tracks in both directions so far that damage is ruled out when backfiring occurs in the intake manifold. A fixed film resistor is included in series with the wiper to prevent damage in the case of short-circuit.

Full-load enrichment
The engine delivers its maximum power at full load, when the air-fuel mixture must be enriched compared with that at part load.
In contrast to part load, where the calibration is for minimum fuel consumption and low emissions, at full load it is necessary to enrich the air-fuel mixture. This enrichment is programmed

to be engine-speed dependent. It provides maximum possible torque over the entire engine-speed range, and this ensures optimum fuel-economy figures during full-load operation.
At full load, e.g. in the engine-speed ranges between 1,500 and 3,000 min^{-1} and above 4,000 min^{-1}, the KE-Jetronic enriches the air-fuel mixture. The full-load signal is delivered by a full-load switch on the throttle valve, or by a microswitch on the accelerator-lever linkage. The information on engine speed is taken from the ignition. From this data, the ECU calculates the additional fuel quantity needed, and this is put into effect by the pressure actuator.

Throttle-valve switch

The throttle-valve switch communicates the "idle" and "full load" throttle positions to the ECU.
The throttle-valve switch (Figure 26) is mounted on the throttle body and actuated by the throttle-valve shaft. Attached to this shaft is a contoured switch guide that closes the idle contacts at one end of its travel and the full-load contacts at the other. Recognition of these two engine-operating modes is essential for the engine-management system.

Fig. 26

Throttle-valve switch

1 Full-load contact, **2** Contoured switching plate, **3** Throttle-valve shaft, **4** Idle contact, **5** Electrical connection.

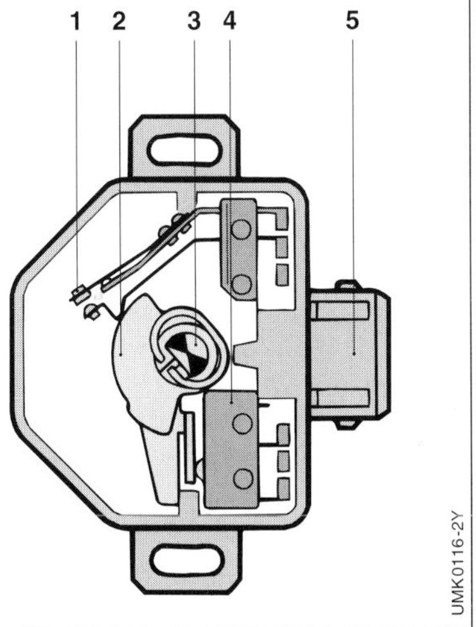

Fig. 25

Potentiometer for determining sensor-plate position

1 Pickoff brush, **2** Main brush, **3** Wiper lever, **4** Potentiometer plate (shifted out of the focal plane), **5** Air-flow sensor housing, **6** Sensor-plate shaft.

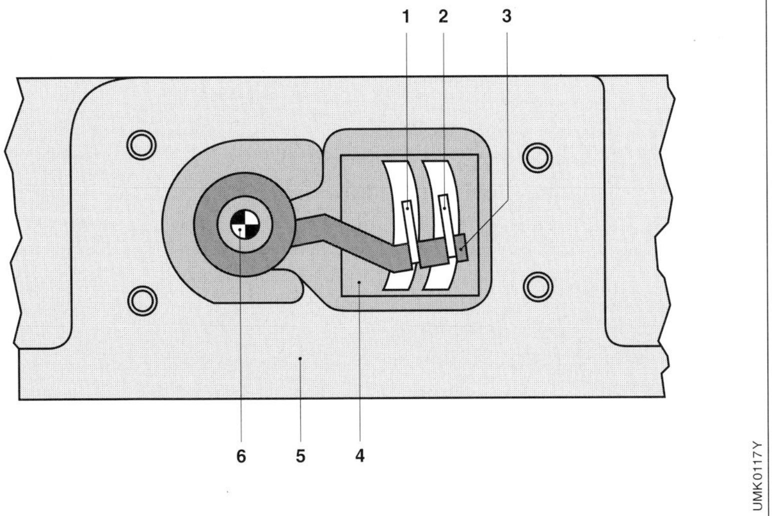

Fig. 27

Auxiliary-air device (section)

Top:
Air passage partially opened by the perforated plate.
Bottom:
Perforated plate blocks off the air passage since the engine has reached the appropriate operating temperature.

1 Plate opening,
2 Pivot,
3 Electrical heating,
4 Air passage,
5 Perforated plate.

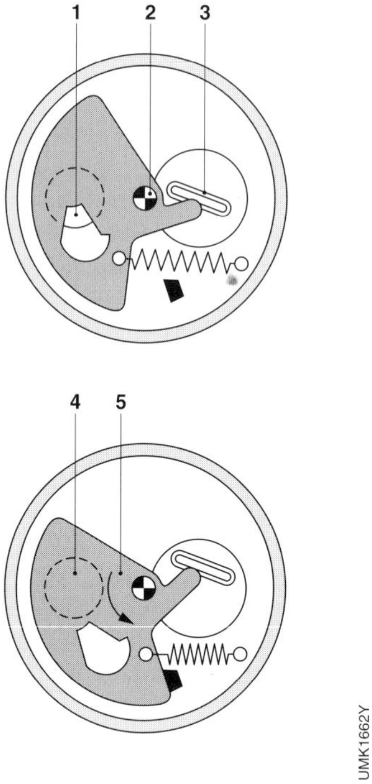

Fig. 28

Electrically heated auxiliary-air device

1 Electrical connection,
2 Electrical heating,
3 Bimetal strip,
4 Perforated plate.

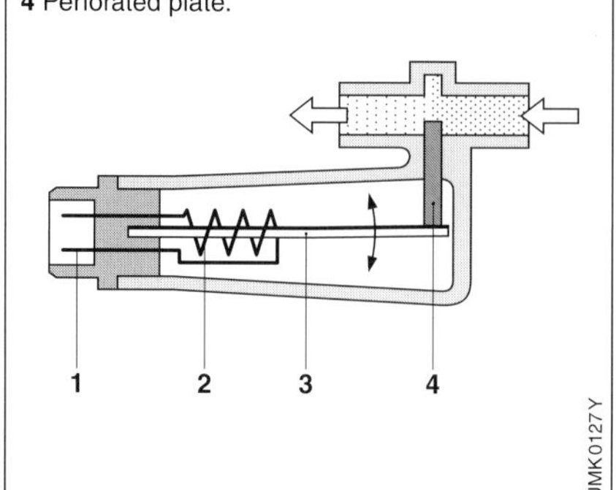

Controlling the idle speed by means of the auxiliary-air device

In order to achieve smoother running at idle, the idle-speed is increased. This also leads to a more rapid warm-up of the engine. Depending upon engine temperature, an auxiliary-air device in the form of a bypass around the throttle plate allows the engine to draw in more air. This auxiliary air is also measured by the air-flow sensor, and leads to the KE-Jetronic providing the engine with more fuel. Precise adaptation is by means of the electrical heating facility. The engine temperature then determines how much auxiliary air is fed in initially through the bypass, and the electrical heating is mainly responsible for subsequently reducing the auxiliary air as a function of time.

Auxiliary-air device

The auxiliary-air device incorporates a perforated plate (Figures 27, 28 and 30) which is actuated by the bimetallic strip and which controls the cross-section of the bypass passage. Initially, the bypass cross-section opened by the perforated plate is determined by the engine temperature, so that during a cold start, the bypass opening is adequate for the auxiliary air required. The opening closes steadily along with increasing engine temperature until, finally, it is closed completely. The bimetal strip is electrically heated and this limits the opening time, starting from the initial setting which is dependent upon engine temperature. The auxiliary-air device is fitted in the best possible position on the engine for it to assume engine temperature. It does not function when the engine is warm.

Closed-loop idle-speed control by rotary idle actuator

The air quantity or charge is the best correcting variable for closed-loop idle-speed control. Closed-loop idle-speed control via the charge (also termed idle-mixture control) permits a stable, low and, thus, economical idle speed which does not vary throughout the service life of the vehicle.
Excessive idle speed increases the fuel consumption at idle and, as a result, the vehicle's overall fuel consumption. This problem is solved by the closed-loop idle-speed control which always provides exactly the right amount of mixture in order to maintain the idle speed, regardless of the engine load (e.g. cold engine with increased frictional resistance). Furthermore, the emission figures remain constant in the long term without having to adjust the idle speed. To a certain extent, closed-loop idle-speed control also compensates for changes in the engine which are attributable to aging. It also stabilizes the idle speed throughout the entire service life of the engine (Fig. 29).

The rotary idle actuator opens a bypass around the throttle valve. Depending upon the signal applied to it, the actuator adjusts a given opening in the bypass. Due to the fact that KE-Jetronic registers the resulting extra air with its sensor plate, the injected fuel quantity changes accordingly. By contrast with other idle-speed controls on the market, this closed-loop idle-speed control controls the idle speed efficiently due to it actually carrying out a comparison between the desired and actual values and, in case of deviation, correcting accordingly.

Rotary idle actuator

Depending upon the deviation of the idle speed from the set value, the rotary idle actuator supplies the engine with more or less intake air through a bypass around the throttle valve. It assumes the function of the auxiliary-air device which is no longer needed.
The rotary idle actuator of the KE-Jetronic (Figures 30 and 31) receives its control signal from the ECU. This control signal depends upon engine speed and temperature, and causes the rotating plate in the idle actuator to change the bypass opening.
The rotary idle actuator is powered by a rotary-magnet drive comprising a winding and a magnetic circuit. Its rotational range is limited to 60 degrees. The rotating slide is attached to the armature shaft and opens the bypass passage far enough for the specified idle speed to be maintained independent of the engine loading. The closed-loop control circuit in the ECU, which is provided by the engine-speed sensor with the necessary

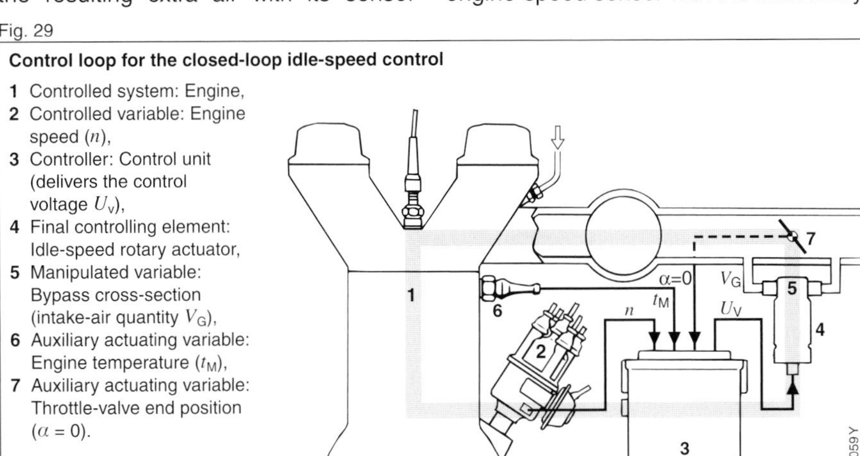

Fig. 29
Control loop for the closed-loop idle-speed control

1 Controlled system: Engine,
2 Controlled variable: Engine speed (n),
3 Controller: Control unit (delivers the control voltage U_V),
4 Final controlling element: Idle-speed rotary actuator,
5 Manipulated variable: Bypass cross-section (intake-air quantity V_G),
6 Auxiliary actuating variable: Engine temperature (t_M),
7 Auxiliary actuating variable: Throttle-valve end position ($\alpha = 0$).

Rotary idle actuator (left) for closed-loop idle-speed control and auxiliary-air device with temperature sensor (right) for idle-speed control

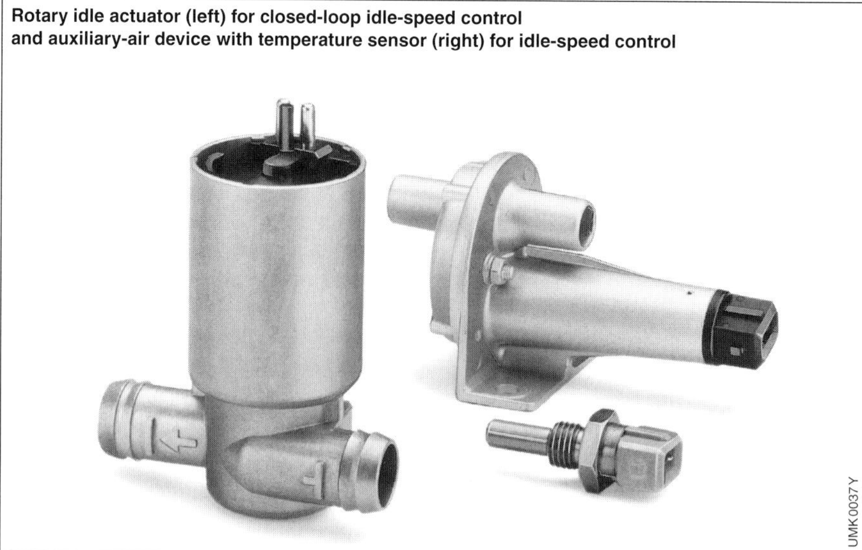

Fig. 30

information concerning engine speed, compares this with the programmed set idle speed and adjusts the air through-flow by means of the idle actuator until the actual idle speed coincides with the set idle speed. With the engine warm and unloaded, the bypass opening is very near to its lower limit.

Further inputs from the ECU, such as temperature and throttle-valve position, ensure that errors do not occur at low temperatures or due to accelerator-pedal movements. The ECU transforms the engine-speed signal into a voltage signal which it compares with a voltage corresponding to the set value. The ECU generates a control signal from the difference voltage, and inputs this to the rotary idle actuator. A pulsating DC is applied to the winding of the coil and causes a torque at the rotating armature which acts against the return spring. The resulting bypass opening depends upon the strength of the current.

In the absence of current (vehicle malfunction), the return spring forces the rotating slide against an adjustable stop and provides an emergency opening. At the maximum on/off ratio of the applied pulsating DC, the bypass is fully opened.

Fig. 31

Rotary idle actuator (single-winding rotary actuator)

1 Electrical connection, **2** Housing, **3** Return spring, **4** Winding, **5** Rotating armature, **6** Air passage as bypass around the throttle plate, **7** Adjustable stop, **8** Rotating slide.

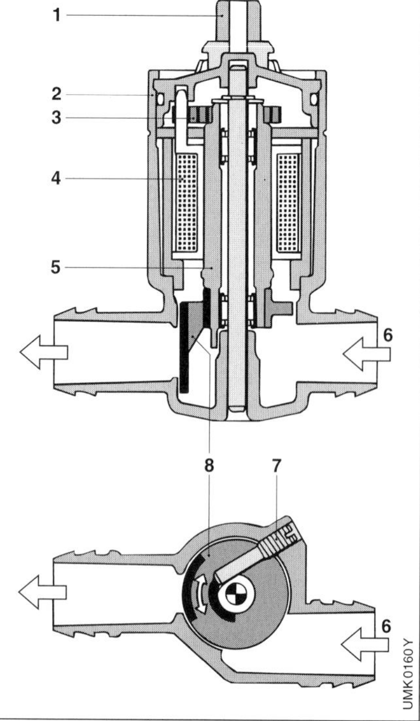

Supplementary functions

Overrun fuel cutoff

Overrun fuel cutoff is the complete interruption of the supply of fuel to the engine during deceleration in order to reduce the fuel consumption and exhaust gas emission when driving downhill and decelerating (i.e. even in town traffic). Since no fuel is burned, there are no emissions.

When the driver takes his foot off the accelerator pedal while driving, the throttle valve returns to the zero position. The throttle-valve switch reports the "throttle-valve closed" condition to the ECU, which, at the same time, receives from the ignition the data concerning the engine speed. If the actual engine speed is within the operating range of the overrun cutoff (i.e. above idle), the ECU reverses the current in the electro-hydraulic pressure actuator. The pressure drop at the actuator is then practically zero. This means that in the fuel distributor, the differential-pressure valves are closed by the springs in their lower chambers (Figure 32) and interrupt the flow of fuel to the injection valves.

Since the injection valves inject continuously, the overrun fuel cutoff operates perfectly smoothly. Its response is also dependent upon the coolant temperature (Figure 33), and in order to avoid continuous switching in and out at a given engine speed, a different switching point is specified depending upon whether the engine speed is decreasing

Fig. 32

Fuel distributor with overrun fuel cutoff

1 Fuel distributor, **2** Fuel inlet, **3** and **5** Inlets to the injection valves (injectors), **4** To the cold-start valve, **6** To the primary-pressure regulator, **7** Upper chamber, **8** Diaphragm (closes inlets 3, 5 to the injection valves), **9** Lower chamber, **10** Nozzle, **11** Winding, **12** Baffle plate.

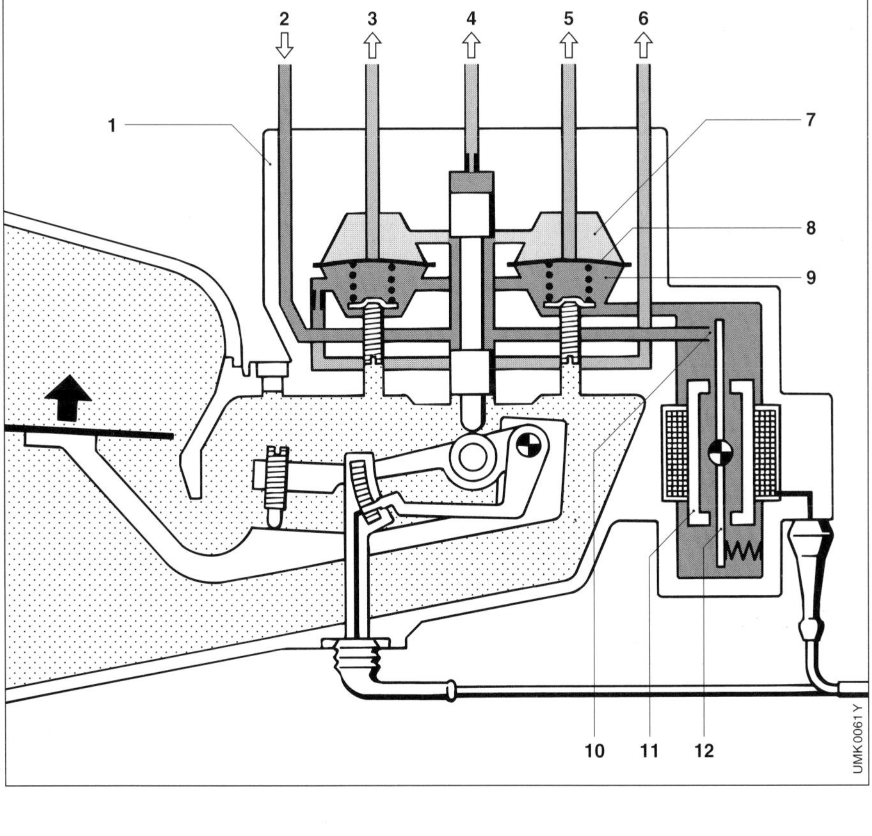

or increasing. The switching thresholds are chosen to be as low as possible for the warm engine in order that maximum fuel savings are achieved. On the other hand, with a cold engine, the thresholds are somewhat higher so that the engine does not stop when the clutch pedal is suddenly pressed.

Engine-speed limiting

Engine-speed limiting shuts off the fuel supply to the injection valves when the maximum permissible engine speed is reached. Conventional mechanical engine-speed limiters have a distributor rotor arranged to short-circuit the ignition when the maximum permissible engine speed is reached.
Today, for reasons of exhaust-gas emissions and fuel economy, this method is no longer entirely satisfactory. It has been superseded by electronic engine-speed limiting using fuel-injection cutoff. When the current through the electro-hydraulic actuator is reversed, the baffle plate is pulled away from the nozzle. The pressure drop approaches zero and the diaphragms in the differential-pressure valves stop the flow of fuel to the injection valves. This involves the same process as on the overrun fuel-cutoff facility. In the ECU, the actual engine speed is compared with the programmed maximum speed n_o. If the maximum speed is exceeded, the ECU suppresses the injection pulses. This function operates within a bandwidth of 80 min^{-1} above and below the permissible maximum speed (Figure 34). This electronic engine-speed limiting facility prevents the engine from "overspeeding" and, at the same time, limits fuel consumption and exhaust-gas emission.

Adaptation of the air-fuel mixture at high altitudes

At high altitudes, due to the lower air density, the volumetric flow measured by the air-flow sensor corresponds to a lower air-mass flow. Depending upon the facilities incorporated in the particular KE-Jetronic, this error can be compensated for by correcting the injection time. Over-enrichment is avoided and, therefore, excessive fuel consumption at high altitudes.
The altitude compensation is provided by a sensor which measures the air pressure. In accordance with the prevailing air pressure, the sensor inputs a signal to the ECU which changes the pressure-actuator current accordingly. This alters the lower-chamber pressure, and therefore the pressure difference at the metering slits (this thus changes the

Fig. 33

The lowest engine speed for the overrun fuel-cutoff facility depends upon coolant temperature

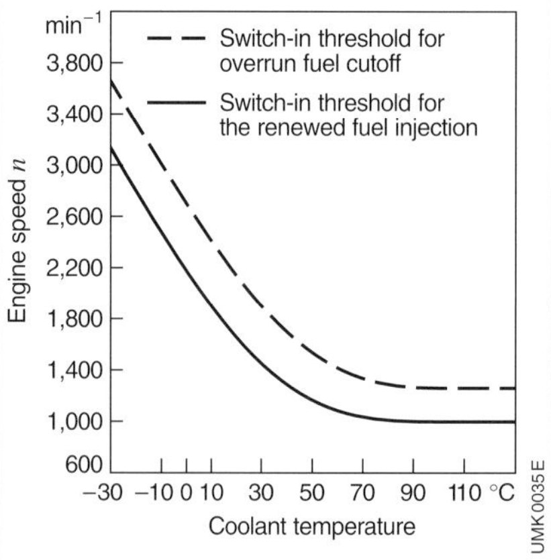

Fig. 34

Limiting the maximum engine speed n_o by cutting off the flow of fuel to the injectors

1 Fuel injection "Off", **2** Fuel injection "On", **3** Engine speed limiting "On".

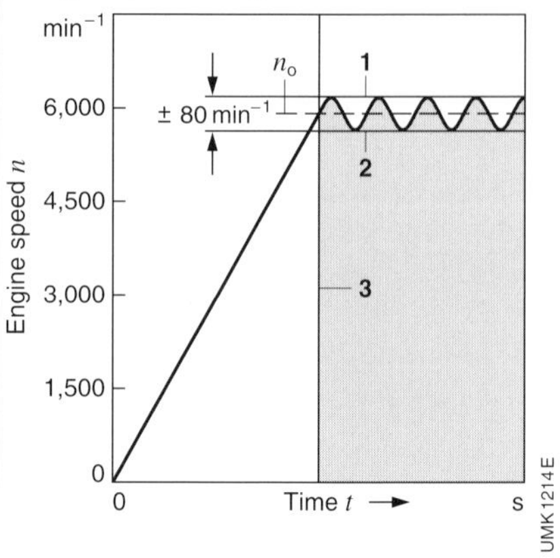

injected-fuel quantity). It is also possible to incorporate continuous adaptation of the injected-fuel quantity according to the changing air pressure.

Lambda closed-loop control

The lambda closed-loop control permits the excess-air factor to be maintained very precisely at $\lambda = 1$. The lambda closed-loop control is an add-on facility which can basically be retrofitted to any electronically controllable fuel-management system.

The lambda closed-loop control is also particularly suitable for use in conjunction with the KE-Jetronic. The signal of

Fig. 35

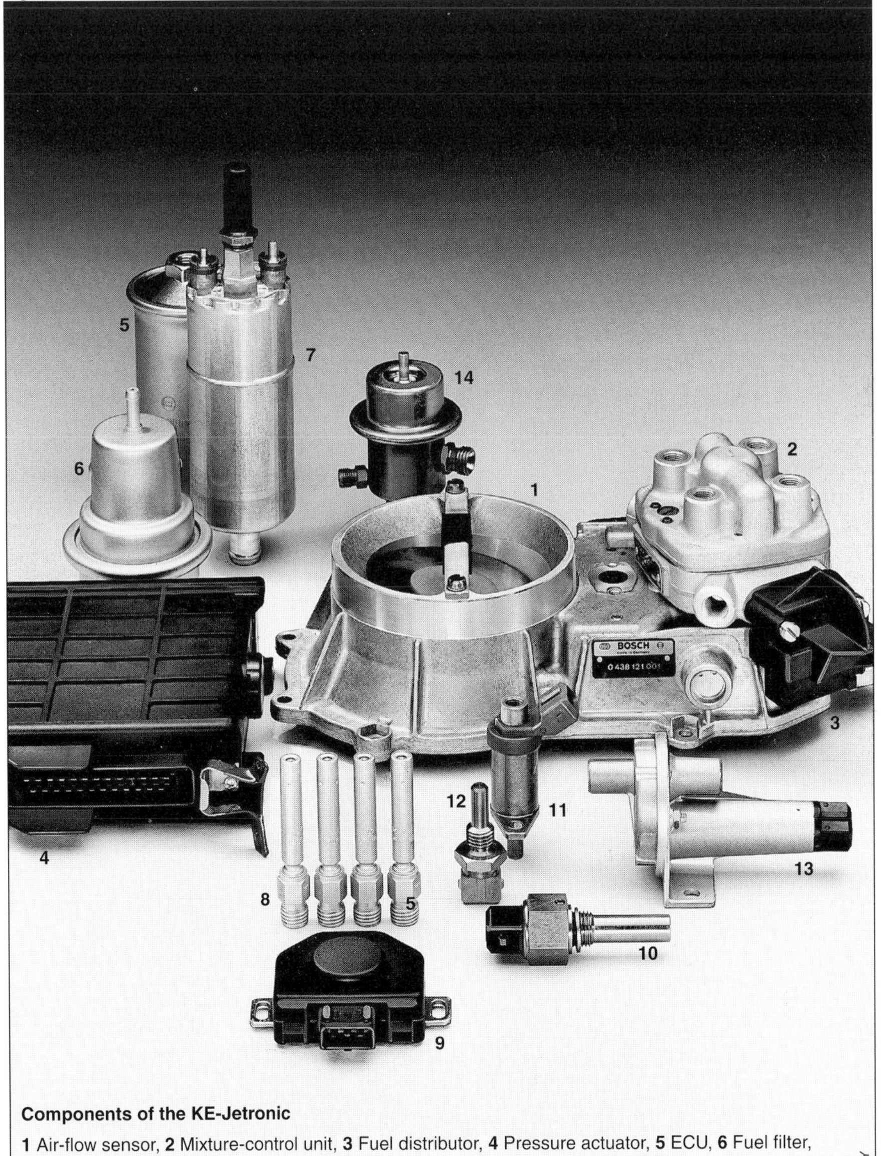

Components of the KE-Jetronic

1 Air-flow sensor, **2** Mixture-control unit, **3** Fuel distributor, **4** Pressure actuator, **5** ECU, **6** Fuel filter, **7** Fuel accumulator, **8** Electric fuel pump, **9** Fuel-injection valves (injectors), **10** Throttle-valve switch, **11** Thermo-time switch, **12** Cold-start valve, **13** Engine-temperature sensor, **14** Auxiliary-air device, **15** Primary-pressure regulator.

the lambda sensor is processed in the control unit which is already fitted, and the required control intervention for correction of the fuel allocation is carried out via the pressure actuator.

Electrical circuitry

If the engine stops but the ignition remains switched on, the electrical fuel pump is switched off by a safety circuit. The KE-Jetronic system is equipped with a number of electrical components such as electric fuel pump, auxiliary-air device, cold-start valve and thermo-time switch. All these components are controlled by the control relay which, itself, is switched by the ignition and starting switch. Apart from its switching functions, the control relay also has a safety function. A commonly used circuit is described below.

Function

When cold-starting the engine, voltage is applied to the cold-start valve and the thermo-time switch through terminal 50 (Figs. 36 and 37). If the cranking process takes longer than between 8 and 15 seconds, the thermo-time switch switches off the start valve in order that the engine does not “flood”. In this case, the thermo-time switch performs a time-switch function.

Fig. 36

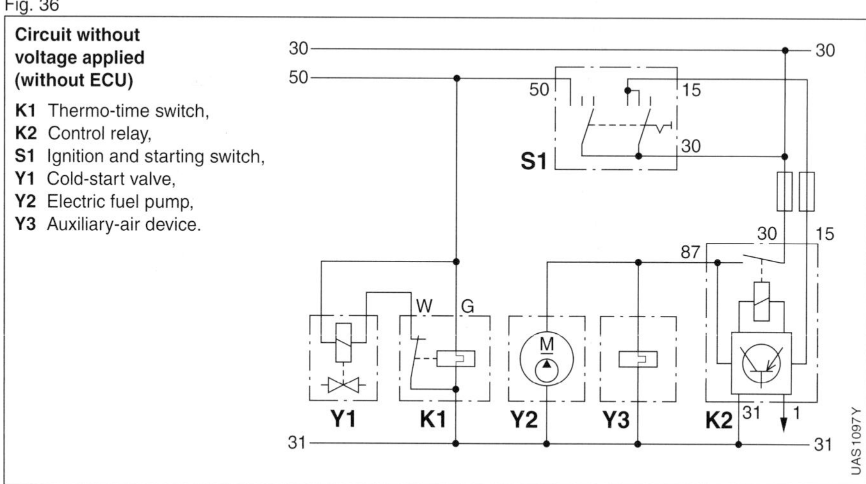

Circuit without voltage applied (without ECU)

K1 Thermo-time switch,
K2 Control relay,
S1 Ignition and starting switch,
Y1 Cold-start valve,
Y2 Electric fuel pump,
Y3 Auxiliary-air device.

Fig. 37

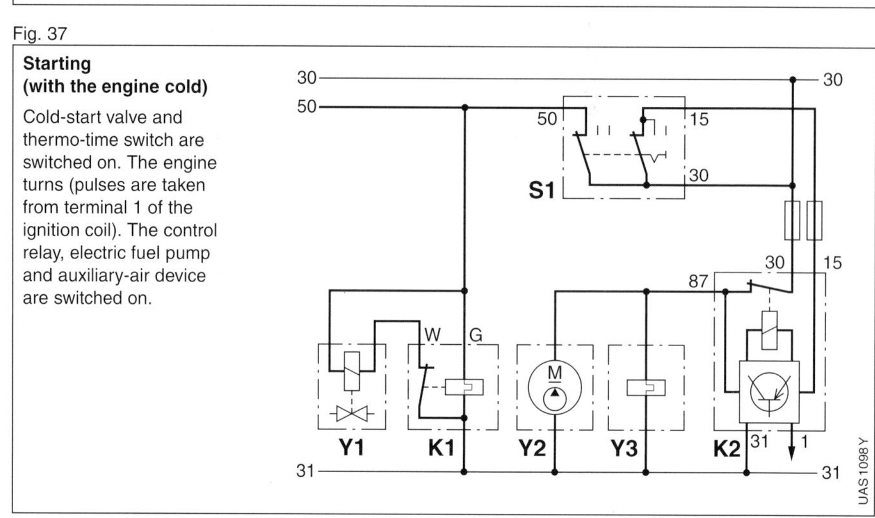

Starting (with the engine cold)

Cold-start valve and thermo-time switch are switched on. The engine turns (pulses are taken from terminal 1 of the ignition coil). The control relay, electric fuel pump and auxiliary-air device are switched on.

If the temperature of the engine is above approximately +35 °C when the starting process is commenced, the thermo-time switch will have already open-circuited the connection to the start valve which, as a result, does not inject extra fuel. In this case, the thermo-time switch functions as a temperature switch.
Voltage from the ignition and starting switch is still present at the control relay which switches on as soon as the engine runs. The engine speed reached when the starting motor cranks the engine is high enough to generate the "engine running" signal which is taken from the ignition pulses coming from terminal 1 of the ignition coil.

An electronic circuit in the control relay evaluates these pulses. After the first pulse, the control relay switches on and applies voltage to the electric fuel pump and the auxiliary-air device. The control relay remains switched on providing the ignition is switched on and the engine is running (Fig. 38).
If the pulses from terminal 1 of the ignition coil stop because the engine has stopped turning (for instance in the case of an accident), the control relay switches off approximately 1 second after the last pulse is received. This safety circuit prevents the fuel pump from pumping fuel when the ignition is switched on but the engine is not turning (Fig. 39).

Fig. 38

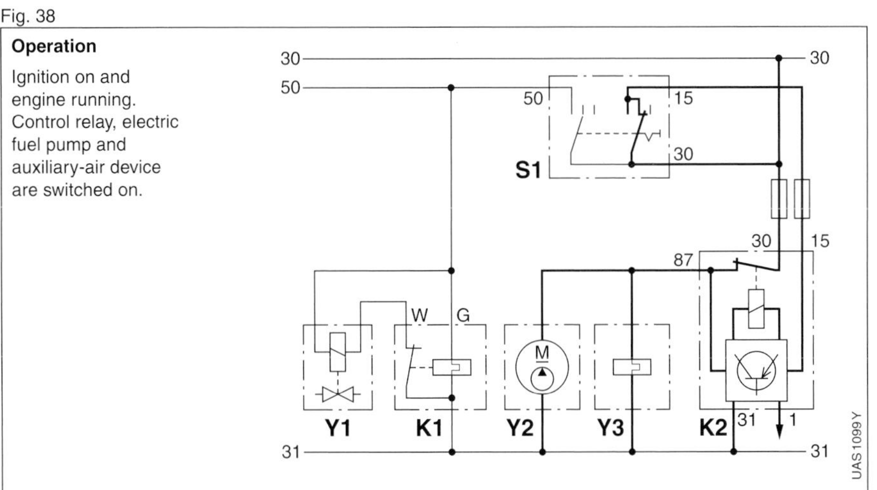

Operation
Ignition on and engine running. Control relay, electric fuel pump and auxiliary-air device are switched on.

Fig. 39

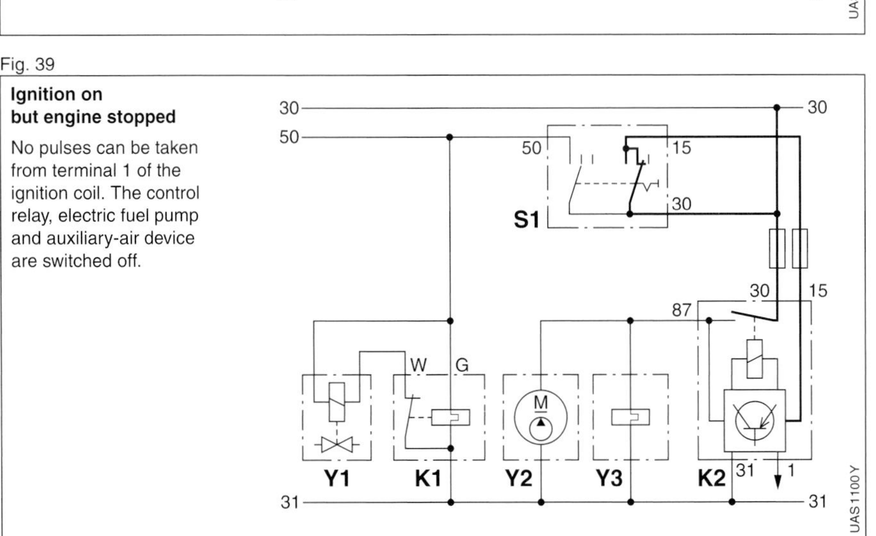

Ignition on but engine stopped
No pulses can be taken from terminal 1 of the ignition coil. The control relay, electric fuel pump and auxiliary-air device are switched off.

Workshop testing techniques

Bosch customer service

Customer service quality is also a measure for product quality. The car driver has more than 10,000 Bosch Service Agents at his disposal in 125 countries all over the world. These workshops are neutral and not tied to any particular make of vehicle. Even in sparsely populated and remote areas of Africa and South America the driver can rely on getting help very quickly. Help which is based upon the same quality standards as in Germany, and which is backed of course by the identical guarantees which apply to customer-service work all over the world. The data and performance specs for the Bosch systems and assemblies of equipment are precisely matched to the engine and the vehicle. In order that these can be checked in the workshop, Bosch developed the appropriate measurement techniques, test equipment, and special tools and equipped all its Service Agents accordingly.

Testing techniques for KE-Jetronic

Apart from the regular replacement of the fuel filter as stipulated by the particular vehicle's manufacturer, the KE-Jetronic gasoline-injection system requires no special maintenance work.
Basically speaking, in case of malfunctions, the workshop expert has the following test equipment, together with the appropriate test specs, at his disposal:

- Injector tester,
- Injected-quantity comparison tester,
- Pressure-measuring device,
- Lambda closed-loop-control tester (only if Lambda control is fitted),
- Universal test adapter,
- Universal multimeter.

Together with the relevant Test Instructions and Test Specifications in a variety of different languages, this uniform testing technology is available throughout the world at the Bosch Service Agent workshops and at the workshops belonging to the vehicle manufacturers. Efficient trouble-shooting and technically correct repairs cannot be performed at a reasonable price without this equipment. It is therefore inadvisable for the vehicle owner to attempt to carry out his own repairs.

Injector tester

The injector tester is specifically for testing the K- and KE-Jetronic injectors when removed from the engine. The tester checks all the functions of the injector which are essential for correct engine running:

- Opening pressure,
- Leakage integrity,
- Spray shape,
- Chatter.

Those injectors whose opening pressure is outside tolerance are replaced. For the leak test, the pressure is slowly increased up to 0.5 bar below the opening pressure and held at this point. Within 60 secs, no droplet of fuel is to form at the injector. During the chatter test, the injector must generate a "chattering" noise without a fuel droplet being formed. Serviceable injectors generate a fully atomized spray pattern. "Pencil" jets and "bundled" jets are not to form.

Injected-quantity comparison tester

Without removing the fuel distributor from the vehicle, a comparitive measurement is made to determine the differences in the delivered quantities from the various fuel-distributor outlets (this applies to all engines of up to maximum eight cylinders). And since the test is performed using the original injectors it is possible to ascertain at the same time whether any scatter in the figures results from the fuel distributor itself or from the injectors.
The tester's small measuring tubes serve for idle measurement and its larger measuring tubes for part-load or full-load measurement.
Connection to the fuel distributor is by means of eight hoses. The injectors are

pulled out of their mountings on the engine and inserted in the automatic couplings at the ends of the hoses. Each automatic coupling incorporates a push-up valve which prevents fuel escaping from hoses which are not connected (e.g. on 6-cylinder systems. Fig. 40). A further hose returns the fuel to the tank.

Pressure-measuring device
This is used to measure all the pressures which are important for correct KE-Jetronic operation:
- Primary (system) pressure: Provides information on the performance of the fuel-supply pump, on fuel-filter flow resistance, and on the condition of the primary-pressure regulator.
- Control pressure: Important for assessment of all operating conditions (for instance: Cold/warm engine; part load/full load; fuel-enrichment functions).
- Leakage integrity of the complete system. This is particularly important with regard to the cold-start and hot-start behavior.

Lambda closed-loop-control tester
On KE-Jetronic systems with Lambda closed-loop control, this tester serves to check the duty factor of the Lambda-sensor signal (using simulation of the "rich"/"lean" signal), and the "open-loop/closed-loop control function". Special adapter lines are available for connection to the Lambda-sensor cable of the various vehicle models. Measured values are shown on an analog display.

Universal test adapter
In the case of KE-Jetronic versions without (or with only restricted) self-diagnosis, the universal test adapter serves for rapid and reliable system testing.

Universal multimeter
The universal multimeter is used to measure the pressure-actuator currents in all operating modes. It also serves for voltage and resistance measurements at the various components (e.g. potentiometer for the air-flow sensor).

Fig. 40
Injected-quantity comparison tester (connected to a 6-cylinder installation)
1 Fuel-distributor injection lines,
2 Injectors,
3 Automatic couplings,
4 Comparison-tester hoses,
5 Small measuring tubes,
6 Large measuring tubes,
7 Return line to fuel tank.

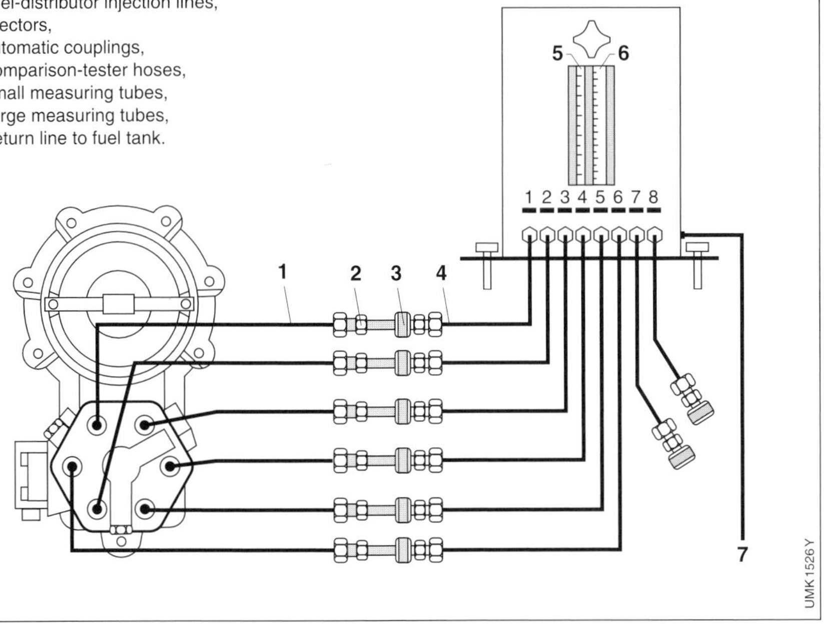

1 2 3 4 5 6 7 8
UMK1526Y

L-Jetronic

System overview

The L-Jetronic is an electronically controlled fuel-injection system which injects fuel intermittently into the intake ports. It does not require any form of drive. It combines the advantages of direct air-flow sensing and the special capabilities afforded by electronics.

As is the case with the K-Jetronic system, this system detects all changes resulting from the engine (wear, deposits in the combustion chamber and changes in valve settings), thus guaranteeing a uniformly good exhaust gas quality.
The task of the gasoline injection system is to supply to each cylinder precisely the correct amount of fuel as is necessary for the operation of the engine at that particular moment. A pre-requisite for this, however, is the processing of as many influencing factors as possible relevant to the supply of fuel. Since, however, the operating condition of the engine often changes quite rapidly, a speedy adaptation of the fuel delivery to the driving situation at any given moment is of prime importance. Electronically controlled gasoline injection is particularly suitable here. It enables a variety of operational data at any particular location of the vehicle to be registered and converted into electrical signals by sensors.

Fig. 1

Principle of the L-Jetronic (simplified)

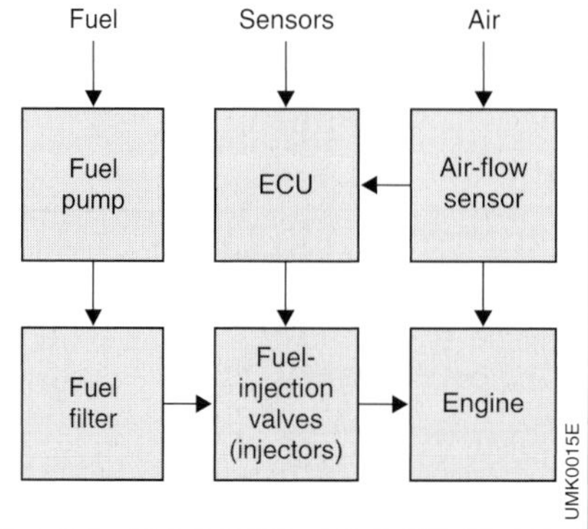

These signals are then passed on to the control unit of the fuel-injection system which processes them and calculates the exact amount of fuel to be injected. This is influenced via the duration of injection (Fig. 1).

Function

A pump supplies the fuel to the engine and creates the pressure necessary for injection. Injection valves inject the fuel into the individual intake ports and onto the intake valves. An electronic control unit controls the injection valves.
The L-Jetronic consists principally of the following function blocks:
- Fuel supply system,
- Operating-data sensing system and
- Fuel-metering system.

Fuel-supply system

The fuel system supplies fuel from the fuel tank to the injection valves, creates the pressure necessary for injection and maintains it at a constant level.

Operating-data sensing system

The sensors register the measured variables which characterize the operating mode of the engine.
The most important measured variable is the amount of air drawn in by the engine and registered by the air-flow sensor. Other sensors register the position of the throttle, the engine speed, the air temperature and the engine temperature.

Fuel-metering system

The signals delivered by the sensors are evaluated in the electronic control unit (ECU) where they are used to generate the appropriate control pulses for the injection valves.

Advantages of the L-Jetronic system

Low fuel consumption

In carburetor systems, due to segregation processes in the intake manifold, the individual cylinders of the engine do not all receive the same amount of air-fuel

mixture. Optimum fuel distribution cannot be achieved if a mixture is created which is suitable for supplying sufficient fuel even to the worst-fed cylinder. This would result in high fuel consumption and unequal stressing of the cylinders.
In L-Jetronic systems (Fig. 2), each cylinder has its own injection valve. The injection valves are controlled centrally; this ensures that each cylinder receives precisely the same amount of fuel, the optimum amount, at any particular moment and under any particular load.

Adaptation to operating conditions

The L-Jetronic system adapts to changing load conditions virtually immediately since the required quantity of fuel is computed by the control unit (ECU) within milliseconds and injected by the injection valves onto the engine intake valves.

Low-pollution exhaust gas

The concentration of pollutants in the exhaust gas is directly related to the air-fuel ratio. If the engine is to be operated with the least pollutant emission, then a fuel-management system is necessary which is capable of maintaining a given air-fuel ratio. The L-Jetronic works so precisely that the precise mixture formation necessary for observing the present-day exhaust regulations is guaranteed.

Higher power output per litre

The fact that there is no carburetor enables the intake passages to be designed aerodynamically in order to achieve optimum air distribution and cylinder charge and, thus, greater torque. Since the fuel is injected directly onto the intake valves, the engine receives only air through the intake manifold. This results in a higher power output per litre and a torque curve appropriate to practice.

Fig. 2

Schematic diagram of an L-Jetronic system with lambda closed-loop control

1 Fuel tank, **2** Electric fuel pump, **3** Fuel filter, **4** ECU, **5** Fuel-injection valve (injector), **6** Fuel rail and fuel-pressure regulator, **7** Intake manifold, **8** Cold-start valve, **9** Throttle-valve switch, **10** Air-flow sensor, **11** Lambda sensor, **12** Thermo-time switch, **13** Engine-temperature sensor, **14** Ignition distributor, **15** Auxiliary-air device, **16** Battery, **17** Ignition and starting switch.

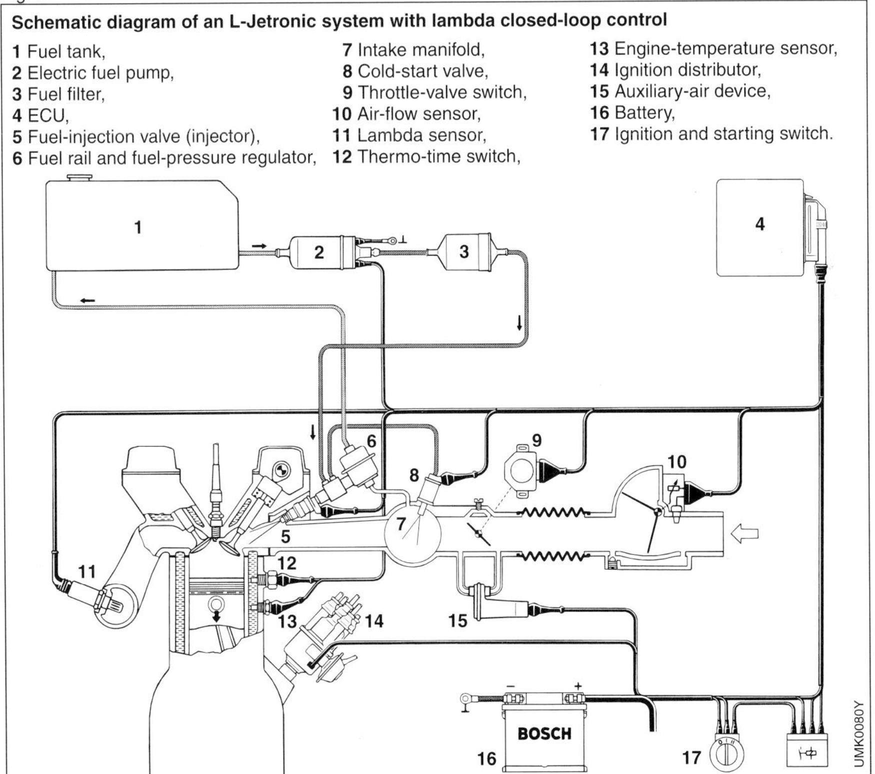

Fuel supply system

The fuel supply system comprises the following components:

- Electric fuel pump,
- Fine filter,
- Fuel rail,
- Pressure regulator and
- Fuel-injection valves.

An electrically driven roller-cell pump pumps the fuel from the fuel tank at a pressure of approximately 2.5 bar through a filter into the fuel rail. From the fuel rail, fuel lines diverge to the injection valves. At the end of the fuel rail is a pressure regulator which maintains the injection pressure at a constant level (Figure 3). More fuel circulates in the fuel system than is needed by the engine even under the most extreme conditions. The excess fuel is returned to the fuel tank by the pressure regulator but not under pressure. The constant flushing through of the fuel system enables it to be continually supplied with cool fuel. This helps to avoid the formation of fuel vapour bubbles and guarantees good hot-starting characteristics.

Electric fuel pump

The electric fuel pump (Fig. 4) is a roller-cell pump driven by a permanent-magnet electric motor. The rotor plate which is eccentrically mounted in the pump housing is fitted with metal rollers in notches around its circumference which are pressed against the pump housing by centrifugal force and act as seals. The fuel is carried in the cavities which form between the rollers. The pumping action takes place when the rollers, after having closed the inlet port, force the trapped fuel around in front of them until it can escape from the pump through the outlet port (Figure 5). The fuel flows directly around the electric motor. There is no danger of explosion, however, because there is never an ignitable mixture in the pump housing.

The electric fuel pump delivers more fuel than the maximum requirement of the engine so that the pressure in the fuel system can be maintained under all operating conditions. A check valve in the pump disconnects the fuel system from the fuel tank by preventing return flow of fuel to the fuel tank.

The electric fuel pump starts immediately when the ignition and starting switch is operated and remains switched on continuously after the engine has started. A safety circuit prevents fuel from being delivered when the ignition is switched on, but when the engine is stationary (e.g. after an accident). The fuel pump is located in the direct vicinity of the fuel tank and requires no maintenance.

Fig. 3

Fuel supply system

1 Fuel tank,
2 Electric fuel pump,
3 Fuel filter,
4 Fuel rail,
5 Fuel-pressure regulator,
6 Fuel-injection valve (injector),
7 Cold-start valve.

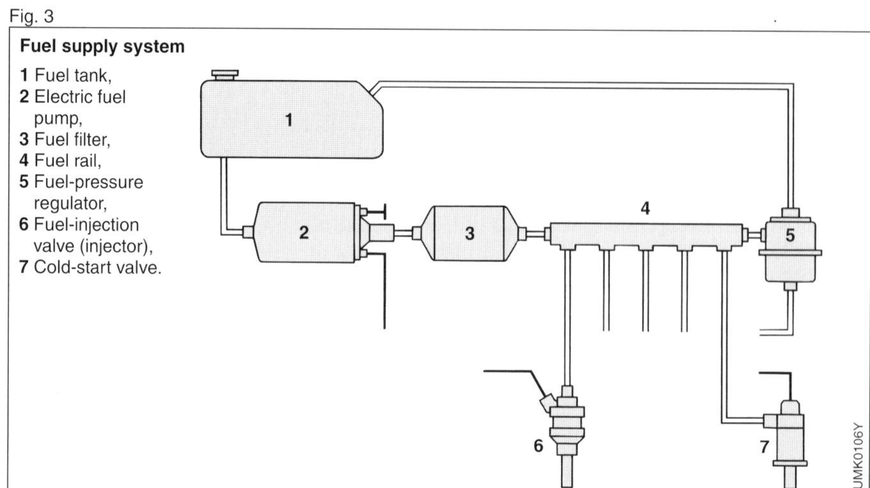

Fuel filter

The fuel filter filters off impurities in the fuel which could impair the function of the injection system. The filter contains a paper element with an average pore size of 10 µm, which is backed up by a fluff strainer (Figure 6). This combination ensures a high degree of filtration. A support plate secures the filter in its metal housing. The filter is installed in the fuel line downstream of the fuel accumulator. When the filter is changed, it is imperative that the throughflow direction as indicated by the arrow on the housing be observed.

Fuel rail

The fuel rail supplies all injection valves with an equal quantity of fuel and ensures the same fuel pressure at all injection valves.

The fuel rail has a storage function. Its volume, compared with the amount of fuel injected during each working cycle of the engine, is large enough to prevent variations in pressure. The injection valves connected to the fuel rail are therefore subjected to the same fuel pressure. The fuel rail also facilitates easy fitting of the injection valves.

Pressure regulator

The pressure regulator keeps the pressure differential between the fuel pressure and manifold pressure constant. Thus, the fuel delivered by the electromagnetic injection valve is determined solely by the valve opening time.

The pressure regulator is a diaphragm-controlled overflow pressure regulator which controls pressure at 2.5 or 3 bar, dependent upon the system in question. It is located at the end of the fuel rail and consists of a metal housing, divided into two spaces by a flanged diaphragm: a chamber for the spring that preloads the diaphragm, and a chamber for the fuel (Figure 7). When the preset pressure is exceeded, a valve operated by the diaphragm opens the return line for the excess fuel to flow back, not under pressure, to the fuel tank. The spring chamber is connected by a tube with the intake manifold downstream of the throttle valve. This has the effect that the pressure in the fuel system is dependent upon the absolute manifold pressure; therefore, the pressure drop across the fuel-injection valves is the same for any throttle position.

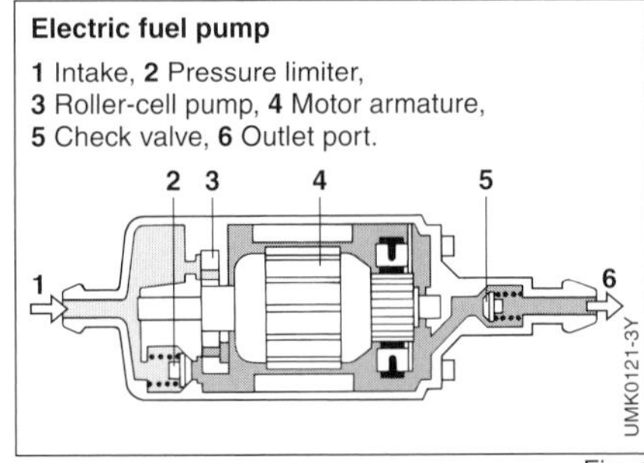

Electric fuel pump

1 Intake, **2** Pressure limiter, **3** Roller-cell pump, **4** Motor armature, **5** Check valve, **6** Outlet port.

Fig. 4

Fig. 5

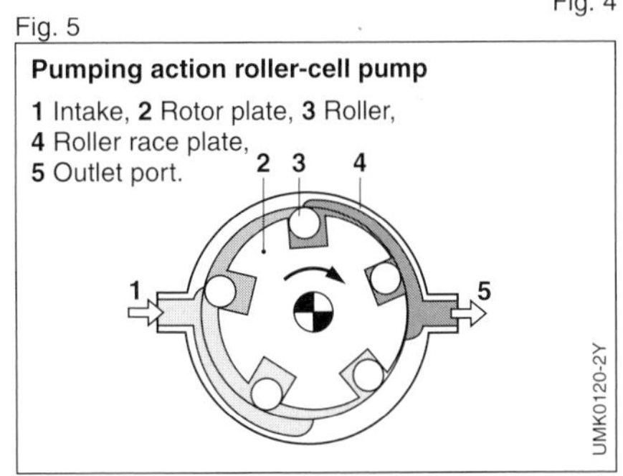

Pumping action roller-cell pump

1 Intake, **2** Rotor plate, **3** Roller, **4** Roller race plate, **5** Outlet port.

Fig. 6

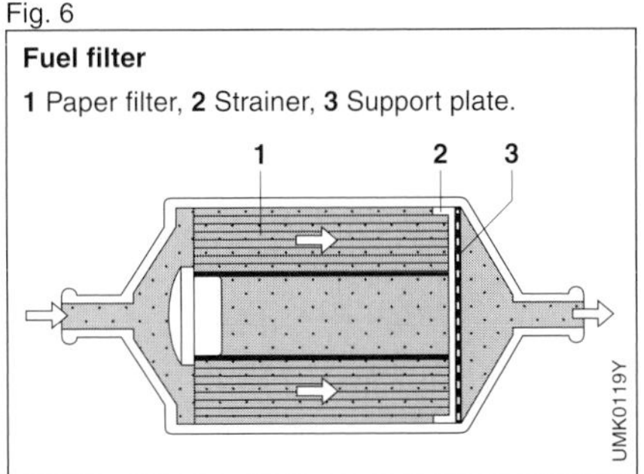

Fuel filter

1 Paper filter, **2** Strainer, **3** Support plate.

Fuel-injection valves

The electronically controlled fuel-injection valves inject precisely metered fuel into the intake ports and onto the intake valves.

Each engine cylinder has its own fuel-injection valve. The valves are solenoid-operated and are opened and closed by means of electric pulses from the electronic control unit. The fuel-injection valve consists of a valve body and the needle valve with fitted solenoid armature. The valve body contains the solenoid winding and the guide for the needle valve. When there is no current flowing in the solenoid winding, the needle valve is pressed against its seat on the valve outlet by a helical spring. When a current is passed through the solenoid winding, the needle valve is lifted by approximately 0.1 mm from its seat and the fuel can be injected through the precision annular orifice. The front end of the needle valve has a specially ground pintle for atomizing the fuel (Figure 8). The pickup and release times of the valve lie in the range of 1 to 1.5 ms. To achieve good fuel distribution together with low condensation loss, it is necessary that wetting of the intake-manifold walls be avoided.

The means that a particular spray angle in conjunction with a particular distance of the injection valve from the intake valve must therefore be maintained, specific to the engine concerned. The fuel-injection valves are fitted with the help of special holders and are mounted in rubber mouldings in these holders. The insulation from the heat of the engine thereby achieved prevents the formation of fuel-vapour bubbles and guarantees good hot-starting characteristics. The rubber mouldings also ensure that the fuel-injection valves are not subjected to excessive vibration.

Fuel-pressure regulator

1 Intake-manifold connection, **2** Spring, **3** Valve holder, **4** Diaphragm, **5** Valve, **6** Fuel inlet, **7** Fuel return.

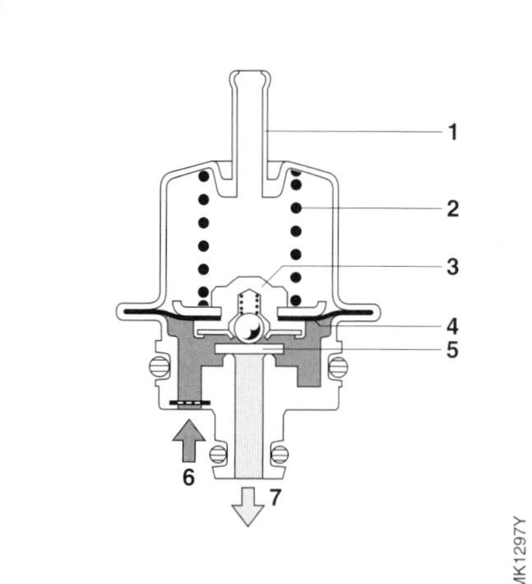

Fig. 7

Fig. 8

Solenoid-operated fuel-injection valve (injector)

1 Filter in fuel inlet, **2** Electrical connection, **3** Solenoid winding, **4** Valve housing, **5** Armature, **6** Valve body, **7** Valve needle.

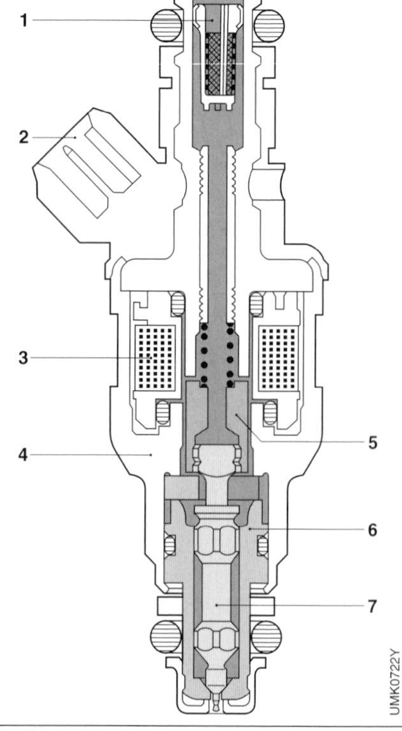

Operating-data sensing system

Sensors detect the operating mode of the engine and signal this condition electrically to the control unit. The sensors and ECU form the control system. The sensors are described in conjunction with the relevant main function or compensation function.

Measured variables

The measured variables characterizing the operating mode of the engine are as follows:

- Main measured variables,
- Measured variables for compensation,
- Measured variables for precision compensation.

The ECU evaluates all measured variables together so that the engine is always supplied with exactly the amount of fuel required for the instantaneous operating mode. This achieves optimum driveability.

Main measured variables

The main measured variables are the engine speed and the amount of air drawn in by the engine. These determine the amount of air per stroke which then serves as a direct measure for the loading condition of the engine.

Measured variables for compensation

For operating conditions such as cold start and warm-up and the various load conditions which deviate from normal operation, the mixture must be adapted to the modified conditions. Starting and warm-up conditions are detected by sensors which transmit the engine temperature to the control unit. For compensating various load conditions, the load range (idle, part-load, full-load) is transmitted to the control unit via the throttle-valve switch.

Measured variables for precision compensation

In order to achieve optimum driving behavior, further operating ranges and influences can be considered: the sensors mentioned above detect the data for transition response when accelerating, for maximum engine-speed limitation and during overrun. The sensor signals have a particular relationship to each other in these operating ranges. The control unit recognizes these relationships and influences the control signals of the injection valves accordingly.

Calculating engine speed

Information on engine speed and the start of injection is passed on to the L-Jetronic ECU in breaker-triggered ignition systems by the contact-breaker points in the ignition distributor, and, in breakerless ignition systems, by terminal 1 of the ignition coil (Fig. 9).

Measuring the air flow

The amount of air drawn in by the engine is a measure of its loading condition. The air-flow measurement system allows for all changes which may take place in the engine during the service life of the vehicle, e.g. wear, combustion-chamber deposits and changes to the valve settings.

Since the quantity of air drawn in must first pass through the air-flow sensor before entering the engine, this means that, during acceleration, the signal leaves the sensor before the air is actually drawn into the cylinder. This permits correct mixture adaptation at any time during load changes.

Fig. 9

Calculating engine speed with a breaker-triggered ignition system

1 Ignition distributor, **2** ECU.
n Engine speed.

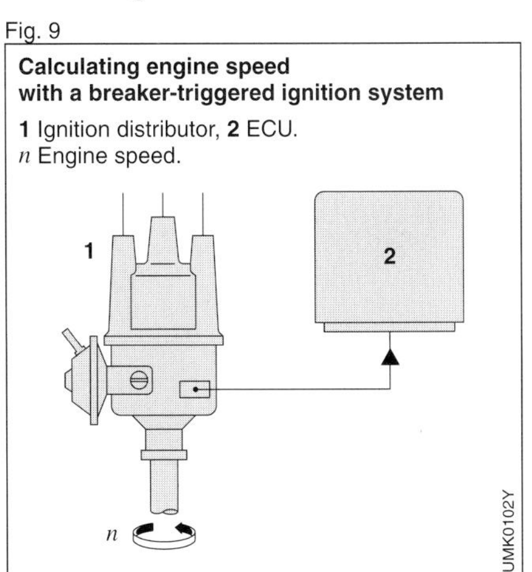

Air-flow sensor in the intake system

1 Throttle valve,
2 Air-flow sensor,
3 Intake air temperature signal to the ECU,
4 ECU,
5 Air-flow sensor signal to the ECU,
6 Air filter,
Q_L Intake-air quantity,
α Deflection angle.

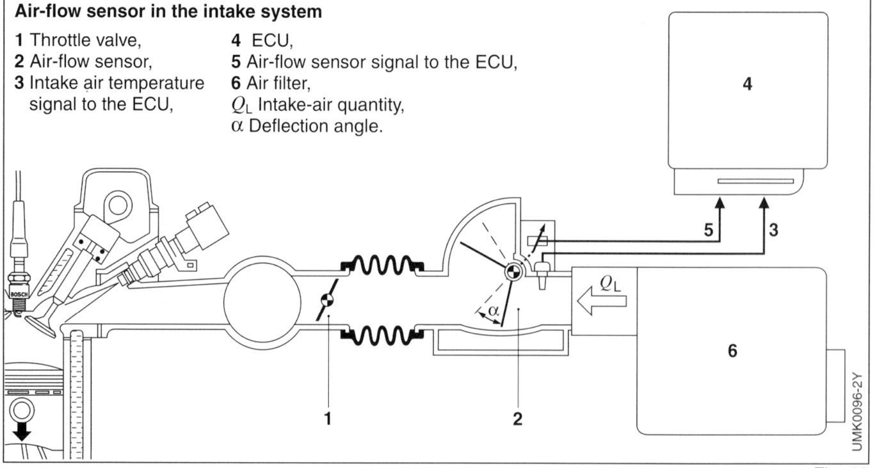

Fig. 10

Fig. 11

Air-flow sensor (air side)

1 Compensation flap,
2 Damping volume,
3 Bypass,
4 Sensor flap,
5 Idle-mixture adjusting screw (bypass).

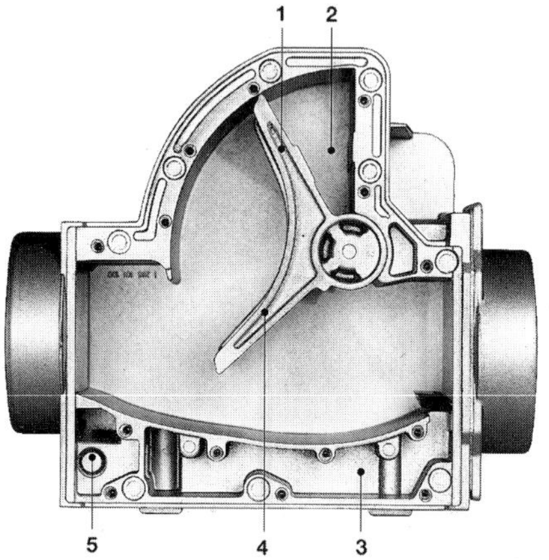

Fig. 12

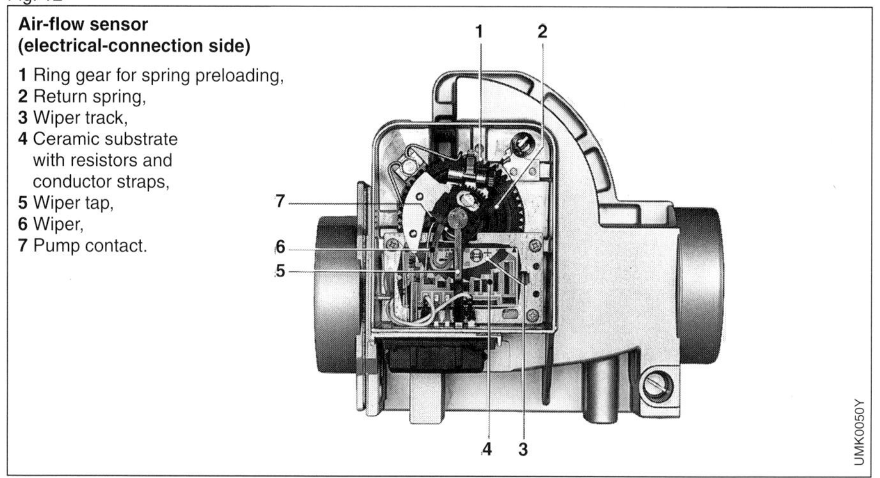

Air-flow sensor (electrical-connection side)

1 Ring gear for spring preloading,
2 Return spring,
3 Wiper track,
4 Ceramic substrate with resistors and conductor straps,
5 Wiper tap,
6 Wiper,
7 Pump contact.

The sensor flap in the air-flow sensor measures the entire air quantity inducted by the engine, thereby serving as the main controlled variable for determining the load signal and basic injection quantity.

Air-flow sensor

The principle is based on the measurement of the force emanating from the stream of air drawn in by the engine. This force has to counteract the opposing force of a return spring acting upon the air-flow sensor flap. The flap is deflected in such a manner that, together with the profile of the measurement channel, the free cross-section increases along with the rise in the quantity of air passing through it (Figs. 10, 11 and 12).

The change in the free air-flow sensor cross-section, depending on the position of the sensor flap, was selected so that a logarithmic relationship results between flap angle and air throughput. The result is that, at low air throughput, where measurement precision must be particularly high, the sensitivity of the air-flow sensor is also high. In order to prevent the oscillations in the intake system caused by the engine intake strokes from having more than a minimum effect upon the sensor-flap position, a compensation flap is attached rigidly to the sensor flap. The pressure oscillations have the same effects upon both flaps and the moments of force therefore cancel each other out so that the measurement is not affected. The angular position of the sensor flap is transformed by a potentiometer into a voltage. The potentiometer is calibrated such that the relationship between air throughput and voltage output is inversely proportional. In order that aging and the temperature characteristic of the potentiometer have no effect upon the accuracy, only resistance values are evaluated in the ECU. In order to set the air-fuel ratio at idle, an adjustable bypass channel is provided.

Fuel metering

As the central unit of the system, the ECU evaluates the data delivered by the sensors on the operating mode of the engine. From this data, control pulses for the injection valves are generated, whereby the quantity to be injected is determined by the length of time the injection valves are opened.

Electronic control unit (ECU)

Configuration

The L-Jetronic ECU is in a splash-proof sheet-metal housing which is fitted where it is not affected by the heat radiated from the engine. The electronic components in the ECU are arranged on printed-circuit boards; the output-stage power components are mounted on the metal frame of the ECU thus assuring good heat dissipation. By using integrated circuits and hybrid modules, it has been possible to reduce the number of parts to a minimum. The reliability of the ECU was increased by combining functional groups into integrated circuits (e.g. pulse shaper, pulse divider and division control multivibrator, Fig. 13) and by combining components into hybrid modules.

A multiple plug is used to connect the ECU to the injection valves, the sensors and the vehicle electrical system. The input circuit in the ECU is designed so that the latter cannot be connected with the wrong polarity and cannot be short-circuited. Special Bosch testers are available for carrying out measurements on the ECU and on the sensors. The testers can be connected between the wiring harness and the ECU with multiple plugs.

Operating-data processing

Engine speed and inducted air quantity determine the basic duration of injection. The timing frequency of the injection pulses is determined on the basis of the engine speed.

The pulses delivered by the ignition system for this purpose are processed by the ECU. First of all, they pass through a pulse-shaping circuit which generates square-wave pulses from the signal "delivered" in the form of damped oscillations, and feeds these to a frequency divider.

The frequency divider divides the pulse frequency given by the ignition sequence in such a manner that two pulses occur for each working cycle regardless of the number of cylinders. The start of the pulse is, at the same time, the start of injection for the injection valves. For each turn of the crankshaft, each injection valve injects once, regardless of the position of the intake valves. When the intake valve is closed, the fuel is stored and the next time it opens the fuel is drawn into the combustion chamber together with the air. The duration of injection depends on the amount of air measured by the air-flow sensor and the engine speed.

The ECU also evaluates the signal supplied by the potentiometer. Fig. 14 shows the interrelationships between intake air quantity, flap angle, potentiometer voltage and injected quantity. Assuming a specific intake-air quantity Q_L flowing through the air-flow sensor (point Q), we thus obtain the theoretically required injection quantity Q_K (point D). In addition, a specific flap angle (point A) is established as a function of the air intake quantity. The potentiometer actuated by the air-flow sensor flap supplies a voltage signal US to the ECU (point B) which controls the injection valves, whereby point C represents the injected fuel quantity V_E. It can be seen that the fuel quantity injected in practice and the theoretically required injection quantity are identical (line C–D).

Fig. 14

Interrelationships between intake-air quantity, sensor-flap angle, voltage at the potentiometer and injected fuel quantity

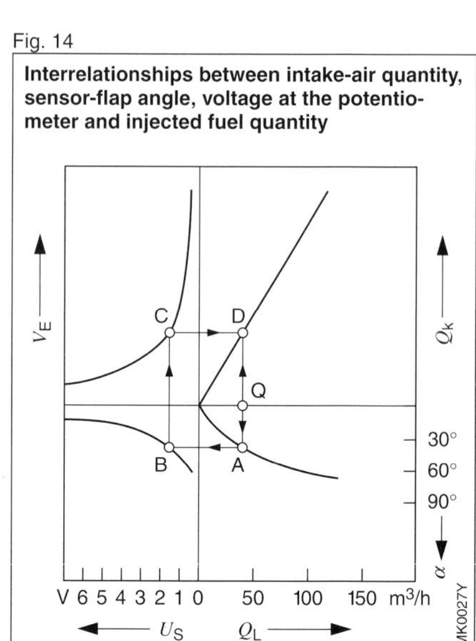

Fig. 13

Block diagram of the ECU

T_i Injection pulses, corrected, T_p Basic injection duration, n Engine speed.

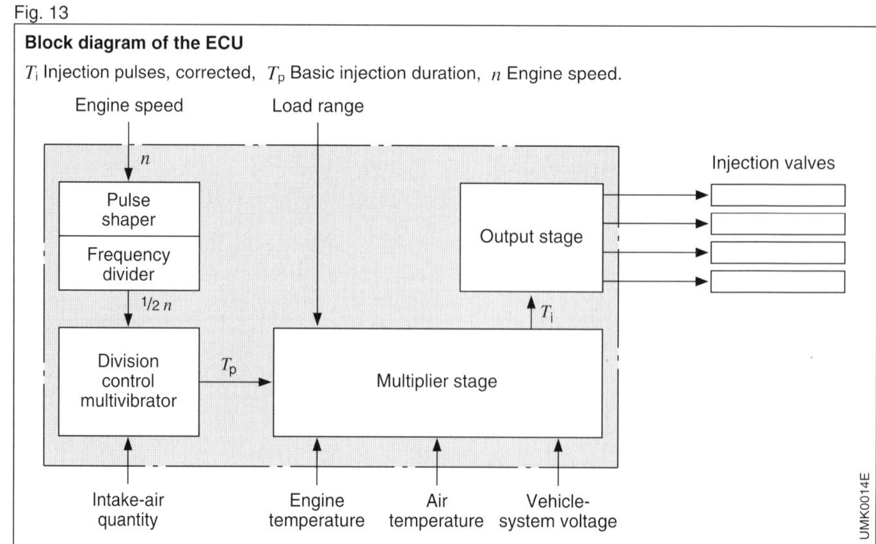

Generation of injection pulses (Fig. 15)

The generation of the basic injection duration is carried out in a special circuit group in the ECU, the division control multivibrator.

The division control multivibrator (DSM) receives the information on speed n from the frequency divider and evaluates it together with the air-quantity signal U_S. For the purpose of intermittent fuel injection, the DSM converts the voltage U_S into square-wave control pulses. The duration T_p of this pulse determines the basic injection quantity, i.e. the quantity of fuel to be injected per intake stroke without considering any corrections. T_p is therefore regarded as the "basic injection duration". The greater the quantity of air drawn in with each intake stroke, the longer the basic injection duration.

Two border cases are possible here: if the engine speed n increases at a constant air throughput Q_L, then the absolute pressure sinks downstream of the

Fig. 15

Complete schematic pulse-timing diagram of the L-Jetronic for 4-cylinder engines

f Ignition pulse, frequency or sparking rate,
n Engine speed,
T_p Basic duration of injection,
T_m Pulse duration extension resulting from corrections,
T_u Pulse duration extension resulting from voltage compensation,
T_i Pulse control time. The actual injection duration per cycle differs from the pulse control time since both a response delay and a release delay change the injection duration.

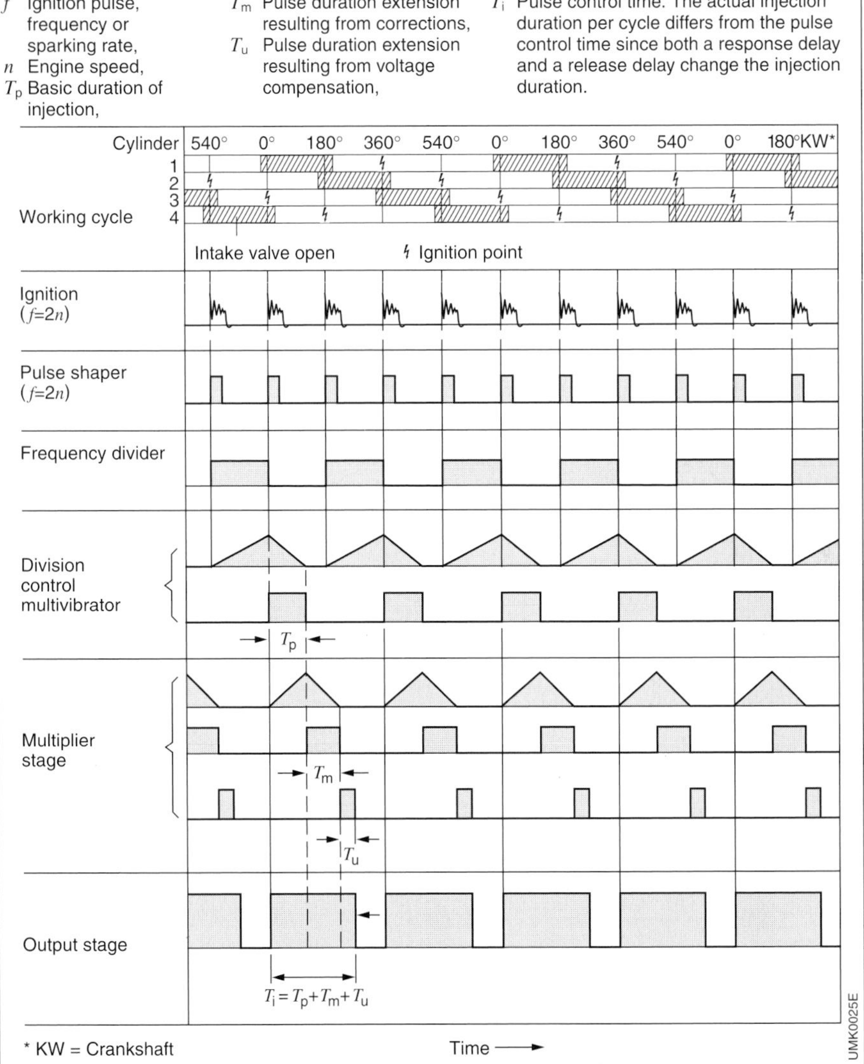

throttle valve and the cylinders draw in less air per stroke, i.e. the cylinder charge is reduced. As a result, less fuel is needed for combustion and the duration of the pulse T_p is correspondingly shorter. If the engine output and thereby the amount of air drawn in per minute increase and providing the speed remains constant, then the cylinder charge will improve and more fuel will be required: the pulse duration T_p of the DSM is longer (Figs. 15 and 16).
During normal driving, engine speed and output usually change at the same time, whereby the DSM continually calculates the basic injection duration T_p. At a high speed, the engine output is normally high (full load) and this results ultimately in a longer pulse duration T_p and, therefore, more fuel per injection cycle.
The basic injection duration is extended by the signals from the sensors depending on the operating mode of the engine.
Adaptation of the basic injection duration to the various operating conditions is carried out by the multiplying stage in the ECU. This stage is controlled by the DSM with the pulses of duration T_p. In addition, the multiplying stage gathers information on various operating modes of the engine, such as cold start, warm-up, full-load operation etc. From this information, the correction factor k is calculated. This is multiplied by the basic injection duration T_p calculated by the division control multivibrator. The resulting time is designated $T_m \cdot T_m$ is added to the basic injection duration T_p, i.e. the injection duration is extended and the air-fuel mixture becomes richer. Tm is therefore a measure of fuel enrichment, expressed by a factor which can be designated "enrichment factor". When it is very cold, for example, the valves inject two to three times the amount of fuel at the beginning of the warm-up period (Figures 13 and 15).

Fig. 16

Signals and controlled variables at the ECU

Q_L Intake air quantity, ϑ_L Air temperature, n Engine speed, P Engine load range, ϑ_M Engine temperature, V_E Injected fuel quantity, Q_{LZ} Auxiliary air, V_{ES} Excess fuel for starting, U_B Vehicle-system voltage.

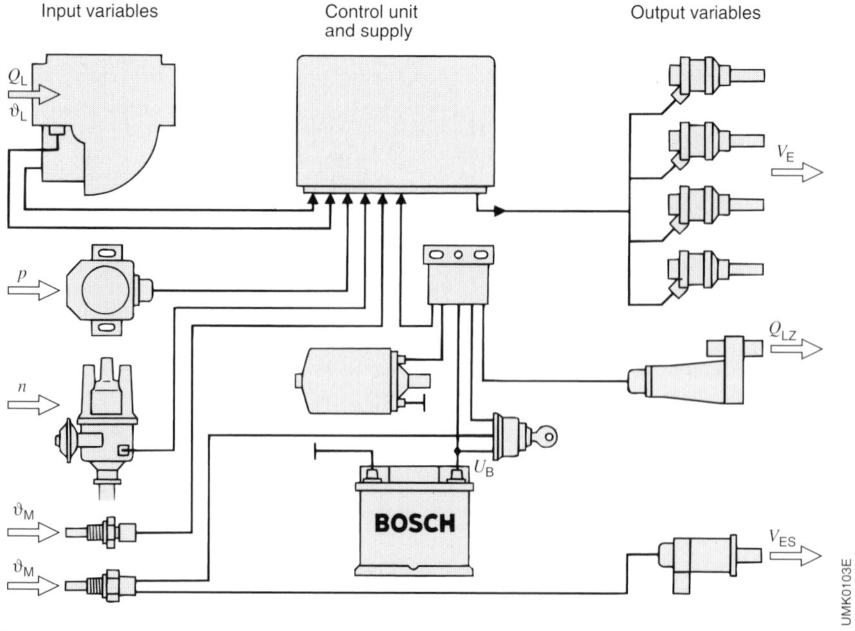

Voltage correction

The pickup time of the fuel-injection valves depends very much on the battery voltage. Without electronic voltage correction, the response delay which results from a low-voltage battery would cause the injection duration to be too short and, as a result, insufficient fuel would be injected. The lower the battery voltage, the less fuel the engine would receive. For this reason, a low battery voltage, i.e. after starting with a heavily discharged battery, must be compensated for with an appropriate extension T_u of the pre-calculated pulse time in order that the engine receives the correct fuel quantity. This is known as "voltage compensation". For voltage compensation, the effective battery voltage is fed into the control unit as the controlled variable. An electronic compensation stage extends the valve control pulses by the amount T_u which is the voltage-dependent pickup delay of the injection valves. The total duration of the fuel-injection pulses T_i is thus the sum of T_p, T_m and T_u (Fig. 15).

Amplification of the injection pulses

The fuel-injection pulses generated by the multiplying stage are amplified in a following output stage. The injection valves are controlled with these amplified pulses.

All the fuel-injection valves in the engine open and close at the same time. With each valve, a series resistor is wired into the circuit and functions as a current limiter. The output stage of the L-Jetronic supplies 3 or 4 valves simultaneously with current. Control units for 6 and 8-cylinder engines have two output stages with 3 and 4 injection valves respectively. Both output stages operate in unison. The injection cycle of the L-Jetronic is selected so that for each revolution of the camshaft (= 1 working cycle) half the amount of fuel required by each working cylinder is injected twice.

In addition to controlling the fuel-injection valves through the series resistors, some control units have a regulated output stage. In these control units, the fuel-injection valves are operated without series resistors. Control of the fuel-injection valve takes place then as follows: as soon as the valve armatures have picked up at the beginning of the pulse, the valve current is regulated for the rest of the pulse duration to a considerably reduced current, the holding current. Since these valves are switched on at the start of the pulse with a very high current, short response times are the result. By means of the reduction in current strength after switching on, the output stage is not subjected to such heavy loading. In this way, up to 12 fuel-injection valves can be switched with only one output stage.

Mixture formation

Intermittent injection onto the engine intake valve.

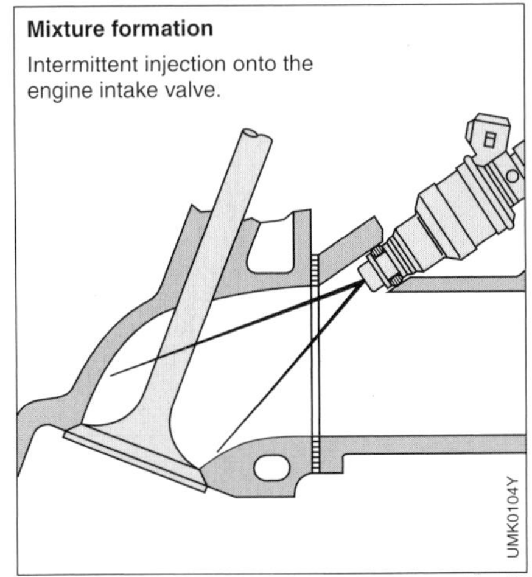

Fig. 17

Mixture formation

Mixture formation is carried out in the intake ports and in the engine cylinder.

The fuel-injection valve injects its fuel directly onto the engine intake valve and, when this opens, the cloud of fuel is entrained along with the air which is drawn in by the engine and an ignitable mixture is formed by the swirling action which takes place during the intake cycle (Fig. 17).

Adaptation to operating modes

In addition to the basic functions described up to now, the mixture has to be adapted during particular operating modes.
These adaptations (corrections) are necessary in order to optimize the power delivered by the engine, to improve the exhaust-gas composition and to improve the starting behavior and driveability. With additional sensors for the engine temperature and the throttle-valve position (load signal), the L-Jetronic ECU can perform these adaptation tasks. The characteristic curve of the air-flow sensor determines the fuel-requirement curve, specific to the particular engine, for all operating ranges.

Cold-start enrichment

When the engine is started, additional fuel is injected for a limited period depending on the temperature of the engine. This is carried out in order to compensate for fuel condensation losses in the inducted mixture and in order to facilitate starting the cold engine.
This extra fuel is injected by the cold-start valve into the intake manifold. The injection duration of the cold-start valve is limited by a thermo-time switch depending upon the engine temperature.
This process is known as cold-start enrichment and results in a "richer" air-fuel mixture, i.e. the excess-air factor λ is temporarily less than 1.
There are two methods of cold-start enrichment:
– Start control with the aid of the ECU and injection valves (Figure 18) or
– Cold-start enrichment via thermo-time switch and cold-start valve (Figure 19).

Start control

By extending the period during which the fuel-injection valves inject, more fuel can be supplied during the starting phase. The electronic control unit controls the start procedure by processing the signals from the ignition and starting switch and the engine-temperature sensor (Figure 18). The construction and method of operation of the temperature sensor are described in Chapter "Warm-up enrichment".

Cold-start valve

The cold-start valve (Figure 20) is a solenoid-operated valve. The solenoid winding is located in the valve. In neutral position, a helical spring presses the movable solenoid armature against a seal, thereby shutting off the valve.
When a current is passed through the solenoid, the armature, which now rises from the valve seat, allows fuel to flow along the sides of the armature to a nozzle where it is swirled. The swirl nozzle atomizes the fuel very finely and as a result enriches the air in the intake

Fig. 18

Cold-start enrichment by start control

1 Engine-temperature sensor, **2** ECU,
3 Fuel-injection valves (injectors),
4 Ignition and starting switch.

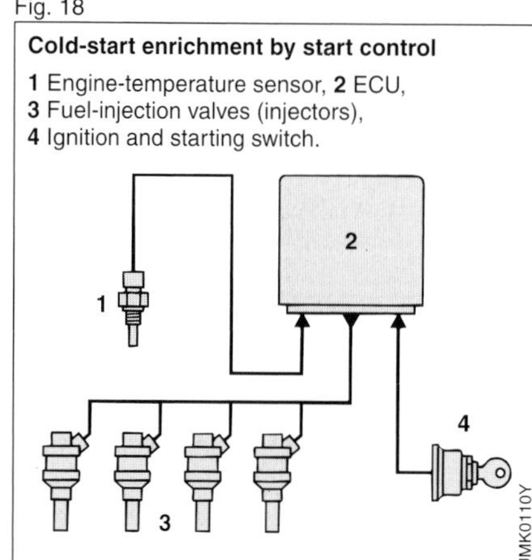

Fig. 19

Cold-start enrichment by cold-start valve

1 Cold-start valve, **2** Thermo-time switch,
3 Relay combination,
4 Ignition and starting switch.

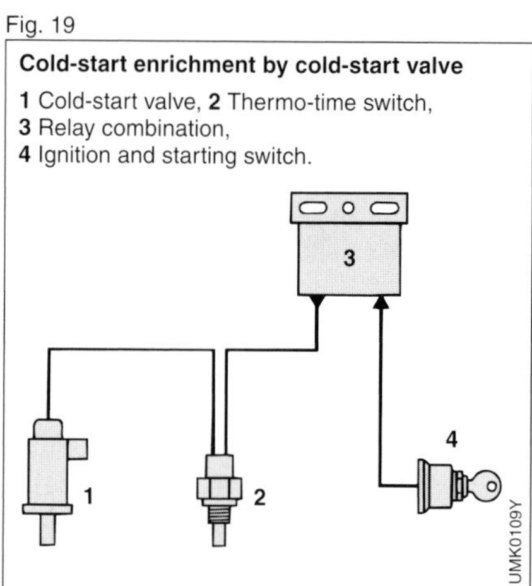

Cold-start valve operated

1 Electrical connection, **2** Fuel inlet with filter strainer, **3** Valve (solenoid armature), **4** Solenoid winding, **5** Swirl nozzle, **6** Valve seat.

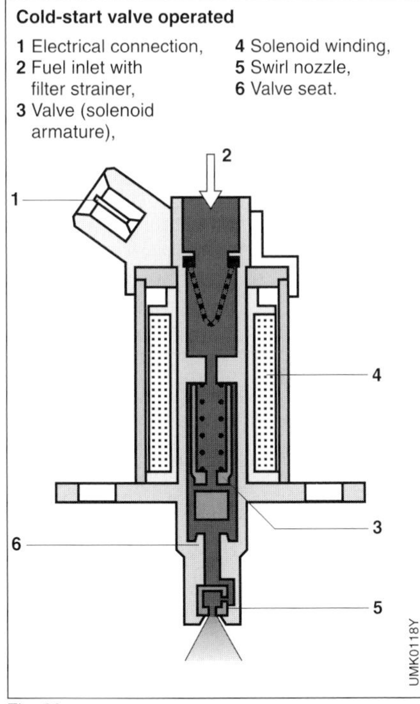

Fig. 20

Thermo-time switch

1 Electrical connection, **2** Housing, **3** Bimetal, **4** Heating winding, **5** Electrical contact.

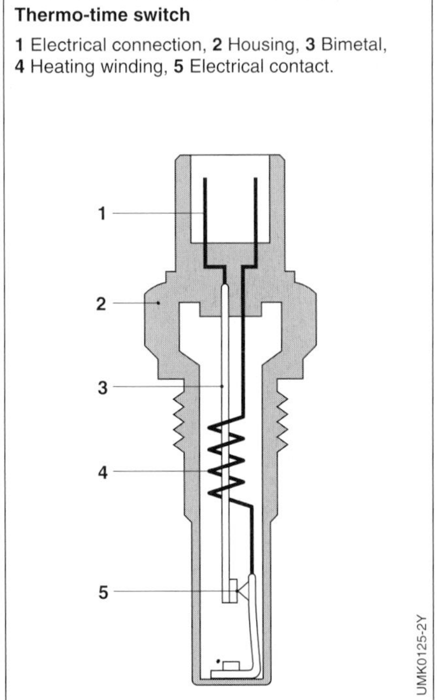

Fig. 21

manifold downstream of the throttle valve with fuel. The cold-start valve is mounted on the intake manifold in such a way as to provide a favourable distribution of the air-fuel mixture to all engine cylinders.

Thermo-time switch

The thermo-time switch limits the duration of injection of the cold-start valve depending on the temperature of the engine.

The thermo-time switch (Figure 21) is an electrically heated bimetal switch which opens or closes a contact depending on its temperature. It is con-trolled via the ignition and starting switch. The thermo-time switch is attached in a position representative of the engine temperature. During a cold start, it limits the "on" period of the cold-start valve. In the case of repeated start attempts, or when starting takes too long, the cold-start valve ceases to inject.

The "on" period is determined by the thermo-time switch which is heated by the heat of the engine as well as by its own built-in electric heater. The electrical heating is necessary in order to ensure that the "on" period of the cold-start valve is limited under all conditions, and engine flooding is prevented. During an actual cold start, the heat generated by the built-in heating winding is mainly responsible for the "on" period (switch-off, for instance, at –20 °C after approx. 7.5 s). With a warm engine, the thermo-time switch has already been heated so far by engine heat that it remains open and prevents the cold-start valve from going into action.

Post-start and warm-up enrichment

During warm-up, the engine receives extra fuel.

The warm-up phase follows the cold-start phase of the engine. During this phase, the engine needs substantially more fuel since some of the fuel condenses on the still cold cylinder walls. In addition, without supplementary fuel enrichment during the warm-up period, a major drop in engine speed would be noticed after the additional fuel from the cold-start valve has been cut off.

For example, at a temperature of –20 °C, depending on the type of engine, two to three times as much fuel must be injected immediately after starting compared with when the engine is at normal operating temperature. In this first part of the warm-up phase (post-start), there must be an enrichment dependent on time. This is the so-called post-start enrichment. This enrichment has to last about 30 s and, dependent upon temperature, results in between 30 % and 60 % more fuel.

When the post-start enrichment has finished, the engine needs only a slight mixture enrichment, this being controlled by the engine temperature. The diagram (Figure 22) shows a typical enrichment curve with reference to time with a starting temperature of 22 °C. In order to trigger this control process, the electronic control unit must receive information on the engine temperature. This comes from the temperature sensor.

Fig. 22

Warm-up enrichment curve

Enrichment factor F as a function of time:
a Proportion mainly dependent on time,
b Proportion mainly dependent on engine temperature.

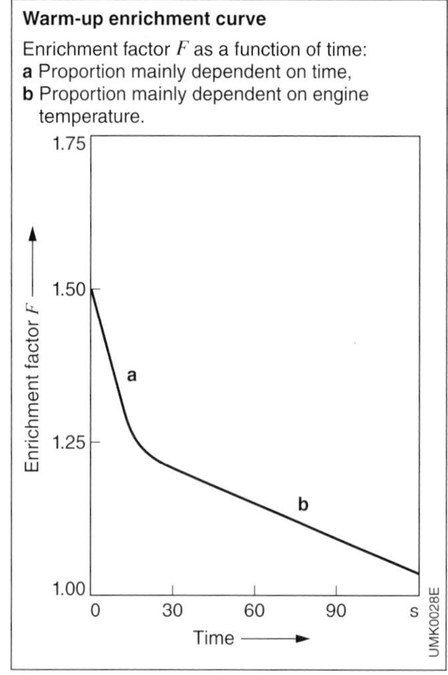

Engine-temperature sensor

The engine-temperature sensor (Figure 23) measures the temperature of the engine and converts this into an electrical signal for the ECU.

It is mounted on the engine block on air-cooled engines. With water-cooled engines, it projects into the coolant.

The sensor "reports" the particular engine temperature to the ECU in the form of a resistance value. The ECU then adapts the quantity of fuel to be injected during post-start and during warm-up. The temperature sensor consists of an NTC resistor embedded in a threaded sleeve.

NTC stands for Negative Temperature Coefficient, the decisive characteristic of this resistor. When the temperature increases, the electrical resistance of the semiconductor resistor decreases.

Part-load adaptation

By far the major part of the time, the engine will be operating in the part-load range. The fuel-requirement curve for this range is programmed in the ECU and determines the amount of fuel supplied. The curve is such that the fuel con-

Fig. 23

Engine-temperature sensor

1 Electrical connection, **2** Housing, **3** NTC resistor.

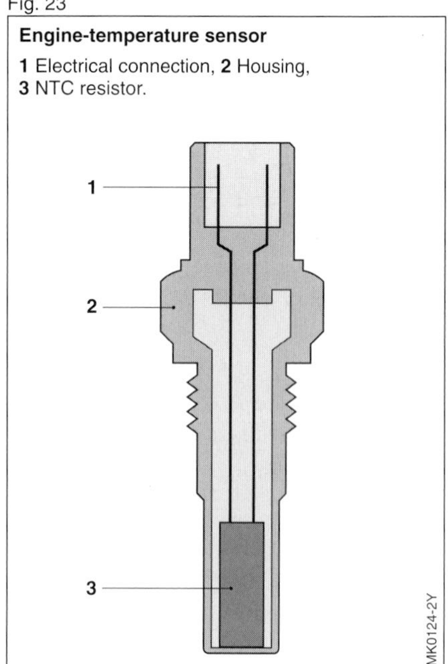

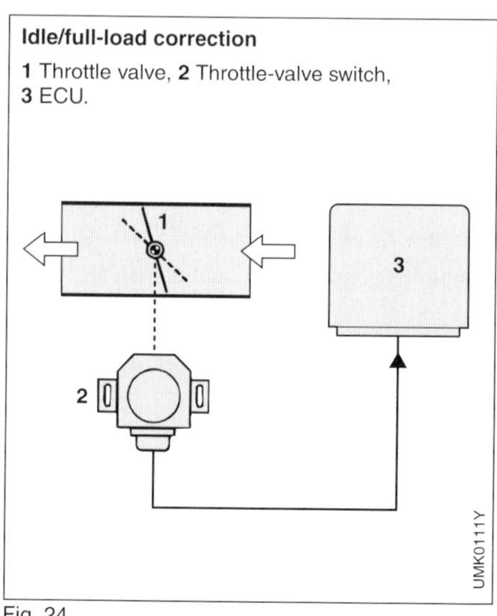

Idle/full-load correction
1 Throttle valve, **2** Throttle-valve switch, **3** ECU.

Fig. 24

sumption of the engine is low in the part-load range.

Acceleration enrichment

During acceleration, the L-Jetronic meters additional fuel to the engine.
If the throttle is opened abruptly, the air-fuel mixture is momentarily leaned-off, and a short period of mixture enrichment is needed to ensure good transitional response.
With this abrupt opening of the throttle valve, the amount of air which enters the combustion chamber, plus the amount of air which is needed to bring the manifold pressure up to the new level, flow through the air-flow sensor. This causes the sensor plate to "overswing" past the wide-open-throttle point. This "overswing" results in more fuel being metered to the engine (acceleration enrichment) and ensures good acceleration response.
Since this acceleration enrichment is not adequate during the warm-up phase, the control unit also evaluates a signal representing the speed at which the sensor flap deflects during this operating mode.

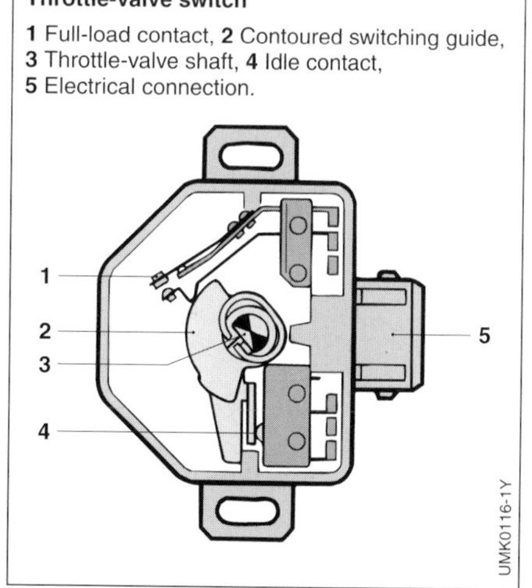

Throttle-valve switch
1 Full-load contact, **2** Contoured switching guide, **3** Throttle-valve shaft, **4** Idle contact, **5** Electrical connection.

Fig. 25

Full-load enrichment

The engine delivers its maximum torque at full load, when the air-fuel mixture must be enriched compared to that at part-load.
In contrast to part-load where the calibration is for minimum fuel consumption and low emissions, at full load it is necessary to enrich the air-fuel mixture. This enrichment is programmed in the electronic control unit, specific to the particular engine. The information on the load condition is supplied to the control unit by the throttle-valve switch.

Throttle-valve switch

The throttle-valve switch (Figures 24 and 25) communicates the "idle" and "full load" throttle positions to the control unit.
It is mounted on the throttle body and actuated by the throttle-valve shaft. A contoured switching guide closes the "idle" contact at one end of switch travel and the "full-load" contact at the other.

Controlling the idle speed

The air-flow sensor contains an adjustable bypass via which a small quantity of air can bypass the sensor flap.

The idle-mixture-adjusting screw in the bypass permits a basic setting of the air-fuel ratio or mixture enrichment by varying the bypass cross-section (Figure 26).

In order to achieve smoother running even at idle, the idle-speed control increases the idle speed. This also leads to a more rapid warm-up of the engine. Depending upon engine temperature, an electrically heated auxiliary-air device in the form of a bypass around the throttle plate allows the engine to draw in more air (Figure 26).

This auxiliary air is measured by the air-flow sensor, and leads to the L-Jetronic providing the engine with more fuel. Precise adaptation is by means of the electrical heating facility. The engine temperature then determines how much auxiliary air is fed in initially through the bypass, and the electrical heating is mainly responsible for subsequently reducing the auxiliary air as a function of time.

Auxiliary-air device

The auxiliary-air device incorporates a perforated plate (Figure 27) which is actuated by the bimetallic strip and which controls the cross-section of the bypass passage.

Initially, the bypass cross-section opened by the perforated plate is determined by the engine temperature, so that during a cold start the bypass opening is adequate for the auxiliary air required. The opening closes steadily along with increasing engine temperature until, finally, it is closed completely. The bimetal strip is electrically heated and this limits the opening time, starting from the initial setting which is dependent upon the engine temperature.

The auxiliary-air device is fitted in the best possible position on the engine for it to assume engine temperature. It does not function when the engine is warm.

Adaptation to the air temperature

The quantity of fuel injected is adapted to the air temperature. The quantity of air necessary for combustion depends upon the temperature of the air drawn in. Cold air is denser. This means that with the same throttle-valve position the volumetric efficiency of the cylinders drops as the temperature increases.

To register this effect, a temperature sensor is fitted in the intake duct of the air-flow sensor. This sensor measures the temperature of the air drawn in and passes this information on to the control unit which then controls the amount of fuel metered to the cylinders accordingly.

Fig. 26

Idle-speed control

1 Throttle valve, **2** Air-flow sensor,
3 Auxiliary-air device,
4 Idle-mixture-adjusting screw.

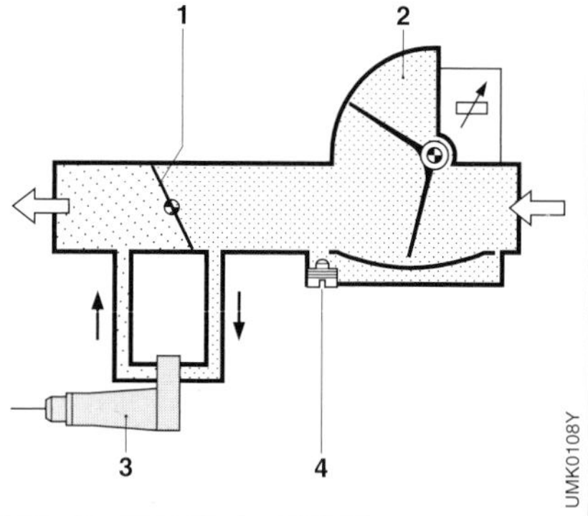

Fig. 27

Electrically heated auxiliary-air device

1 Electrical connection,
2 Electric heating element,
3 Bimetal strip,
4 Perforated plate.

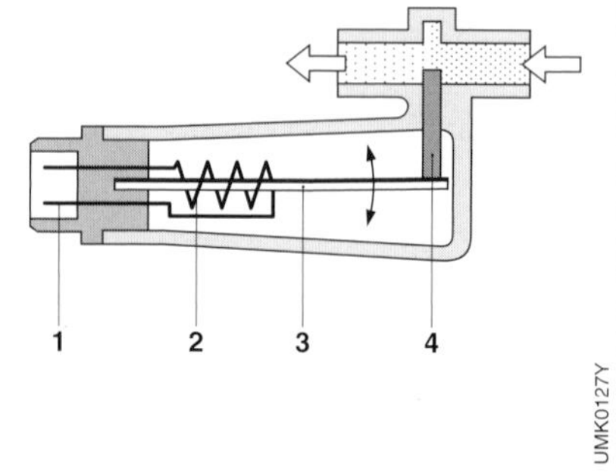

Supplementary functions

Lambda closed-loop control

By means of the lambda closed-loop control, the air-fuel ratio can be maintained precisely at $\lambda = 1$. In the control unit, the lambda-sensor signal is compared with an ideal value (setpoint), thus controlling a two-position controller. The intervention in fuel-metering is accomplished through the opening time of the fuel-injection valves.

Overrun fuel cutoff

Overrun fuel cutoff is the interruption of the supply of fuel to the engine in order to reduce consumption and emissions during downhill driving and braking. When the driver takes his foot off the accelerator pedal while driving, the throttle-valve switch signals "throttle-valve closed" to the ECU and fuel injection is interrupted. The engine-speed switching threshold for injection-pulse cutoff, as well as that for the resumption of fuel injection, depend upon engine temperature.

Engine-speed limiting

When the maximum permissible engine speed is reached, the engine-speed limiting system suppresses the injection signals and interrupts the supply of fuel to the injection valves.

Fig. 28

Lambda closed control loop of the L-Jetronic

The lambda closed control loop is superimposed upon the open-loop air-fuel mixture control. The injected fuel quantity, as determined by the A/F mixture control, is modified by the lambda closed-loop control in order to provide optimum combustion. U_L Intake-air quantity signal, U_λ Lambda-sensor signal.

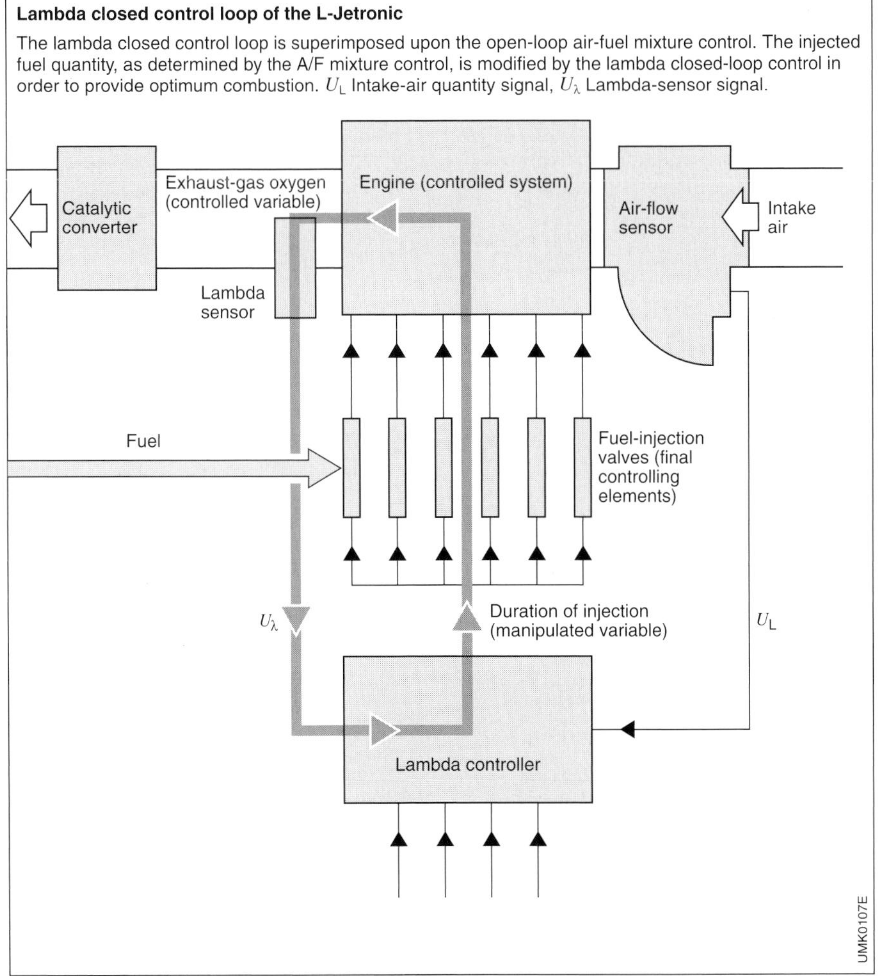

Fig. 29

Components of the L-Jetronic

1 Air-flow sensor, **2** ECU, **3** Fuel filter, **4** Fuel pump, **5** Fuel pressure regulator, **6** Auxiliary-air device, **7** Thermo-time switch, **8** Temperature sensor, **9** Throttle-valve switch, **10** Cold-start valve, **11** Fuel-injection valves (injectors).

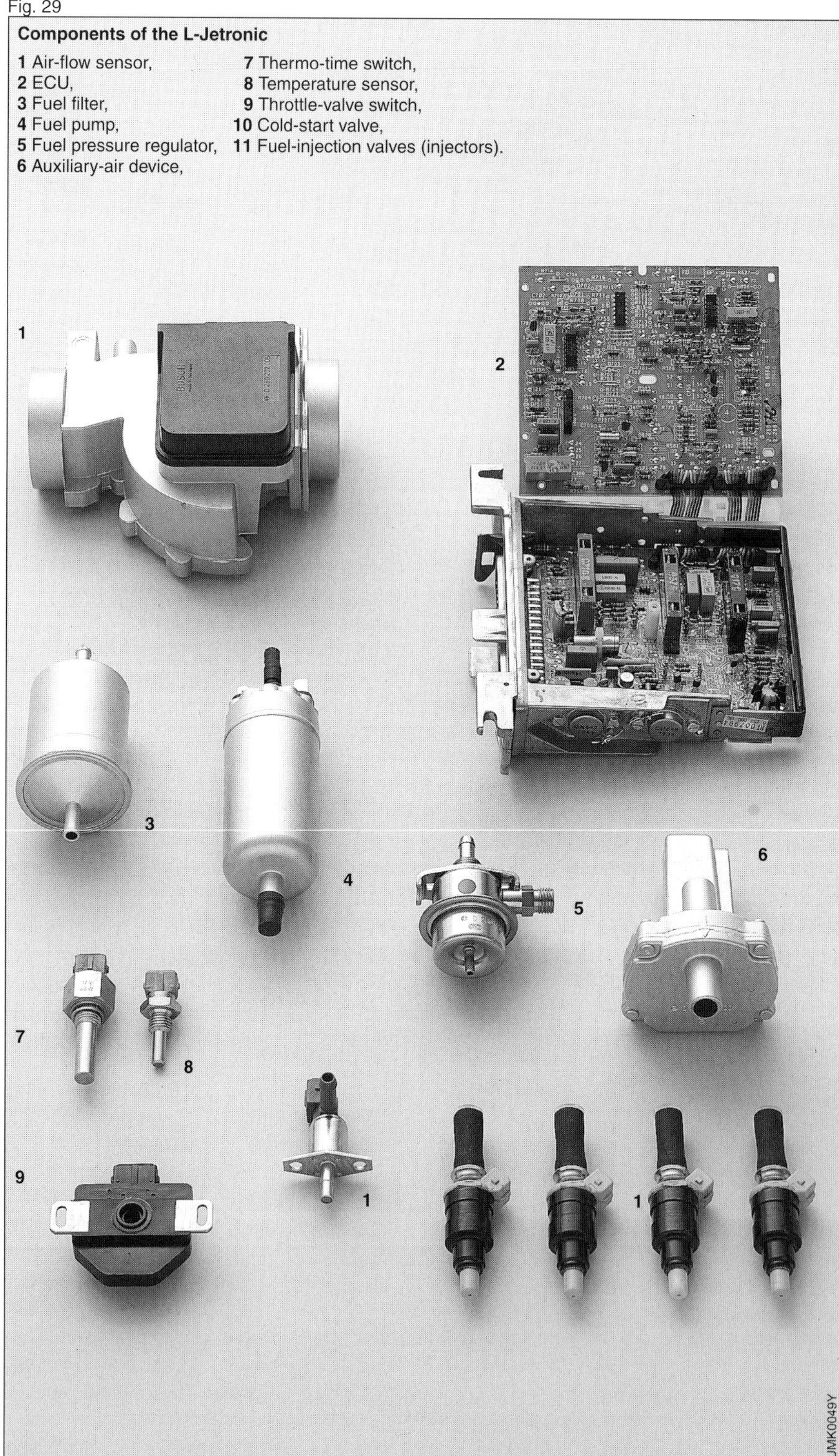

Electric circuitry

The complete circuitry of the L-Jetronic (Fig. 29) has been designed so that it can be connected to the vehicle electrical system at a single point.
At this point, you will find the relay combination which is controlled by the ignition and starting switch, and which switches the vehicle voltage to the control unit and the other Jetronic components.
The relay combination has two separate plug connections, one to the vehicle electrical system and one to the Jetronic.

Safety circuit

In order to prevent the electric fuel pump from continuing to supply fuel following an accident, it is controlled by means of a safety circuit. When the engine is running, the air passing through the air-flow sensor causes a switch to be operated. This switch controls the relay combination which in turn switches the electric fuel pump. If the engine stops with the ignition still on, air is no longer drawn in by the engine and the switch interrupts the power supply to the fuel pump. During starting, the relay combination is controlled accordingly by terminal 50 of the ignition and starting switch.

Terminal diagram

The example shown here is a typical terminal diagram for a vehicle with a 4-cylinder engine. Please note with the wiring harness that terminal 88z of the relay combination is connected directly and without a fuse to the positive pole (terminal post) of the battery in order to avoid trouble and voltage drops caused by contact resistances. Terminals 5, 16 and 17 of the control unit, as well as terminal 49 of the temperature sensor, must be connected with separate cables to a common ground point (Fig. 30).

Fig. 30

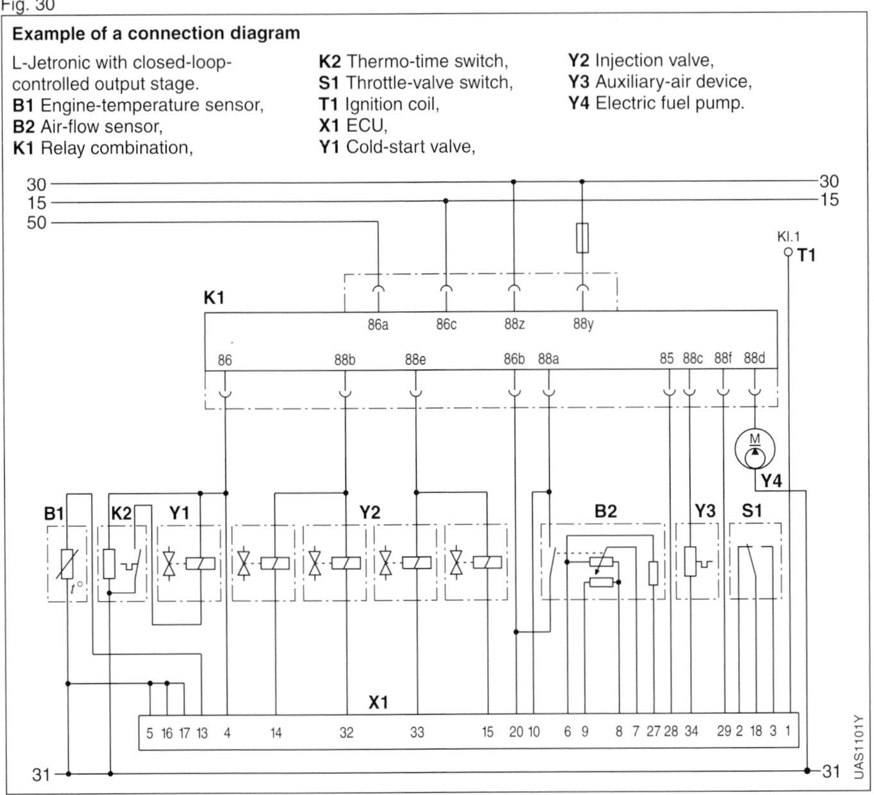

Example of a connection diagram

L-Jetronic with closed-loop-controlled output stage.
B1 Engine-temperature sensor,
B2 Air-flow sensor,
K1 Relay combination,
K2 Thermo-time switch,
S1 Throttle-valve switch,
T1 Ignition coil,
X1 ECU,
Y1 Cold-start valve,
Y2 Injection valve,
Y3 Auxiliary-air device,
Y4 Electric fuel pump.

L3-Jetronic

Specific systems for specific markets have in the meantime been developed on the basis of the L-Jetronic. These systems include the LE-Jetronic without lambda closed-loop control for Europe and the LU-Jetronic system with lambda closed-loop control for countries with strict exhaust gas emission legislation (e.g. the USA). The most recent stage of development is the L3-Jetronic which differs from its predecessors in respect of the following details:
– The control unit, which is suitable for installation in the engine compartment, is attached to the air-flow sensor and thus no longer requires space in the passenger compartment,
– The combined unit of control unit and air-flow sensor with internal connections simplifies the cable harness and reduces installation expense,
– The use of digital techniques permits new functions with improved adaptation capabilities to be implemented as compared with the previous analog techniques used.
The L3-Jetronic system is available both with and without lambda closed-loop control. Both versions have what is called a "limp-home" function which enables the driver to drive the vehicle to the nearest workshop if the microcomputer fails. In addition the input signals are checked for plausibility, i.e. an implausible input signal (e.g. engine temperature lower than –40°C) is ignored and a default value stored in the control unit is used in its place.

Fuel supply

On this system, the fuel is supplied to the injection valves in the same way as on the L-Jetronic system via an electric fuel pump, fuel filter, fuel rail and pressure regulator.

Fig. 31

Schematic diagram of an L3-Jetronic system with lambda closed-loop control

1 Fuel tank, **2** Electric fuel pump, **3** Fuel filter, **4** Fuel-injection valve (injector), **5** Fuel rail, **6** Fuel-pressure regulator, **7** Intake manifold, **8** Throttle-valve switch, **9** Air-flow sensor, **10** ECU, **11** Lambda sensor, **12** Engine-temperature sensor, **13** Ignition distributor, **14** Auxiliary-air device, **15** Battery, **16** Ignition and starting switch.

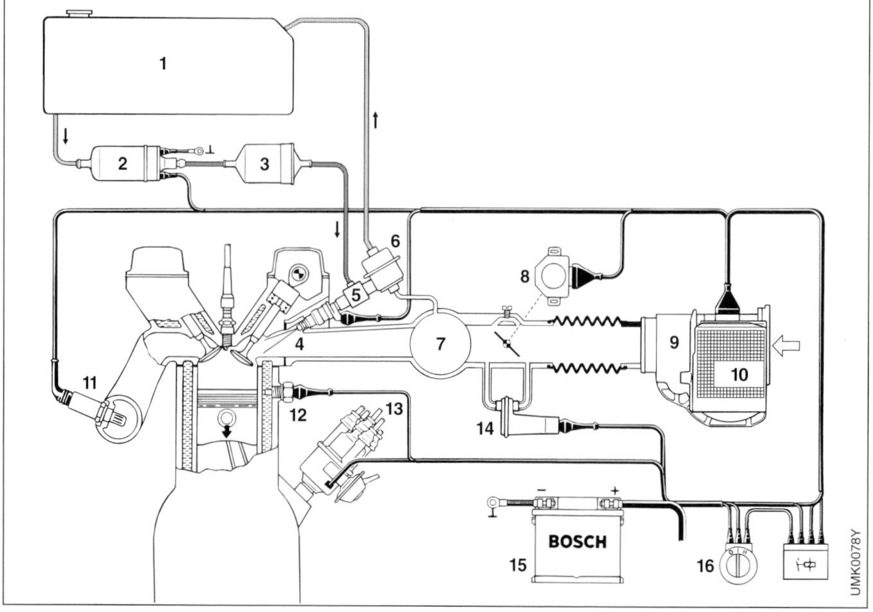

BOSCH

UMK0078Y

Operating-data sensing system

The ignition system supplies the information on engine speed to the control unit. A temperature sensor in the coolant circuit measures the engine temperature and converts it to an electrical signal for the control unit. The throttle-valve switch signals the throttle-valve positions "idle" and "full load" to the control unit for controlling the engine in order to allow for the different optimization criteria in the various operating conditions. The control unit senses the fluctuations in the electrical vehicle supply and compensates for the resultant response delays of the fuel-injection valves by correcting the duration of injection.

Air-flow sensor

The air-flow sensor of the L3-Jetronic system measures the amount of air drawn in by the engine using the same measuring principle as the air-flow sensor of the conventional L-Jetronic system. Integrating the control unit with the air-flow sensor to form a single measuring and control unit requires a modified configuration however. The dimensions of the potentiometer chamber in the air-flow sensor and of the control unit have been reduced to such an extent that the overall height of the entire unit does not exceed that of the previous air-flow sensor alone. Other features of the new air-flow sensor include the reduced weight due to the aluminum used in place of the zinc material for the housing, the extended measuring range and the improved damping behavior in the event of abrupt changes in the intake air quantity. Thus, the L3-Jetronic incorporates clear improvements both in respect of electronic components and in respect of mechanical components whilst requiring less space (Figs. 32 and 33).

Fuel metering

Fuel is injected onto the intake valves of the engine by means of solenoid-operated injection valves. One solenoid valve is assigned to each cylinder and is operated once per crankshaft revolution. In order to reduce the circuit complexity, all valves are connected electrically in

Fig. 32

Integration of ECU and air-flow sensor of the L3-Jetronic to form a single measuring and control unit

1 ECU, **2** Air-flow sensor with potentiometer.

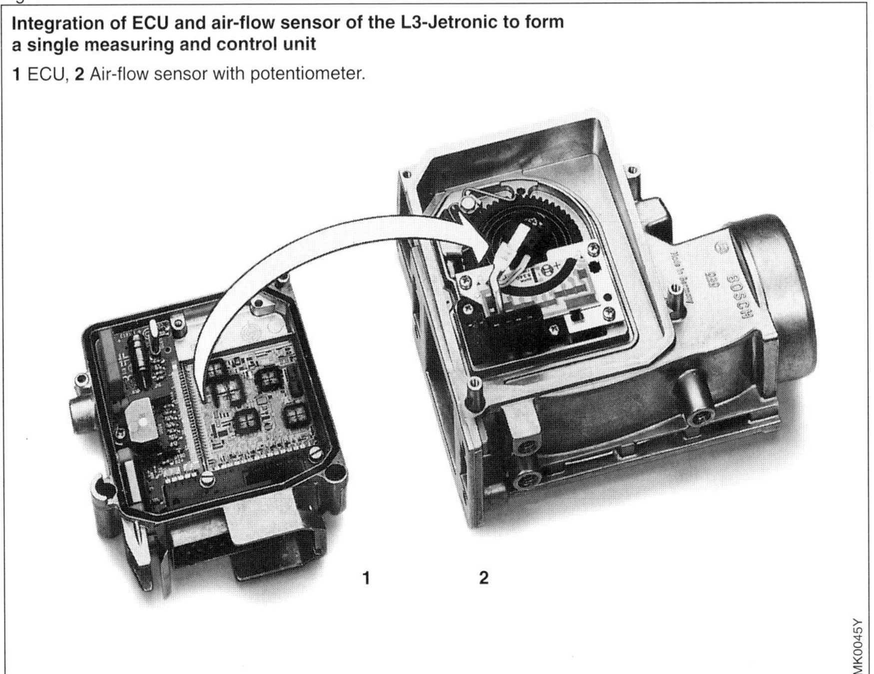

parallel. The differential pressure between the fuel pressure and intake-manifold pressure is maintained constant at 2.5 or 3 bar so that the quantity of fuel injected depends solely upon the opening time of the injection valves. For this purpose, the control unit supplies control pulses, the duration of which depends upon the inducted air quantity, the engine speed and other actuating variables which are detected by sensors and processed in the control unit.

Electronic control unit (ECU)
By contrast with the L-Jetronic system, the digital control unit of this system adapts the air-fuel ratio by means of a load/engine-speed map. On the basis of the input signals from the sensors, the control unit computes the injection duration as a measure of the amount of fuel to be injected. The microcomputer system of the control unit permits the required functions to be influenced. The control unit for attachment to the air-flow sensor must be very compact and must have very few plug connections in addition to being resistant to heat, vibration and moisture. These conditions are met by the use of a special-purpose hybrid circuit and a small PC board in the control unit. In addition to accommodating the microcomputer, the hybrid circuit also accommodates 5 other integrated circuits, 88 film resistors and 23 capacitors. The connections from the ICs to the thick-film board comprise thin gold wires which are a mere 33 thousandths of a millimeter in thickness.

Adaptation to operating conditions

During certain operating conditions (cold start, warm-up, acceleration, idle and full load), the fuel requirement differs greatly from the normal value so that it is necessary to intervene in mixture formation.

Throttle-valve switch
This switch is operated by the throttle-valve shaft and has a switching contact for each of the two end positions of the throttle valve. When the throttle valve is closed (idle) or fully open (full load), the

Air-flow sensor of the L3-Jetronic
1 Sensor flap, **2** Compensation flap, **3** Damping volume.

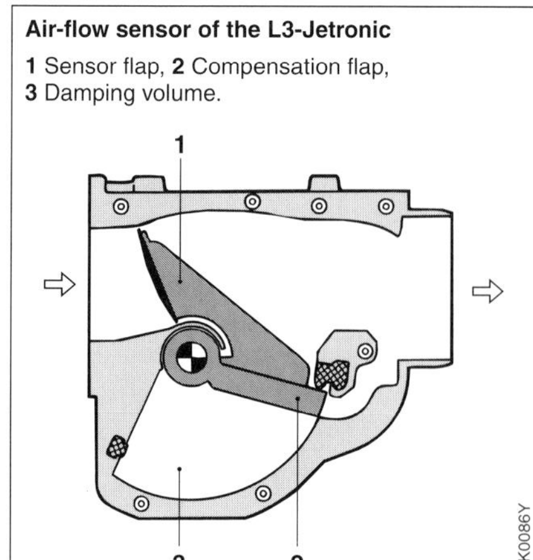

Fig. 33

switch issues a switching signal to the control unit.

Auxiliary-air device
A plate which is moved by a bimetallic spring or expansion element supplies extra air to the engine during the warm-up phase. This results in the higher idle speed which is required during the warm-up phase for smooth running of the engine.
A closed-loop idle-speed control system, in the form of a separate system, can be used instead of the auxiliary-air device to control the idle speed.

Engine-temperature sensor
The engine-temperature sensor, a temperature-dependent resistor, controls warm-up enrichment. The overrun fuel cutoff function, and the speed limiting function at maximum permissible engine speed, permit fuel economy and a reduction in pollutant emission.

Lambda closed-loop control

In the control unit, the signal from the lambda sensor is compared with an ideal value (setpoint), thus controlling a two-position controller. Dependent upon the result of the comparison, either an excessively lean air-fuel mixture is enriched or an excessively rich mixture is leaned. Fuel metering is influenced via the opening time of the injection valves.

LH-Jetronic

The LH-Jetronic (Fig. 34) is closely related to the L-Jetronic. The difference lies in the hot-wire air-mass meter which measures the air mass inducted by the engine. The result of measurement is thus independent of the air density which is itself dependent upon temperature and pressure.

Fuel supply

The fuel is supplied to the injection valves through the same components as with the L-Jetronic.

Operating-data sensing system

The information on engine speed is supplied to the control unit by the ignition system. A temperature sensor in the coolant circuit measures the engine temperature and converts it to an electronic signal for the control unit. The throttle-valve switch signals the throttle-valve positions “idle” and “full load” to the control unit for engine control in order to allow for the different optimization criteria in the various operating conditions. The control unit detects the fluctuations in the vehicle electrical supply and compensates for the resultant response delays of the injection valves by correcting the duration of injection.

Air-mass meters

The hot-wire and hot-film air-mass meters are “thermal” load sensors. They are installed between the air filter and the throttle valve and register the air-mass flow [kg/h] drawn in by the engine. Both sensors operate according to the same principle.

Hot-wire air-mass meter

With the hot-wire air-mass meter, the electrically heated element is in the form of a 70 µm thick platinum wire. The intake-air temperature is registered by a temperature sensor. The hot wire and the intake-air temperature sensor are part of

Fig. 34

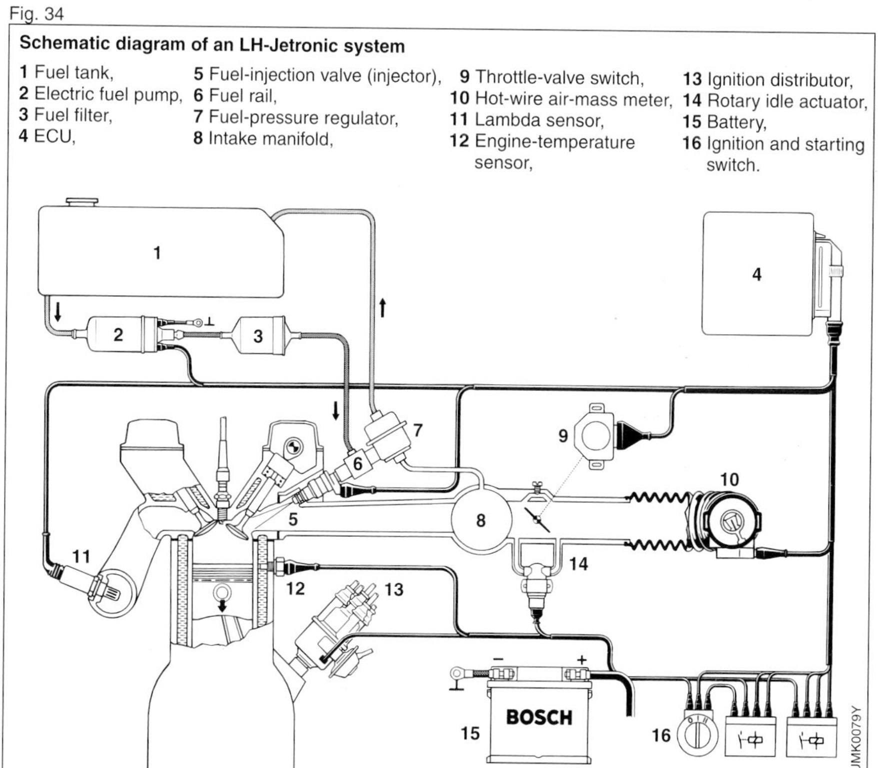

Schematic diagram of an LH-Jetronic system

1 Fuel tank, **2** Electric fuel pump, **3** Fuel filter, **4** ECU, **5** Fuel-injection valve (injector), **6** Fuel rail, **7** Fuel-pressure regulator, **8** Intake manifold, **9** Throttle-valve switch, **10** Hot-wire air-mass meter, **11** Lambda sensor, **12** Engine-temperature sensor, **13** Ignition distributor, **14** Rotary idle actuator, **15** Battery, **16** Ignition and starting switch.

a bridge circuit in which they function as temperature-dependent resistances. A voltage signal which is proportional to the air-mass flow is transmitted to the ECU (Figs. 35 and 36).

Hot-film air-mass meter

With the hot-film air-mass meter, the electrically heated element is in the form of a platinum film resistance (heater). The heater's temperature is registered by a temperature-dependent resistor (through-flow sensor). The voltage across the heater is a measure for the air-mass flow. It is converted by the hot-film air-mass meter's electronic circuitry into a voltage which is suitable for the ECU (Fig. 37).

Fuel metering

Fuel is injected by means of solenoid-operated injection valves onto the intake valves of the engine. A solenoid valve is assigned to each cylinder and is operated once per crankshaft revolution. In order to reduce the circuit complexity, all valves are connected electrically in parallel. The differential pressure between the fuel pressure and intake-manifold pressure is maintained constant at 2.5 or 3 bar so that the quantity of fuel injected depends solely upon the opening time of the injection valves. For this purpose, the control unit supplies control pulses, the duration of which are dependent upon the inducted air quantity, the engine speed and other actuating variables which are detected by sensors and processed in the control unit.

Electronic control unit (ECU)

By comparison with the L-Jetronic system, the digital control unit of this system adapts the air-fuel ratio by means of a load/engine-speed map. On the basis of the input signals from the sensors, the control unit computes the injection duration as a measure of the quantity of fuel to be injected. The microcomputer system of the control unit permits the required functions to be influenced.

Fig. 35

Hot-wire air-mass meter.

The 70 µm thin platinum wire is suspended inside the measuring venturi.

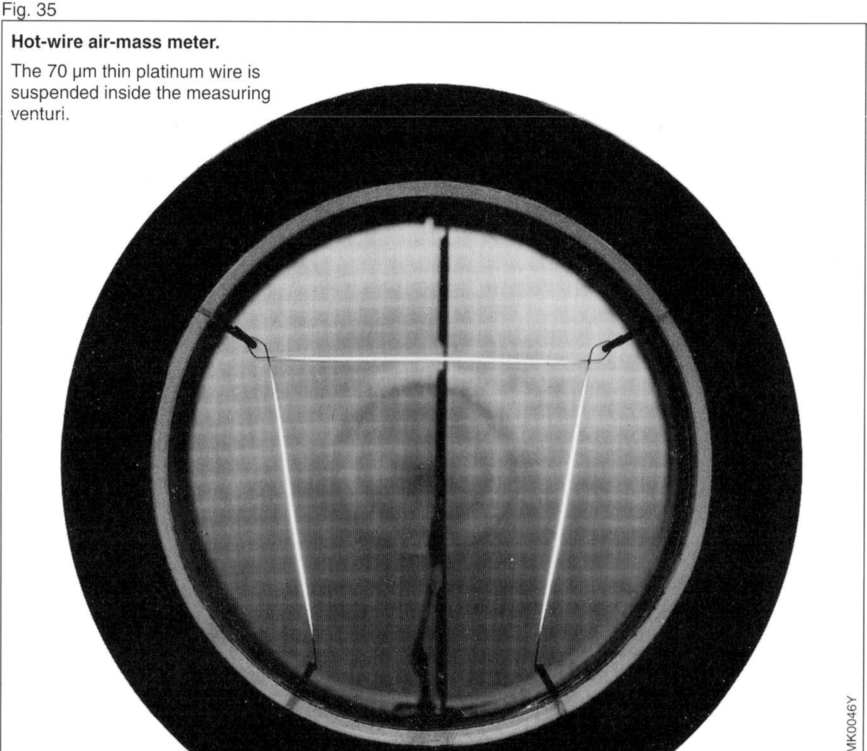

Adaptation to operating conditions
During certain operating conditions (cold start, warm-up, acceleration, idle and full load), the fuel requirement differs greatly from the normal value, thus necessitating an intervention in mixture formation.

Throttle-valve switch
This switch has a switching contact for each of the two end positions of the throttle valve. It issues a switching signal to the control unit when the throttle valve is closed (idle) or fully open (full load).

Rotary idle actuator
The idle speed can be reduced and stabilized with the idle-speed control function. For this purpose, the rotary idle actuator opens a bypass line to the throttle valve and supplies the engine with more or less air. Since the hot-wire air-mass meter senses the extra air, the injected fuel quantity also changes as required.

Engine-temperature sensor
The engine-temperature sensor, a temperature-dependent resistor, controls warm-up enrichment.

Supplementary functions
The overrun fuel cutoff function, and the speed limiting function at maximum permissible engine speed, permit fuel economy and a reduction in pollutant emission.

Lambda closed-loop control
The lambda sensor supplies a signal which represents the instantaneous mixture composition. In the control unit, the signal of the lambda sensor is compared with an ideal value (setpoint), thus controlling a two-position controller. Dependent upon the result of comparison, either an excessively lean air-fuel mixture is enriched or an excessively rich mixture is leaned. Fuel metering is influenced via the opening time of the injection valves.

Fig. 36
Hot-wire air-mass meter
1 Hybrid circuit, **2** Cover, **3** Metal insert, **4** Venturi with hot wire, **5** Housing, **6** Screen, **7** Retaining ring.

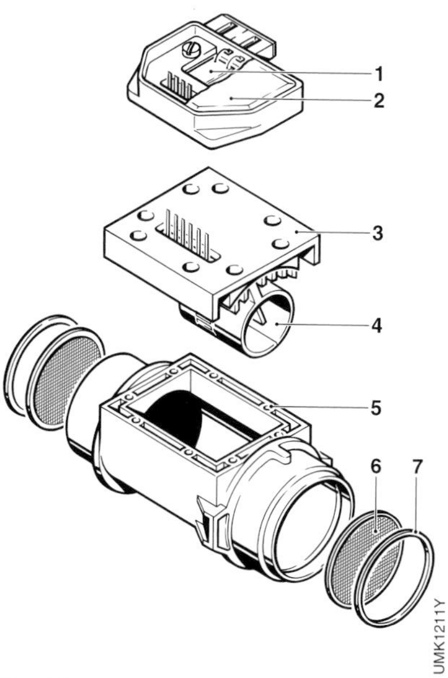

Fig. 37
Hot-film air-mass meter
a Housing, **b** Hot-film sensor (fitted in the middle of the housing).
1 Heat sink, **2** Intermediate module, **3** Power module, **4** Hybrid circuit, **5** Sensor element.

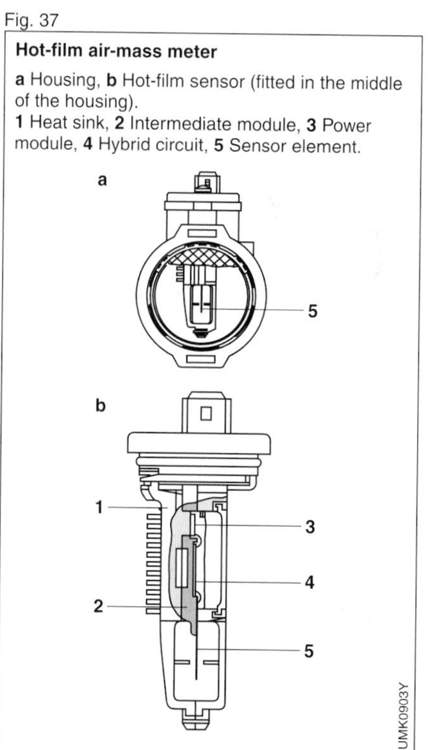

Workshop testing techniques

Bosch customer service

Customer service quality is also a measure for product quality. The car driver has more than 10,000 Bosch Service Agents at his disposal in 125 countries all over the world. These workshops are neutral and not tied to any particular make of vehicle. Even in sparsely populated and remote areas of Africa and South America the driver can rely upon getting help quickly. Help which is based upon the same quality standards as in Germany, and which is backed of course by the identical guarantees which apply to customer-service work all over the world.
The data and performance specs for the Bosch systems and assemblies of equipment are precisely matched to the engine and the vehicle. In order that these can be checked in the workshop, Bosch developed the appropriate measurement techniques, test equipment, and special tools, and equipped its Service Agents accordingly.

Testing techniques for L-Jetronic

Apart from the regular replacement of the air filter and fuel filter as stipulated by the particular vehicle's manufacturer, the L-Jetronic fuel-injection system requires no special maintenance work.
Essentially, in case of malfunctions the workshop expert has the following test equipment and the appropriate test specs at his disposal:

- Universal test adapter with system adapter line, and multimeter or engine analyzer,
- Jetronic set (hydraulic case with pressure-measuring device),
- Injected-quantity comparison tester with triggering unit,
- Lambda closed-loop-control tester (only needed if Lambda control is fitted),
- Diagnosis tester for more modern Jetronic systems.

Fig. 38

Test set-up with universal test adapter, system adapter line, and multimeter

Test components: **1** Universal test adapter, **2** Multiple-contact switch, **3** Plug-in connection, **4** Adapter line, 4.1 Y-version adapter line (depending on system), **5** Measurement lines, **6** Multimeter, Vehicle components: **7** ECU, **8** System wiring harness.

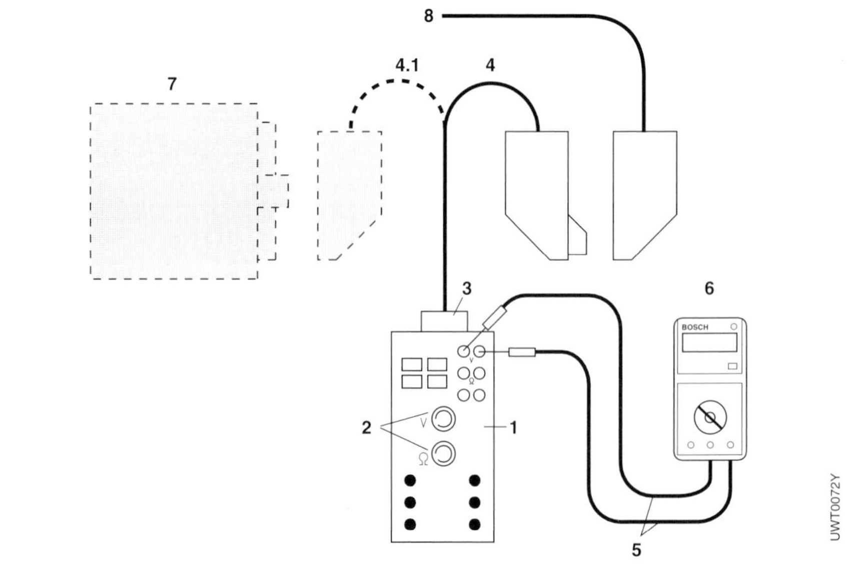

Together with the relevant Test Instructions and Test Specifications in a variety of different languages, this uniform testing technology is available throughout the world at the Bosch Service Agent workshops and at the majority of the workshops belonging to the vehicle manufacturers. Purposeful trouble-shooting and technically correct repairs cannot be performed at a reasonable price without this equipment. It is therefore inadvisable for the vehicle owner to attempt his own repairs.

Universal test adapter, system adapter line, and multimeter or engine tester
The universal test adapter (Fig. 38) was specifically developed for testing electronic fuel-injection (EFI) systems such as the L-, LE-, LU-, L2-, L3-, LH-Jetronic and Motronic. Using this test adapter, all the system's important components and functions which are essential for optimal engine running can be tested:
- Air-flow or air-mass sensor,
- Intake-air temperature sensor,
- Engine-temperature sensor,
- Throttle-valve switch or potentiometer,
- Electric fuel pump, and
- Lambda oxygen sensor.

The system-specific adapter line is used to connect the universal test adapter to the ECU wiring-harness plug. By means of 2 multiple-contact switches, it is then a simple matter to quickly select the various line to the components, and to measure the voltage and resistance values using the multimeter or engine analyzer.
Using the Y-version adapter line, supplementary ECU functions such as overrun fuel cutoff, and full-load and warm-up enrichment can be tested with the engine running.

Jetronic set
The fuel pressures at idle and with the manifold-pressure hose disconnected can be measured using the pressure-measuring device from the Jetronic set. This permits testing of:
- Electric fuel-pump delivery,
- Fuel-filter permeability,
- Return-line restriction,
- Fuel metering by the injectors, and
- Pressure-regulator function.

In addition, the pressure-measuring device can be used to check the sealing integrity of the complete fuel system. This is vitally important for starting behavior.

Injected-quantity comparison tester and triggering unit
A comparative measurement can be made to determine whether there are differences in the delivered quantities from the individual injectors. To do so, the injectors are pulled out of the intake manifold and connected to the comparison tester. The triggering unit generates the electrical pulses needed to operate the injectors. The injected-fuel quantity from up to eight injectors can be compared in this manner.

Lambda closed-loop-control tester
On L-Jetronic systems with Lambda closed-loop control, this tester serves to check the integrator voltage, the lambda-sensor signal (with simulation of the "rich/lean" signal), and the "open-loop/closed-loop control" function. Special adapter lines are available for connection to the Lambda-sensor cable of the various vehicle models. Measured values are shown on an analog display.

Diagnosis tester
Newer Jetronic systems using digital circuitry are equipped with an on-board diagnosis (OBD) facility with fault memory. Using the diagnosis tester fault codes and actual values can be read out, and actuator tests performed.

Mono-Jetronic fuel-injection system

System overview

Mono-Jetronic is an electronically controlled, low-pressure, single-point injection (SPI) system for 4-cylinder engines. While port injection systems such as KE and L-Jetronic employ a separate injector for each cylinder, Mono-Jetronic features a single, centrally-located, solenoid-controlled injection valve for the entire engine.

The heart of the Mono-Jetronic is the central injection unit (described in the following). It uses a single solenoid-operated injector for intermittent fuel injection above the throttle valve.

The intake manifold distributes the fuel to the individual cylinders.

A variety of different sensors are used to monitor engine operation and furnish the essential control parameters for optimum mixture adaptation. These include:

- Throttle-valve angle,
- Engine speed,
- Engine and intake-air temperature,
- Throttle-valve positions (idle/full-throttle),
- Residual oxygen content of exhaust gas, and (depending on the vehicle's equipment level):
- Automatic transmission, air-conditioner settings, and a/c compressor clutch status (engaged-disengaged).

Fig. 1

Mono-Jetronic schematic diagram

1 Fuel tank, **2** Electric fuel pump, **3** Fuel filter, **4** Fuel-pressure regulator, **5** Solenoid-operated fuel injector, **6** Air-temperature sensor, **7** ECU, **8** Throttle-valve actuator, **9** Throttle-valve potentiometer, **10** Canister-purge valve, **11** Carbon canister, **12** Lambda oxygen sensor, **13** Engine-temperature sensor, **14** Ignition distributor, **15** Battery, **16** Ignition-start switch, **17** Relay, **18** Diagnosis connection, **19** Central injection unit.

UMK0114Y

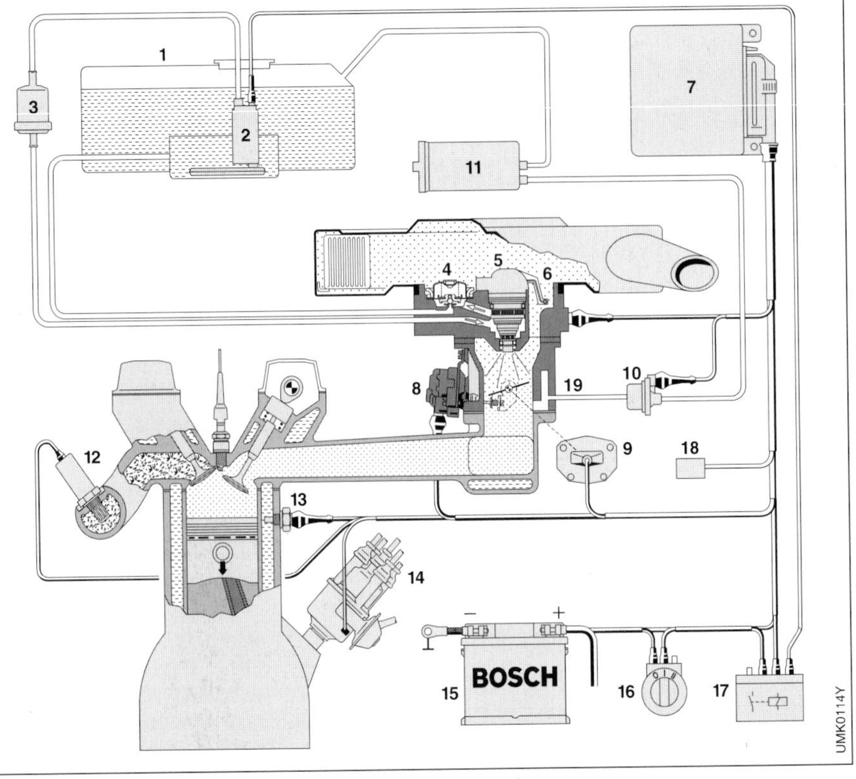

Input circuits in the ECU convert the sensor data for transmission to the microprocessor, which analyzes the operating data to determine current engine operating conditions; this information, in turn, provides the basis for calculating control signals to the various final-control elements (actuators). Output amplifiers process the signals for transmission to the injector, throttle-valve actuator and canister-purge valve.

Versions

The following text and illustrations describe a typical Mono-Jetronic installation (Figure 1). Other versions are available to satisfy any specific individual requirements that the manufacturers define for fuel-injection systems.

The Mono-Jetronic system discharges the following individual functions (Fig. 2):
- Fuel supply,
- Acquisition of operating data, and
- Processing of operating data.

Basic function

Mono-Jetronic's essential function is to control the fuel-injection process.

Supplementary functions

Mono-Jetronic also incorporates a number of supplementary closed-loop and open-loop control functions with which it monitors operation of emissions-relevant components. These include idle-speed control, Lambda closed-loop

Fig. 2

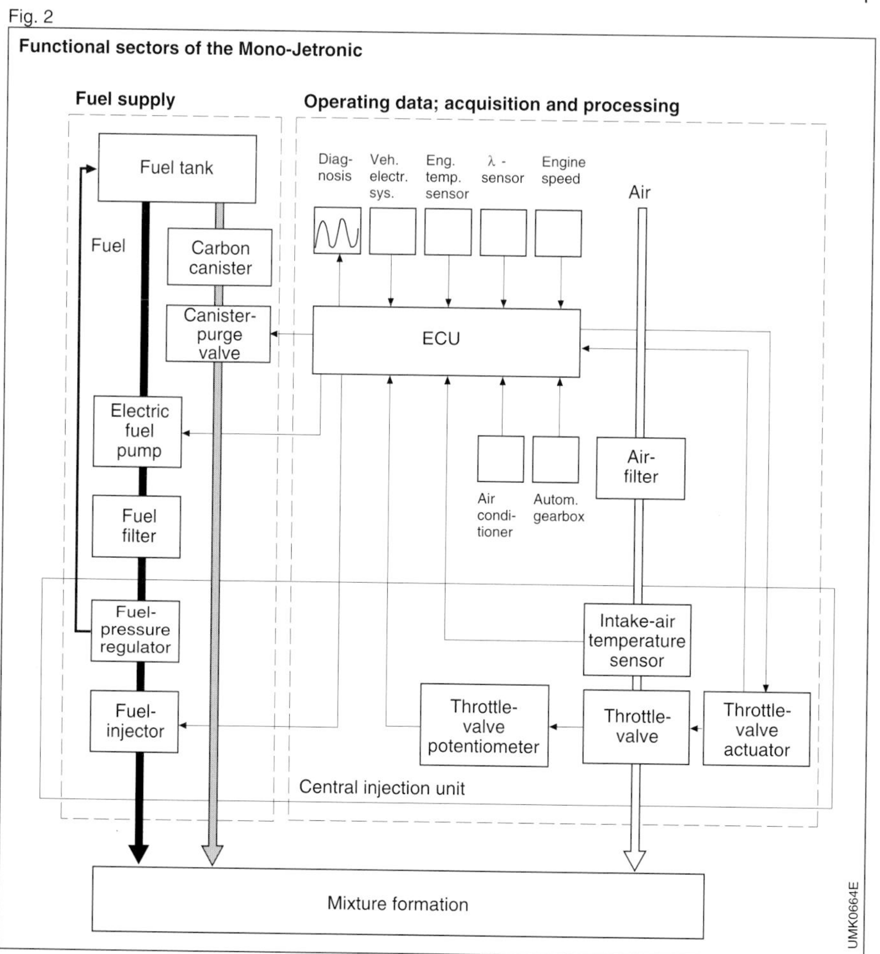

control, and open-loop control of the evaporative-emissions control system.

Fuel supply

The fuel system supplies fuel from the tank to the solenoid-operated injector.

Fuel delivery

The electric fuel pump continuously pumps fuel from the tank and through the fuel filter to the central injection unit. Fuel pumps are available as in-line or in-tank versions.
In-line fuel pumps are situated outside the fuel tank, in the fuel line on the vehicle body platform between the fuel tank and the fuel filter.
Usually, the Mono-Jetronic has an in-tank fuel pump. As the name implies, this pump is located inside the tank. It is held in a special holder which normally incorporates an additional fuel filter on the intake side, a fuel-level gauge, a fuel-swirl pot which serves as a fuel reservoir, and the electrical and hydraulic connections to the outside (Figures 3 and 5).

Electric fuel pump

The electric motor and the pump itself are encased in a common housing. Fuel circulates around both pump and motor for continuous cooling. No complex sealing arrangements are required between pump and motor, so the motor's performance is better. There is no danger of explosion because an ignitable mixture can never form in the electric motor. The pump's output-end cover contains the electrical connections, the check valve, and the hydraulic pressure connection. The check valve also maintains primary pressure in the system for some time after the fuel pump is deactivated; this inhibits formation of vapor bubbles in warm fuel. In addition, the output-end cover can also incorporate the interference-suppression devices (Fig. 4).
The electric fuel pump described above is the most common type used in Mono-Jetronic applications. It provides ideal performance with this system's low primary pressure. It is a two-stage flow-type pump: a side-channel pump being used as the preliminary stage (pre-stage) and a peripheral pump as the main stage. Both stages are integrated in a single impeller wheel.

Fig. 3

Mono-Jetronic fuel supply

1 Fuel tank, **2** Electric fuel pump, **3** Fuel filter, **4** Fuel-pressure regulator, **5** Fuel injector, **6** Throttle valve.

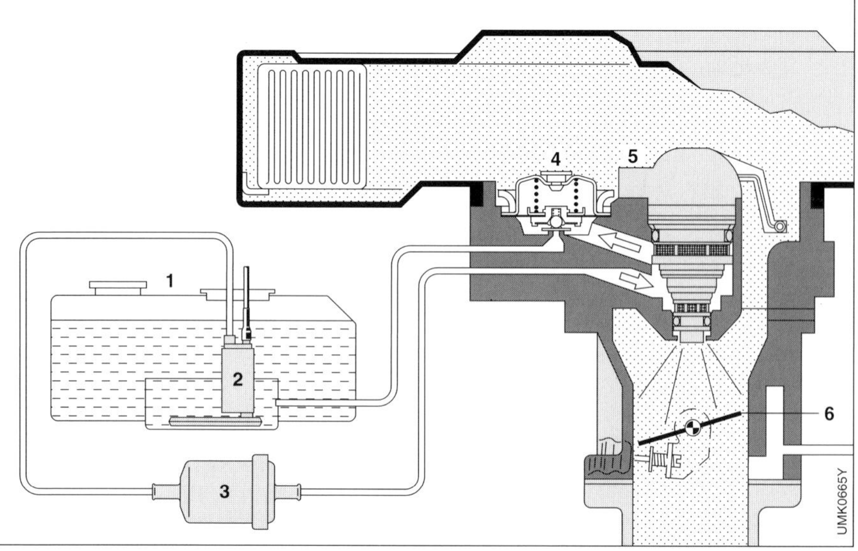

The pre-stage circuit features a side channel on each side of the impeller blade ring, that is, one in the pump cover and one in the pump housing. The fuel, which is accelerated by the blade ring of the rotating impeller, converts its velocity energy into pressure energy in these two side channels, at the end of which the fuel is then transferred to the outside (as viewed from the end along the center axis) main pump stages. A degassing vent is provided in the overflow channel between the preliminary stage and the main stage. Through it, excess fuel, together with any vapor bubbles which may have formed, is continuously returned to the fuel tank.

The main stage and pre-stage function identically. The main difference is in the design of the impeller wheel and the shape of the channel which encloses the blade ring at the side and around its periphery (peripheral principle). There is a device for rapid venting of the main stage at the end of the peripheral channel. This takes the form of a diaphragm plate which closes an opening in the intake cover and functions as a discharge valve (Fig. 6).

With the discharge valve closed, the fuel is forced into the pump's electric-motor chamber from where it flows through the check valve into the fuel-supply line.

At high fuel temperatures, since the fuel-vapor bubbles have already been

Fig. 5

In-tank fuel pump with noise encapsulation

1 Electric fuel pump, **2** Rubber hose, **3** Rubber collar, **4** Plastic housing, **5** Swirl pot, **6** Fuel filter.

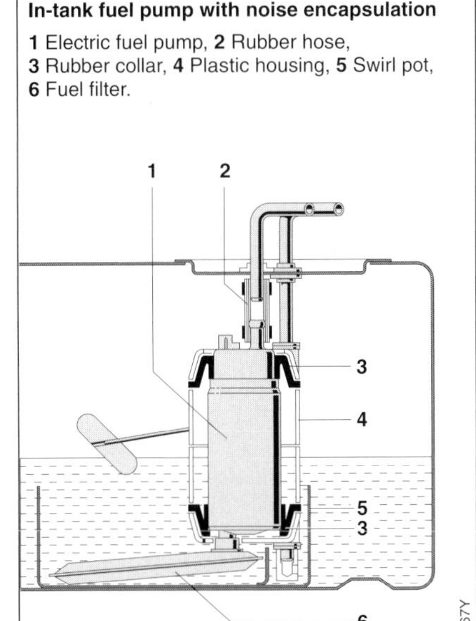

Fig. 4

Two-stage electric fuel pump, in-tank version. With side-channel pump (pre-stage) and peripheral pump (main stage)

1 Input-end cover with intake connection, **2** Impeller wheel, **3** Side-channel pump, **4** Peripheral pump, **5** Pump housing, **6** Armature, **7** Check valve, **8** Output-end cover with pressure connection.

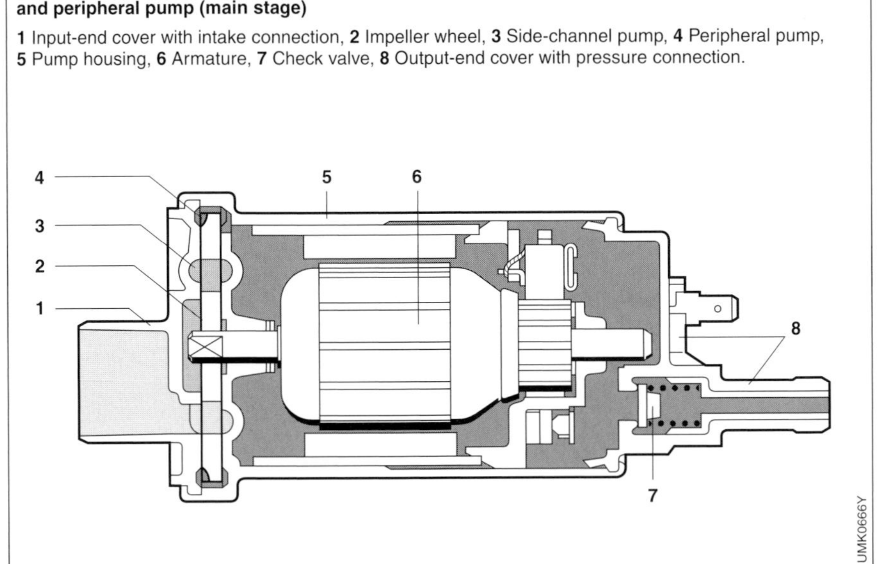

removed, this pump is distinguished by its excellent delivery characteristic and outstanding quietness.
A further advantage of the flow-pump principle lies in its fuel delivery, which is almost completely free of pressure pulsations. This fact also contributes to the pump's quiet operation.

Fuel filtration

Fuel-borne contaminants and impurities can prevent the injectors and fuel-pressure regulator from operating correctly. A filter is therefore fitted in the fuel line between the electric fuel pump and the central injection unit, preferably in a position underneath the vehicle protected against stone-throw.

Fuel filter

The filter's paper element has a mean pore size of 10 µm and comprises a paper tube with sealing rim. In order to completely separate the filter element's contaminated side from its clean side, the sealing rim is welded to the inside of the housing, which is made of impact-resistant plastic material. The paper tube is held axially by a plug and by support ribs in the filter cover (Fig. 7).
Depending on fuel-contamination levels and filter volume, filter service life is usually between 30,000 and 80,000 km.

Fuel-pressure control

It is the task of the fuel-pressure regulator to maintain the differential between line pressure and the local pressure at the injector nozzle constant at 100 kPa. In the Mono-Jetronic, the pressure regulator is an integral part of the central injection unit's hydraulic circuit.

Fig. 6

Components of the two-stage electric fuel pump

a Input-end cover (seen from the impeller wheel), **b** Impeller wheel, **c** Pump housing (seen from the impeller wheel).
1 Venting valve, **2** Degassing vent, **3** Inlet port for the side channel, **4** Side channel (pre-stage), **5** Peripheral channel (main stage), **6** Blade ring for side-channel pump (pre-stage), **7** Blade ring for side-channel pump (main stage), **8** Outlet port for peripheral channel

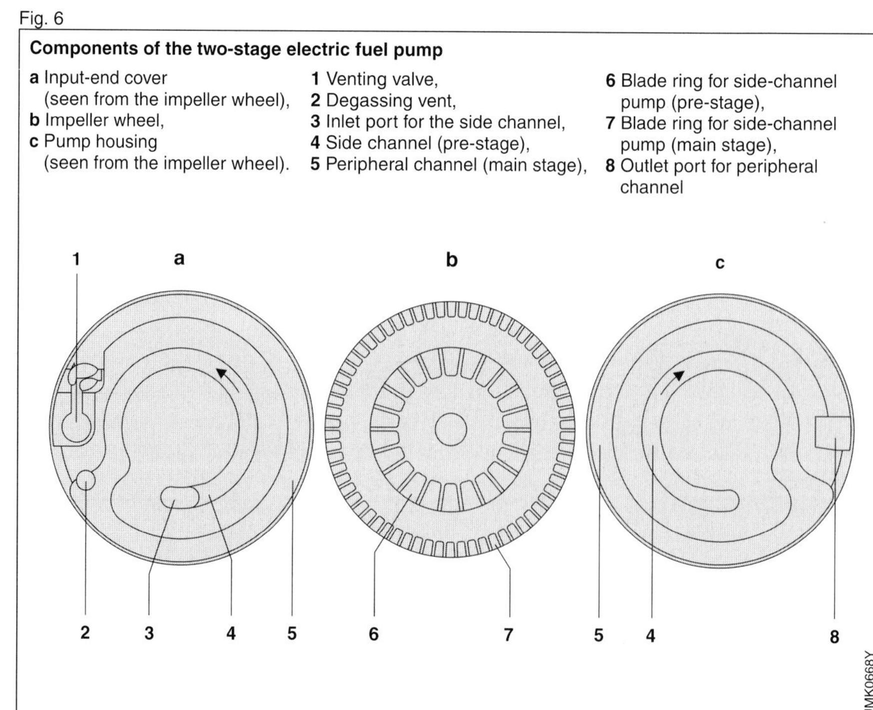

Fuel-pressure regulator

A rubber-fabric diaphragm divides the fuel-pressure regulator into the lower (fuel) chamber and the upper (spring) chamber. The pressure from the helical spring is applied to the diaphragm. A movable valve plate which is connected to the diaphragm through the valve holder is pressed onto the valve seat by spring pressure (flat-seat valve).

When the force generated by the fuel pressure against the diaphragm's area exceeds the opposing spring force, the valve plate is lifted slightly from its seat and the excess fuel can flow back to the tank through the open valve. In this state of equilibrium, the differential pressure between the upper and lower chambers of the pressure regulator is 100 kPa.

Vents maintain the spring chamber's internal pressure at levels corresponding to those at the injector nozzle. The valve-plate lift varies depending upon the delivery quantity and the actual fuel quantity required.

The spring characteristics and the diaphragm area have been selected to provide constant, narrow-tolerance pressure over a broad range of fuel-delivery rates. When the engine is switched off, fuel delivery also stops. The check valve in the electric fuel pump then closes, as does the pressure-regulator valve, maintaining the pressure in the supply line and in the hydraulic section for a certain period (Fig. 8).

Because this design configuration effectively inhibits the formation of vapor bubbles which can result from fuel-line heat build-up during pauses in operation, it helps ensure trouble-free warm starts.

Fig. 7

Fuel filter

1 Filter cover, **2** Sealing rim, **3** Filter housing, **4** Plug, **5** Support ribs, **6** Paper element, **7** Paper tube.

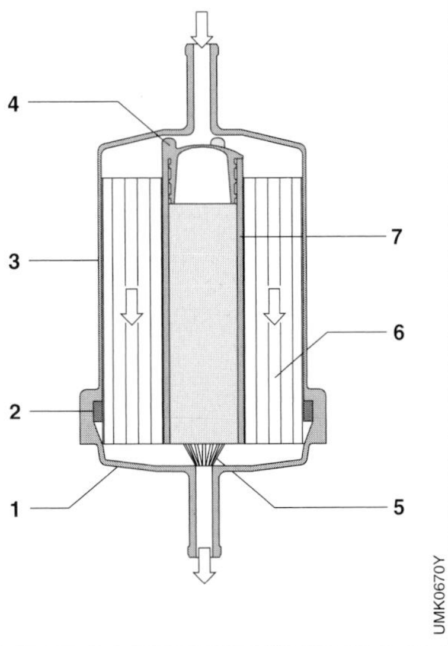

UMK0670Y

Fig. 8

Fuel-pressure regulator

1 Venting ports, **2** Diaphragm, **3** Valve holder, **4** Compression spring, **5** Upper chamber, **6** Lower chamber, **7** Valve plate.

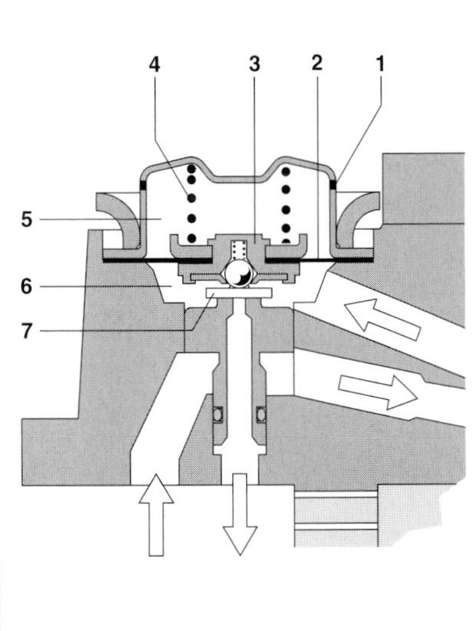

Evaporative-emissions control

In order to reduce the emissions of hydrocarbon compounds, legislation in a number of countries forbids the escape of fuel vapors from the fuel tank into the atmosphere.

Vehicles must be equipped with an evaporative-emissions control system in which the fuel tank is connected to a carbon canister. The carbon in the canister absorbs the fuel contained in the vapors passed through it. In order to transport the fuel from the canister to the engine, the fresh air drawn in by the engine is passed through the canister to reabsorb the fuel. From the canister, the intake air which is now enriched with hydrocarbons is passed through the intake manifold to the engine for combustion (Fig. 9).

Carbon canister

The dimensions of the carbon canister have been selected to maintain a state of equilibrium between the quantity of fuel absorbed and the quantity removed by the fresh air passing through the canister. To minimize canister size, the regenerative air stream is maintained at the highest possible level under all operating conditions (from idle to full throttle).

The level of the regeneration air flow is essentially determined by the difference between intake-manifold and ambient pressure. Because the differential is considerable at idle, the flow volume must be minimized to prevent driveability problems. At higher engine load factors, conditions are exactly the opposite because the regeneration-gas flow may be quite high, although the available pressure difference is only slight.

An ECU-controlled canister-purge valve provides precise metering of the fuel vapor flow.

Fig. 9

Evaporative-emissions control system

1 Line from fuel tank to carbon canister, **2** Carbon canister, **3** Fresh air, **4** Canister-purge valve, **5** Line to intake manifold, **6** Throttle valve.

p_S Intake-manifold pressure, p_u Atmospheric pressure, Δp Difference between intake-manifold pressure and atmospheric pressure.

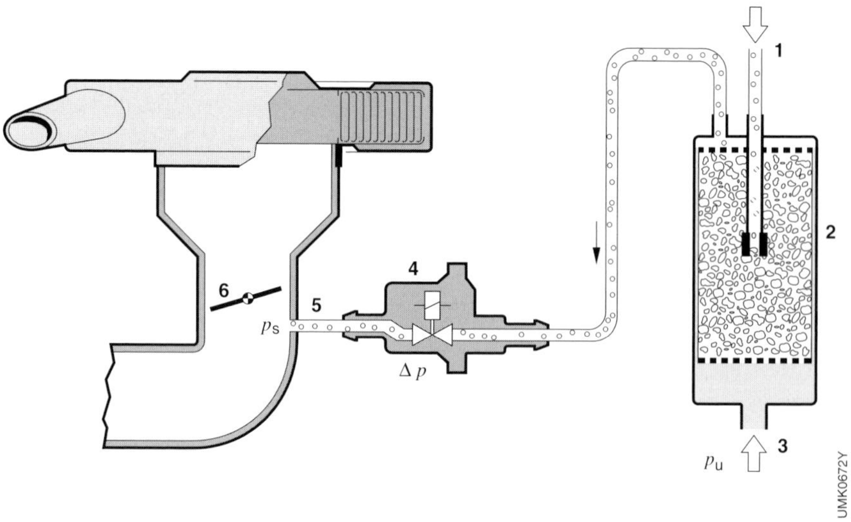

Acquisition of operating data

Sensors monitor all essential operating data to furnish instantaneous information on current engine operating conditions. This information is transmitted to the ECU in the form of electric signals, which are converted to digital form and processed for use in controlling the various final-control elements.

Air charge

The system derives the data required for maintaining the required air-fuel mixture ratio by monitoring the intake-air charge for each cycle. Once this air mass (referred to as air charge in the following) has been measured, it is possible to adapt the injected fuel quantity by varying injection duration.
With the Mono-Jetronic, the air charge is determined indirectly, using coordinates defined by the throttle-valve angle α and engine speed n. For this design to function, the relationship between throttle-valve angle and flow area within the throttle-valve housing must be maintained within very close tolerances on all production units.
The driver controls the engine's intake-air stream with the accelerator pedal, this determines throttle opening and load factor. The throttle-valve angle α is registered by the throttle-valve potentiometer. Engine speed n and intake-air density supplement the throttle-valve position α as additional variables for determining the intake-air mass.

Air charge as a function of α and n is determined for a given engine on the engine dynamometer. Figure 10 shows a typical set of curves for an engine program; it illustrates the relative air-charge factors for various throttle-valve apertures α and engine speeds n. If the engine response data are already available, and assuming constant air density, air charge can be defined precisely using α and n exclusively (α/n system).

The Mono-Jetronic throttle-valve assembly is an extremely precise air-measuring device and provides the ECU with a very accurate signal for the

Fig. 10

Engine map

Relative air charge as a function of engine speed n and throttle-valve angle α.
x Relative change in air-charge.

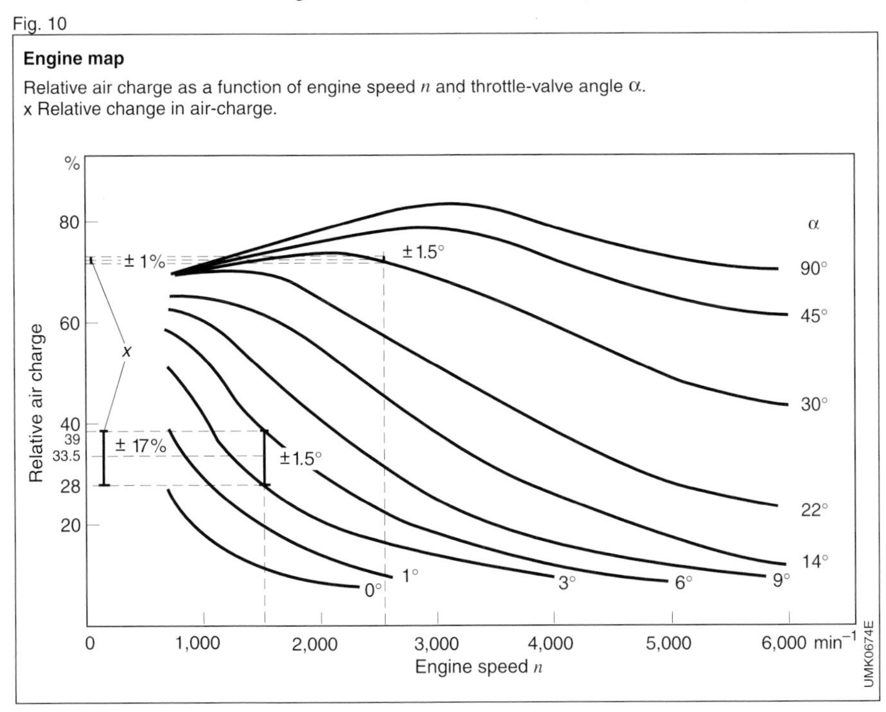

throttle-valve angle. The engine-speed information is provided by the ignition system. Because the differential between internal fuel pressure within the injector and air pressure at the nozzle is maintained at a constant level, injected fuel quantities are determined solely by the length of time (injection duration) the injector remains open for each triggering pulse.
This injection duration must be proportional to the monitored air charge to maintain a specific air-fuel ratio. In other words: The injection duration is a direct function of α and n. In the Mono-Jetronic, this relationship is governed by a Lambda program map with input variables α and n. The system is programmed to compensate for fluctuations in air density, which is a function of the intake-air temperature and the air pressure. The intake-air temperature, measured as the air enters the central injection unit, provides the ECU with the basic data required to determine the corresponding correction factor.

To enable it to comply with the stringent US exhaust-gas regulations, the Mono-Jetronic is always equipped with the Lambda closed-loop control, designed to maintain the air-fuel ratio at precisely λ = 1. In addition, the Lambda closed-loop control is used to implement adaptive mixture adaptation, in other words, the system learns to adapt itself to the changing operating conditions.
Correction factors for variations in atmospheric pressure (especially those associated with altitude changes) are supplemented by correction factors designed to compensate for differences in production tolerances and ongoing wear. When the engine is switched off, the system stores the correction factors so that they are effective immediately once the engine is started again.
This system of indirect determination of intake-air mass – with control based on the α/n control parameters – operates with adaptive mixture control and superimposed Lambda closed-loop control to accurately maintain a constant mixture, without any need for direct measurement of air mass.

Throttle-valve angle

The ECU employs the throttle-valve angle α to calculate the throttle-valve's position and angular velocity. Throttle-valve position is an important input parameter for determining intake-air volume, the basic factor in calculating injection duration. When the throttle is closed, the idle switch provides the throttle-valve actuator with a supplementary position signal.
Data on the throttle valve's angular velocity are used mainly for the transition compensation. The resolution of the α signal is determined by the air-charge measurement. To ensure good driveability and low emissions, the resolution of the air-charge measurement and injection duration must take place in the smallest-possible digital steps (quantization), to maintain the air-fuel ratio within a tolerance range of 2 %.

The program range that displays the largest intake-charge variations relative to α is defined by minimal apertures α and low engine speeds n, i.e., at idle and low part load. Within this range – as illustrated in Figure 10 – a change of ±1.5° in throttle-valve angle shifts the air-charge/lambda factor by ±17 %, whereas outside this range, with higher throttle-valve angles, a similar increment has an almost negligible effect. This means that a high angular resolution is necessary at idle and low part load.

Throttle-valve potentiometer

The potentiometer wiper arm is fastened to the throttle-valve shaft. The potentiometer resistance tracks and the electrical connection are on a plastic plate screwed onto the underside of the fuel-injection assembly. Power supply is from a stabilized 5 V source.
The required high level of signal resolution is achieved by distributing the throttle-valve angle for the range between idle and full-throttle between two resistance tracks.

Voltage drop is linear across each section of track. In parallel with each resistance track is a second (collector) track. The resistance tracks and the collector tracks are manufactured using thick-film techniques.
The wiper arm carries four wipers, each of which contacts one of the potentiometer tracks. The wipers for each resistance and collector-track pair are connected, the signal from the resistance track is thus transmitted to the collector track (Fig. 11).

Track 1 covers the angular range from 0° ... 24°, and track 2 the range from 18° ... 90°. The angle signals (α) from each track are converted separately in the ECU, each in its own analog/digital converter circuit. The ECU also evaluates the voltage ratios, using this data to compensate for wear and temperature fluctuations at the potentiometer. The potentiometer plate features a seal ring in a circumferential groove to prevent moisture and contamination from entering the unit. The potentiometer chamber is connected to the atmosphere by means of a venting device.

Engine speed
The engine-speed information required for α/n control is obtained by monitoring the periodicity of the ignition signal. These signals from the ignition system are then processed in the ECU. The ignition signals are either T_D pulses already processed by the ignition trigger box, or the voltage signal available at Terminal 1 (U_S) on the low-voltage side of the ignition coil. At the same time, these signals are also used for triggering the injection pulses, whereby each ignition pulse triggers an injection pulse (Fig. 12).

Engine temperature
Engine temperature has a considerable influence on fuel consumption. A temperature sensor in the engine coolant circuit measures the engine temperature and provides the ECU with an electrical signal.

Engine-temperature sensor
The engine-temperature sensor consists of a threaded sleeve with an integral NTC semiconductor resistor (NTC = Negative Temperature Coefficient).

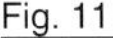
Fig. 11

Throttle-valve potentiometer

a Housing with wiper, **b** Housing cover with potentiometer tracks.
1 Bottom section of the central injection unit, **2** Throttle-valve shaft, **3** Wiper arm, **4** Wiper,
5 Resistance track 1, **6** Collector track 1, **7** Resistance track 2, **8** Collector track 2, 9 Toroidal seal ring.

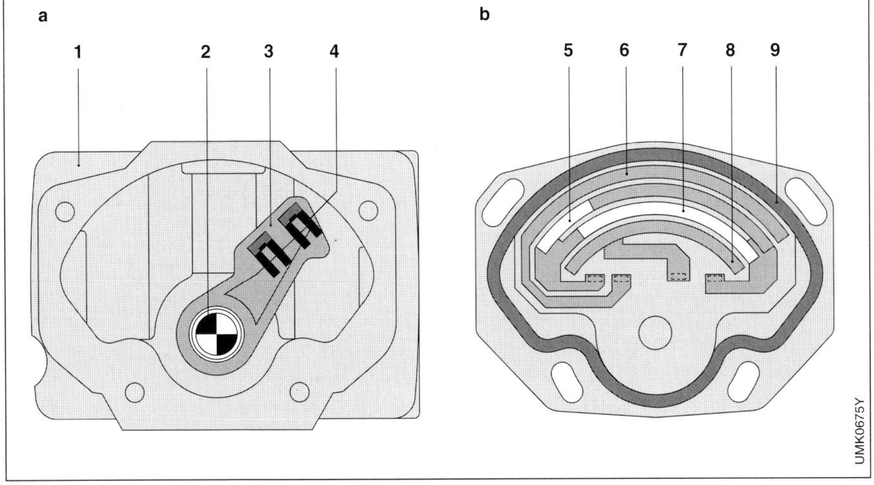

The sensor's resistance changes with temperature, and the ECU evaluates this change (Fig. 13).

Intake-air temperature

Intake-air density varies according to temperature. To compensate for this influence, a temperature sensor on the intake side of the throttle body monitors the temperature of the intake air for transmission to the ECU.

Air-temperature sensor

The air-temperature sensor incorporates a NTC resistor element. So that changes in intake-air temperature can be registered immediately, the resistor protrudes from the end of a trunk-shaped moulding into the area of maximum air-flow speed. The 4-pole plug-and-socket connection includes the plug for the fuel injector (Fig. 14).

Operating modes

Idle and full-throttle must be registered accurately so that the injected fuel quantity can be optimized for these operating modes, and in order that full-load enrichment and overrun fuel cutoff can function correctly.

The idle operating mode is registered from the actuated idle contact of a switch in the throttle-valve actuator, which is closed by a small plunger in the actuator shaft (Fig. 15). The ECU recog-nizes full-throttle operation based on the electrical signal from the throttle-valve potentiometer.

Battery voltage

The solenoid fuel injector's pickup and release times vary according to battery voltage. If system voltage fluctuates during operation, the ECU adjusts the injection timing to compensate for delays in injector response times.

In addition, the ECU responds to the low system voltages encountered during starting at low temperatures by extending injection duration.

The extended duration compensates for voltage-induced variations in the pumping characteristic of the electric fuel pump, which does not achieve maximum system pressure under these conditions.

The ECU receives the battery voltage in the form of a continuous signal transmitted to the microprocessor via the A/D converter.

Fig. 12

Engine-speed signals from the ignition system

1 Ignition distributor, **2** Trigger box, **3** Ignition coil, n Engine speed, T_D Conditioned pulses from ECU, U_S Voltage signal.

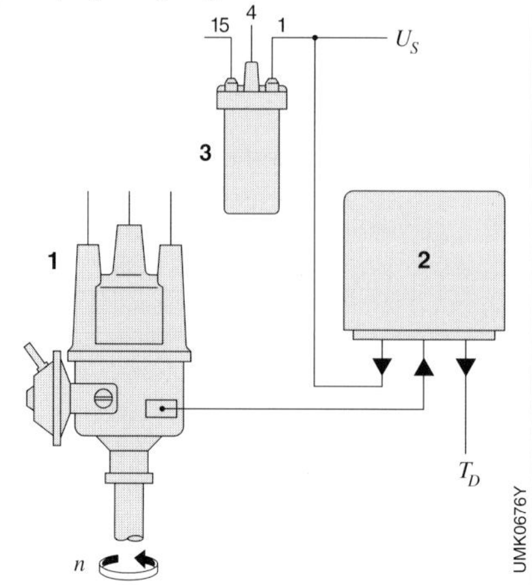

Fig. 13

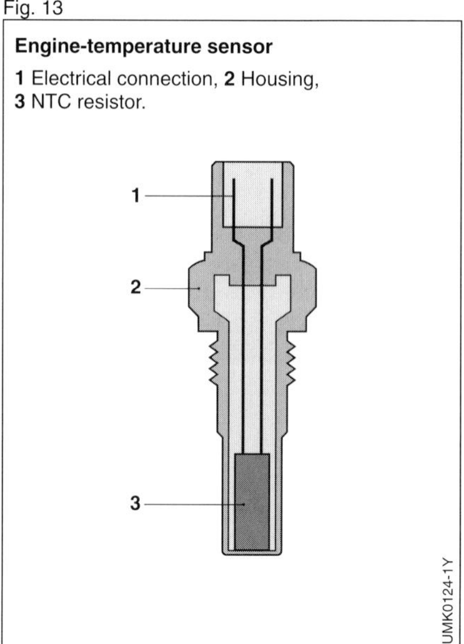

Engine-temperature sensor

1 Electrical connection, **2** Housing, **3** NTC resistor.

Control signals from the air-conditioner and/or automatic transmission

When the air conditioner is switched on, or the automatic transmission is placed in gear, the resulting engine load would normally cause the idle speed to drop. To compensate, the air-conditioner modes "air-conditioner ON" and "compressor ON," as well as the "Drive" position on the automatic gearbox, are registered by the ECU as switching signals. The ECU then modifies the idle-speed control signal to compensate. It may be necessary to increase the idle speed to ensure that the air conditioner continues to operate effectively, and reductions in idle speed are often required when "Drive" is selected on automatic-transmission vehicles.

Mixture composition

Due to the exhaust-gas aftertreatment using a three-way catalytic converter, the correct air-fuel mixture composition must be precisely maintained. A Lambda oxygen sensor in the exhaust-gas flow provides the ECU with an electric signal indicating the current mixture composition. The ECU then uses this signal for the closed-loop control of the mixture composition to obtain a stoichiometric ratio. The Lambda oxygen sensor is installed in the engine's exhaust system at a position which ensures that it is kept at the temperature required for correct functioning across the complete engine operating range.

Lambda oxygen sensor

The Lambda oxygen sensor protrudes into the exhaust-gas stream and is designed so that the outer electrode is surrounded by the exhaust gas while the inner electrode remains in contact with the atmosphere (Fig. 16).
The Lambda sensor consists of a special-ceramic housing, the surfaces of which are covered by gas-permeable platinum electrodes. The sensor's operation is based on the fact that the ceramic material is porous and permits the diffusion of the oxygen present in the air (solid electrolyte). At higher temperatures the ceramic becomes conductive, and if the oxygen concentration at one of the electrodes differs from that at the other, a voltage is generated between them. At a stoichiometric air-fuel ratio $\lambda = 1.0$, a jump takes place in the sensor's output voltage. This voltage is used as the measuring signal (Fig. 17).

Fig. 14

Intake-air temperature sensor

1 Intake air, **2** Trunk-shaped moulding, **3** Guard, **4** NTC resistor, **5** Fuel injector.

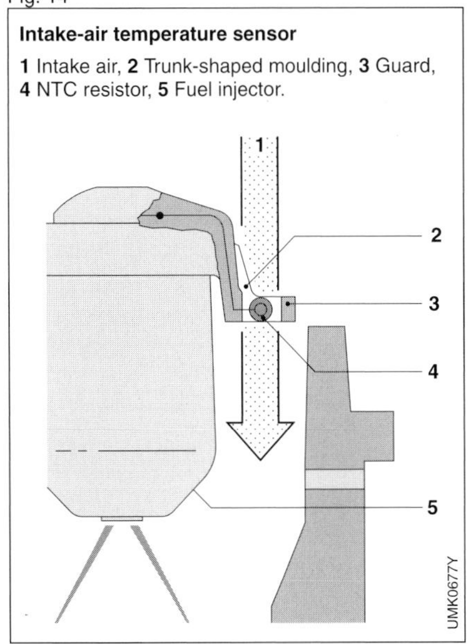

Fig. 15

Idle switch

1 Actuation by throttle-valve lever, **2** Idle contact, **3** Electrical connections.

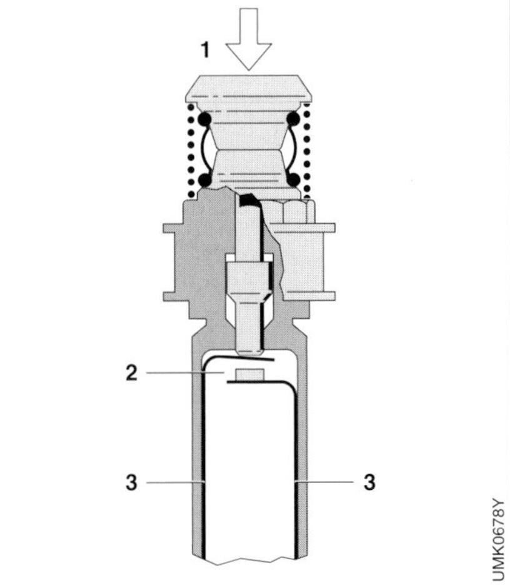

The sensor ceramic body is held in a threaded mounting and provided with a protective tube and electrical connections. The sensor-ceramic surface is covered by a microporous platinum layer which on the one hand has a decisive effect on the sensor's characteristic while on the other serving as an electrical contact. A highly adhesive and highly porous ceramic coating has been applied over the platinum layer at the end of the sensor ceramic which is in contact with the exhaust gas. This serves to protect the platinum layer against erosion due to the solid particles in the exhaust gas.
The exhaust-gas end of the sensor is provided with a protective tube to keep combustion deposits away from the sensor ceramic. This tube is slotted, so that the exhaust gases and the combustion particles they contain cannot reach the sensor ceramic. In addition to the mechanical protection provided by this tube, it also effectively reduces the temperature changes at the sensor during transitions from one operating mode to the other.
A protective metal sleeve is fitted over the electrical contact end of the sensor. As well as serving as the support for a disc spring, it has a bore which ensures pressure compensation in the sensor interior. The electrical connection cable exits the sensor through an insulating sleeve.
The voltage and internal resistance of the sensor depend upon its temperature. Efficient closed-loop control is possible with exhaust-gas temperatures above 350 °C (unheated sensor) and above 200 °C (heated sensor).

Heated Lambda oxygen sensor

To a large extent, the design principle of the heated Lambda sensor (Fig. 18) is identical to that of the unheated version. The sensor's active ceramic body is heated internally by a ceramic heating element to maintain its temperature above the 350°C operating limit, even if the exhaust gas is cooler. The heated Lambda sensor features a protective tube with a reduced flow opening. Among other things, this prevents the sensor ceramic cooling off when the exhaust gas is cold.
The advantages of the heated Lambda sensor are reliable closed-loop control at low exhaust-gas temperatures (e.g., at idle); almost complete independence from variations in exhaust-gas tem-

Fig. 16

Installation of Lambda oxygen sensor in exhaust pipe (schematic)

1 Sensor ceramic, **2** Electrodes, **3** Contact, **4** Housing contact, **5** Exhaust pipe, **6** Ceramic protective layer (porous), **7** Exhaust gas, **8** Air.

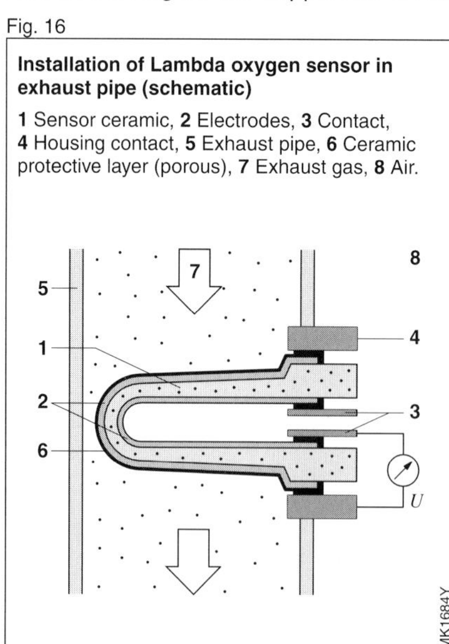

Fig. 17

Voltage characteristic of the Lambda oxygen sensor at a working temperature of 600 °C

a Rich mixture (air deficiency), **b** Lean mixture (excess air).

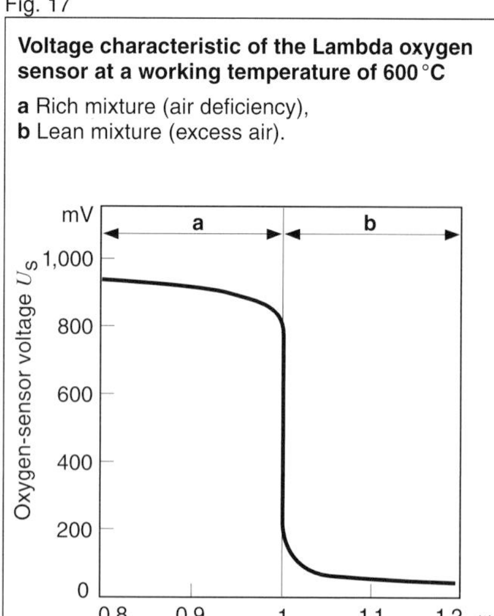

perature; rapid response to minimize delays before effective Lambda control can be implemented after cold starts; low exhaust emissions due to improved sensor reaction time; and greater flexibility in selecting the installation location (because the sensor does not depend upon the exhaust gas to reach operating temperature).

Processing of operating data

The ECU processes the engine operating data received from the sensors. From this data, it uses the programmed ECU functions to generate the triggering signals for the fuel injector, the throttle-valve actuator, and the canister-purge valve.

Electronic control unit (ECU)

The ECU is housed in a fiberglass-reinforced, polyamide plastic casing. To insulate it from the engine's heat, it is located either in the passenger compartment or the ventilation plenum chamber between the engine and passenger compartments.

All of the ECU's electronic components are installed on a single printed-circuit board. The output amplifiers and the voltage regulator responsible for maintaining 5V supply to the electronic components are installed on heat sinks for better thermal dissipation.

A 25-pin plug connects the control unit to the battery, the sensors and the actuators (final-control elements).

Fig. 18

Heated Lambda oxygen sensor

1 Sensor housing, **2** Ceramic support tube, **3** Electrical connections, **4** Protective tube with slots, **5** Active sensor ceramic, **6** Contact element, **7** Protective sleeve, **8** Heating element, **9** Clamp-type connections for heating element, **10** Disc spring.

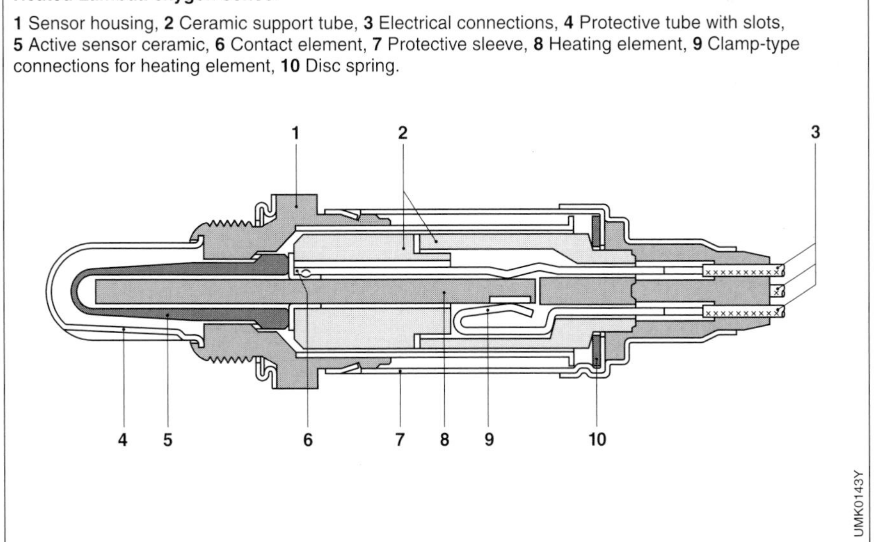

Analog-digital converter (A/D)
The signals from the throttle-valve potentiometer are continuous analog signals; as are the Lambda-sensor signal; the engine-temperature signal; the intake-air signal; the battery voltage; and a voltage-reference signal generated in the ECU. These analog signals are converted to data words by the analog-digital converter (A/D) and transmitted to the microprocessor via the data bus. An analog-digital input is used so that, depending upon the input voltage, various data records in the read memory can be addressed (data coding). The engine-speed signal provided by the ignition is conditioned in an integrated circuit (IC) and transmitted to the microprocessor. The engine-speed signal is also used to control the fuel-pump relay via an output stage.

Microprocessor
The microprocessor is the heart of the ECU (Fig. 19). It is connected through the data and address bus with the programmable read-only memory (EPROM) and the random-access memory (RAM). The read memory contains the program code and the data for the definition of operating parameters. In particular, the random-access memory serves to store the adaptation values (adaptation: adapting to changing conditions through self-learning). This memory module remains permanently connected to the vehicle's battery to maintain the adaptation data contained in the random-access memory when the ignition is switched off.

A 6 MHz quarz oscillator provides the stable basic clock rate needed for the arithmetic operations. A signal interface adapts the amplitude and shape of the

Fig. 19

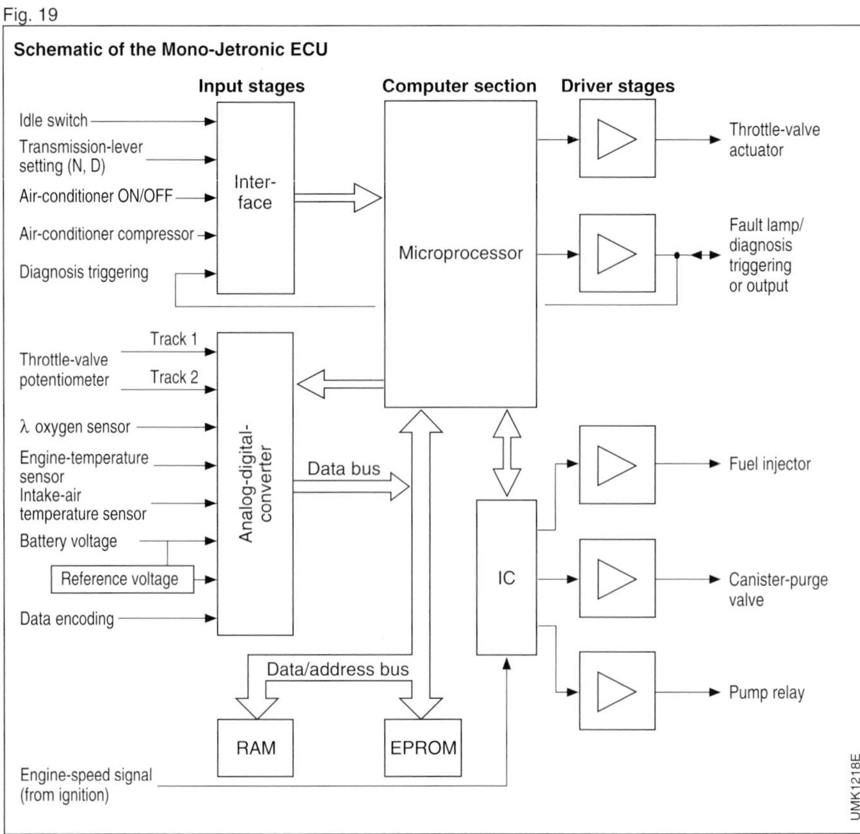

control signals before transmitting them to the microprocessor for processing. These signals include the idle-switch setting, diagnosis activation, the position of the selector lever on automatic-transmission vehicles (Neutral, Drive), and a signal for "Air-conditioner ON" and "Compressor ON/OFF" in vehicles with air-conditioning.

Driver stages
A number of different driver stages generate the control signals for the fuel injector, the throttle-valve actuator, the canister-purge valve, and the fuel-pump relay. A fault lamp is installed in some vehicles to warn the driver in case of sensor or actuator faults. The fault lamp's output also serves as an interface for diagnosis activation and read-outs.

Lambda program map
The Lambda map is used for precise adaptation of the air-fuel ratio at all static operating points once the engine is warm. This map is stored electronically in the digital circuit section of the ECU; the reference data are determined empirically through tests on the engine dynamometer. For a Lambda closed-loop-controlled engine-management concept such as the Mono-Jetronic, this testing is employed to determine optimum injection timing and duration for a specific engine under all operating conditions (idle, part or full-throttle). The resulting program consistently maintains an ideal (stoichiometric) air-fuel mixture throughout the operating range.
The Mono-Jetronic Lambda program map consists of 225 control coordinates; these are assigned to 15 reference coordinates for the parameters throttle-valve angle α and n for engine speed n. Because the α/n curves are extremely non-linear, necessitating high resolution accuracy at idle and in the lower part-load range, the data points are situated very closely together in this area of the map (Fig. 20). Control coordinates located between these reference coordinates are determined in the ECU using linear interpolation.

Lambda map
Injection duration as a function of engine speed and throttle-valve angle.

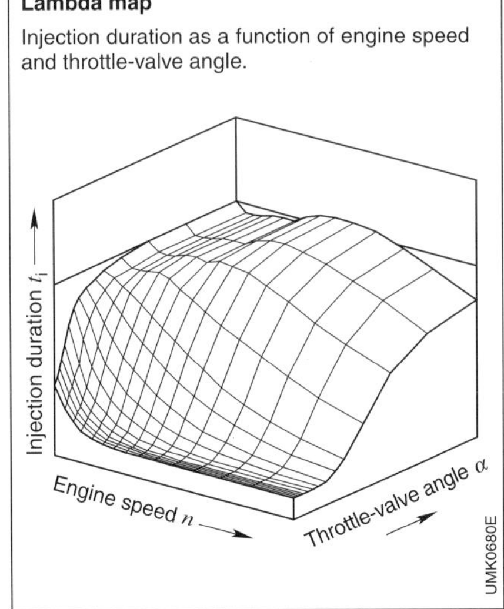

Fig. 20

Because the Lambda map is designed for the engine's normal operating and temperature range, it becomes necessary to correct the basic injection timing when engine temperatures deviate or when special operating conditions are encountered.

If the ECU registers deviations from λ = 1 in the signals from the Lambda sensor, and as a result is forced to correct the basic injection duration for an extended period of time, it generates mixture correction values and stores them in an internal adaptation process. From then on, these values are effective for the complete map and are continually up-dated.
This layout ensures consistent compensation for individual tolerances and for permanent changes in the response characteristics of engine and injection components.

Fuel injection
The fuel-injection system must be able to accurately meter to the engine the minimum amount of fuel required (at idle or zero-load), as well as the maximum (at full throttle). The control coordinates for these conditions must be situated within the linear range on the injector curves (Fig. 21).

One of the Mono-Jetronic's most important assignments is the uniform distribution of the air-fuel mixture to all cylinders. Apart from intake-manifold design, the distribution depends mainly upon the fuel injector's location and position, and the quality of its air-fuel mixture preparation. The ideal fuel-injector posi-tion within the Mono-Jetronic housing was determined in the research and development phase. Special adaptations for operation in individual engines are not required.

The central injection unit's housing is centered in the intake-air flow by a special bracket, and has been designed for maximum aerodynamic efficiency. The housing contains the injector, which is installed directly above the throttle valve for intensive mixing of the injected fuel with the intake air. To this end, the injector finely atomizes the fuel and injects it in a cone-shaped jet between the throttle valve and the throttle-valve housing into the area with the highest airflow speed.

The fuel injector is sealed-off to the outside by seal rings. The central injection unit is closed off at the top by a semicircular plastic cap which not only contains the injector's electrical connections but also ensures correct axial positioning.

Injector

The fuel injector (Fig. 22) comprises the valve housing and the valve group. The valve housing contains the solenoid winding and the electrical connections. The valve group includes the valve body which holds the valve needle and its solenoid armature.

When no voltage is applied to the solenoid winding, a helical spring assisted by the primary system pressure forces the valve needle onto its seat. With voltage applied and the solenoid energized, the needle lifts about 0.06 mm (depending upon valve design) from its seat so that fuel can exit through an annular gap. The pintle on the front end of the valve needle projects from the valve-needle bore, and its shape ensures excellent fuel atomization.

The size of the gap between the pintle and the valve body determines the valve's static fuel quantity. In other words, the maximum fuel flow with the valve permanently open. The dynamic fuel

Fig. 21

Fuel-injector characteristic curve

At an engine speed of 900 min^{-1} (corresponds to a 33 ms injection pulse train).
1 Voltage-dependent fuel-injector delay time,
2 Non-linear characteristic section,
3 Injection-duration section at idle or no-load operation.

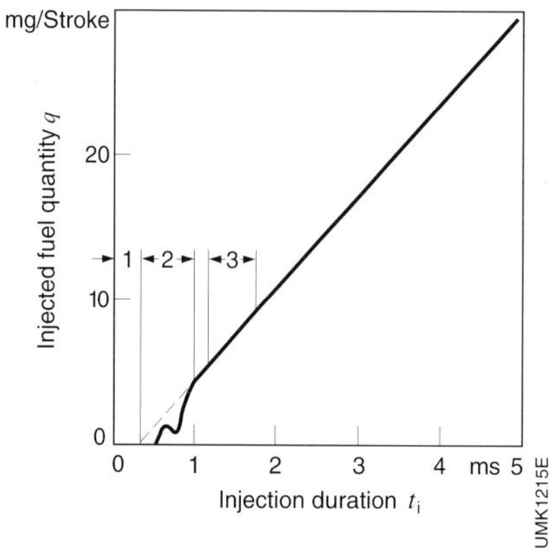

mg/Stroke
Injected fuel quantity q
20
10
0
1 2 3
0 1 2 3 4 ms 5
Injection duration t_i
UMK1215E

Fig. 22

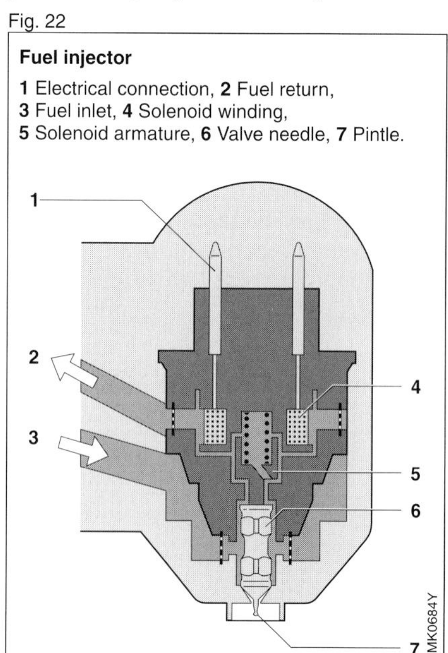

Fuel injector

1 Electrical connection, **2** Fuel return,
3 Fuel inlet, **4** Solenoid winding,
5 Solenoid armature, **6** Valve needle, **7** Pintle.

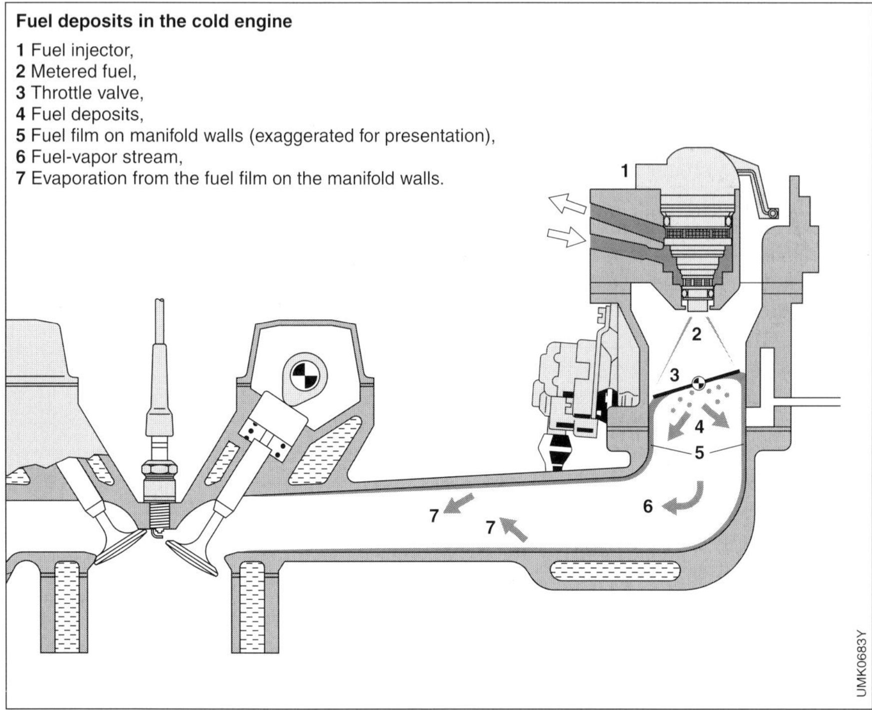

Fuel deposits in the cold engine
1 Fuel injector,
2 Metered fuel,
3 Throttle valve,
4 Fuel deposits,
5 Fuel film on manifold walls (exaggerated for presentation),
6 Fuel-vapor stream,
7 Evaporation from the fuel film on the manifold walls.

Fig. 23

quantity injected during intermittent operation depends also on the valve spring, the valve-needle mass, the magnetic circuit and the ECU driver stage. Because the fuel pressure is constant, the amount of fuel actually injected by the fuel injector depends solely upon the valve's opening time (injection duration). Due to the frequency of the injection pulse train (every ignition pulse triggers an injection pulse) the fuel injector must feature extremely short switching times. Its pick-up and release times are kept to below one millisecond by the low mass of armature and needle, as well as by the optimized magnetic circuit. Precise metering of even the smallest amount of fuel is therefore guaranteed.

Mixture adaptation

Starting phase

When the engine is cold, effective fuel vaporization is inhibited by the following factors:

- cold intake air,
- cold manifold walls,
- high manifold pressure,
- low air-flow velocity in the intake manifold, and
- cold combustion chambers and cylinder walls.

These factors mean that part of the fuel metered to the engine condenses on the cold manifold walls and covers them with a film of fuel (Fig. 23).

To ensure that the fuel leaving the injector reaches the engine for combustion, this condensation process must be terminated as soon as possible. To do so, more fuel is metered to the engine when starting than would otherwise be needed for the combustion of the intake air. Being as the amount of fuel condensation mainly depends upon the intake-manifold temperature, the injection times for starting are specified by the ECU according to engine temperature (Fig. 24a).

The fuel condensation rate not only depends upon the manifold-wall temperature but also on the air velocity in the intake manifold. The higher the air-flow velocity, the less fuel is deposited on the manifold walls. The injection duration is reduced as engine speed increases (Fig. 25a).

In order to achieve very short starting times, the build-up of the fuel film on the manifold walls must take place as quickly as possible, in other words a large quantity of fuel must be metered to the engine in a short period. At the same time, precautions must be taken to prevent flooding the engine.

These demands are in opposition to each other, but are complied with by reducing the injection duration the longer engine cranking continues (Fig. 25b). The engine is considered to have started as soon as the so-called end-of-starting speed, which is dependent upon engine temperature, is exceeded (Fig. 24b).

Fig. 24

Corrections as a function of engine temperature. During the cranking phase, the post-start phase, and the warm-up phase

a Injection duration at start of cranking,
b Engine speed at end of cranking,
c Post-start factor, **d** Warm-up factor.

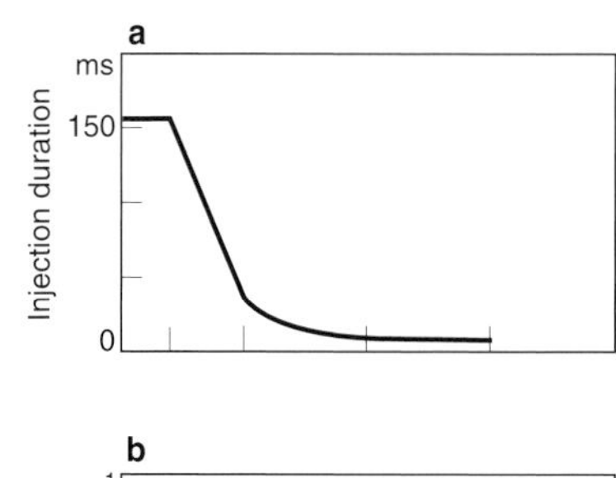

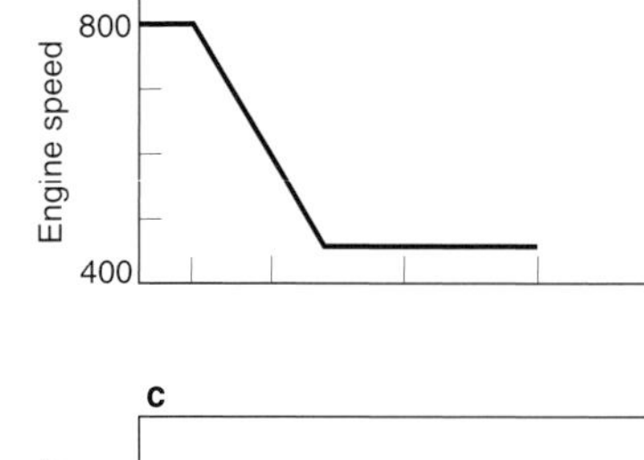

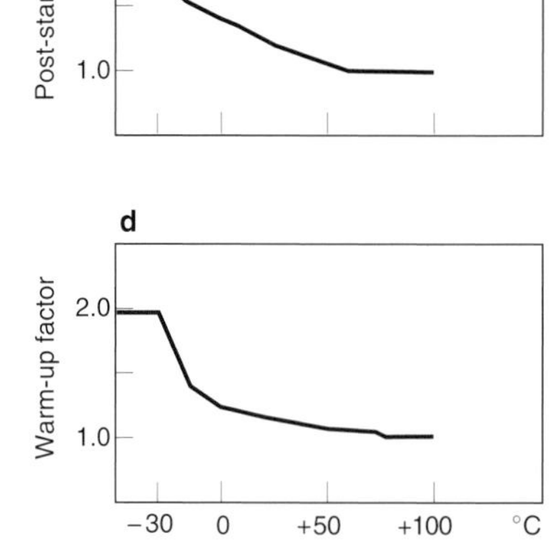

Post-start and warm-up phase

As soon as the engine has started, and depending upon throttle-valve position and engine speed, the fuel injector is triggered with the injection durations

Fig. 25

Reduction of injected-fuel quantity during starting

a Reduction as a function of speed,
b Reduction as a function of time.

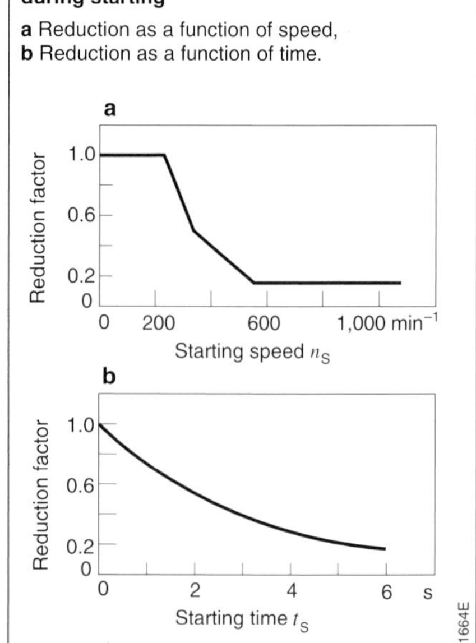

stored in the Lambda map. From this point on, and until engine operating temperature has been reached, fuel enrichment continues to be necessary due to the fuel condensation on the combustion chamber and manifold walls, which are still cold.
Immediately following a successful start, the enrichment must be increased briefly, but this increase is then followed by normal enrichment which depends solely on the engine temperature.
Two functions determine the engine's fuel requirement in the phase between end-of-start and reaching operating temperature:

– Post-start enrichment is stored as an engine-temperature-dependent correction factor. This post-start factor is used to correct the injection durations calculated from the Lambda map. The reduction of the post-start factor to the value 1 is time-dependent (Fig. 24c).
– The warm-up enrichment is also stored as an engine-temperature-dependent correction factor. Reduction of this factor to value 1 depends solely on engine temperature (Fig. 24d).
Both functions operate simultaneously; the injection durations calculated from the Lambda map are adapted with the post-start factor as well as with the warm-up factor.

Mixture adaptation as a function of intake-air temperature

The air mass required for the combustion is dependent upon the intake air's temperature. Taking a constant throttle-valve setting, this means that cylinder charge reduces along with increasing air temperature. The Mono-Jetronic central injection unit therefore is equipped with a temperature sensor which reports the intake-air temperature to the ECU which then corrects the injection duration, e.g., the injected fuel quantity, by means of an intake-air-dependent enrichment factor (Fig. 26).

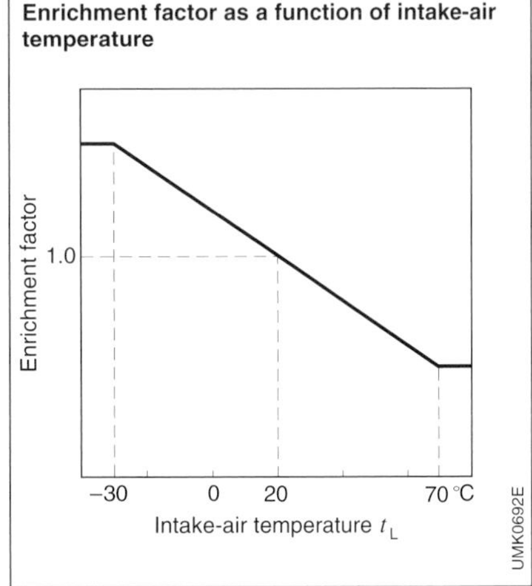

Fig. 26

Transition compensation

The dynamic mixture correction which is necessary to compensate for the load changes due to throttle-valve movements is provided by the transition compensation. This facility is required in order to achieve best-possible driveability and exhaust-gas behavior, and in the case of a single-point fuel-injection system its functional complexity is somewhat higher than with a multipoint system. This is due to the fact that mixture distribution is via the intake manifold in a single-point system, which means that for transition compensation three different factors must be taken into consideration with regard to the transportation of the fuel:

– Fuel vapor in the central injection unit or in the intake manifold, or which forms on the manifold walls due to the evaporation of the fuel film. This vapor is transported very quickly at the same speed as the intake air.
– Fuel droplets, which are transported at different speeds, but nevertheless at about the same speed as the intake-air. Some of these droplets are flung against the intake-manifold walls though, where they contribute to the evaporation of the fuel film.
– Liquid fuel, transported to the combustion chamber at reduced speed, stemming from the fuel film on the intake-manifold walls. There is a time lag in the

availability of this portion of the fuel for combustion.

At low intake-manifold pressures, that is at idle or lower part load, the fuel in the intake manifold is almost completely in vaporous form and there is practically no fuel film on the manifold walls. When the manifold pressure increases though, i.e. when the throttle valve is opened (or when speed drops), the proportion of fuel in the wall film increases. The result is that when the throttle valve is operated during a transition, the balance between the increase and decrease in the amount of fuel in the wall film is disturbed. This means that when the throttle valve is opened, some form of compensation is necessary in order to prevent the mixture leaning-off due to the increase in the amount of fuel deposited on the manifold walls. This compensation is provided by the acceleration-enrichment facility. Correspondingly, when the throttle valve is closed, the wall film reduces by releasing some of its fuel, and without some form of compensation this would lead to mixture enrichment at the cylinders during the transitional phase. Here, the deceleration lean-off facility comes into effect.

In addition to the tendency of the fuel to evaporate as a result of the intake-manifold pressure, the temperature also plays an important role. Therefore, when the intake manifold is still cold, or with low intake-air temperatures, there is a further increase in the amount of fuel held in the wall film.

In the Mono-Jetronic, complex electronic functions are applied to compensate for these dynamic mixture-transportation effects, and these functions ensure that the air-fuel ratio remains as near as possible to $\lambda = 1$ during transition modes. The acceleration enrichment and deceleration lean adjustment functions are based on throttle-valve angle, engine speed, intake-air temperature, engine temperature and the throttle-valve's angular velocity.

Acceleration enrichment and trailing-throttle lean adjustment are triggered when the throttle valve's angular velocity exceeds the respective trigger threshold. For acceleration enrichment, the trigger threshold is in the form of a characteristic curve based on the throttle valve's angular velocity, and for deceleration lean adjustment it is constant (Fig. 27).

Also based on the throttle valve's rate of travel are the two dynamic-response correction factors for acceleration enrichment and trailing-throttle lean adjustment. Both of these dynamic mixture-correction factors are stored as characteristic curves (Fig. 28).

To reduce the tendency for fuel to condense along the intake tracts, the intake manifold is heated by including it in the engine's cooling circuit. Meanwhile, a heat-riser warms the intake air to

Fig. 27

Trigger thresholds for transition compensation

1 Acceleration enrichment, **2** Deceleration lean-off. No triggering.

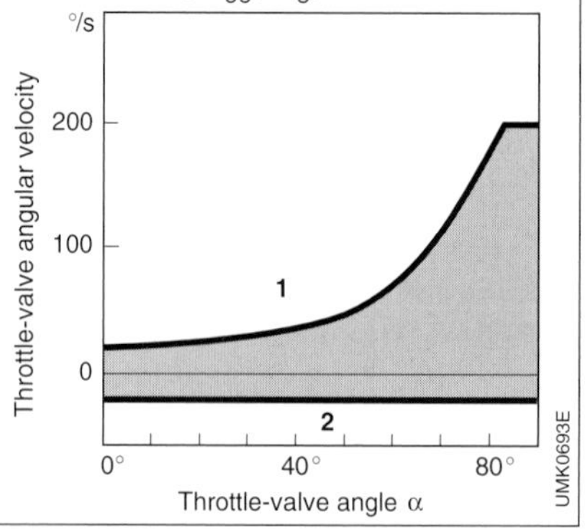

Fig. 28

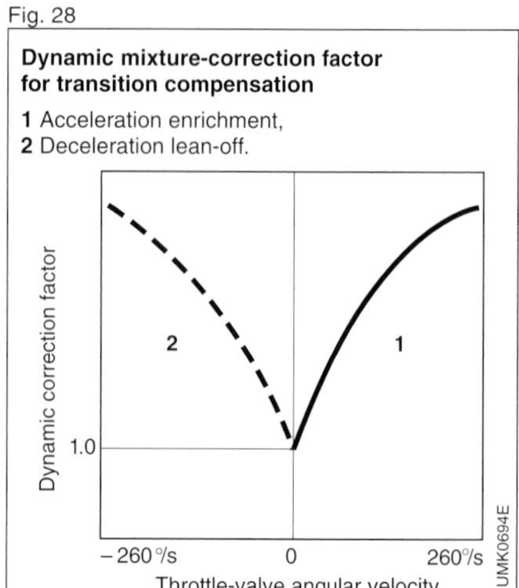

Dynamic mixture-correction factor for transition compensation

1 Acceleration enrichment,
2 Deceleration lean-off.

facilitate formation of a homogeneous air-fuel mixture. These influences are taken into account by using correction curves which modify the dynamic mixture-correction factors to compensate for variations in engine and intake-air temperature (Figures 29 and 30a).
To compensate for the variations in manifold vacuum and their effect on the rate at which fuel condenses along the intake tracts, yet another program containing supplementary evaluation factors adapts the dynamic correction factors in response to changes in throttle-valve angle and engine speed (Fig. 31).

When the throttle valve's angular velocity drops below one of the control thresholds, or if the ECU responds to the current sensor signals by prescribing a series of radical reductions in mixture-correction factor, the system adapts by phasing out the last dynamic correction factors for acceleration enrichment and trailing-throttle lean correction. The process periodicity coincides with that of the ignition pulse, and is based on an engine-temperature factor of less than 1. Each of the reduction factors, that is, acceleration enrichment and trailing-throttle lean correction, is defined in a separate program curve (Fig. 30b). The program's compensation mode thus functions as a comprehensive transition factor for adapting both injection timing and duration.

Fig. 29

Transition-compensation map

Weighting factors as a function of engine speed and throttle-valve angle.

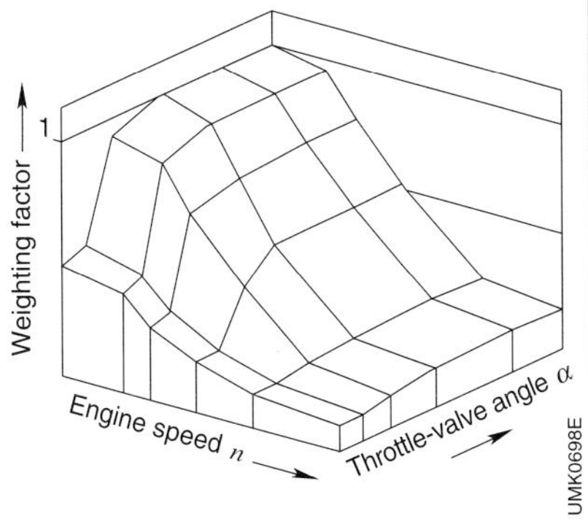

Because changes in load factor can be quite rapid relative to injection periodicity, the system is also capable of generating a supplementary injection pulse to provide additional compensation.

Fig. 30

Transition-compensation factors related to engine temperature

a Weighting factor, **b** Reduction factor.
1 Acceleration enrichment,
2 Deceleration lean-off.

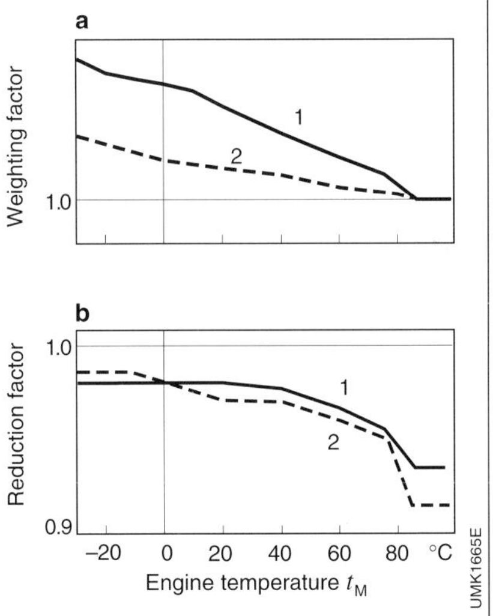

Fig. 31

Transition-compensation weighting factor related to intake-air temperature

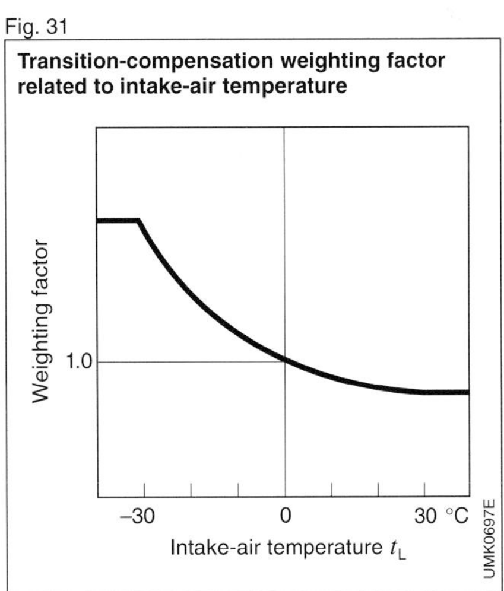

Lambda closed-loop control

The Lambda closed-loop control circuit maintains an air-fuel mixture of precisely $\lambda = 1$. The ECU receives continuous signals from the Lambda oxygen sensor located in the exhaust-gas stream. Using this signal, the ECU monitors the instantaneous level of the residual oxygen in the exhaust gas. The ECU employs these signals to monitor the current mixture ratio and to adjust injection duration as required.
The Lambda closed-loop control is superimposed on the basic mixture-control system. It ensures that the system is always optimally matched to the 3-way catalytic converter (Fig. 32).

Lambda closed control loop
1 Fuel, **2** Air, **3** Central injection unit, **4** Fuel injector, **5** Engine, **6** Lambda oxygen sensor, **7** Catalytic converter, **8** ECU with Lambda closed-loop control, **9** Exhaust gas.
U_{λ} Sensor voltage, U_{V} Injector triggering pulse.

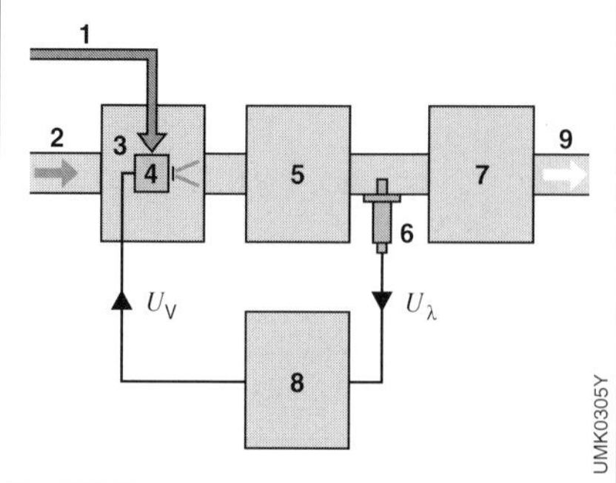

Fig. 32

Lambda closed-loop control circuit

Deviations from the stoichiometric air-fuel ratio are detected and corrected with the aid of the Lambda oxygen sensor. The control principle is based on the Lambda sensor measuring the level of residual oxygen in the exhaust gas. This residual oxygen is a measure for the composition of the air-fuel mixture supplied to the engine. The Lambda sensor is installed so that it extends into the exhaust gas and acts as a probe which delivers information as to whether the mixture is richer or leaner than $\lambda = 1$. If the mixture deviates from $\lambda = 1$, the sensor output voltage changes abruptly and this change is evaluated by the ECU control circuit. A high (approx. 800 mV) sensor output voltage indicates a mixture which is richer than $\lambda = 1$, and a low output signal (approx. 200 mV) a mixture which is leaner. The Lambda sensor's signal periodicity is illustrated in Figure 33. The Lambda control stage receives a signal for every transition from rich to lean mixture and from lean to rich.

The Lambda correction factor is applied to adjust injection duration. At Lambda values above 1.0 (low sensor voltage) the injection quantity is increased, at values below 1.0, it is decreased. When a sensor voltage jump occurs, in order to generate a correction factor as soon as possible, the air-fuel mixture is changed immediately by a given amount. The manipulated variable then follows a programmed adaptation function until the next Lambda-sensor volt-age jump takes place. The result is that within a range very close to $\lambda = 1$, the air-fuel mixture permanently changes its composition either in the rich or in the lean direction. If it were possible to adapt the Lambda map to the ideal value $\lambda = 1$, the manipulated variable for the Lambda control stage (the Lambda correction factor) would control permanently to the neutral value $\lambda = 1$.

Unavoidable tolerances make this impossible though, so the Lambda closed-loop control follows the deviations from the ideal value and controls each map point to $\lambda = 1$. In this manner, the fuel is metered so precisely that the air-fuel ratio is an optimum for all operating conditions. The system compensates for the effects of production tolerances and engine wear. Without this continuous, practically instantaneous adjustment of the mixture to $\lambda = 1$, efficient treatment of the exhaust gas by the downstream catalytic converter would be impossible.
The Lambda oxygen sensor needs a temperature above about 350 °C before it delivers a signal which can be evaluated. There is no closed-loop control below this temperature.

Mixture adaptation

The mixture-adaptation function provides separate, individual mixture adjustment for specific engine-defined operating environments. It also furnishes the mixture-control circuits with reliable compensation for variations in air density. The mixture adaptation program is designed to compensate for the effects of production tolerances and wear on engine and injection-system components.

The system must compensate for three variables:
– Influences due to air-density changes when driving at high altitudes (air-flow multiplicative influence).
– Influences related to vacuum leakage in the intake tract. Leakage downstream from the throttle valve can also vary as deposits reseal the affected areas (air-flow additive influence).
– Influences due to individual deviations in the injector response delay (injection-duration additive influence).

As there are certain map sectors where these influences have a very marked effect, the map is subdivided into three mixture-adaptation sectors:
– Air-density changes which have the same effect over the complete map.
The mixture-adaptation sector for the adaptation variable which takes the air density into account (air-flow multiplicative value) therefore applies to the complete map.
– Changes in the leakage-air rate have a pronounced effect at low air-flow rates (e.g. in the vicinity of idle). An additional adaptation value is therefore calculated in a second sector (airflow-additive value).
– Changes in the injected fuel quantity for every injection pulse are particularly effective when the injection duration is very short. A further adaptation value is therefore calculated in a third sector (injection-duration additive value).

The mixture-adaptation variable is calculated as follows:
The Lambda control-stage manipulated variable is a defined quantity. When a mixture fault is detected, this variable is changed until the mixture has been corrected to $\lambda = 1$. Here the mixture correction for the Lambda controller is defined by the deviation from the Lambda control value. For mixture adaptation, these Lambda controller parameters are evaluated using a weighting factor before being added to the adaptation variables for the individual ranges. The adaptation variable thus varies in fixed increments, with the size of the increments being proportional to the current Lambda

Fig. 33

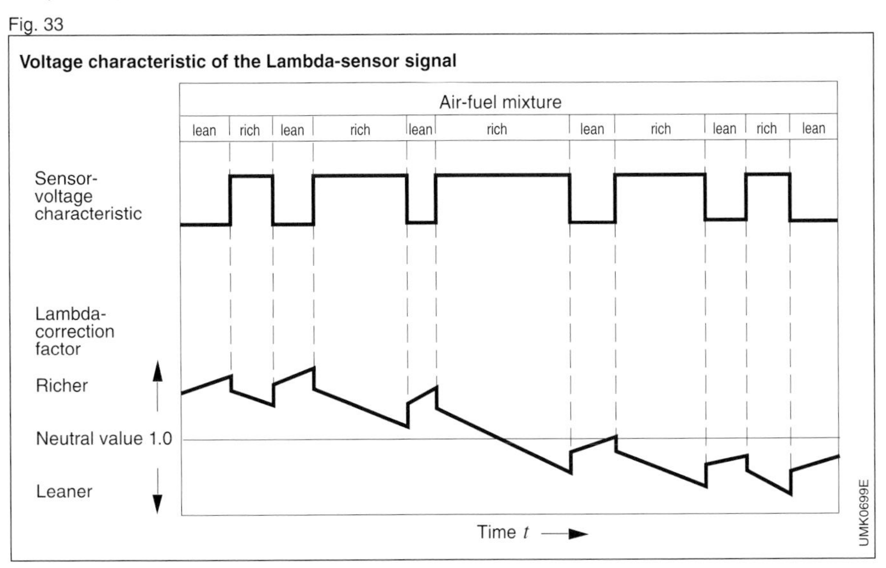

mixture-correction factor. Each increment provides compensation for another segment of the mixture correction (Figure 34).

Depending upon the engine load and speed, the incremental periodicity is between 1 s and a few 100 ms. Adaptation variables are updated so quickly that any effects of tolerance and drift on driveability and exhaust-gas composition are compensated for completely.

Idle-speed control

The idle-speed (closed-loop) control is used to reduce and stabilize the idle speed, and maintains a consistent idle speed throughout the whole of the vehicle's service life. The Mono-Jetronic system is maintenance-free, because no idle speed or idle mixture adjustments are required. In this type of idle-speed control, the throttle-valve actuator, which opens the throttle valve by means of a lever, is so controlled that the stipulated idle speed is maintained exactly under all operating conditions. This applies no matter whether the vehicle electrical system is heavily loaded, or the air-conditioner is switched on, or the automatic gearbox is at "Drive", or the power-assisted steering is at full lock etc. Engine temperature also has no effect, nor do high altitudes, where larger throttle-valve angles are required to compensate for lower barometric pressure.

The idle-speed control adapts the idle speed to the engine operating condition. In most cases, the idle speed is reduced, which is a decisive contribution to a reduction in fuel consumption and exhaust-gas emissions.

For the idle-speed control facility there are two engine-temperature-dependent curves stored in the ECU (Fig. 35a):

– Curve 1 for automatic-transmission vehicles when "Drive" is engaged.
– Curve 2 for manually-shifted transmissions, or automatic gearboxes if "Drive" is not engaged (Neutral).

The idle speed is usually reduced on automatic vehicles in order to reduce their tendency to creep when "Drive" is engaged. When the air-conditioner is switched on, the idle speed is often increased to maintain a defined minimum idle and ensure adequate cooling (Curve 3). In order to prevent engine-speed fluctuations when the air-conditioner compressor engages and disengages, the high idle speed is maintained even when the compressor is not engaged.

Fig. 34

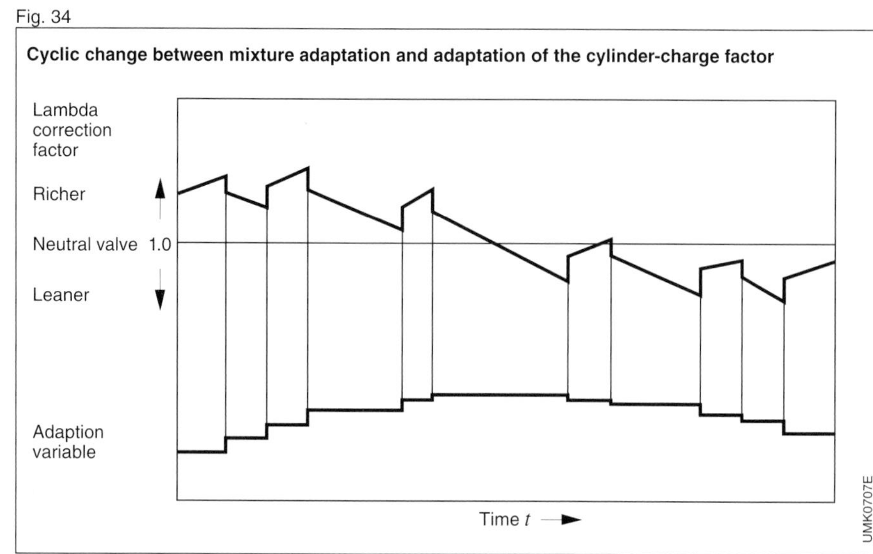

From the difference between the actual engine speed and the set speed (n_{set}), the engine-speed control stage calculates the appropriate throttle-valve setting.

With the idle switch closed, the control signal to the throttle-valve actuator comes from a position controller. This generates the control signal from the difference between the actual throttle-valve setting as registered by the throttle-valve potentiometer, and the desired throttle-valve setting as calculated by the engine-speed control stage.

In order to avoid sudden drops in engine speed during transitions, for instance, in the transition from overrun to idle, the throttle valve must not be closed too far. This is ensured by applying pilot control characteristics which electronically limit the minimum correcting range of the throttle-valve actuator. This necessitates the ECU containing a temperature-dependent throttle-valve pilot-control characteristic for "Drive" and for "Neutral" (Fig. 35b).

In addition, a number of different pilot-control corrections come into effect when the air-conditioner is switched on, dependent on whether the compressor is engaged or not. So that the pilot control is always at optimum value, pilot-control corrections are also adapted to cover all possible combinations resulting from the input signals for Transmission setting (Drive/Neutral), "Air-conditioner ready" (YES/NO), and "Compressor engaged" (YES/NO). It is the object of this adaptation to select the overall effective pilot-control value so that at idle this is at a specified number of degrees in advance of the actual throttle-valve setting.

In order that the correction of the pilot-control values becomes effective before the first idle phase when driving at high-altitudes, an additional air-density-dependent pilot-control correction is applied. The possibility of also being able to operate the throttle-valve outside the idle range (if the driver is not pressing the accelerator pedal) is taken advantage of and applied as a vacuum-limiting function. During overrun, this function opens the throttle valve in accordance with an engine-speed-dependent characteristic curve (Fig. 35b), just far enough to avoid operating points which have a very low cylinder charge (incomplete combustion).

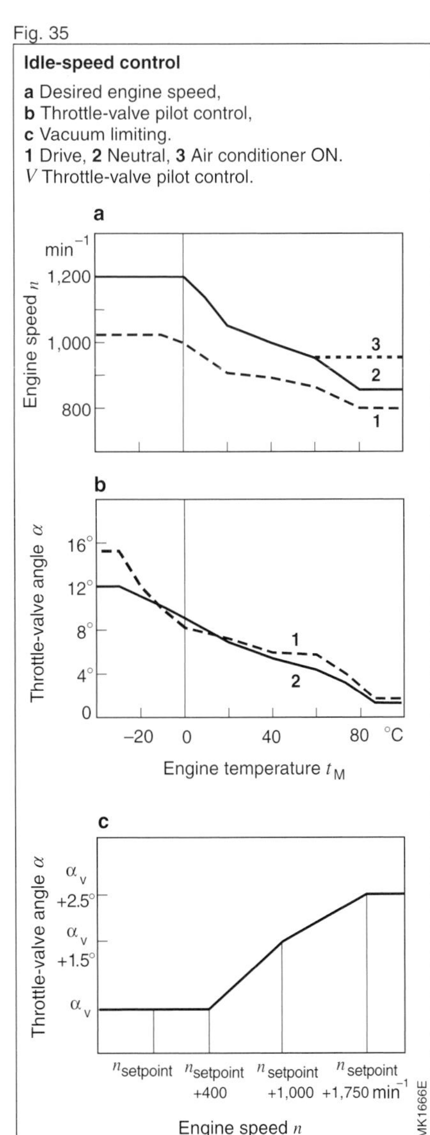

Fig. 35
Idle-speed control
a Desired engine speed,
b Throttle-valve pilot control,
c Vacuum limiting.
1 Drive, **2** Neutral, **3** Air conditioner ON.
V Throttle-valve pilot control.

Throttle-valve actuator

Through its control shaft, the throttle-valve actuator can adjust the throttle-valve lever and thereby influence the amount of air available to the engine. The actuator is powered by a DC motor which drives the setting shaft through a worm and roller gearset. Depending upon the motor's direction of rotation (which in turn depends upon the polarity applied to it), the setting shaft either extends and opens the throttle valve or retracts and reduces the throttle-valve angle. The control shaft incorporates a switching contact which closes when the shaft abuts against the throttle-valve lever and provides the ECU with the idle signal.
A rubber bellows device between the control shaft and the throttle-valve actuator housing prevents the intrusion of dirt and damp (Fig. 36).

Full-load enrichment

The driver presses the accelerator pedal to the floor to obtain maximum power from the engine. An IC engine develops maximum power with an air-fuel mixture which is about 10...15 % richer than the stoichiometric air-fuel ratio. The degree of enrichment is stored in the ECU in the form of a factor which is used to multiply the injection duration figures calculated from the Lambda map. Full-load enrichment comes into effect as soon as a specified throttle-valve angle is exceeded (this is a few degrees before the throttle-valve stop).

Engine-speed limiting

Extremely high rotational speeds can destroy the engine (valve gear, pistons). The limiting circuit prevents the engine from exceeding a given maximum speed. This speed n_0 can be defined for every engine, and as soon as it is exceeded the ECU suppresses the injection pulses. When the engine speed drops below n_0, the injection pulses are resumed again. This cycle is repeated in rapid succession within a tolerance band centered around the maximum permitted engine speed (Fig. 37).

The driver notices a sharp reduction in engine response, which also signals that an upshift is due.

Fig. 36

Throttle-valve actuator

1 Housing with electric motor, **2** Worm, **3** Wormwheel, **4** Setting shaft, **5** Idle contact, **6** Rubber bellows.

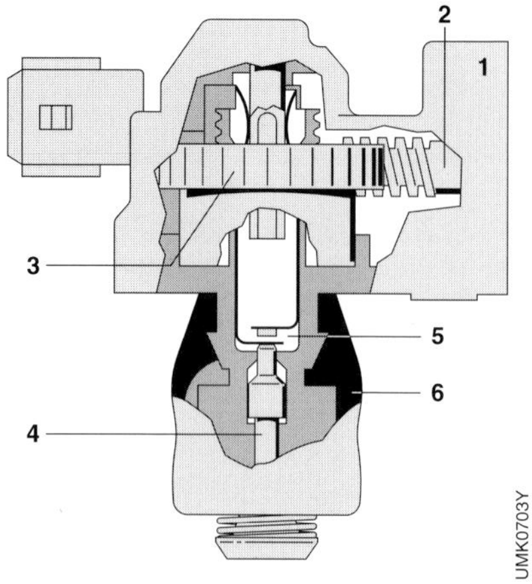

Fig. 37

Limitation of engine speed n_0 by means of injection-pulse suppression

a Area of fuel cutoff.

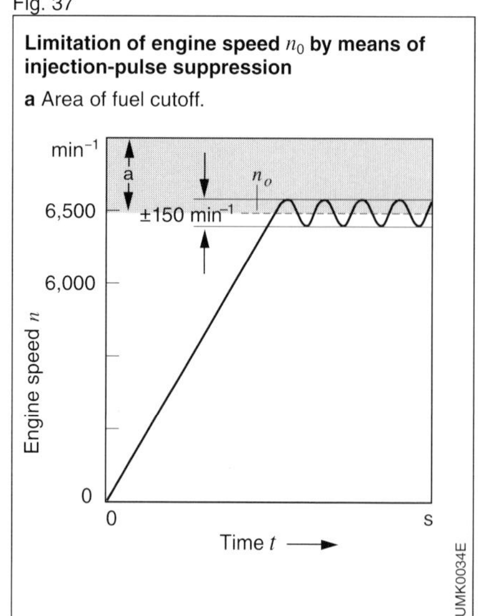

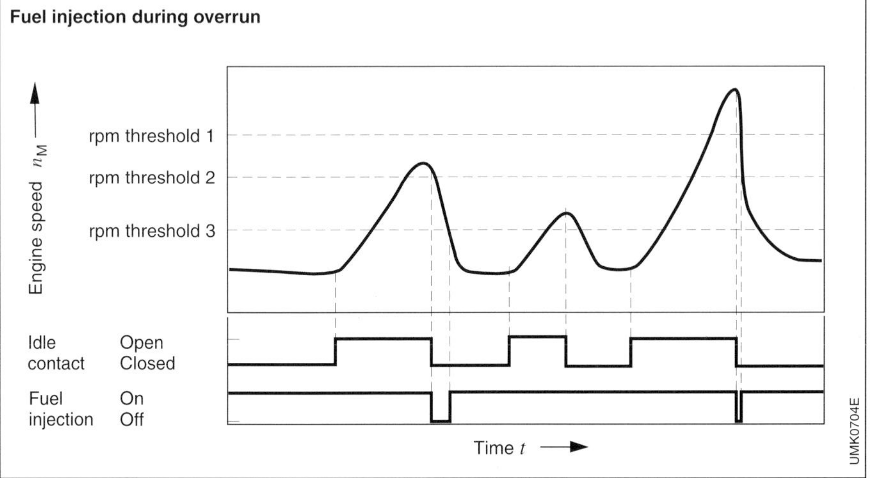

Fig. 38

Overrun (trailing throttle)

When the driver takes his foot off of the accelerator pedal while driving the vehicle, the throttle valve closes completely, and the vehicle is driven by its own kinetic energy. This operating mode is termed overrun or trailing throttle.

In order to reduce exhaust emissions and to improve driveability, a number of functions come into operation during overrun:

If the engine speed is above a given threshold (rpm threshold 2) when the throttle valve is closed, injector triggering stops, the engine receives no more fuel, and its speed falls. Once the next threshold is reached (rpm threshold 3), fuel injection is resumed. If the engine speed drops abruptly during overrun, as can occur when the clutch is pressed, fuel injection is resumed at a higher engine speed (rpm threshold 1), otherwise the engine speed could fall below idle or the engine could even stall (Fig. 38).

If the throttle valve is closed at high engine speeds, on the one hand the vehicle is decelerated abruptly due to the braking effect of the motored engine. On the other, the emission of hydrocarbons increases because the drop in manifold pressure causes the fuel film on the manifold walls to evaporate. This fuel though cannot combust completely because there is insufficient combustion air. To counteract this effect, during overrun the throttle-valve actuator opens the throttle valve slightly as a function of engine speed. This is de-scribed above under engine-speed limiting. If on the other hand the engine-speed drop is very abrupt during overrun, the throttle-valve opening angle is no longer a function of the fall in engine speed but instead the reduction in open-ing angle is slower and follows a time function.

During overrun the film of fuel deposited on the intake-manifold walls evaporates completely and the intake manifold dries-out. When the overrun mode is completed, fuel must be made available to build up this fuel film on the manifold walls again. This results in a slightly leaner air-fuel mixture in the transitional period until equilibrium returns. Fuel-film buildup is aided by an additional injection pulse, the length of which depends upon the overrun duration.

System voltage compensation

Fuel-injector voltage compensation

A feature of the solenoid-operated fuel injector is that due to self-induction it tends to open more slowly at the beginning of the injection pulse and to close more slowly at the end. Opening and closing times are in the order of 0.8 ms. The opening time depends heavily on the battery voltage, whereas the closing time depends only slightly on this factor. Without electronic voltage correction, the resulting delay in injector pickup would mean a far too short injection duration and therefore insufficient fuel would be injected.

In other words, the lower the battery voltage, the less fuel is metered to the engine. Therefore, a reduction in battery voltage must be counteracted by a voltage-dependent increase in injection duration, the so-called additive injector correction factor (Fig. 39a). The ECU registers the actual battery voltage and increases the injector-triggering pulse by the voltage-dependent injector pickup delay.

Fig. 39

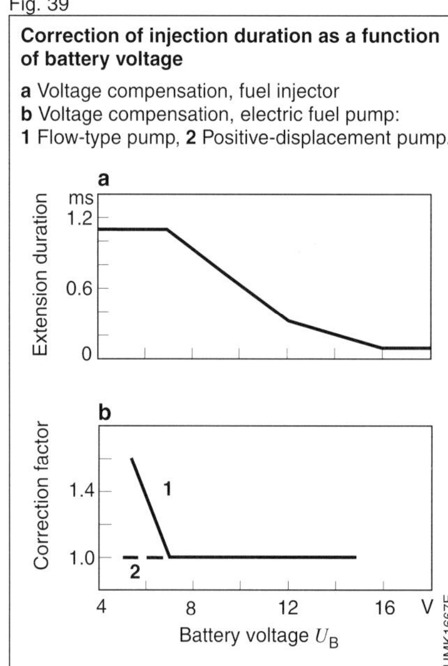

Correction of injection duration as a function of battery voltage

a Voltage compensation, fuel injector
b Voltage compensation, electric fuel pump:
1 Flow-type pump, **2** Positive-displacement pump.

Fuel-pump voltage compensation

The speed of the fuel-pump motor is very sensitive to variations in battery voltage. For this reason, if the battery voltage is low (for instance during a cold start), the fuel pump, which functions according to the hydrodynamic principle, is unable to bring the primary pressure up to its specified level. This would result in insufficient fuel being injected. To compensate for this effect, particularly at very low battery voltages, a voltage-correction function is applied to correct the injection duration (Fig. 39b).

If a positive-displacement electric fuel pump is used, this voltage correction function is unnecessary. The ECU is provided with a coding input, therefore, which enables the voltage-correction function to be activated depending upon pump type.

Controlling the regeneration-gas flow

When fresh air is drawn through the carbon in the carbon canister (purging), it transports the fuel trapped in the carbon to the engine for combustion.

The canister-purge valve between the carbon canister and the central injection unit controls the regeneration-gas flow. The control ensures that under all operating conditions as much of the trapped fuel as possible is transported away to the engine for burning. In other words, the regeneration-gas flow is kept as high as possible, without driveability being impaired. In general, the limit for the regeneration-gas flow is reached when the fuel contained in the regeneration-gas flow is about 20 % of the fuel required by the engine at the given operating point.

In order to ensure the correct functioning of the mixture adaptation it is imperative that a cyclic change is made between normal operation, which makes mixture adaptation possible, and regeneration operation. Furthermore, it is necessary during the regeneration phase to ascertain the fuel content in the regeneration gas and to use this value for adaptation. The same as with the mixture adaptation, this takes place by evaluating

the Lambda control stage's deviation from the $\lambda = 1$ value. Once the fuel content is known, the injection duration can be increased or decreased accordingly when the cycle changes so that during transition the mixture is maintained within tight limits around $\lambda = 1$.
To specify the fuel content in the regeneration-gas stream as a function of the engine operating mode, as well as to adapt the proportion of fuel in the regeneration-gas stream, the relationship between the regeneration-gas stream and the amount of air drawn in by the engine via the throttle valve must be determined. Both partial streams are practically proportional to the open cross-section areas through which they flow.
Whereas it is relatively simple to calculate the cross-section area in the throttle housing from the throttle-valve angle, the open cross-section area of the canister-purge valve changes according to the applied differential pressure.
The differential pressure applied at the canister-purge valve depends upon the engine's operating point and can be derived from the injection durations stored in the Lambda map.

Fig. 40

Canister-purge valve

1 Hose connection, **2** Non-return valve, **3** Leaf spring, **4** Sealing element, **5** Solenoid armature, **6** Seal seat, **7** Solenoid winding.

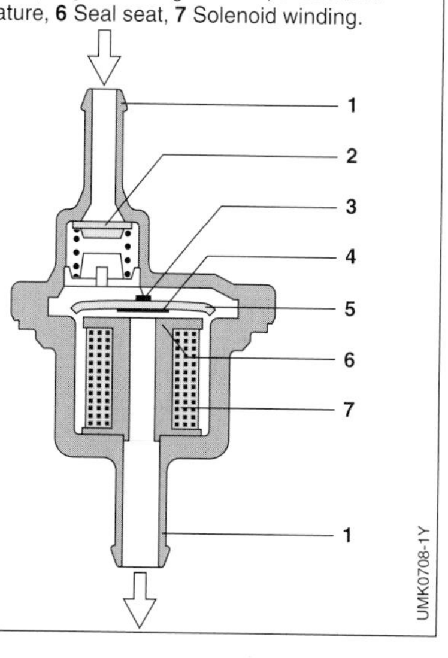

The ratio of regeneration-gas stream to air stream drawn in by the engine can be calculated for each engine operating point as defined by the throttle-valve angle and the engine speed. The regeneration-gas stream can be further reduced by cycling the canister-purge valve and in this manner adjusted precisely to the required ratio necessary to ensure acceptable driveability.

Canister-purge valve

Due to the canister-purge valve's throughflow characteristic, it is possible to have a large regeneration stream for relatively small pressure differentials (operation near to WOT), and a small regeneration stream for large pressure differentials (idle operation). With switched-mode operation, the throughflow values can be further reduced by increasing the valve's on-off ratio. It is provided with two hose fittings for connection to the carbon canister and to the intake manifold (Fig. 40).
With a signal applied, the solenoid winding draws in the armature so that its sealing element (rubber seal) is pressed against the seal seat and closes the valve's outlet. The armature is fastened with a thin leaf spring which is secured at one end. When no current flows through the solenoid winding, this spring lifts the armature and seal element from the seal seat so that the throughflow cross section is open.
When the differential pressure between valve inlet and valve outlet increases, the forces acting on the leaf spring cause it to bend in the direction of the stream flow and in doing so bring the seal element nearer to the seal seat so that the effective throughflow cross section is reduced.
A check valve is provided in the valve-inlet area to prevent fuel vapors enter-ing the intake manifold from the carbon canister when the engine is switched off.

Limp-home and diagnosis

All sensor signals are continuously checked for their plausibility by monitoring functions incorporated in the ECU. If one of the sensor signals deviates from its defined, plausible range, this means that either the sensor itself is defective or there is a fault in one of its electrical connections.

If one of the sensor signals fails or is no longer plausible, it must be replaced with a substitute signal in order that the vehicle does not break down completely but can still safely reach the next specialist workshops under its own power even though with some restrictions in driveability and comfort.

For instance, if temperature signals fail, signals are substituted which normally prevail when the engine is at its operating temperature: for the intake air 20°C and for the engine coolant 100°C. If a fault occurs in the Lambda closed-loop control, the complete Lambda control facility is closed down, that is, the injection durations from the Lambda map are only corrected with the mixture-adaptation value (if available).

If the signals from the throttle-valve potentiometer are not plausible, this means that one of the main controlled variables is missing. In other words, access is no longer possible to the injection durations stored in the Lambda map. In this case. the fuel injector is triggered with defined constant-length pulses. Two different injection durations are available, and switching between them depends upon engine speed.

In addition to the sensors, the throttle-valve actuator (idle-speed control) is also monitored continuously.

Fig. 41

Central injection unit (Section drawing)

1 Fuel-pressure regulator, **2** Intake-air temperature sensor, **3** Fuel injector, **4** Upper section (hydraulic section), **5** Fuel-inlet channel, **6** Fuel-return channel, **7** Thermal-insulation plate, **8** Throttle valve, **9** Lower section.

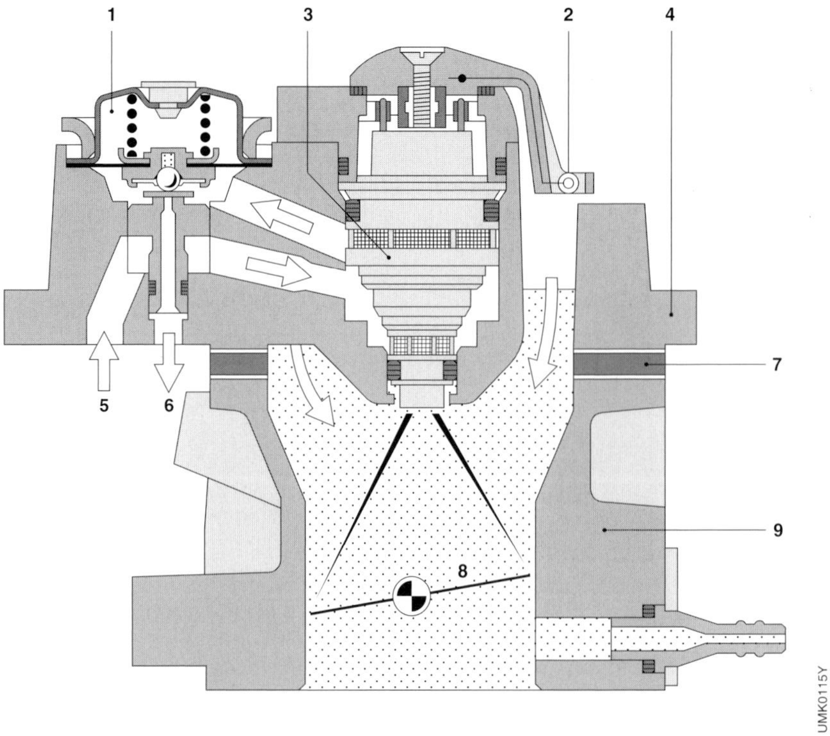

Fault memory

As soon as a sensor failure or throttle-valve actuator malfunction is registered, a corresponding entry is made in the fault memory. This entry remains accessible for a number of operating cycles (that is, even when the engine is switched off and the vehicle is left standing overnight a number of times) so that the workshop is able to localize the fault. This also applies to sporadic faults such as loose contacts etc.

Fault diagnosis

When diagnosis is triggered, the contents of the fault memory are made available to the workshop in the form of a flashing code, or with the aid of a diagnostic tester. As soon as the fault has been repaired, the Mono-Jetronic system operates again normally.

Central injection unit

The Mono-Jetronic central injection unit is bolted directly to the intake manifold. It supplies the engine with finely atomized fuel, and is the heart of the Mono-Jetronic system. Its design is dictated by the fact that contrary to multipoint fuel injection (e.g., L-Jetronic), fuel-injection takes place at a central point, and the intake-air quantity is measured indirectly through the two factors throttle-valve angle α and engine speed n, as already described (Figures 41 and 42).

Lower section

The lower section of the central injection unit comprises the throttle valve together with the throttle-valve potentiometer for measuring the throttle-valve angle. The throttle-valve actuator for the idle-speed

Fig. 42

Central-injection unit (part sectional drawing)

1 Fuel injector, **2** Air-temperature sensor, **3** Throttle valve, **4** Fuel-pressure regulator, **5** Fuel return, **6** Fuel inlet, **7** Throttle-valve potentiometer (on throttle-valve shaft extension, not shown), **8** Throttle-valve actuator.

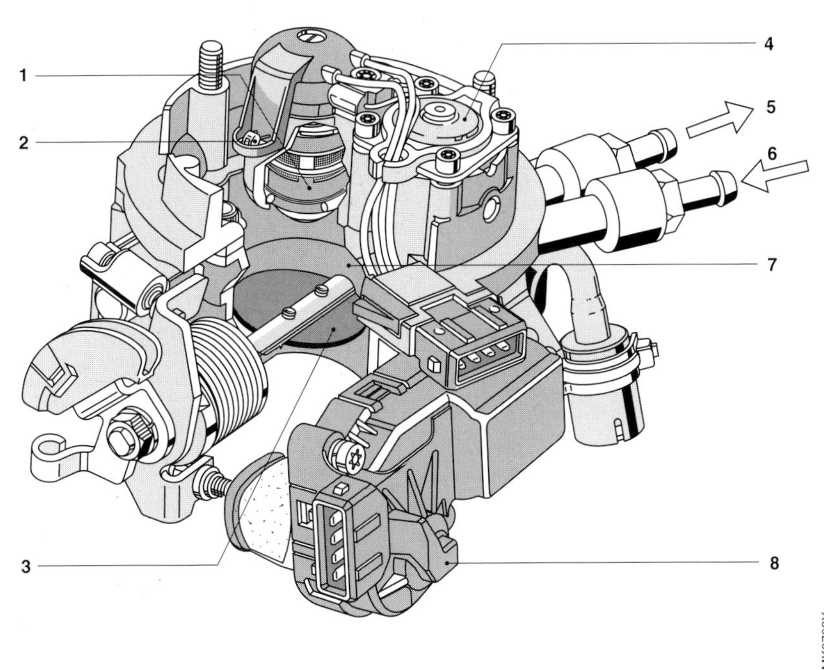

control is mounted on a special bracket on the lower section.

Upper section

The complete fuel system for the central injection unit is in the upper section, consisting of the fuel injector, the fuel-pressure regulator, and the fuel channels to the fuel injector which are incorporated in the injector bracket. Two passages descend through the housing to supply the injector with fuel. The fuel flows to the injector via the lower passage.

The upper passage is connected to the lower chamber of the pressure regulator. From the fuel-pressure regulator, excess fuel enters the fuel-return line via the plate valve. This fuel-channel arrangement was selected so that even when the formation of fuel-vapor bubbles increases (as can occur for instance when the fuel-injection assembly heats up after the engine is switched off), enough fuel collects at the injector metering area to ensure trouble-free starting.
The open cross-section area between the return channel and the inlet channel is limited to a defined dimension by a shoulder on the fuel-injector strainer.
This ensures that excessive fuel that has not been injected is divided into two partial streams, one of which flows through the injector and the other around it. This form of intensive flushing cools the fuel injector, and the fuel-channel arrangement with fuel circulating around the injector as well as through it, results in the Mono-Jetronic's excellent hot-starting characteristics. The temperature sensor for measuring the intake-air temperature is fitted in the central injection unit's cover cap.

Power supply

Battery

The battery supplies the complete electrical system with electrical energy.

Ignition/starter switch

The ignition switch is a multipurpose switch. It is used for starting the engine, and from here, electric power is switched to most of the vehicle's electrical circuits, including the ignition and the fuel-injection system.

Relay

The relay is controlled by the ignition switch, and switches the vehicle system voltage to the ECU and the other electrical components.

Electrical circuitry

The ECU is connected through a 25-pole plug-and-socket connection to all the Mono-Jetronic components as well as to the vehicle electrical system (Fig. 43).
The ECU is connected to the vehicle system voltage through two connections:
– One connection is used to permanently connect the ECU to the battery positive pole (Terminal 30). This permanent connection to the battery maintains the stored data (adaptation values, diagnostic fault memory) even when the ignition is switched off.
– The other connection connects the ECU to the battery when the ignition is switched on. In order to avoid voltage peaks which can be caused by the ignition coil's self-inductance, instead of connecting the ECU directly to Terminal 15 it may be necessary to connect it to the battery via a relay (main relay) which is controlled by Terminal 15 on the ignition switch.

ECU ground
Two separate lines are also used for connecting the ECU to ground:
– The ECU electronic circuitry requires a separate ground connection in order to correctly process the sensor signals (Lambda oxygen sensor, potentiometer, NTC sensors).
– The high currents from the ECU driver stages for triggering the actuators flow through the second ground line.

Lambda oxygen sensor connection
In order to screen the Lambda oxygen sensor line from interference due to voltage peaks, it is surrounded by a wire-mesh sheath in the wiring harness.

Fuel-pump safety cicuit
The fuel-pump relay is directly controlled from the ECU. This ensures that the fuel-pump stops delivering fuel if the engine has ceased running, for instance in case of an accident. Switching on the ignition activates the fuel pump for about 1 second and so does each ignition pulse. This is termed dynamic pump control. If the engine stops with the ignition still switched on, the fuel-pump relay releases and open-circuits the power supply to the fuel pump.

Fig. 43
Mono-Jetronic diagram

B1 Intake-air temperature sensor, **B2** Lambda oxygen sensor, **B3** Engine-temperature sensor, **B4** Throttle-valve potentiometer, **F1, F2** Fuses, **H1** Diagnosis lamp and tester connection, **K1** Pump relay, **K2** Main relay, Kl.1/TD engine-speed information, **R1** Dropping resistor, S1 Air-conditioner ON, S2 Air-conditioner compressor, S3 Gearbox switch, W1 t_v-coding, W2 pump coding, **X1** ECU, **Y1** Canister-purge valve, **Y2** Electric fuel pump, **Y3** Fuel injector, **Y4** Throttle-valve actuator with idle switch.

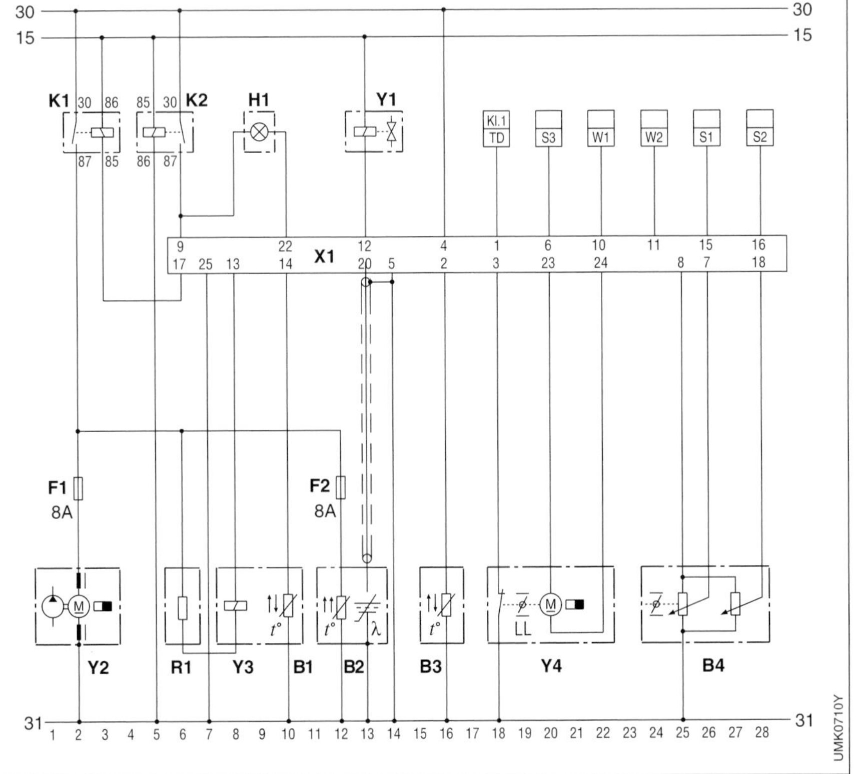

Workshop testing techniques

Bosch customer service

Customer service quality is also a measure for product quality. The car driver has more than 10,000 Bosch Service Agents at his disposal in 125 countries all over the world. These workshops are neutral and not tied to any particular make of vehicle. Even in sparsely populated and remote areas of Africa and South America the driver can rely on getting help very quickly. Help which is based upon the same quality standards as in Germany, and which is backed of course by the identical guarantees which apply to customer-service work all over the world. The data and performance specs for the Bosch systems and assemblies of equipment are precisely matched to the engine and the vehicle. In order that these can be checked in the workshop, Bosch developed the appropriate measurement techniques, test equipment, and special tools and equipped all its Service Agents accordingly.

Testing techniques for Mono-Jetronic

Apart from the regular replacement of the fuel filter as stipulated by the particular vehicle's manufacturer, the Mono-Jetronic gasoline-injection system requires no special maintenance work. Essentially, in case of malfunctions, the workshop expert has the following test equipment, together with the appropriate test specs, at his disposal:

- Universal test adapter, system adapter line, and multimeter or engine analyzer,
- Jetronic set (hydraulic case with pressure-measuring device),
- Lambda closed-loop-control tester,
- Pocket system-diagnosis tester, KTS 300 or evaluation unit for blink code KDAW 9975 or KDAW 9980.

Universal test adapter, system adapter line, and multimeter or engine analyzer

The universal test adapter (Fig. 44) was specifically developed for testing electronic fuel-injection systems, such as practically all the Jetronic systems and

Fig. 44

Test set-up with universal test adapter, system adapter line, and multimeter

a Test set-up with KTS 300 system tester, **b** Test set-up with universal test adapter.
1 System tester KTS 300, **2** Diagnosis plug on vehicle, **3** ECU, **4** Diagnosis adapter line for KTS 300, **5** System wiring harness, **6** Universal test adapter, **7** Multiple-contact switch, **8** Plug-in connection, **9** System adapter line, **10** Measurement lines, **11** Multimeter.

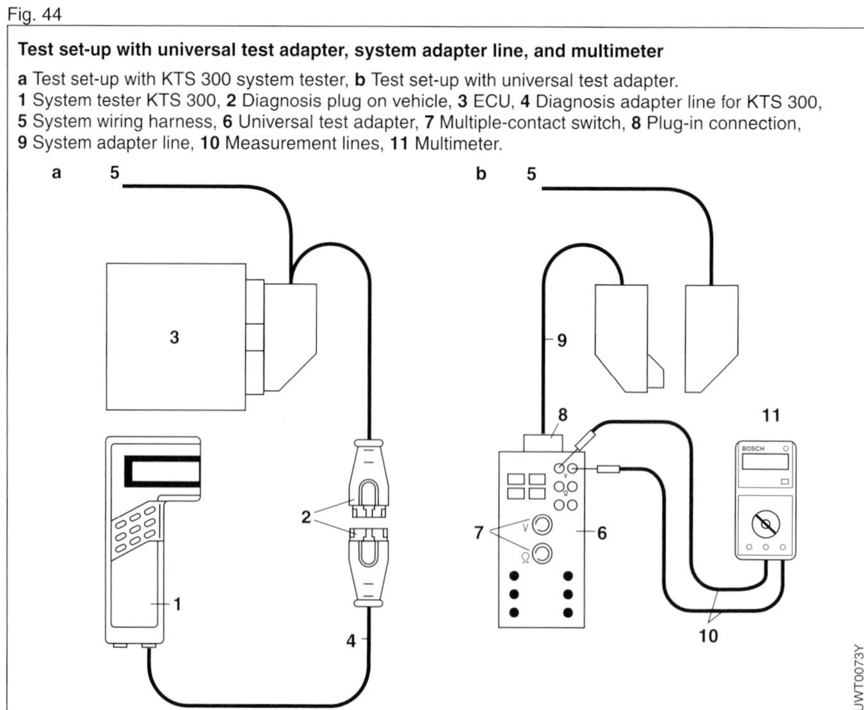

various Motronic systems. Using this test adapter, all the Mono-Jetronic's components and functions which are essential for correct engine running can be tested:

- Throttle-valve switch or potentiometer (load registration),
- Throttle-valve actuator,
- Injector,
- Ignition-coil signal (injection triggering),
- Engine-temperature sensor,
- Intake-air temperature sensor,
- Electric fuel pump,
- Lambda oxygen sensor, and
- Canister-purge valve.

The system adapter line is used to connect the universal test adapter to the ECU wiring-harness plug. By means of two multiple-contact switches, it is then a simple matter to quickly select the various lines to the components, and to measure the voltage and resistance values using the multimeter or engine analyzer.

Jetronic set

The fuel-system pressure can be measured using the pressure-measuring device from the Jetronic set.
Fuel-pressure measurement permits the assessment of such measured quantities and characteristics as:

- Electric fuel-pump delivery,
- Fuel-filter permeability,
- Restriction presented by the return line,
- Pressure-regulator functioning, and
- The sealing integrity of the complete fuel system, this being of vital importance for starting behaviour.

Lambda closed-loop-control tester

This tester is used for testing the Lambda-sensor signal (and for simulation of the "rich/lean" signal).
Special adapter lines are available for connection to the Lambda-sensor lines found in the various vehicle models. The measured values are shown on an analog display.

Diagnosis tester and evaluation unit

The Mono-Jetronic ECU features digital circuitry and includes a self-diagnosis facility with error storage. The error storage can be read out with suitable evaluation equipment.
The following testers are suitable for this purpose and are available on the market:

Pocket system-diagnosis tester
KTS 300:
The KTS 300 can display one or more faults in the system by means of a corresponding error code in connection with a text which provides information on faulty components or their lines and plugs.

Blink-code evaluation unit KDAW 9975 or KDAW 9980:
The ECU is designed to perform self-diagnosis by means of a blink code, the faults being read out in the form of blink pulses. On some vehicles, it is possible to read out the blink code directly through the check lamp on the vehicle's instrument panel.

Important: In the case of the Mono-Jetronic, an option is not provided for both diagnosis methods on one and the same vehicle. The automaker decides which diagnosis method can be used.
If the ECU is configured for self-diagnosis using blink code, it is impossible to read out the error storage using a self-diagnosis tester such as the KTS 300.

Test technology for the Mono-Motronic

The Mono-Motronic's engine-management system is a further development of the Mono-Jetronic. It combines the ignition and fuel-injection sub-systems. Similar to the Mono-Jetronic, this system features an ECU. The system test is performed using the KTS 300. On some older Mono-Motronic systems, the error storage can also be read out by means of blink code.

Mono-Motronic engine management

System overview

The Mono-Motronic engine-management system is a low-pressure single-point injection system (fuel-injection subsystem) with integrated electronic map-controlled ignition (ignition subsystem). Combination of the two subsystems enables joint optimization of fuel-metering and ignition-control.
The heart of the Mono-Motronic system is an electronic control unit (ECU) which incorporates the high-performance microcomputer required for the processing not only of the fuel injection but also of the ignition control. Since, for instance, only one power supply and one case are needed, the outlay required for the single ECU is less than that which would be required for separate fuel-injection and ignition systems. This leads to increased reliability of the system as a whole and a more favorabe cost/benefit ratio.

The fact that the ECU utilises sensor signals for controlling the ignition and fuel-injection functions leads to the following Mono-Motronic advantages:
– Precise metering of the injected fuel quantity, and temperature-dependent adaptive ignition angle lead to warm-running adaptation with attendant fuel-consumption reduction,
– Minimization of fuel consumption together with favorable emissions figures as a result of precision ignition-angle adaptation covering the complete ignition map and under all operating conditions.
– Idle-speed stabilization due to dynamic ignition timing,
– Increase in driving comfort by ignition-timing intervention during acceleration and deceleration,
– With automatic transmissions, ignition-timing intervention for improved shifts.

Subsystem: Fuel injection

The intermittent electronically controlled single-point injection system is based on the familiar Mono-Jetronic. Functions have been added to improve driving comfort and to further improve the limp-home facilities in case of sensor failure.
The fuel is pumped by the electric fuel pump from the fuel tank, through the fuel filter, and to the central injection unit which is located directly on the intake manifold. The solenoid-operated injector, which injects a finely-atomized jet of fuel just above the throttle plate, and the fuel-pressure regulator, are situated in the injection unit's hydraulic stage.
The ECU calculates the basic injection quantity from the throttle-plate angle and the engine-speed signal. More or less fuel is injected to take into account such operating conditions as cold-start, post-start, warm-up, and overrun, as well as engine-speed limitation.
The vapors from the fuel tank are also returned to the running engine from the carbon canister (canister-purge).

Subsystem: Ignition

The ignition map stored in the ECU supersedes the centrifugal (mechanical) and vacuum advance mechanisms located in the ignition distributor. The ignition (advance) angle as a function of engine speed and load is stored in the ignition map. In addition, the ignition angle can be changed depending upon engine and intake-air temperature, throttle-valve setting, and throttle-valve angular velocity.

Rotating voltage distribution

If an ignition system with rotating high-voltage (HT) distribution is used, the distributor only retains the Hall triggering function for engine-speed measurement and the actual HT distributor. The ECU, which triggers the ignition's external driver stage, is responsible for the engine-speed and load-dependent ignition-timing adjustment and dwell-angle control. The HT distributor is responsible for assigning the ignition spark to the correct cylinder.

Distributorless (stationary) voltage distribution

If a distributorless ignition system (Fig. 1) is used, the mechanically driven HT distributor is dispensed with. The ECU distributes the primary voltage to the ignition coils which generate the HT and pass it directly to the spark plug of the respective cylinder. For example, a 4-cylinder engine is equipped with 2 dual-spark ignition coils which are triggered by the ECU through external power output stages. A speed sensor picks off the engine-speed information and the reference-mark signal for cylinder 1 (or 4) at a sensor wheel which is attached to the crankshaft.

Knock control

The Mono-Motronic can also be equipped with knock-control. This utilises the signal from the knock sensor on the engine block to adjust the ignition advance angle in order to take full advantage of the available fuel quality. The result is a reduction in fuel consumption while at the same time ensuring that the engine cannot be damaged by combustion knock.

Self-diagnosis

The ECU continually checks all the signals which are required for correct operation, and stores the fault type as soon as one of the parameters leaves its defined range. Using a diagnostic tester, faults can be read out from the fault store during servicing.

Supplementary functions

Exhaust-gas recirculation (EGR) and secondary-air injection are further possibilities for reducing toxic emissions.

Fig. 1

Schematic diagram of the Mono-Motronic

1 Injector, **2** Air-temperature sensor, **3** Fuel-pressure regulator, **4** Ignition coil, **5** Canister-purge valve, **6** Throttle actuator, **7** Carbon canister, **8** Pressure actuator, **9** Throttle-valve potentiometer, **10** ECU, **11** Fuel filter, **12** EGR valve, **13** Knock sensor, **14** Engine-speed sensor, **15** Engine-temperature sensor, **16** Lambda sensor, **17** Electric fuel pump.

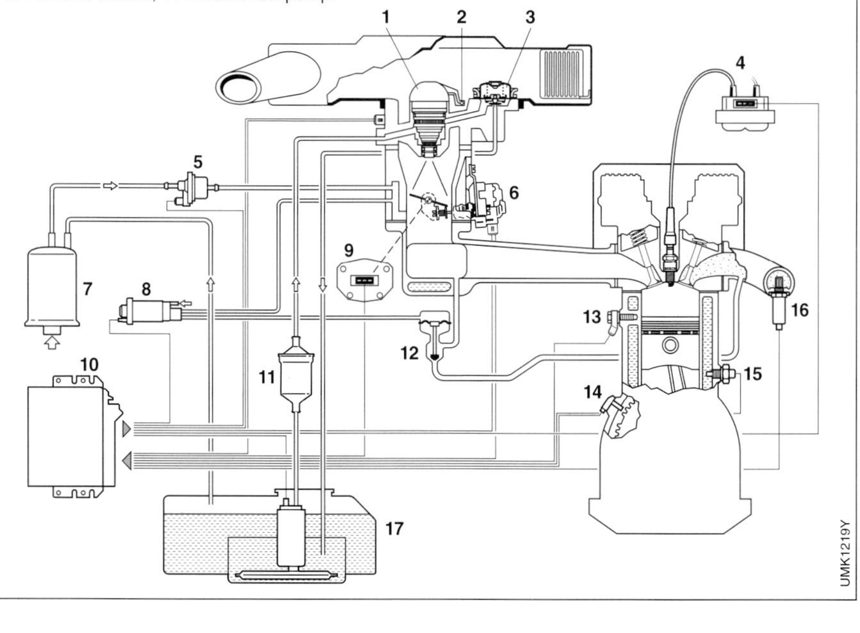

UMK1219Y

Ignition

Ignition in the gasoline engine

The design of the ignition system for the spark-ignition engine varies according to how ignition is triggered, how the timing is adjusted and how the high voltage is distributed.
Table 1 shows a classification of various ignition systems.

Ignition point (timing)

The ignition point is essentially dependent on the variables "engine speed" and "load." It is dependent upon the engine speed since the time taken for complete combustion of the mixture at constant charge and air-fuel ratio is constant and, thus, the ignition point (timing) must be shifted forward as engine speed increases, in other words it must take place earlier. The dependence upon load is due to the leaner mixtures, higher proportions of residual gas and less dense cylinder charges that accompany low load factors. This influence causes a longer ignition delay and lower combustion rates in the mixture; the timing must be advanced to compensate (Fig. 1).

Spark advance

Variable ignition timing allows the ignition to respond to variations in engine speed and load factor. On simple systems, timing is adjusted by a centrifugal advance mechanism and a vacuum control unit. Manifold vacuum provides a reasonably accurate index of engine load.
Semiconductor ignition systems also allow for other influences of the engine, e.g., temperature or changes in the mixture composition. The values of all ig-

Table 1

Definition of the ignition system.
An ignition system must perform at least the following functions:

Function	**Ignition system**			
	CI	**TI**	**SI**	**DLI**
	Coil ignition	Transistorized ignition	Semiconductor ignition	Distributorless semiconductor ignition
Ignition triggering (pulse generator)	mechanical	electronic	electronic	electronic
Determining the ignition angle on the basis of the speed and load condition of the engine	mechanical	mechanical	electronic	electronic
High-tension generation	inductive	inductive	inductive	inductive
Distribution and assignment of the ignition spark to the correct cylinder	mechanical	mechanical	mechanical	electronic
Power section	mechanical	electronic	electronic	electronic

nition timing functions are linked either mechanically or electronically in order to determine the ignition point. The energy storage device must be fully charged before the actual ignition point. This requires the formation of a dwell period or dwell angle in the ignition system. The energy is generally stored in an inductive storage device, and, in rare cases, in a capacitive storage device. High voltage is generated by disconnecting the primary inductor from the power supply followed by transformation. The high voltage is applied to the cylinder currently performing the working stroke. When an ignition distributor is used, the crankshaft position information required for this is provided by an appropriate mechanism via the ignition distributor drive.
In the case of stationary voltage distribution, an electrical signal from the crankshaft or the camshaft provides the position signal. The connecting elements (plugs and high-tension cable) convey the high voltage to the spark plug. The spark plug must function reliably in all engine operating ranges in order to ensure consistent ignition of the mixture.

Fig. 1

Pressure curve in combustion chamber for different ignition points

1 Ignition (Z_a) at correct time
2 Ignition (Z_b) too soon (ignition knock)
3 Ignition (Z_c) too late

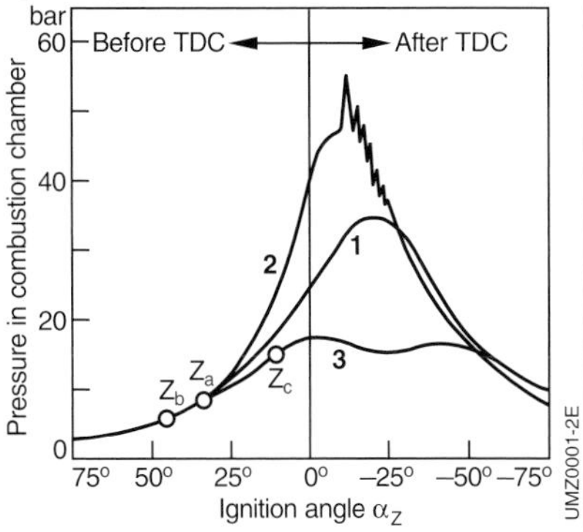

Ignition voltage

The excess-air factor λ and the cylinder pressure which is determined by charge and compression have, together with the spark-plug electrode gap, a crucial influence upon the required ignition voltage and, thus, upon the required secondary available voltage of the ignition system.

Ignition of the mixture

Ignition energy

Approximately 0.2 mJ of energy is required per individual ignition for igniting an air-fuel mixture by electric spark, providing the mixture (static, homogeneous) has a stoichiometric composition. Rich and lean mixtures (turbulent) require over 3 mJ. This amount of energy is but a fraction of the total energy contained in the ignition spark, the ignition energy.
If insufficient ignition energy is available, ignition does not occur; the mixture cannot ignite and there are combustion misses. This is why adequate ignition energy must be provided to ensure that, even under worst-case external conditions, the air-fuel mixture always ignites. It may suffice for a small cloud of explosive mixture to move past the spark. The cloud of mixture ignites and, in turn, ignites the rest of the mixture in the cylinder, thus initiating fuel combustion.

Influences on ignition characteristics

Good induction and easy access of the mixture to the ignition spark improve the ignition characteristics as do long spark duration and a long spark or large electrode gap. Intense turbulence of the mixture also has a similarly favourable effect providing that adequate ignition energy is available. The spark position and spark length are determined by the dimensions of the spark plug. The spark duration is determined by the type and design of ignition system and the instantaneous ignition conditions. The spark position and accessibility of the mixture to the spark plug influence the

exhaust gas, especially in the idle range. Particularly high ignition energy and a long spark duration are favourable in the case of lean mixtures. This can be demonstrated using an engine at idle. During idle, the mixture may be very inhomogeneous. Valve overlap results in a high residual exhaust-gas component.
If we compare a normal breaker-triggered coil ignition system and a high-energy transistorized ignition system, we can see that the spark of the transistorized ignition system clearly reduces and stabilizes HC emission. Smooth running of the engine is also stabilized at the same time.
Fouling of the spark plug is also an important factor. If spark plugs are badly fouled, energy is discharged from the ignition coil via the spark-plug shunt path during the period in which the high voltage is being built up. This shortens the spark duration, thus affecting the exhaust gas and, in critical cases (if the spark plugs are badly fouled or wet) may result in complete misfiring. A certain amount of misfiring is normally not noticed by the driver but does result in higher fuel consumption and may damage the catalytic converter.

Pollutant emission

The ignition angle a_z or the ignition point has an important influence on the exhaust-gas values, the torque, the fuel consumption and the driveability of the spark-ignition engine. The most important pollutants in the exhaust gas are the unburned hydrocarbons (HC), oxides of nitrogen (NO_X) and carbon monoxide (CO).
The emission of unburned hydrocarbons increases the more the ignition is advanced.
NO_x emission also increases when the ignition is advanced across the entire air-fuel ratio range. The reason for this is the rise in combustion-chamber temperature associated with increasing ignition advance.
CO emission is practically independent of the ignition point and is virtually exclusively a function of the air-fuel ratio.

Fuel consumption

The influence of the ignition point on fuel consumption conflicts with the influence on pollutant emission. With increasing excess-air factor λ, ignition must occur earlier in order to compensate for the lower combustion rate and thus maintain an optimum combustion process. Thus, an advanced ignition point means lower fuel consumption and high torque but only if the mixture is controlled accordingly.

Knocking tendency

One further important interrelationship is that between ignition point and the engine's tendency to knock. This is demonstrated by the effect of a too early or too late ignition point (by comparison with the correct ignition point) on the pressure in the combustion chamber (Fig. 1). If the ignition point is too early, mixture at various points in the combustion chamber also ignites owing to the ignition pressure wave. This means that the mixture burns irregularly and intense pressure fluctuations occur with high combustion-pressure peaks. This effect, called knocking, can be heard clearly at low engine speeds. At high engine speeds, the noise is smothered by the engine noise. But it is precisely in this range that knocking may lead to engine damage and it must thus be avoided by finding an optimum combination of suitable fuel and ignition point.

Conventional coil ignition CI

The conventional coil-ignition system is breaker-triggered. This means that the current flowing through the ignition coil is switched on and off mechanically via a contact in the ignition distributor (contact breaker).
The breaker-triggered coil-ignition system is the simplest version of an ignition system in which all functions are implemented. In addition to the ignition distributor, there are a whole number of other components, as shown in Table 2 together with their functions.

Operating principle

Synchronization and high-voltage distribution

Synchronization with the crankshaft and, thus, with the position of the individual pistons is guaranteed by the mechanical coupling of the ignition distributor to the camshaft or to another shaft which turns at only half the speed of the crankshaft. Thus, if the ignition distributor is turned, this will also result in a shift in the ignition point or, in other words, turning the ignition distributor permits setting a prescribed ignition point.
The mechanical rotor arm which is also permanently coupled to the upper section of the ignition-distributor shaft ensures correct distribution of the high voltage, in conjunction with routing of the high-voltage cables to the individual spark plugs.

Ignition sequence

When the system is operating, voltage from the battery (1) flows through the starter/ignition switch (2) and to terminal 15 on the ignition coil (3) (Figures 1 and 2). When the contact breaker (6) is closed, the current flows through the primary winding of the ignition coil to ground. This builds up a magnetic field in the ignition coil, thus storing the ignition energy. The current rise is exponential owing to the inductance and the primary resistance of the primary winding. The charging time is determined by the dwell angle. In turn, the dwell angle is determined by the design of the cam which actuates the contact breaker via the cam follower. At the end of the dwell period, the ignition-distributor cam opens the ignition contact and thus interrupts the coil current.

Table 2
Components of the conventional ignition system
A coil-ignition system is composed of various components and subassemblies, the actual design and construction of which depend mainly on the engine with which the system is to be used.

Component	Function
Ignition coil	Stores the ignition energy and delivers it in the form of a high voltage surge through the HT ignition cable to the distributor
Ignition and starting switch	A switch in the primary circuit of the coil, manually operated with the ignition key
Ballast resistor	This is shorted during starting for starting-voltage boosting
Contact breaker	Opens and closes the primary circuit of the ignition coil for the purposes of energy storage and voltage conversion
Ignition condenser/ capacitor	Provides for low-loss interruption of the primary current and suppresses most of the arcing between the contact points
Ignition distributor	At the instant of ignition, distributes the firing voltage to the spark plugs in a pre-set sequence
Centrifugal advance mechanism	Automatically shifts the ignition timing depending on the engine speed
Vacuum advance mechanism	Automatically shifts the ignition timing depending on the engine load
Spark plug	Contains the electrodes which are the most important parts required to generate the ignition spark and seals off the combustion chamber

Fig. 2

Circuit diagram of the coil ignition system

1 Battery, **2** Ignition and starting switch, **3** Ignition coil, **4** Ignition distributor, **5** Ignition capacitor, **6** Contact breaker, **7** Spark plugs, R_V ballast resistor.

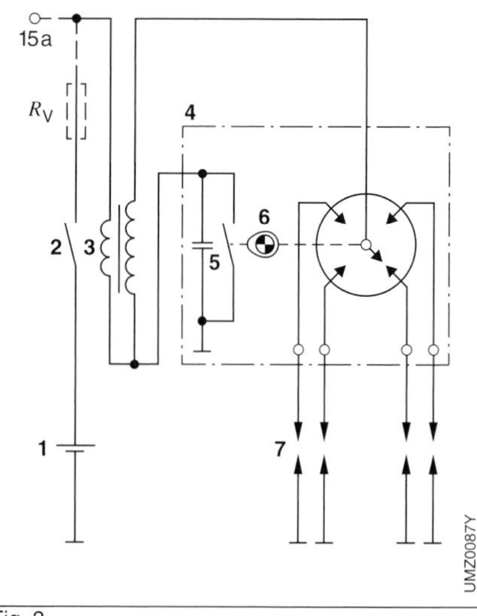

The current, the off time and the number of windings in the secondary circuit of the ignition coil essentially determine the ignition voltage induced in the secondary circuit. Since the current tends to carry on flowing, an arc would form at the ignition contact. In order to prevent this, the ignition capacitor (5) is connected in parallel with the contact breaker. This means that the primary current flows to the capacitor and charges it until the ignition voltage discharges disruptively. This means that voltages of a few hundred V occur briefly at terminal 1 of the ignition coil in the primary circuit (Figures 1 and 2).

The high tension generated in the secondary circuit charges the connection to the center tower of the ignition distributor, causes a disruptive discharge between the rotor arm and outer electrode, then charges the high-voltage cable to the relevant spark plug and finally causes disruptive discharge at the spark plug, i.e. causes the ignition spark.

After this, the magnetic energy stored in the ignition coil is discharged constantly as electrical energy to the spark. This causes a spark voltage of approximately 400 V at the spark plug. The spark duration is generally 1 to 2 ms. After the ignition coil is discharged, the cam of the ignition distributor switches the contact breaker back on and the ignition coil is recharged.

The rotor arm which carries on moving in

Fig. 1

Ignition system with conventional coil ignition

1 Battery, **2** Ignition and starting switch, **3** Ignition coil, **4** Ignition distributor, **5** Ignition capacitor, **6** Contact breaker, **7** Spark plugs, R_V ballast resistor for boosting the starting voltage (not always fitted).

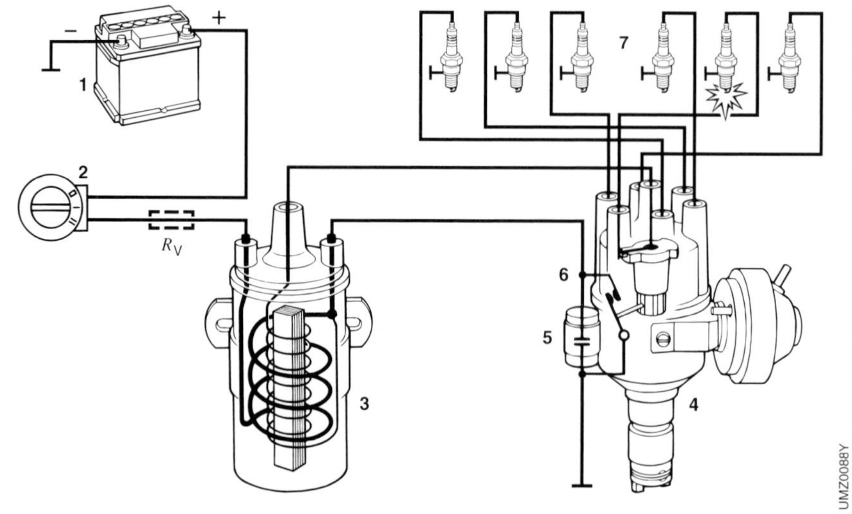

the meantime transfers the high tension to the next spark plug during the next ignition.

Ignition coil

Construction

The ignition coil consists of a metal case which accommodates metal plate jackets for reducing stray magnetic fields. The secondary winding is wound directly onto the laminated iron core and connected electrically to the center tower in the cap of the ignition coil via the core.

Since the high voltage is applied to the iron core, the core must be insulated by the cap and an additional insulator inserted in the base. The primary winding is located near to the outside around the secondary winding (Fig. 3).

The insulated ignition coil cap contains the terminals 15 and 1 for the battery voltage and the connection to the contact breaker, arranged symmetrically with the high-tension tower, terminal 4. The windings are insulated and mechanically locked in position by potting with asphalt. Oil-filled ignition coils are also available.

The power loss occurs chiefly in the primary winding. The heat is dissipated through the metal plate jackets to the case. This is why the ignition coil is secured to the bodywork with such a wide clamp so that as much heat as possible is dissipated via this metal band.

Function

The primary current which is switched on and off by the ignition distributor flows through the ignition-coil primary winding. The magnitude of the current is determined by the battery voltage at terminal 15 and the ohmic resistance of the primary winding. The primary resistance may lie between 0.2 and 3 Ω, dependent upon use of the ignition coil. The primary inductance L_1 is a few mH. The following formula applies to the energy stored in the magnetic field of the ignition coil

$$W_{Sp} = \frac{1}{2}\ L_1 \cdot i_1^2$$

W_{Sp} stored energy, L_1 inductance of the primary winding, i_1 current which flows in the ignition distributor at the moment at which the contact breaker opens.

At the ignition point, the voltage at terminal 4 (high-voltage tower of the ignition coil) rises approximately sinusoidally. The rate of rise is determined by the capacitive load at terminal 4. When the spark-plug breakdown voltage is reached, the voltage drops to the spark voltage of the spark plug and the energy stored in the ignition coil flows to the ignition spark. As soon as the energy no

Fig. 3

Section through an ignition coil

1 High-tension connection on the outside, **2** Winding layers with insulating paper, **3** Insulating cap, **4** High-tension connection on the inside via spring contact, **5** Case, **6** Mounting bracket, **7** Metal plate jacketing (magnetic), **8** Primary winding, **9** Secondary winding, **10** Sealing compound, **11** Insulator, **12** Iron core.

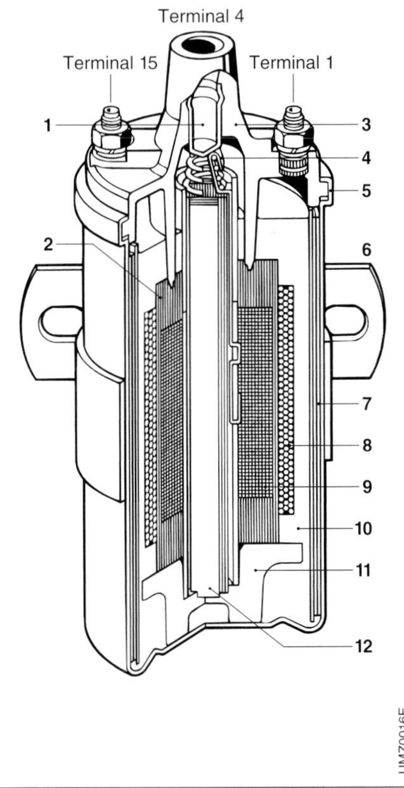

longer suffices to maintain the glow discharge, the spark breaks down and the remaining energy decays in the secondary circuit of the ignition coil.
The high voltage is polarized such that the center electrode of the spark plug is negative with respect to the vehicle chassis or ground. If the polarity were the reverse, this would mean a slightly higher required voltage. The ignition coil is designed as an auto-transformer such that the secondary side is connected to terminal 1 or 15.
In the same way as the primary inductance and the primary resistance determine the stored energy, the secondary inductance determines the high-voltage and spark characteristic. A typical turns-ratio of primary to secondary winding is 1:100. The induced voltage, the spark current and the spark duration are dependent upon both the stored energy and the secondary inductance.

Internal resistance

One further important value is the ignition-coil internal resistance since it is one of the factors which determine the rate of voltage rise and is thus a measure of how much energy is dis-charged from the ignition coil via shunt resistances at the moment of spark discharge. A low internal resistance is advantageous in the case of fouled or wet spark plugs. The internal resistance is dependent upon the secondary inductance.

Fig. 4
Contact breaker (schematic diagram)
a Contact closed,
b Large point gap, small dwell angle,
c Small point gap, large dwell angle.

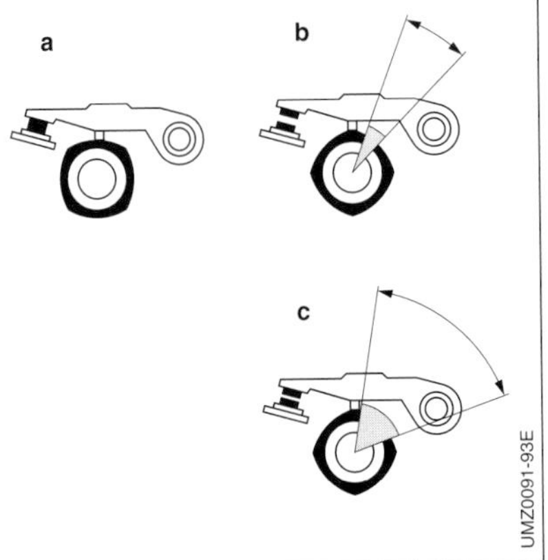

Contact breaker

The contact breaker is triggered via the breaker cam which has as many lobes as the number of cylinders in the engine. The breaker cam can be turned on the ignition distributor shaft. It is adjusted dependent upon the engine-speed-dependent ignition-timing adjustment input from the centrifugal advance mechanism. The cam is configured such that there results a dwell angle corresponding to the ignition coil and the sparking rate (Fig. 4).
This means that the dwell angle is permanently preset for a breaker-triggered ignition system and is invariable throughout the entire engine-speed range. However, the dwell angle does change throughout the service life of the engine owing to wear of the cam follower on the breaker lever. The abrasion which is thus produced means that the contact breaker opens at a later point. The resultant ignition retard generally results in higher fuel consumption. This is one of the reasons why the contact breaker needs to be renewed regularly and the dwell angle checked. Another reason why maintenance is required is contact erosion (pitting). The contact must switch currents of up to 5 A and break voltages of up to 500 V. On a 4-cylinder engine with an engine speed of 6,000 min^{-1} the contact switches 12,000 times per minute, corresponding to a frequency of 200 Hz.
Defective contacts mean inadequate charging of the ignition coil, undefined ignition points and, thus, higher fuel consumption and poorer exhaust-gas values.

Ignition distributor

The ignition distributor is the component of the ignition system which performs most functions. It rotates at half the crankshaft speed. A 4-cylinder distributor has, for instance, 4 outputs which each generate an ignition pulse each time the rotor turns (Fig. 5).

Features

The main exterior features are the pot-shaped ignition-distributor housing and the distributor cap made of insulating material with the towers for the high-voltage connections. On some versions of shaft distributors, the drive shaft projects into the engine. It is then driven via a gearing system or a coupling. Another design, the short-type distributor, simplifies direct attachment to the camshaft. In this case, there is no drive shaft and the drive coupling is located directly at the base of the ignition-distributor housing. The stringent requirements in respect of ignition-distributor accuracy require very good bearing support. On shaft distributors, the shaft itself provides an adequately long bearing section. Short-type dis-tributors require an additional bearing above the triggering system.

Construction

The ignition-distributor housing accommodates the centrifugal advance mechanism, the actuation system for the vacuum advance system and the ignition triggering system. The ignition capacitor and the vacuum control unit are secured on the outside of the ignition-distributor housing. In addition, the outside of the housing also has the catches for securing the distributor cap and the electrical connection. The dust-protection cover protects the triggering system against dirt deposits and moisture. There is a slot on the distributor shaft above the breaker cam. This slot serves to define the installation position of the distributor rotor. This is why, when fitting, it must be ensured that the rotor arm is fitted in the correct position. The distributor rotor and distributor cap are made of a high-grade plastic which is required to meet very stringent requirements in respect of dielectric strength, climatic resistance, mechanical strength and flammability. The high voltage generated in the ignition coil is fed to the ignition distributor via the central tower. There is a small, spring-

Fig. 5

Components of an ignition distributor

1 Distributor cap,
2 Distributor rotor with electrode (E),
3 Dust-protection cover (condensation barrier),
4 Distributor shaft,
5 Breaker cam,
6 Connection for vacuum hose,
7 Vacuum control unit,
8 Ignition condenser/capacitor.

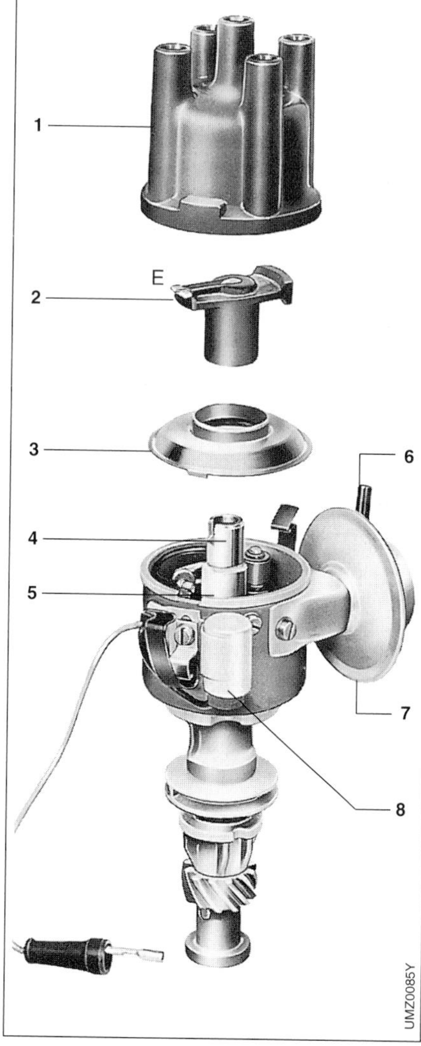

1
E
2
3
6
4
5
7
8
UMZ0085Y

loaded carbon pin between the distributor rotor and central tower. This pin establishes contact between the fixed cap and the rotating distributor rotor. The ignition energy flows from the centerpoint of the distributor rotor through the interference-suppression resistor with a rating of $\geq 1\,k\Omega$ to the distributor-rotor electrode, and, from there, sparks over to the respective outer electrode which is recessed in the outer tower. The voltage required for this is in the kV region. The resistor in the distributor rotor limits the peak currents when the sparks are being built up and thus serves to suppress interference. With the exception of the contact breaker, all parts of the ignition distributor require virtually no maintenance.

Spark-advance mechanism

The centrifugal advance mechanism advances ignition timing as a function of the engine speed. Assuming constant charge and fuel induction, this results in a fixed duration for ignition and complete combustion of the mixture. This fixed duration means that it is necessary to produce the ignition spark correspondingly earlier at high engine speeds. However, the progression of an ignition-distributor characteristic curve is also influenced in practice by the knock limit and the variation in composition of the mixture.

The vacuum advance mechanism takes the load condition of the engine into account since the ignition and combustion rate of the fresh gas in the cylinder is highly dependent upon the charge in the cylinder.

The engine-speed or centrifugal advance mechanism, and the vacuum advance mechanism or load adjustment, are interconnected mechanically in such a way that both adjustments are added (Fig. 6).

Centrifugal advance mechanism

The centrifugal advance mechanism adjusts the ignition point as a function of engine speed. The support plate which rotates with the distributor shaft bears the flyweights. As engine speed increases,

Fig. 6

Example of an overall timing-control system comprising adjustment dependent upon engine-speed and upon intake-manifold pressure

1 Part load on road,
2 Full load.

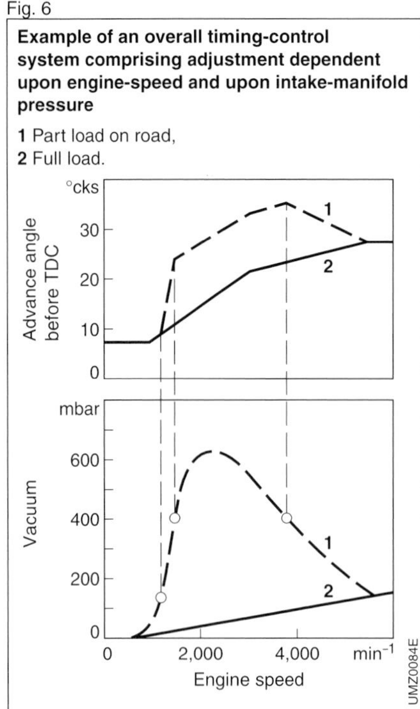

Fig. 7

Centrifugal advance mechanism in rest position (top) and in operating position (below)

1 Support plate,
2 Distributor cam,
3 Rolling contact path,
4 Flyweight,
5 Ignition-distributor shaft,
6 Yoke.

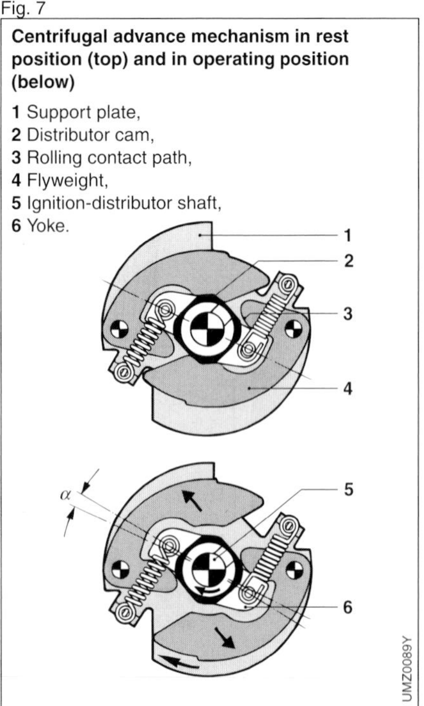

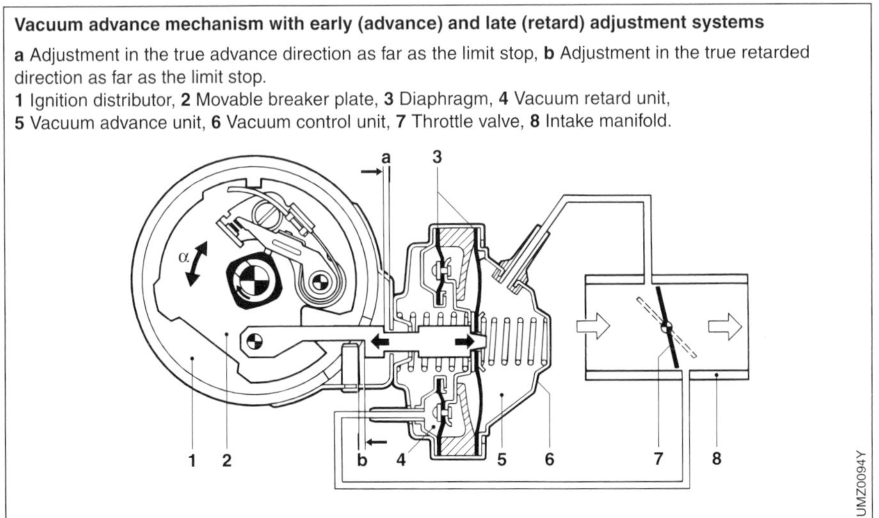

Vacuum advance mechanism with early (advance) and late (retard) adjustment systems

a Adjustment in the true advance direction as far as the limit stop, **b** Adjustment in the true retarded direction as far as the limit stop.
1 Ignition distributor, **2** Movable breaker plate, **3** Diaphragm, **4** Vacuum retard unit, **5** Vacuum advance unit, **6** Vacuum control unit, **7** Throttle valve, **8** Intake manifold.

Fig. 8

the flyweights move outwards. By means of their contact path, they rotate the yoke in the direction of shaft rotation but relative to the shaft. The result is that the distributor cam also rotates relative to the shaft by the same angle, the advance angle α_z, and thus advances the ignition point by this angle (Fig. 7).

Vacuum advance mechanism

The vacuum advance mechanism adjusts the ignition point as a function of the engine power or engine load. The vacuum in the intake manifold near to the throttle valve serves as a measure for this vacuum advance. The vacuum is measured by one or two aneroid capsules (Fig. 8).

"Early" (advance) adjustment system

The lower the load, the earlier the air-fuel mixture needs to be ignited since it burns more slowly. The percentage of burned but non-exhausted residual gases in the combustion chamber increases and the mixture becomes leaner.

The vacuum for adjustment in the advance direction is taken off at the intake manifold. As the engine load decreases, the vacuum in the vacuum advance unit increases, causing the diaphragm and the vacuum advance arm to move to the right (Fig. 8). The vacuum advance arm turns the movable breaker plate in the opposite direction to the direction of rotation of the distributor shaft. The ignition point is advanced further, i.e. in the "early" direction.

"Late" (retard) adjustment system

The vacuum in the intake manifold is, in this case, taken off downstream of the throttle valve. With the aid of the annular "vacuum retard unit", the ignition point is retarded, i.e. moved in the "late" direction, under specific engine conditions (e.g., idle, overrun) in order to improve the exhaust-gas values. The ring diaphragm moves, together with the vacuum advance arm, to the left as soon as there is a vacuum. The vacuum advance arm turns the movable breaker plate, including the contact breaker, in the direction of rotation of the ignition distributor shaft. The late adjustment system is subordinate to the early adjustment system: simultaneous vacuum in both aneroid capsules means an ignition-timing shift in the advance direction for part-load operation.

Breaker-triggered transistorized ignition TI-B

The ignition distributor of the breaker-triggered transistorized ignition system (TI-B) is identical to the ignition distributor of the breaker-triggered coil ignition system (CI). However, the contact breaker no longer needs to switch the primary current but only the control current for the transistorized ignition system. The transistorized ignition system itself plays the role of a current amplifier and switches the primary current via an ignition transistor (generally a Darlington transistor). In order to facilitate understanding, the wiring of the contact and the function of a simple TI-B are compared below to a breaker-triggered coil-ignition system.

Operating principle

Figures 2 and 3 clearly show that the breaker-triggered transistorized ignition system is a further development of the conventional, non-electronic coil-ignition system: the transistor T is used as the circuit breaker in place of the contact breaker and assumes its switching function in the primary circuit of the ignition system. However, since the transistor has a relay characteristic, it must be caused to switch in the same way as the relay. This can be done, for instance, as shown in Fig. 2, with a control switch. Such transistorized ignition systems are thus termed breaker-triggered.

In Bosch transistorized ignition systems, the cam-operated breaker performs the function of this control switch. When the contact is closed, a control current I_S flows to the base B and the transistor is electrically conductive between the emitter E and the collector C. In this condition, it corresponds to a switch in the "On" position and current can flow through the primary winding L_1 of the ignition coil. However, if the contact of the breaker is open, no control current flows through to the base and the transistor is electrically non-conductive. It thus blocks the primary current and, in this condition, corresponds to a switch in the "Off" position.

Advantages

The breaker-triggered transistorized ignition system has two essential advantages over the breaker-triggered coil-ignition system:

– An increase in the primary current and
– Considerably longer service life of the breaker contact.

Fig. 1

Secondary available voltage of the ignition coil for the spark plug as a function of the sparking rate or engine speed (4-cylinder engine).

a Ignition coil with maximum sparking rate 12,000 min^{-1},
b High-performance ignition coil with maximum sparking rate 21,000 min^{-1}.
U_Z Ignition voltage, ΔU Minimum voltage reserve.
Shaded area: Area for operation free of misfiring.

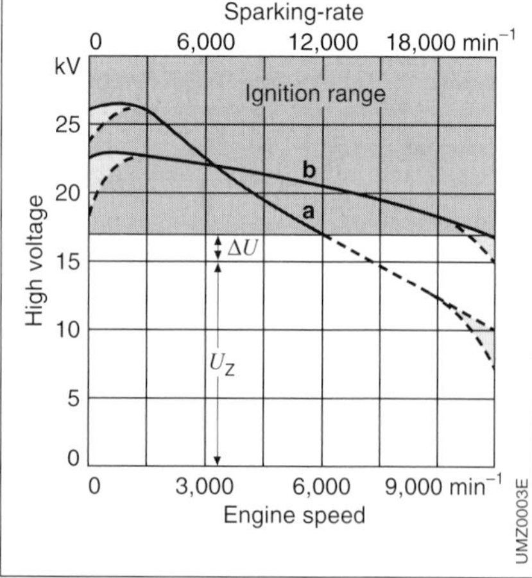

The primary current can be increased if using a switching transistor since a mechanical contact can switch currents of only up to 5 A for long periods and with the required frequency. Since the stored energy is proportional to the square of the primary current, the power of the ignition coil increases and, thus, also all high-voltage data such as secondary available voltage, spark duration and spark current. Thus, a breaker-triggered transistorized ignition system also requires a special ignition coil in addition to the ignition trigger box.

A far longer service life of the TI-B results from the fact that the contact breaker is not required to switch high currents. In addition, the TI-B is also not subject to two other problems which indefinably reduce the secondary available voltage of contact-triggered coil-ignition systems: Contact chatter and the contact-breaking spark which results from the inductance of the ignition coil. The contact-breaking spark reduces the available energy and delays the high-voltage rise, particularly at low engine speed and when starting. Conversely, contact chatter occurs at high engine speeds owing to the high switching frequency of the contact and is a disturbing influence. The contact bounces when closing and thus charges the ignition coil less intensely, precisely at a point in time at which the dwell period is reduced anyway. The first negative characteristic of the contact breaker is not applicable to the breaker-triggered transistorized ignition system, the second is.

Fig. 2

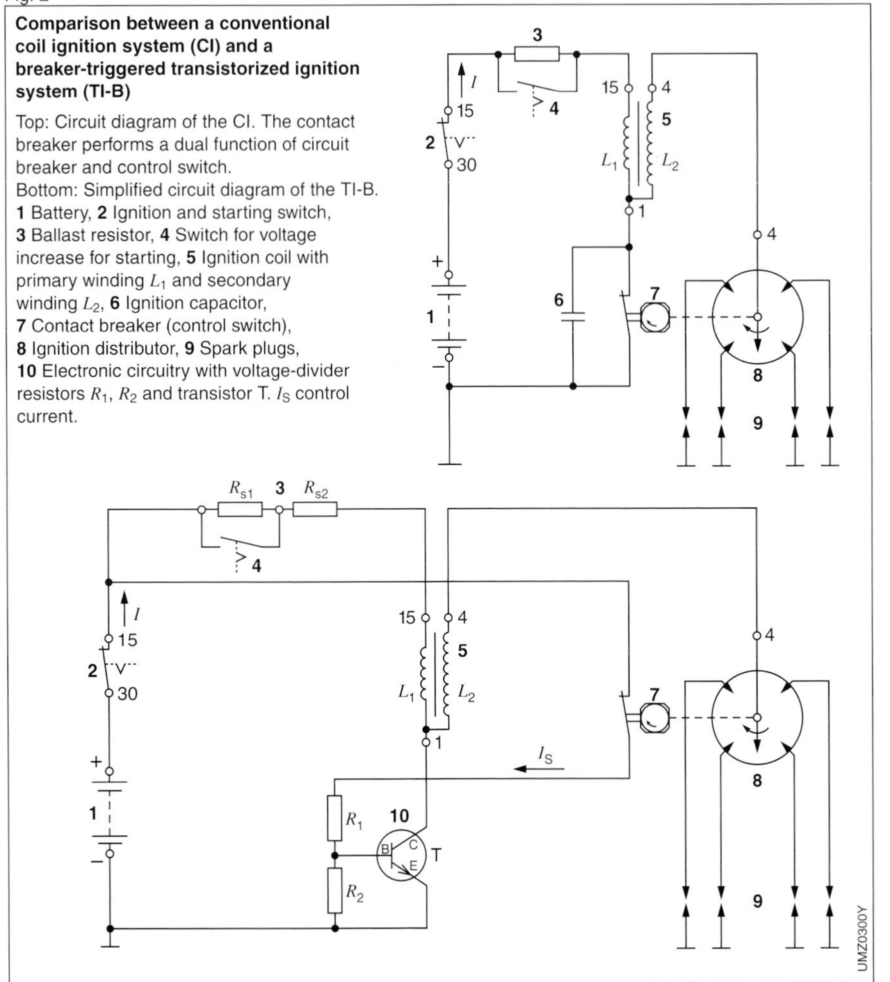

Comparison between a conventional coil ignition system (CI) and a breaker-triggered transistorized ignition system (TI-B)

Top: Circuit diagram of the CI. The contact breaker performs a dual function of circuit breaker and control switch.
Bottom: Simplified circuit diagram of the TI-B.
1 Battery, **2** Ignition and starting switch, **3** Ballast resistor, **4** Switch for voltage increase for starting, **5** Ignition coil with primary winding L_1 and secondary winding L_2, **6** Ignition capacitor, **7** Contact breaker (control switch), **8** Ignition distributor, **9** Spark plugs, **10** Electronic circuitry with voltage-divider resistors R_1, R_2 and transistor T. I_S control current.

Circuit

On a breaker-triggered transistorized ignition system, the ignition trigger box (control unit) is connected between terminal 1 of the ignition distributor (i.e. the contact breaker) and terminal 1 of the ignition coil (Fig. 3). In addition, the ignition trigger box requires one further terminal 15 for its power supply and a ground connection 31. The primary circuit of the ignition coil is powered via a pair of ballast resistors which are normally connected in series. When starting, the left-hand ballast resistor is bypassed by terminal 50 at the starter. This means that a higher supply voltage is applied to the ignition coil via the right-hand ballast resistor. It compensates for the disadvantages which result from the starting operation and the resulting reduction in battery voltage. Ballast resistors serve to limit the primary current in the case of low-resistance, rapidly chargeable ignition coils. They thus prevent overload of the ignition coil, particularly at low engine speeds, and thus protect the ignition contact breaker since the dwell angle is still determined by the distributor cam. Since the ignition coil actually requires a constant period for charging but does not operate with a fixed dwell angle, there is too much time available at low engine speeds for charging and too little at high engine speeds. Ballast resistors and a rapidly chargeable ignition coil permit an optimum situation over the entire operating range.

On older vehicles, the TI-B was an original equipment item. It has now been displaced by the transistorized ignition with maintenance-free trigger systems. However, as a retrofit-equipment set, the TI-B is still well-suited for substantially improving the ignition characteristics on vehicles with breaker-triggered coil-ignition systems fitted as standard. This is why it is advisable to retrofit such a system in the case of general ignition problems, specifically in the case of starting difficulties, and if the vehicle is to be used largely for stop-and-go driving.

Fig. 3

Components and connection diagram of the TI-B

1 Battery, **2** Ignition and starting switch, **3** Ignition trigger box, **4** Ballast resistors, **5** Cable connection to the starter, **6** Ignition coil, **7** Ignition distributor, **8** Spark plugs, I primary current, I_S control current.

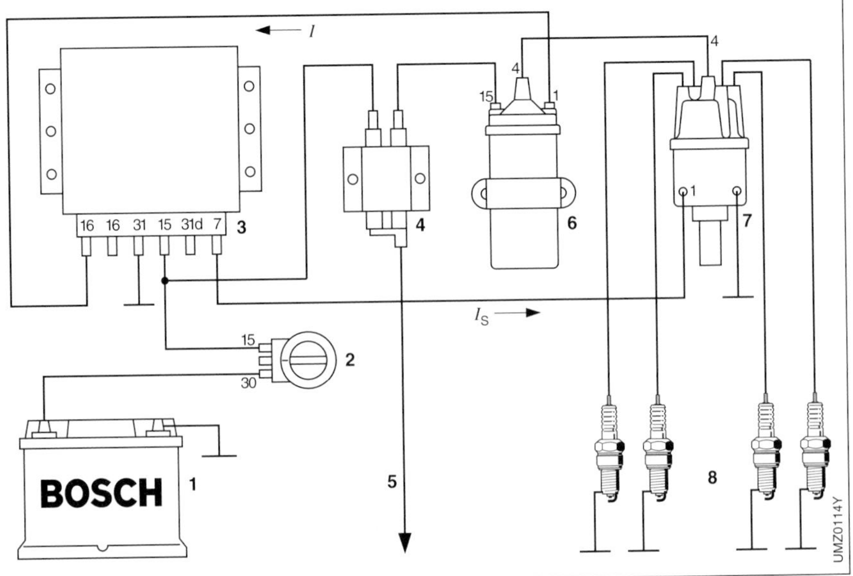

Transistorized ignition with Hall generator TI-H

In addition to the breaker-triggered transistorized ignition system (TI-B), there are two other versions of transistorized ignition with Hall triggering system (TI-H).
On one version, the dwell angle is determined by the shape of the rotor in the ignition distributor. The other version contains a control unit incorporating hybrid circuitry and which automatically regulates the dwell angle. An additional current limiter with a highly efficient ignition coil make this version a particularly high-performance ignition system.

Hall effect

If electrons move in a conductor to which the lines of force of a magnetic field are applied, the electrons are deflected perpendicularly to the current direction and perpendicularly to the direction of the magnetic field: an excess of electrons occurs at A_1 and a deficiency of electrons occurs at A_2, i.e. the Hall voltage occurs across A_1 and A_2. This so-called Hall effect is particularly prominent in the case of semiconductors (Fig. 1).

Hall generator

When the ignition-distributor shaft turns, the vanes of the rotor move through the air gap of the magnetic barrier without touching it. When the air gap is unobstructed the incorporated IC and the Hall layer are subjected to the magnetic field (Fig. 2).
At the Hall layer, the magnetic flux density B is high and the Hall voltage U_H is maximum. The Hall IC is activated. As soon as one of the rotor vanes enters the air gap, most of the magnetic flux runs through the vane area and is thus largely prevented from reaching the Hall layer. The flux density at the Hall layer is reduced to a virtually negligible level, resulting from the leakage field. Voltage U_H is at minimum.

Fig. 1

Hall effect

B Flux density of the magnetic field,
I_H Hall current,
I_V Supply current,
U_H Hall voltage,
d Thickness.

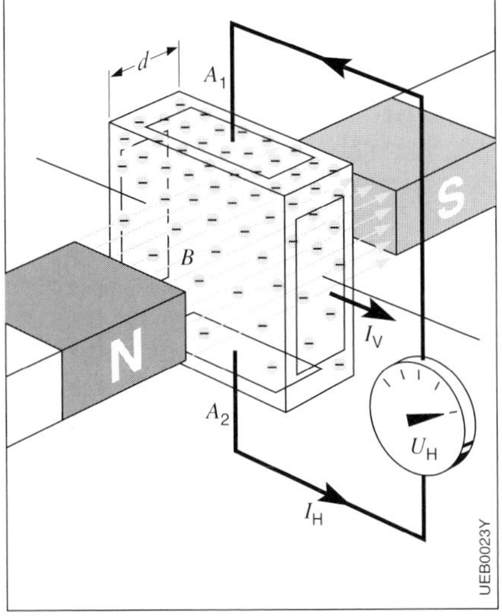

UEB0023Y

Fig. 2

Hall generator in the ignition distributor

Top: Principle. Below: Generator voltage U_G (converted Hall voltage).
1 Vane with width b, **2** Soft magnetic conductive elements with permanent magnet, **3** Hall IC, **4** Air gap.

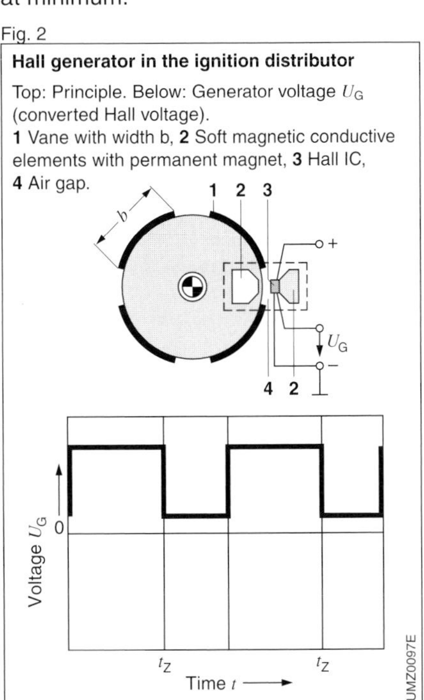

The Hall generator is accommodated in the ignition distributor. The magnetic barrier is mounted on the movable carrying plate.

The Hall IC is located on a ceramic substrate and is potted in plastic together with one of the conductive elements in order to protect it against moisture, dirt and mechanical damage. The conductive elements and trigger wheel are of a soft magnetic material. The trigger wheel and distributor rotor comprise one component on the retrofit version. The number of vanes is equal to the number of cylinders. The width b of the individual vanes can determine the maximum dwell angle of the ignition system, dependent upon the ignition trigger box. The dwell angle consequently remains constant throughout the entire service life of the Hall generator; thus, there is no need to set it. The mode of operation and design of the Hall generator permit the ignition system to be set with the engine switched off providing peak-coil-current cut-off is incorporated. If the technical prerequisites are fulfilled, and the installation instructions are observed precisely, it is an easy matter to convert from conventional ignition to breakerless ignition (Fig. 4). Bosch service stations will be able to provide you with further information.

Current regulation and dwell-angle closed-loop control

High-performance ignition systems operate with ignition coils which charge very rapidly. For this purpose, the ohmic resistance of the primary winding is reduced to less than 1Ω. The information content of the signal of a Hall vane switch in the ignition distributor corresponds to the signal of an ignition contact breaker. In one case, the dwell angle is determined by the distributor cam and, in the other, the pulse duty factor is determined by the rotor vane. A rapidly chargeable ignition coil cannot operate with a fixed dwell angle. This is why two measures

Fig. 3

Varying the dwell angle by shifting the trigger level on the Hall generator

a Dwell angle S_1 correct, **b** Dwell angle S_2 too small, **c** Dwell angle S_3 too large, $t_1 \ldots t_3$ time during which the output stage is conductive, t_1^* current-limiting time correct, t_3^* current-limiting time too long.

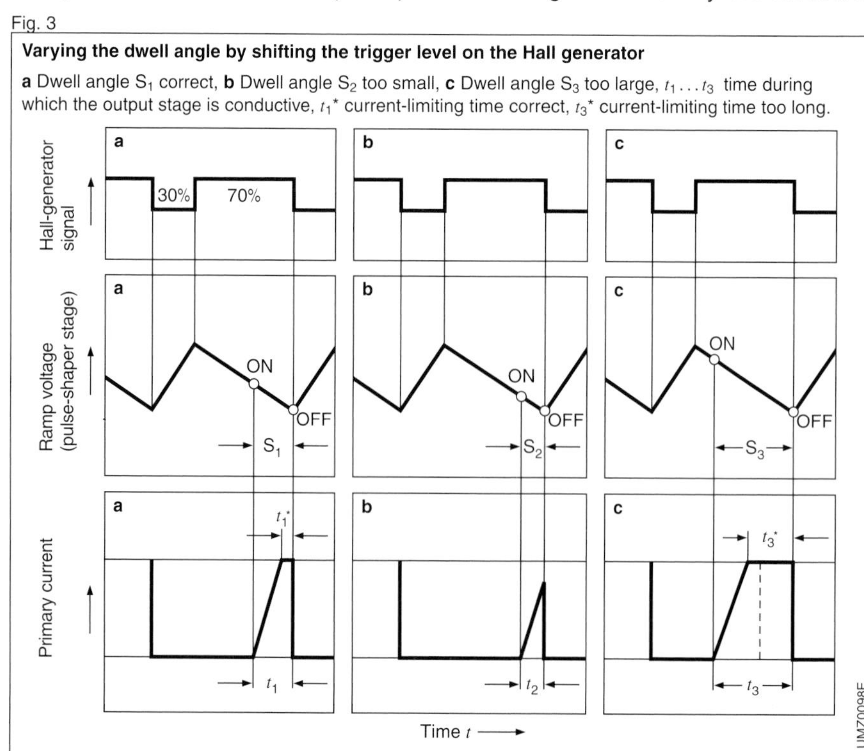

must be taken to protect the ignition coil: a primary-current regulation system and a dwell-angle closed-loop control system (Fig. 3).

Current-regulation function

The primary-current regulation system serves to limit the current through the ignition coil and, thus, to limit the build-up of energy to a specific amount. A certain lead time is required in order to cope with the dynamic conditions applicable when the engine accelerates. This means that the ignition coil should reach its nominal value before the ignition point. During this current regulation phase, the ignition transistor operates in its active range. More voltage than in the pure switch mode drops across the transistor. This means a high power loss which may lie between 20 and 30 W. In order to minimize this, and in order to set the appropriate dwell angle, a dwell-angle closed-loop control system is required (which is actually a dwell-period closed-loop control system since the coil is charged as a function of time).

Function of the dwell-angle closed-loop control system

Since control processes in analog systems are carried out simply by shifting voltage threshold values, the square-wave signal of the Hall generator is first

Fig. 4

Ignition distributor with Hall generator (retrofit version)

1 Vanes, **2** Vane switch, **3** Conductive element, **4** Air gap, **5** Ceramic substrate with Hall IC (potted), **6** 3-core Hall-generator lead, **7** Ignition-distributor shaft, **8** Carrying plate, **9** Distributor housing, **10** Distributor rotor.

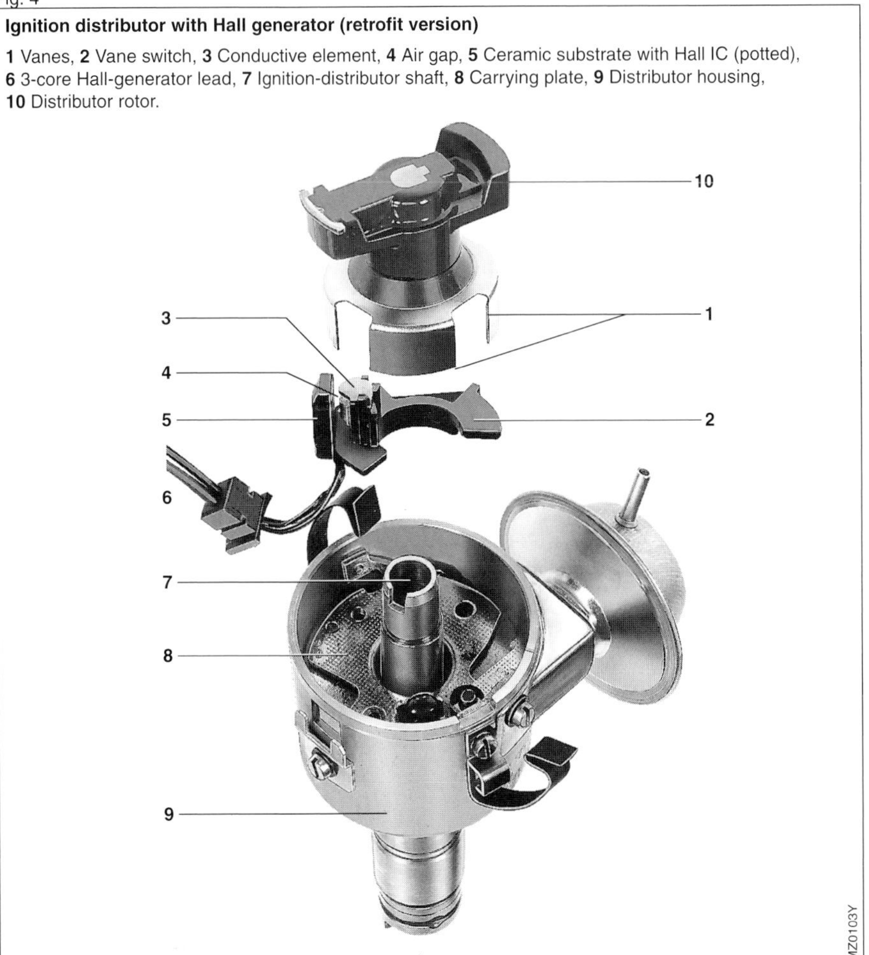

Control unit (trigger box) with output stage for current regulation and dwell-angle closed-loop control

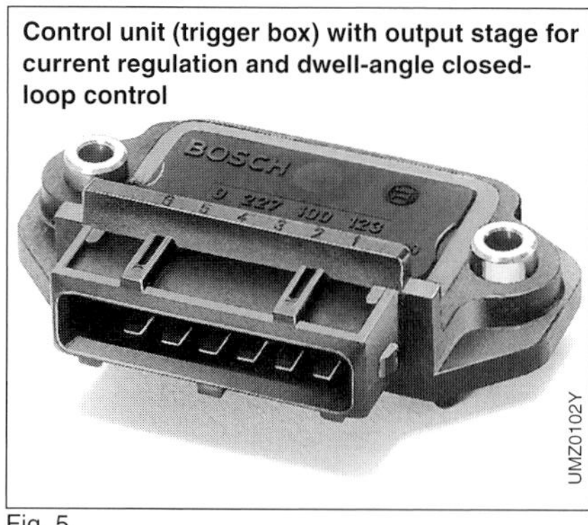

Fig. 5

converted to a ramp signal by charging and discharging capacitors. The pulse duty factor of the Hall generator is 30:70 between two ignition points.

The ignition point determined by adjustment of the ignition distributor lies at the end of the vane width corresponding to 70 %. The closed-loop control system is set such that the current control time t_1 corresponds precisely to the required dynamic lead. A voltage is formed on the basis of value t_1 and is compared with the trailing ramp of the ramp voltage. The primary current is switched on at the intersection point "ON" and the dwell angle starts. In this way, the switch-on point of the dwell angle can be varied as required by shifting the intersection point on the ramp voltage and by varying the voltage derived from the current control time. This means that the correct dwell angle is always formed for every operating range. Since current regulation and dwell-angle closed-loop control are dependent directly upon current and time, this eliminates the effects of varying battery voltage and temperature effects or other ignition-coil tolerances. This renders these ignition systems particularly suitable for cold starting. Since primary current can flow owing to the waveform of the Hall signal with the engine switched off and with the ignition-starting switch switched on, the control units can be equipped with an auxiliary circuit which switches off this "peak-coil current" after a certain period.

Control unit (ECU)

Transistorized ignition systems with current regulation and dwell-angle closed-loop control, virtually all comprise hybrid circuits. This makes it possible to combine the compact and lightweight control units (Fig. 5), for instance with the ignition coil, to form one assembly. Owing to the power loss which occurs in the ignition coil and in the transistorized-ignition control unit, adequate cooling and good thermal contact with the bodywork are required.

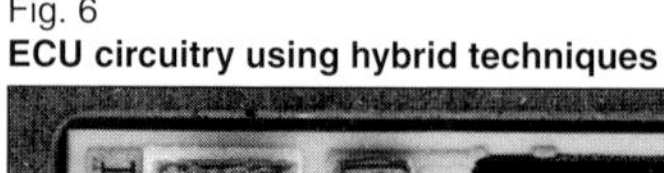

Fig. 6
ECU circuitry using hybrid techniques

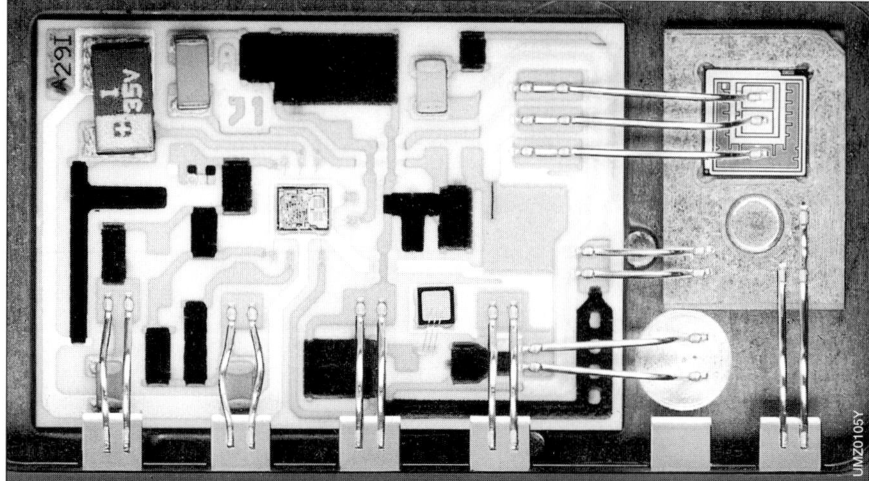

Transistorized ignition with induction-type pulse generator TI-I

The transistorized ignition system with induction-type pulse generator (TI-I) is a high-performance ignition system like the Hall-generator ignition system, and there are only slight differences between the two systems. As compared with the TI-H, the TI-I has a greater phase displacement between the actual ignition point and the off-edge of the pulse-generator voltage at high engine speeds. This is attributable to the induction-type pulse generator in the TI-I which represents an electrical AC generator and has an additional phase displacement owing to the load of the control unit. In certain cases, this effect is even desirable in order to correct the characteristic curves to prevent knocking. Owing to the symmetrical pulse-generator construction, the TI-I is characterized by less spark oscillation in comparison to the Hall barrier of the TI-H which is arranged asymmetrically with respect to the axis of rotation.

Induction-type pulse generator

The permanent magnet, inductive winding and the core of the induction-type pulse generator form a self-contained unit, the "stator". The trigger wheel which is located on the ignition distributor shaft, termed the "rotor" turns with respect to this fixed assembly. The core and rotor are made of soft magnetic steel. They have tooth-shaped extensions (stator teeth, rotor teeth).

The operating principle is based upon the fact that the air gap between the rotor and stator teeth changes periodically when the rotor rotates. This also changes the magnetic flux. The change in flux induces an AC voltage in the inductive winding. The peak voltage $\pm \hat{U}$ is dependent upon the engine speed: approximately 0.5 V at low engine speed and approximately 100 V at high engine speed. The frequency f of this AC voltage corresponds to the sparking rate.

It is as follows,

$$f = z \cdot \frac{n}{2}$$

f Frequency or sparking rate (min^{-1})
z Number of cylinders,
n Engine speed (min^{-1}).

Design features

The induction-type pulse generator is accommodated in the housing of the ignition distributor in place of the contact breaker (Fig. 1). When viewed from the outside, only the plug-in two-core generator lead indicates that this ignition distributor has an induction-type pulse generator. The soft magnetic core of the inductive winding is in the form of a circular disc, called the "pole piece". On the outside, the pole piece has stator teeth which, for instance, may be bent upwards perpendicularly. Consequently the rotor has teeth which are bent downwards.

The trigger wheel, which is comparable with the distributor cam of the contact breaker, is permanently connected to the

Fig. 1

Ignition distributor with induction-type pulse generator

Top: Principle. Below: Induced voltage.
1 Permanent magnet, **2** Inductive winding with core, **3** Variable air gap, **4** Rotor.

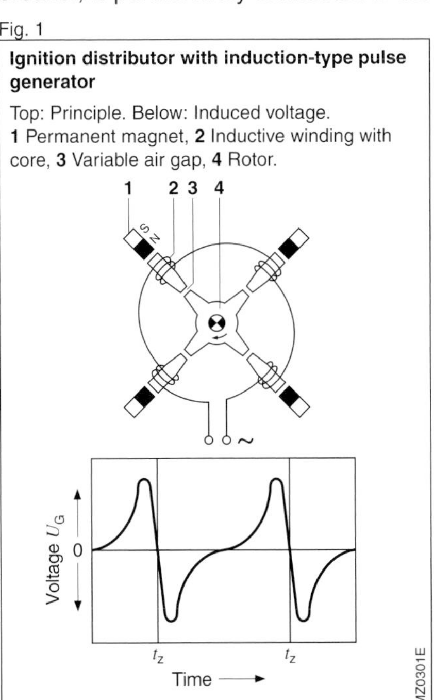

hollow shaft which encloses the distributor shaft. The number of teeth on the trigger wheel and on the pole piece generally correspond to the number of cylinders in the engine. The fixed and moving teeth, when positioned directly opposite each other, have a spacing of approximately 0.5 mm.

Current regulation and dwell-angle closed-loop control

Current regulation and dwell-angle closed-loop control on the TI-I are similar to those on the TI-H. However, they are generally less complex since, normally, it is not necessary to generate a ramp voltage for shifting the on-time of the dwell angle. Instead, the signal of the induction-type pulse generator itself can be used as the voltage ramp on the basis of which the on-time of the dwell angle is determined by comparison with a voltage signal corresponding to the current control time (Fig. 2).

Current regulation function

The current regulation system detects the current by measuring the voltage drop across a low-resistance resistor in the emitter lead of the ignition transistor. The driver stage of the ignition transistor (Darlington transistor) is driven directly via a closed-loop current-limiting circuit.

Function of the dwell-angle closed-loop control system

The dwell-angle closed-loop control system operates with the same measuring-circuit voltage but routes it to its own closed-loop control circuit. Any necessary correction of the dwell angle can be determined by assessing the time during which the current of the transistor is being controlled.

Control unit (ECU)

Control units of TI-I high-performance ignition systems virtually all comprise hybrid circuits since they combine a high packing density with low weight and high reliability.

Fig. 2

Varying the dwell angle by shifting the trigger level on the induction-type pulse generator

a Dwell angle S_1 correct, **b** Dwell angle S_2 too small, **c** Dwell angle S_3 too large, $t_1 \ldots t_3$ time during which the output stage is conductive, t_1^* current-limiting time correct, t_3^* current-limiting time too long.

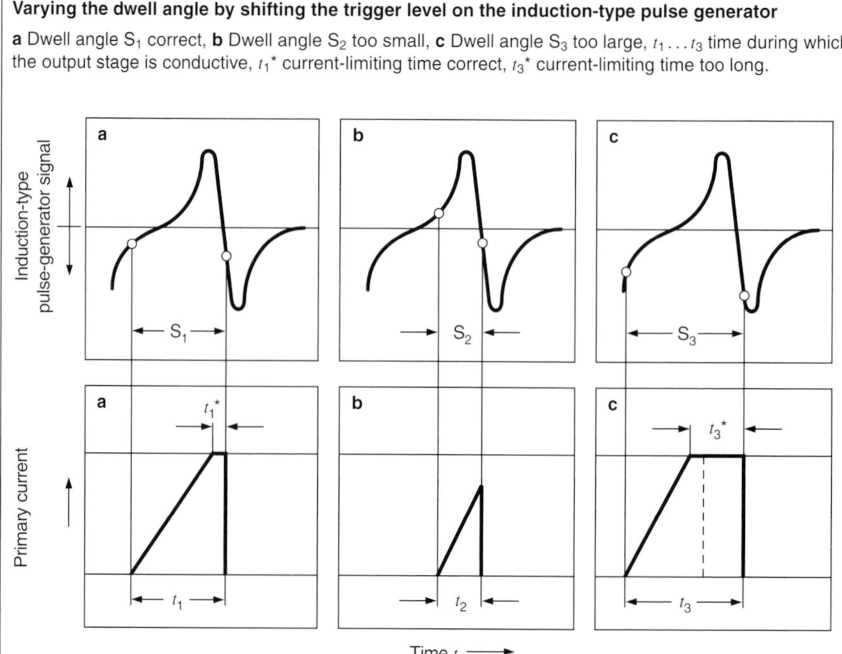

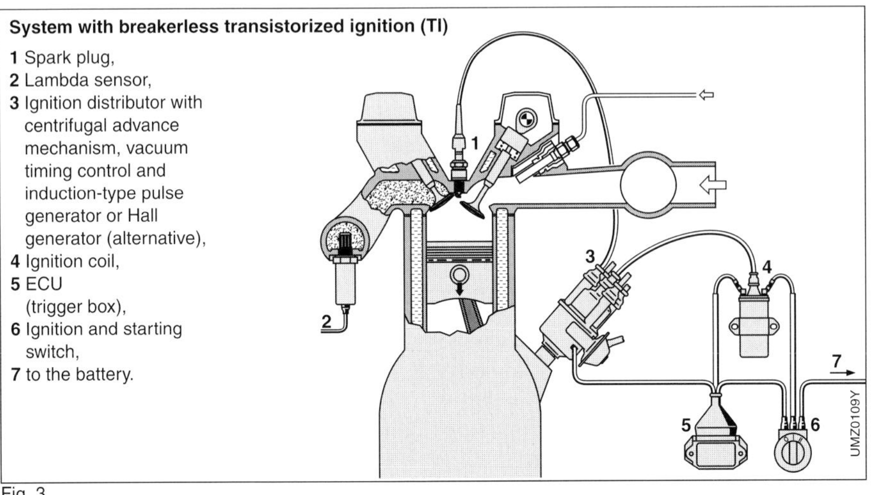

Fig. 3

If less stringent performance data are permitted, it may be possible to dispense with dwell-angle closed-loop control and, possibly even current regulation. Since the control load ratio of the evaluated pulse-generator signal on TI-I systems drops with decreasing engine speed, TI-I control units may be designed more compactly for specific applications and are thus particularly well-suited for direct attachment to the ignition distributor housing. Similar to combining the ECU with the ignition coil, this permits a reduction in the number of components in the ignition system which must be connected with leads (Fig. 4).

Fig. 4

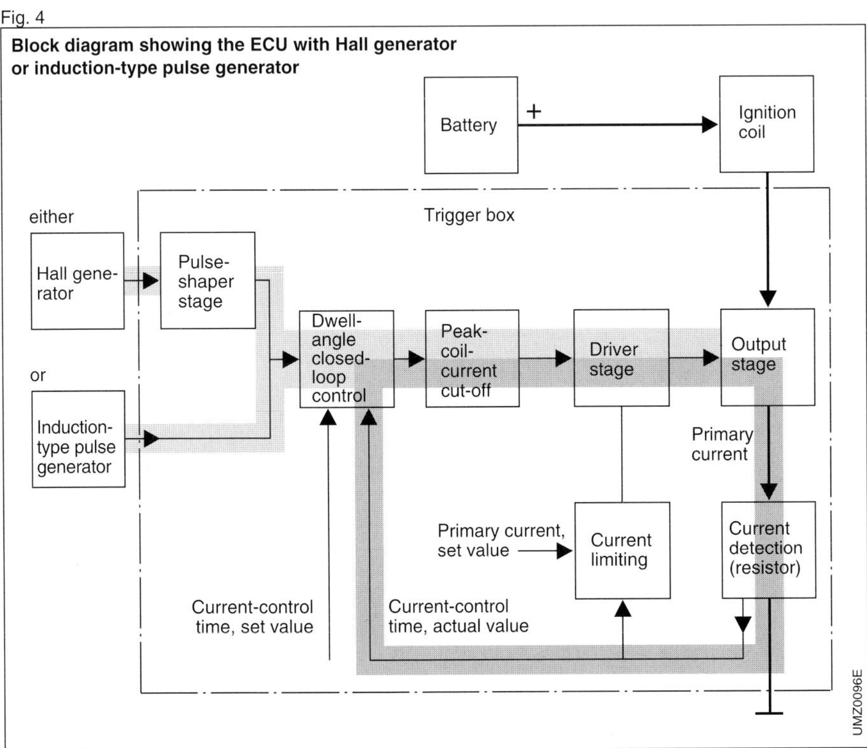

Semiconductor ignition SI

On transistorized ignition systems, the conventional ignition distributor with centrifugal advance and vacuum advance mechanisms is capable of implementing only very simple advance characteristic curves. This means they can only approximately meet the demands im-posed by optimum engine operation.

On the "semiconductor ignition system" (SI, Fig. 1), there is no mechanical spark-advance system in the distributor. Instead, a pulse-generator signal, in the form of an engine-speed signal, is used to trigger the ignition. An additional pressure sensor supplies the load signals. The microcomputer computes the required ignition-point adjustment and modifies the output signal issued to the trigger box accordingly.

Advantages

- Spark advance can be matched better to the individual and diverse requirements made of the engine.
- It is possible to include other control parameters (e.g. engine temperature).
- Good starting behaviour, improved idle control and lower fuel consumption.
- Extended operating-data acquisition.
- It is possible to implement knock control.

The advantages of the semiconductor ignition system are most clearly demonstrated by the ignition-advance map which contains the ignition angle for every given engine operating point. This ignition angle was selected during engine design as the best compromise for every engine speed and for every load condition. The ignition angle for a specific operating point is selected on the basis of the following aspects: fuel consumption, torque, exhaust gas, safety margin from the knock limit, engine temperature and driveability etc. Dependent upon the optimization criterion, one of these

Fig. 1

Semiconductor ignition system (SI)

1 Ignition coil with attached ignition output stage, **2** High-tension distributor, **3** Spark plug, **4** ECU, **5** Engine-temperature sensor, **6** Throttle-valve switch, **7** Rotational-speed sensor and reference-mark sensor, **8** Ring gear, **9** Battery, **10** Ignition and starting switch.

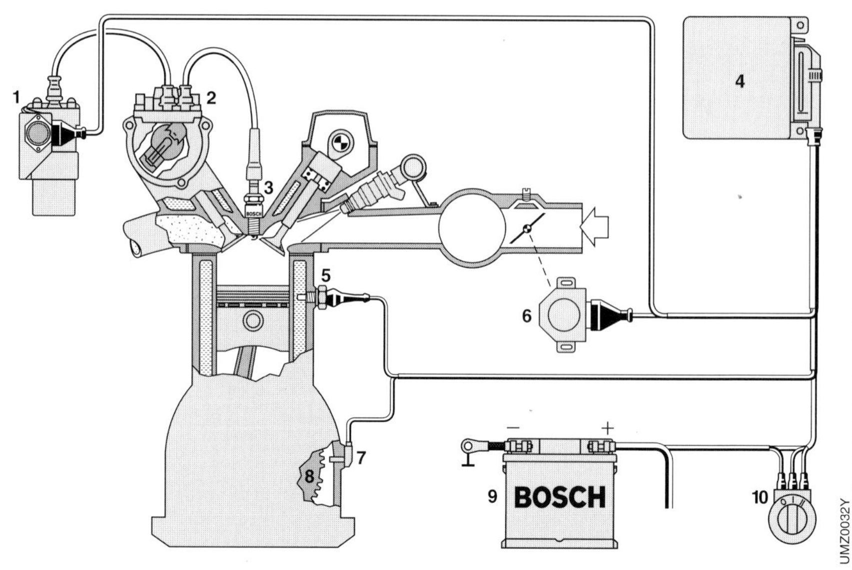

UMZ0032Y

aspects will be more important. This is why the ignition-advance map of a semiconductor spark-advance system frequently appears very rugged and jagged by comparison with the ignition map of a system with centrifugal and vacuum advance mechanisms. If, in addition, the generally non-linear influence of temperature or of another correction function is also to be represented, this would require a four-dimensional ignition map which would be impossible to depict.

Operating principle

The signal issued by the vacuum sensor is used as the load signal for the ignition system. A three-dimensional ignition-advance map is, so to speak, stretched over this signal and the engine speed. This map permits the best ignition point (angle) (in the vertical plane) to be programmed for every engine speed and load condition (horizontal plane) in respect of the exhaust gas and fuel consumption. The entire ignition-advance map contains approximately 1,000 ... 4,000 individual recallable ignition points, dependent upon requirements (Fig. 2).

Fig. 2

Optimized electronic ignition-advance map (top) compared with the ignition-advance map of a mechanical spark-advance system (below)

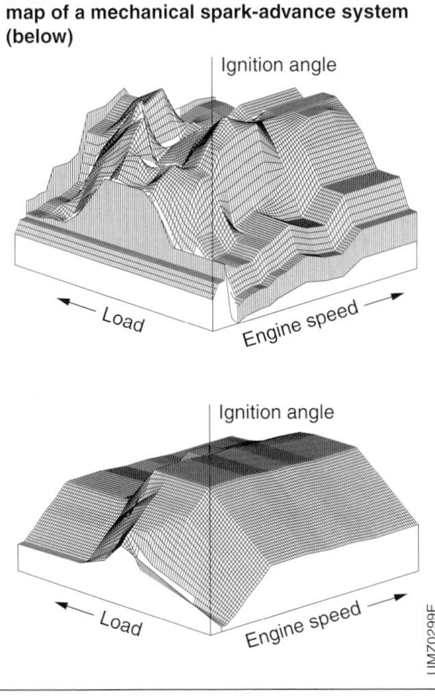

When the throttle valve is closed, the special idle/overrun characteristic curve is selected. The ignition point can be "advanced" for engine speeds below the nominal engine idle speed in order to achieve idle stabilization by increasing the torque. Ignition points matched in respect of exhaust gas, handling and driveability are programmed for overrun operation. At full load, the full-load curve is selected. This curve contains the best programmed ignition parameters allowing for the knock limit.

In the case of specific systems, for starting, a progression of the ignition point, independent of the ignition-advance map, can be programmed as a function of engine speed and engine temperature. This permits high engine torque during starting without the occurrence of countertorques.

Dependent upon requirements, it is possible to implement ignition-advance maps of various degrees of complexity or only a few programmable advance curves. Electronic spark advance is possible within the framework of various semiconductor ignition systems. For instance the Motronic system incorporates fully integrated spark advance. However, it is also possible to implement a spark-advance system as an addition to a transistorized ignition system (in the form of an additional advance system) or as a device with integrated output stage.

Engine-speed sensing

There are two possible methods of engine-speed sensing in order to determine the engine speed and for synchronization with the crankshaft: the signal can be tapped-off directly at the crank-shaft or camshaft, or at an ignition distributor equipped with a Hall ignition vane switch. The advantages afforded by an ignition-advance map with the form already discussed can be utilized to maximum accuracy with an engine-speed sensor on the crankshaft.

Input signals

Engine speed/crankshaft position and intake-manifold pressure are the two main control variables for the ignition point.

Engine speed and crankshaft position

An induction-type pulse generator which scans the teeth of a special-purpose gear wheel on the crankshaft serves to sense the engine speed. The resulting change in magnetic flux induces an AC voltage which is evaluated by the control unit. This gear wheel has a gap which is sensed by the pulse gen-erator and the signal is then processed in a special circuit for clear assignment of the crankshaft position. Triggering with the aid of a Hall generator in the ignition distributor can also be used. In the case of symmetrical engines, it is also possible to trigger pulses inducti-vely via segments on the crankshaft. The number of segments in this case corresponds to half the number of cylinders (Figures 3 to 5).

Load (intake-manifold pressure)

The pressure in the intake manifold acts upon the pressure sensor via a hose.
In addition to the intake-manifold pressure for only indirect load measurement, the air mass or the air quantity per unit of time are also particularly suitable as load signals since they provide a better indication of the charge of the cylinder which is the actual load. On engines equipped with an electronic fuel-injection system, it is thus possible to utilize the load signal not only for the fuel man-agement but also for the ignition as well.

Throttle-valve position

A throttle-valve switch supplies a switching signal during engine idle and full load (Fig. 5).

Temperature

A water-temperature sensor in the engine block (Fig. 5) supplies the control unit with a signal corresponding to the engine temperature. The intake-air temperature can also be sensed by a further sensor either in addition to or instead of the engine temperature.

Battery voltage

The battery voltage is also a correcting quantity which is detected by the control unit.

Signal processing

Intake-manifold pressure, engine temperature and battery voltage, in the form of analog variables, are digitized in the analog-to-digital converter. Engine speed, crankshaft position and throttle-valve stops are digital variables and are

Fig. 3

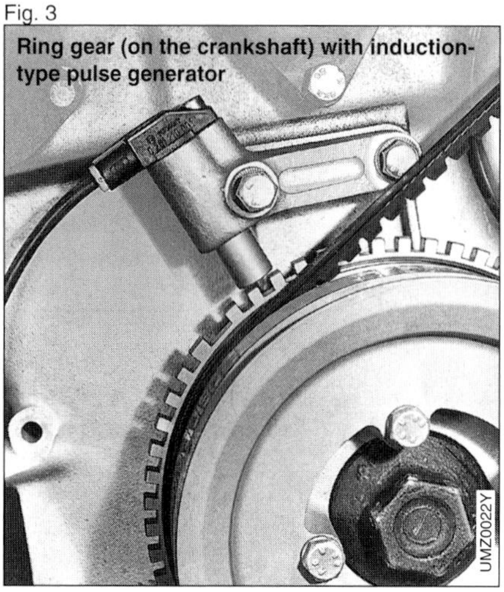

Ring gear (on the crankshaft) with induction-type pulse generator

Fig. 4

Induction voltage curve

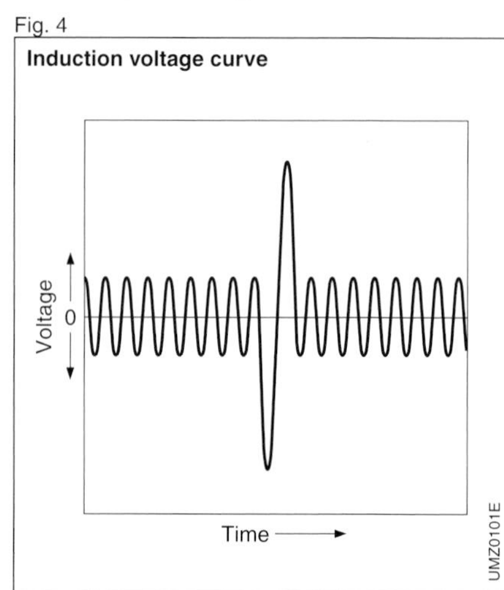

routed directly to the microcomputer. Signal processing is carried out in the microcomputer, which comprises the microprocessor with quartz oscillator crystal for clock-pulse generation. In the computer, the updated values for the ignition angle and the dwell period are calculated anew for each ignition in order that the optimum ignition point is always available to the engine for every operating point.

Ignition output signal

The primary circuit of the ignition coil is switched by means of a power output stage in the electronic control unit. The dwell period is controlled such that the secondary voltage remains practically constant regardless of engine speed and battery voltage.

Since the dwell period or dwell angle is determined anew for each engine speed and battery voltage condition, this requires a further ignition map: the dwell-angle map (Fig. 6). It contains a network of data points between which interpolation is carried out as is the case with the ignition-advance map. Using such a dwell-angle map permits the energy stored in the ignition coil to be metered just as precisely as with a dwell-angle closed-loop control system. However, there are also semiconductor ignition systems in which a dwell-angle closed-loop control is superimposed upon the dwell-angle map. This closed-loop control system optimizes the dwell angle for each cylinder independently.

Dwell-angle map

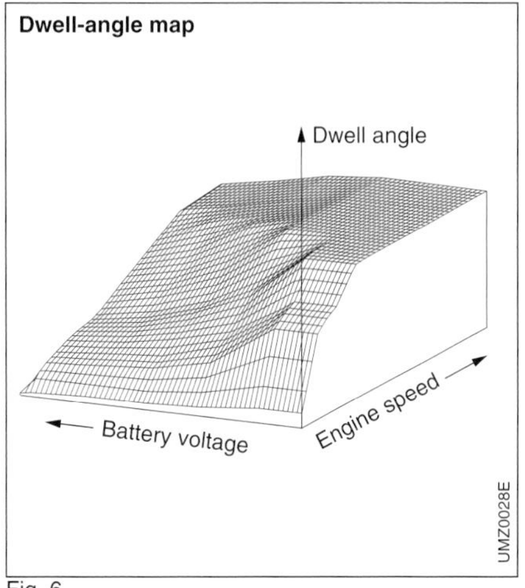

Fig. 6

Control unit (ECU)

As can be seen on the block diagram, the heart of the control unit for a semiconductor ignition system is a microcomputer which contains all data, including the ignition maps, in addition to the programs for detecting the input variables and for computing the output variables. Since the sensors are largely electromechanical components matched to the tough operating conditions of the

Fig. 5

Adaptation components

1 Throttle-valve switch, **2** Pulse generator, **3** Engine-temperature sensor.

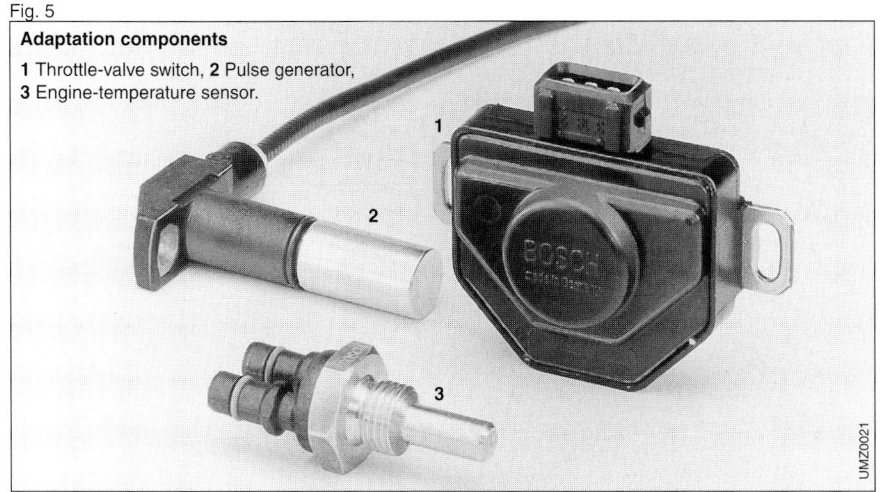

engine, it is necessary to process the signals for the computer.
Pulse-shaping circuits convert the pulsating signals from the sensors (such as those from the rpm sensor) into defined digital signals. As a further example, temperature and pressure sensors frequently transmit their signals in analog form. This analog signal is converted in an analog-to-digital converter and routed to the computer in digital form. The analog-to-digital converter may also be integrated in the microcomputer (Fig. 7).
In order to permit ignition-map data to be changed until shortly before the start of series production, there are control units available with an electronically programmable memory, generally in the form of an EPROM (electronically programmable read-only memory).

Ignition output stage

The ignition output stage may either be incorporated in the control unit itself (as shown in the block diagram) or may be accommodated externally, generally in combination with the ignition coil. In the case of external ignition output stages, the control unit is generally fitted in the passenger compartment. This is also the case less frequently on control units with integrated ignition output stage.
If control units with integrated ignition output stage are accommodated in the engine compartment, they require particularly good heat dissipation. This is achieved by the use of hybrid circuitry. Semiconductor devices and, thus, also the output stage, are then fitted directly to the heat sink which ensures good thermal contact to the bodywork. This means that such control units can be operated at ambient temperatures of over 100°C. Hybrid units have the further advantage of being compact and lightweight.

Fig. 7

Signal processing in the ignition ECU (block diagram)

1 Engine speed, **2** Switch signals, **3** CAN (serial Bus), **4** Intake-manifold pressure, **5** Engine temperature, **6** Intake-air temperature, **7** Battery voltage, **8** A/D converter, **9** Microcomputer, **10** Ignition output stage.

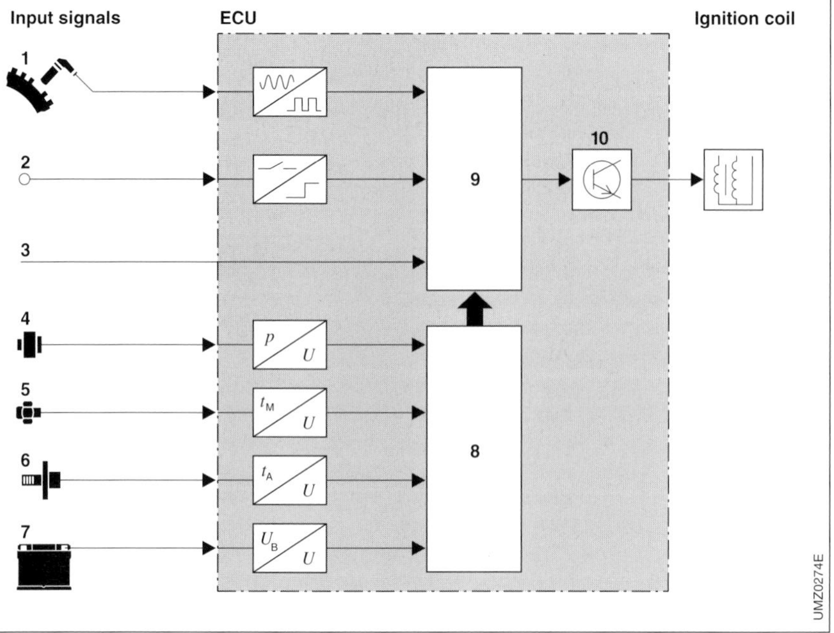

Input signals
ECU
Ignition coil
UMZ0274E

Other output variables

In addition to the ignition output stage, there are controls for further output variables, dependent upon the particular application. Examples of these are outputs for engine-speed signals and status signals for other control units such as injection, diagnostic signals, switching signals for actuating injection pumps or relays etc.

The semiconductor ignition system is particularly suitable for combining with other engine-management functions (Figures 8 and 9). Combined with an electronic fuel-ignition system, this means that the basic version of a Motronic system is realized in a single control unit.

One equally popular form is combining the semiconductor ignition system with a knock-control system. This combination is particularly advantageous since retarding the ignition point is the fastest and most reliable method of intervention to avoid engine knocking.

Fig. 9

Semiconductor-ignition ECU using hybrid techniques

The load sensor is located in the cover.

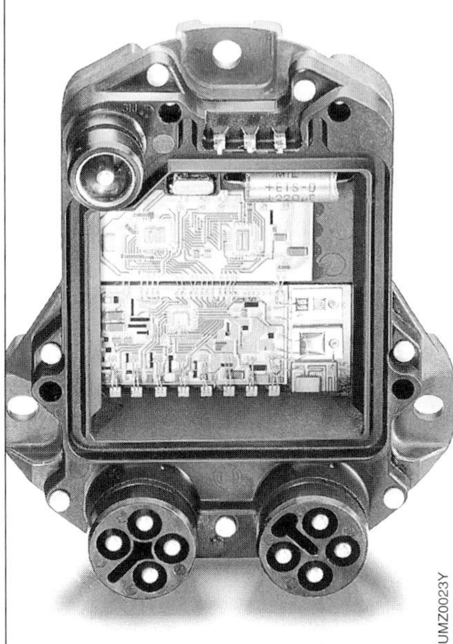

Fig. 8

Semiconductor-ignition ECU with knock control, using printed-circuit-board techniques

The aneroid box **D** serves to measure the intake-manifold pressure.

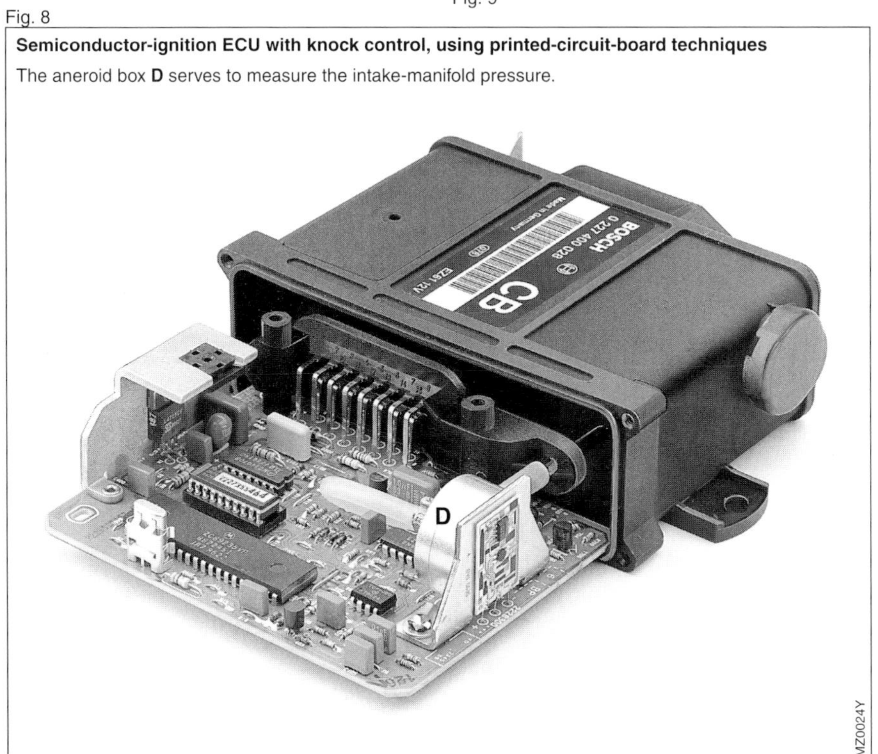

Distributorless semiconductor ignition DLI

The distributorless semiconductor ignition system (DLI, Fig. 1) is characterized by two features: it performs the functions of a semiconductor ignition system, but has no rotating high-voltage distribution system using an ignition distributor.

Advantages

Although it has no advantages regarding weight, stationary high-voltage distribution does have the following advantages:
- Far lower electromagnetic interference level since no open sparks occur,
- no rotating components,
- noise reduction,
- less high-voltage connections and
- design advantages for the engine manufacturer.

The performance data of a distributorless semiconductor ignition system are comparable to those of a conventional semiconductor ignition system.

High-voltage distribution

High-voltage distribution with dual-spark ignition coils

In the simplest case, e.g., on the 4-cylinder engine, two dual-spark ignition coils are used instead of the ignition distributor. These are energized alternately via an ignition output stage. At the ignition point for a given cylinder, which is determined by the microcomputer-controlled ignition map in the same way as with a conventional semiconductor ignition system, the corresponding double-spark ignition coil generates two ignition sparks simultaneously. The two spark plugs at which the sparks are produced are each electrically connected in series with this ignition coil so that one spark plug is connected to each of its high-voltage outputs. These two spark plugs must be arranged so that one spark plug fires in the working stroke of the cylinder in question and the other in the exhaust stroke of the cylinder which is offset by 360°. One rotation of the crankshaft later, these two cylinders are two working strokes further and the spark plugs fire again, but now with reversed roles.

Fig. 1

Distributorless semiconductor-ignition system (DLI)

1 Spark plug, **2** 2 x Dual-spark ignition coils, **3** Throttle-valve switch, **4** Control unit with integrated driver stages, **5** Lambda sensor, **6** Engine-temperature sensor, **7** Engine-speed and reference-mark sensor, **8** Ring gear, **9** Battery, **10** Ignition and starting switch.

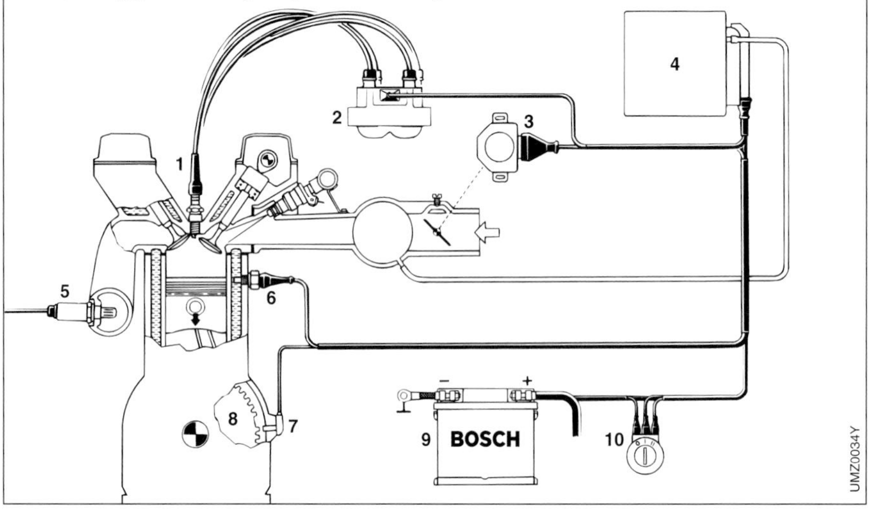

The second dual-spark ignition coil also generates two sparks but offset by 180° (crankshaft) with respect to the first. Using the 4-cylinder engine as an example again, we see that cylinders 1 and 4 always fire simultaneously as do cylinders 3 and 2.

In addition, the dual-spark ignition coil which is the next to be fired requires a signal identifying the start of a revolution. In the example shown, the TDC signal signals that firing must occur in cylinder group 1/4. The computer establishes when the crankshaft has turned a further 180° and then initiates ignition in cylinder group 3/2 with the other dual-spark ignition coil. At the start of the second revolution, the TDC signal is issued again and, once again, causes ignition in cylinder group 1/4.

This forced synchronization system also ensures that the correct firing sequence is maintained even in the event of malfunctions of any kind. Only engines with an even number of cylinders (e.g. 2, 4, 6) are suitable for this type of stationary, or fully-electronic, high-tension distribution. The number of ignition coils required can be calculated in each case by halving the number of cylinders.

The schematic diagram of the distributorless semiconductor-ignition system shows a system with distribution by two double-spark ignition coils. The reference-mark sensor on the crank-shaft also serves to trigger the correct ignition coil, in addition to calculating the ignition angle.

High-voltage distribution with single-spark ignition coils

A distributorless semiconductor-ignition system for an odd number of cylinders (e.g. 3, 5) requires its own ignition coil for each cylinder (single-spark ignition coils are also suitable for engines with an even number of cylinders, in conjunction with distributorless semiconductor ignition systems). The actual distribution of the high voltage to the ignition coils is performed in the low-voltage circuit in a power module with distributor logic. In the case of engines with an odd number of cylinders, one cycle covers two revolutions of the crankshaft. For this reason, a TDC signal from the crankshaft is not sufficient in this case. One signal per camshaft revolution must be triggered by the camshaft for synchronization purposes.

Fig. 2

Firing sequence of a four-stroke engine during two crankshaft revolutions (cycle)

Engines with an even number of cylinders supply a clear signal for ignition of the cylinder group at TDC (0° and 360°).

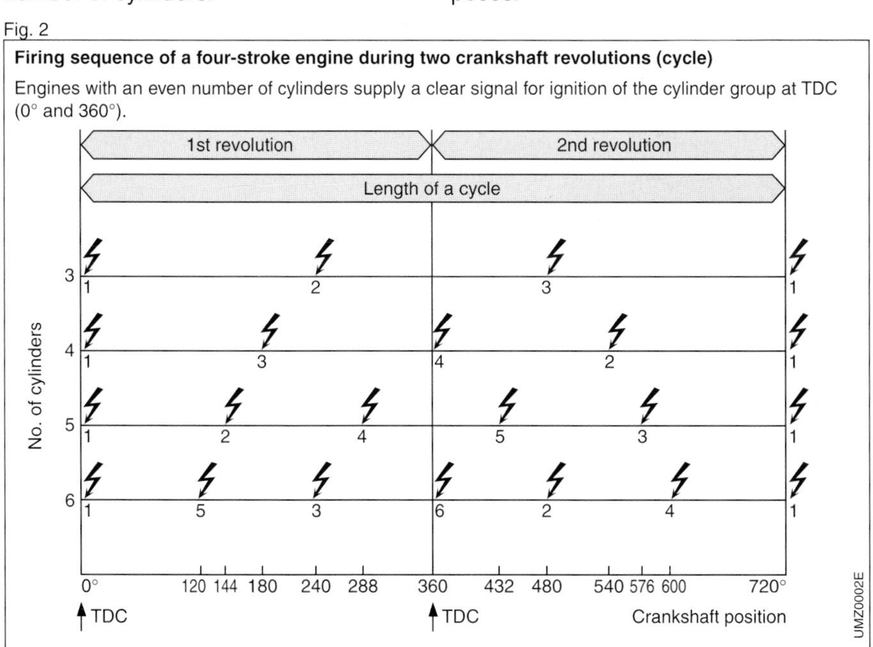

High-voltage distribution with four-spark ignition coils

One further method of stationary high-voltage distribution is a four-spark ignition coil which has two primary windings and one secondary winding. The two primary windings are energized by two ignition output stages. The high-voltage winding has two diodes at each output. From each of these diodes, one high-voltage cable is routed to each spark plug. This means that two sparks are produced alternately, decoupled by the diodes, as is the case on the dual-spark ignition coil.

Required ignition voltage

Since two spark plugs are connected in series on dual-spark and four-spark ignition coils, the required voltage increases by a few kV owing to the spark plug which fires in the low pressure of the exhaust stroke. However, this additional required voltage is compensated for by the fact that there is no ignition-distributor spark gap. In addition, one spark plug is "incorrectly" polarized in each cylinder group. This means that the center electrode is positive and not negative as is normally the case. This also causes the required voltage to be slightly higher.

Ignition coils

Design

Dual-spark and four-spark ignition coils, are designed as plastic-molded coils. The resulting squat and compact design, together with the large area on the upper side, permit two separate high-voltage towers to be provided on these ignition coils. The coil is cooled and secured by the iron core which is led out externally (Fig. 3).

Mode of operation

If we consider the cycle of a single-cylinder four-stroke engine (two revolutions), we can see how the ignition sparks of a dual-spark ignition coil occur during the engine strokes. The first revolution starts shortly after IO (intake valve opens) and lasts until TDC (top dead center). The second revolution starts at TDC and

Fig. 3

Dual-spark ignition coil

UMZ0026Y

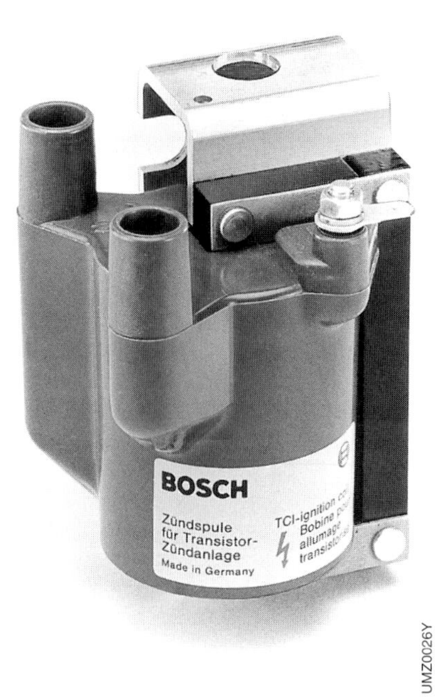

Fig. 4

Occurrence of the ignition sparks of a dual-spark ignition coil in the cycle of a four-stroke engine

1 Switch-on range (start) of the primary current,
2 Ignition range of the first ignition spark,
3 Ignition range of the second ignition spark.
TDC Top dead center, BDC Bottom dead center,
IO Intake valve opens,
IC Intake valve closes, EO Exhaust valve opens,
EC Exhaust valve closes.

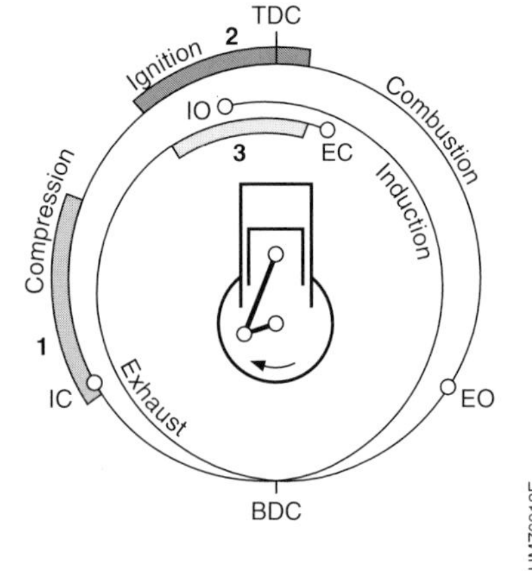

ends shortly before EC (exhaust valve closes). During the working stroke, firing occurs in the area marked red before and shortly after TDC, dependent upon the position of the ignition-advance map point (Fig. 4).

The dwell angle starts in the hatched area, i.e. the ignition-coil primary current is switched on. Depending upon engine speed and battery voltage the switch-on point in this area shifts with the ignition point. At the same time, it also shifts relative to the ignition point in accordance with the dwell-angle map (rotational speed and battery voltage).

Since the two ignition sparks of a double-spark ignition coil are produced simultaneously, i.e. with the same angular crankshaft position, the second ignition spark occurs at the end of the exhaust stroke of the other cylinder (offset by 360° crankshaft) supplied by the given coil. This means that the spark in this cylinder can flash over when the intake valve is starting to open again, and this is critical particularly in the case of large valve overlap (overlap in the open periods of intake and exhaust valves).

Stationary high-voltage distribution with single-spark ignition coils (Fig. 5) requires the same number of ignition output stages and ignition coils as there are cylinders. In such cases, it is practical to combine the power output stage with the ignition coil. This minimizes the number of cables for the high voltage and the medium-high voltage between ignition transistor and ignition coil.

Control unit (ECU)

The electronic control unit of the distributorless semiconductor-ignition system is largely identical to that of the semiconductor ignition system. The ignition output stage can be integrated in the control unit (e.g., in the case of dual-spark or four-spark ignition coils) or can be accommodated externally, in a power module with distributor logic or in combination with the relevant ignition coil (e.g., in the case of single-spark ignition coils).

Fig. 5

Single-spark ignition coil

1 External low-voltage terminal, **2** Laminated iron core, **3** Primary winding, **4** Secondary winding, **5** Internal high-voltage connection (via spring contact), **6** Spark plug.

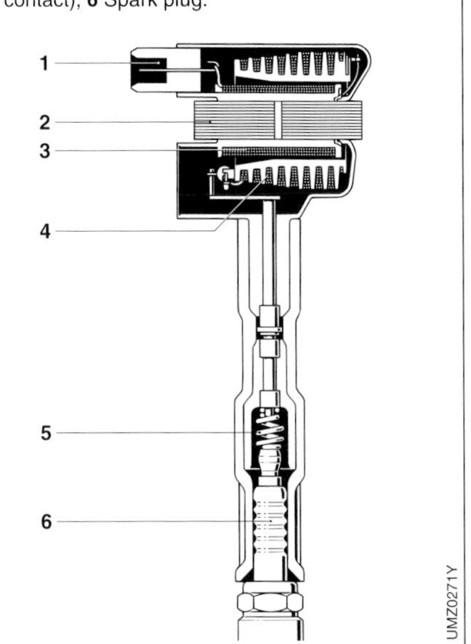

Danger of accident

All electronic ignition systems are inherently dangerous. Before working on these systems, therefore, always switch off the ignition or disconnect the battery. Such work includes:

- Replacing such components as spark plugs, ignition coil or transformer, distributor, high-voltage cables, etc.
- Connecting engine test equipment such as stroboscopic timing light, dwell angle/engine-speed tester, ignition oscilloscope, etc.

Dangerous voltages are present when the ignition is switched on. Testing should therefore be performed by qualified personnel only.

Knock control

Basic functions

Knock limit

Operation with a catalytic converter requires that the engine be operated with unleaded gasoline with an excess-air factor of λ = 1.0. Previously, lead was mixed into gasoline as an antiknock agent, in order to make possible knock-free operation at high compression ratios ε. With unleaded gasoline, a low compression ratio and higher fuel consumption can be expected.

"Knocking" or "pinking", a form of uncontrolled combustion, can lead to engine damage if it occurs too frequently and violently. For this reason, the spark advance is normally designed so that there is always a safety margin before the knock limit is reached. However, since the knock limit is also dependent upon fuel quality, engine con-dition and environmental conditions, the ignition point demanded by this safety margin is too far retarded and the result is a worsening of fuel consumption in the order of several percent. This disadvantage can be avoided if the knock limit is determined continuously during operation, and the ignition angle adjusted to it in a closed loop under the assumption that the ignition angle as specified by the ignition-advance map is already within the knock range. This is carried out by knock control (Fig. 1).

Knock sensor

It is so far impossible to determine the knock limit without the actual occurrence of knocking. Therefore, during closed-loop control along the knock limit there will always be scattered knocking. However, the system is adjusted to the individual type of vehicle concerned so that the knocking is not audible and so that damage is precluded with absolute certainty. The measuring device is the knock sensor which registers the typical noises associated with knocking, turns these into electrical signals and relays them to the electronic control unit (Figures 2 to 4).

The knock sensor is arranged so that knocking from any cylinder can be recognized without difficulty under all conditions. The mounting position is generally on the side of the engine block. With six or more cylinders, one knock sensor is usually inadequate to determine knocking from all cylinders. In such cases, two knock sensors are used per engine, and these are switched corresponding to the firing sequence.

Fig. 1

Schematic diagram of knock control

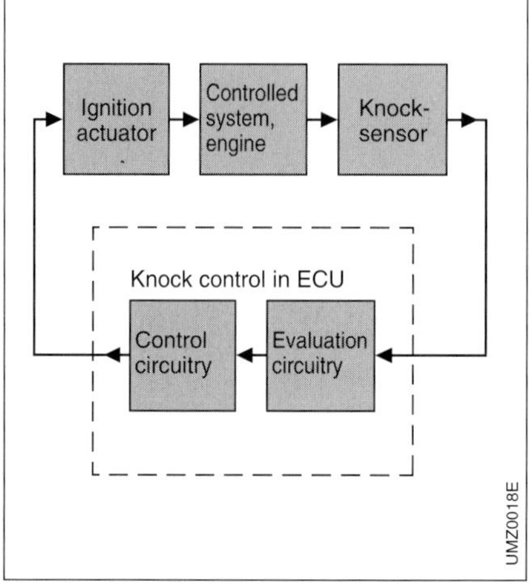

Ignition actuator
Controlled system, engine
Knock-sensor
Knock control in ECU
Control circuitry
Evaluation circuitry
UMZ0018E

Fig. 2

Knock sensor: A wide-band acceleration sensor with a natural frequency above 25 kHz.

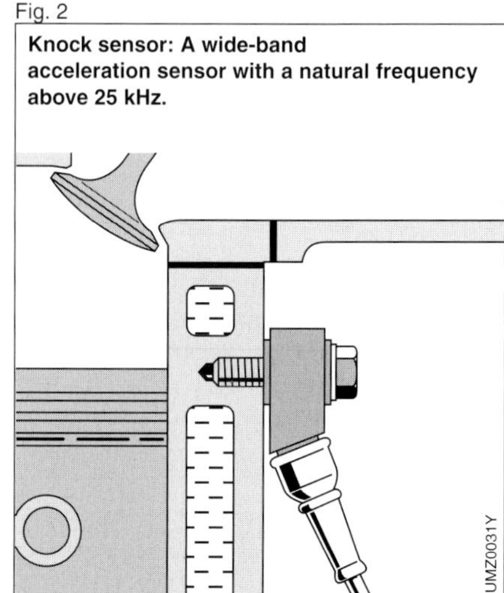

Control unit (ECU)

The sensor signals are evaluated in the electronic control unit. A reference level is individually formed for each cylinder, this level then continuously and automatically adapting itself to operating conditions.

A comparison with the useful signal obtained from the sensor signal after filtering and integration within a crank angle section shows for every combustion process in every cylinder whether knocking is occurring. If this is the case, the ignition point is retarded by a fixed amount, for example 3° crankshaft, for the cylinder involved. This process is repeated for every cylinder for every combustion process recognized as knocking. If there is no more knocking, the ignition point is slowly advanced in small steps until it has returned to its map value.

Since the knock limit varies from cylinder to cylinder within an engine and changes dramatically within the operating range, the result in actual operation at the knock limit is an individual ignition point for every cylinder. This type of "cylinder-selective" knock recognition and control makes possible the best optimization of engine efficiency and fuel consumption.

If the vehicle is designed for operation with unleaded premium fuel, it can also be operated with regular unleaded fuel without damage when provided with knock control. In dynamic operation, knock frequency will increase under such conditions. In order to avoid this, an individual spark-advance map can be stored in the electronic control unit for each of the two fuel types. The engine is then operated after starting with the "premium" map and switched over to the "regular" map if the frequency of knocking exceeds a predetermined limit. The driver is not aware of this switchover; only power and fuel consumption are slightly worsened.

A vehicle designed for premium gasoline and using a conventional ignition system cannot be operated with regular gasoline without danger of knock damage, while a vehicle designed for use with regular gasoline shows no advantages in consumption and power when it is operated with premium gasoline.

Fig. 3

Knock-sensor signals

a Pressure curve in cylinder, **b** Filtered pressure signal, **c** Knock-sensor signal.

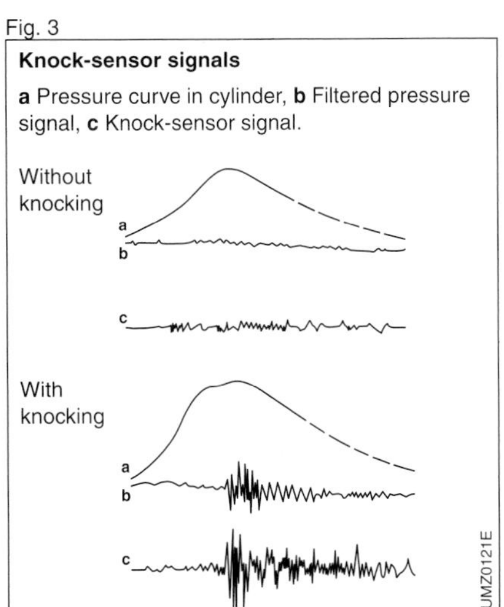

Fig. 4

Knock control

Control algorithm with ignition intervention in a four-cylinder engine.
$K_1 \ldots K_3$ knocking at cylinders 1...3 (no knocking in cylinder 4).
a Retardation, **b** Interval for ignition advance, **c** Advance.

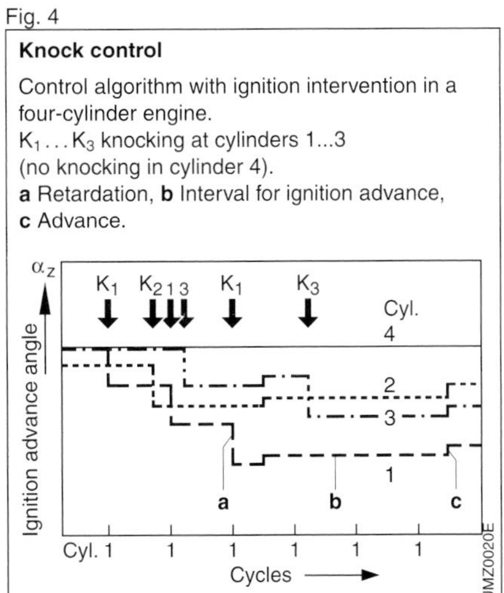

Knock control with turbocharged engines

Boost pressure is controlled via the drive power of the exhaust turbine. Intervention is by the opening cross-section of the exhaust wastegate, which is acted upon with control pressure via a solenoid-operated valve.

A map contains the control values for the solenoid-operated valve. By means of the map, the charge pressure is built up only to the level required by the engine as dictated by the driver (accelerator-pedal position).

The advantages as compared to conventional turbocharged engines are: Less turbocharger work in the part-load range, less exhaust backpressure, less residual exhaust gas in the cylinders, lower charge-air temperature, freely selectable full-load curve of boost pressure as a function of engine speed, softer turbo response, better performance and driveability (Fig. 5).

With closed-loop map control of boost pressure, a control loop is superimposed on the pilot control. A pressure sensor measures intake-manifold pressure which is compared with the values of a stored map. In the case of deviations between the setpoint and actual values the pressure is levelled via the solenoid-operated valve.

Advantages of boost-pressure closed-loop control as compared to open-loop control: Component tolerances and wear, particularly in the exhaust waste-gate and turbocharger, do not affect the level of boost pressure. In addition, if an absolute-pressure sensor is used, the boost pressure can be maintained within a wide range independent of the level of atmospheric pressure (altitude correction).

In case of knocking, ignition timing is retarded for the knocking cylinder just as with a naturally-aspirated engine. In addition, boost pressure is lowered when the ignition retardation of at least 1 cylinder exceeds a prescribed value. This value is stored in the electronic control unit as an engine-speed-dependent characteristic curve. Its quantity is set depending on the maximum allowable exhaust temperature at the turbine inlet. The ignition-timing algorithm (Algorithm: Defined, step-by-step procedure for ob-

Fig. 5

Knock control through combination of semiconductor ignition and boost-pressure control

1 Intake air, **2** Turbocharger, **3** Turbine, **4** Exhaust, **5** Wastegate, **6** Knock sensor, **7** Timing valve, **8** ECU, **9** Ignition coil with attached ignition final stage.

Signals: **a** Throttle-valve position, **b** Intake-manifold pressure, **c** Knock signals, **d** Ignition pulses, **e** Engine temperature, **f** Timing-valve position, **g** Ignition point.

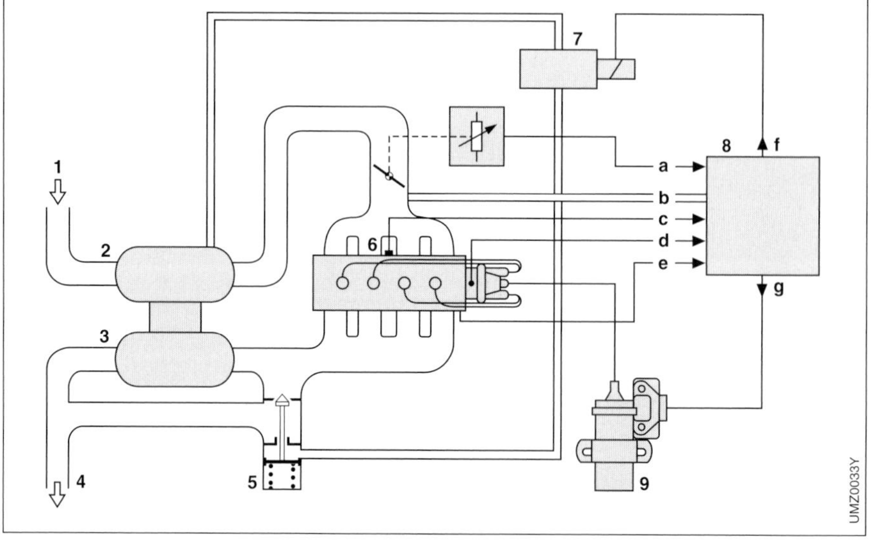

taining mathematical solutions), with fast pressure decrease and slow step-by-step increase up to the nominal value, is similar to the algorithm for ignition timing, but has significantly larger time constants.
The two control algorithms are coordinated taking into consideration the knocking frequency, dynamic behaviour of the engine, exhaust wastegate, and turbocharger, exhaust temperature, driveability, and control stability.
Advantages of this combined control as compared to pure ignition-timing control: Improvement in engine dwell-angle efficiency, reduction in temperature loading of engine and turbocharger, reduction of charge-air temperature.
Advantages compared with pure boost-pressure control: Rapid control response in case of knocking, good dynamic engine behaviour, control stability and driveability.

Special functions

The basic functions of knock-detection and control, ignition-timing, dwell-angle and, where applicable, boost-pressure map, are processed in the control unit. In addition, the intake-manifold pressure as measured by a pressure sensor in the control unit provides information on the load and can be processed in the control unit as can a load signal made available from the gasoline fuel-injection system. Coolant and intake-air temperatures can be taken into account as correction quantities.
If required, overrun fuel cut-off, idle stabilization, and engine-speed limitation can be achieved by switching off ignition or fuel pump and with a fuel-pump control. In addition, should the computer fail – the driver will be informed of this condition – limp-home operation is possible, so that the vehicle must not be left standing.
With turbocharged engines, an engine-speed-dependent full-load signal can be generated and sent to the ignition, in the same way as a signal can be sent to reduce the boost pressure in case of knocking.

Safety and diagnosis

All the functions of knock control, failure of which could lead to engine damage make monitoring a necessity. This must trigger conversion to safe operation should a malfunction arise. The driver can be alerted to the switch to the safety mode by a display on the instrument panel. The exact defect can then be read out via a pulse code when the vehicle is inspected.

The following are monitored:
1. Knock sensor, including wiring harness, continually during operation above a certain engine speed. If a defect is detected, ignition timing in the map range where knock control is active is retarded by a fixed angle; with turbocharged engines, the boost pressure is also decreased.
2. Evaluation circuitry, including the computer, below a certain engine speed. Detection of a defect leads to the same reaction as described above.
3. The load signal, continuously during operation. In the event of defect, the full-load spark advance is used with simultaneous continuous activation of knock control.

Depending upon application, other sensors and signals are monitored and the appropriate reactions determined (e.g., temperature sensor).

Electrical connecting elements

The task of the electrical connecting elements is to reliably transmit the high voltage from the ignition coil through the ignition distributor to the spark plug. Various connection techniques are used, dependent upon the requirements made of the engine and, thus, of the ignition.

Plugs and sockets

Basic versions

One example of the available connection techniques is the plug connection at the high-voltage terminal towers of the ignition distributor. Socket version A (Fig. 1) has only a relatively low high-voltage strength and is thus encountered only in isolated cases in original equipment applications. Versions B and C are the main versions. Both are characterized by having locking pins deep in the terminal tower and by their guaranteeing a substantially higher electric contacting strength owing to the long leakage path. Enlargement in the geometry (as is the case on version C) creates the necessary reserve for guaranteeing the 30 kV required for lean-burn engines. Furthermore, the insertion forces and watertightness are carefully optimized to each other.

Service life

The related mean service life in operating hours is shown by the curves beneath the relevant plug versions. Fig 3 demonstrates their significance: If a voltage U_x is applied to new parts, they initially withstand the stress. However, the insulating capability slowly decreases and, as from time t_1, isolated disruptive discharges must be anticipated. The process advances and, at time t_2, 63% of the parts are destroyed. At low voltages, the parts withstand the stress substantially longer

Fig. 1

Plug and socket versions A, B and C and their high-voltage strength as a function of service life

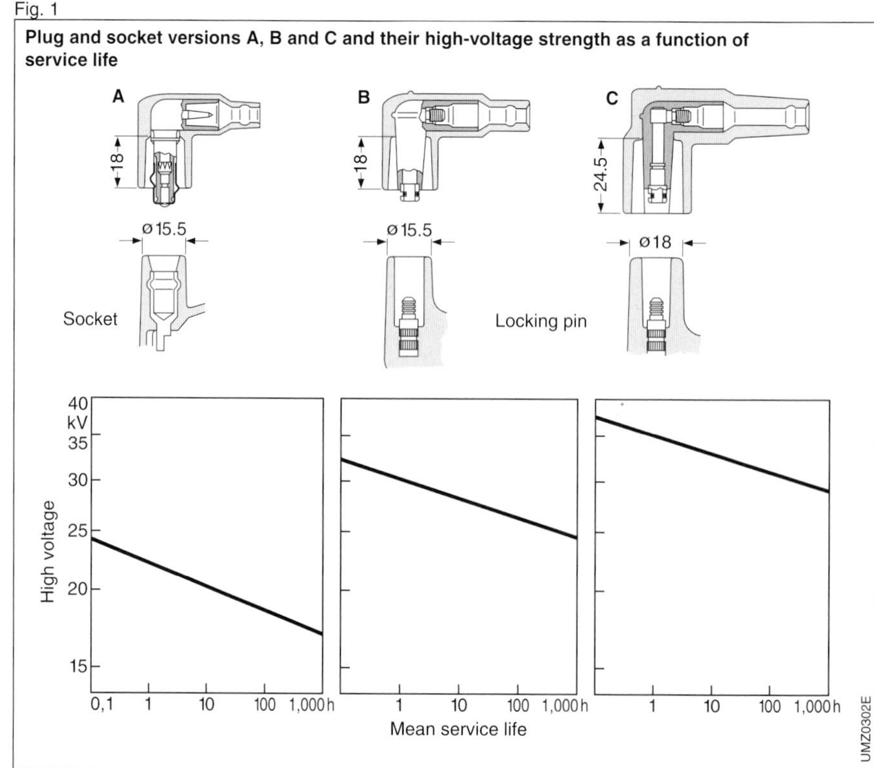

than at high voltages (logorithmic scales). This also approximately corresponds to the statistical distribution of the engine voltage curve. The very high required voltage occurs only rarely, referred to the total number of ignitions. The accumulation point lies at values below 25 kV and this is why versions B and C, in conjunction with a maintenance-free ignition system, sturdy high-voltage cables with metal core and regular spark-plug replacement, provide an ignition system which poses no problems throughout the vehicle's service life.

Special versions

One particularly carefully designed connection technique comprises watertight spark-plug connectors, high-quality ignition cables, watertight ignition-distributor and ignition-coil connectors and protective hoods for the ignition distributor and the ignition coil. These protective hoods provide additional protection against hose water and against dirt. In addition, the ignition-distributor hood improves the interference suppression (Fig. 2).

Insulating capability of plug connectors as a function of time

U_x Voltage,
t_1 Time with isolated disruptive discharges,
t_2 Time with many disruptive discharges.

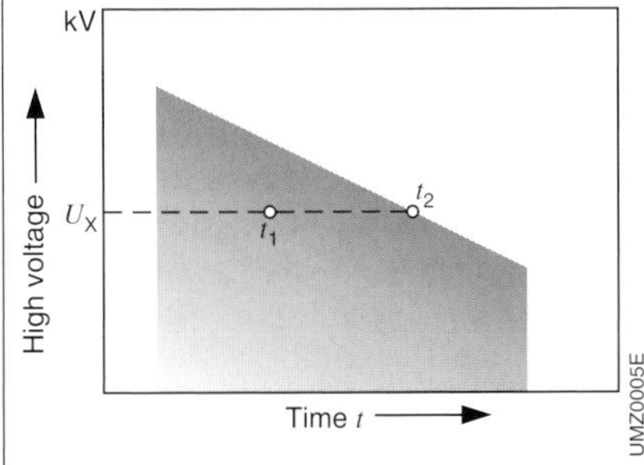

Fig. 3

Fig. 2

Ignition-system cable connections

Protective caps prevent the penetration of dirt and moisture.

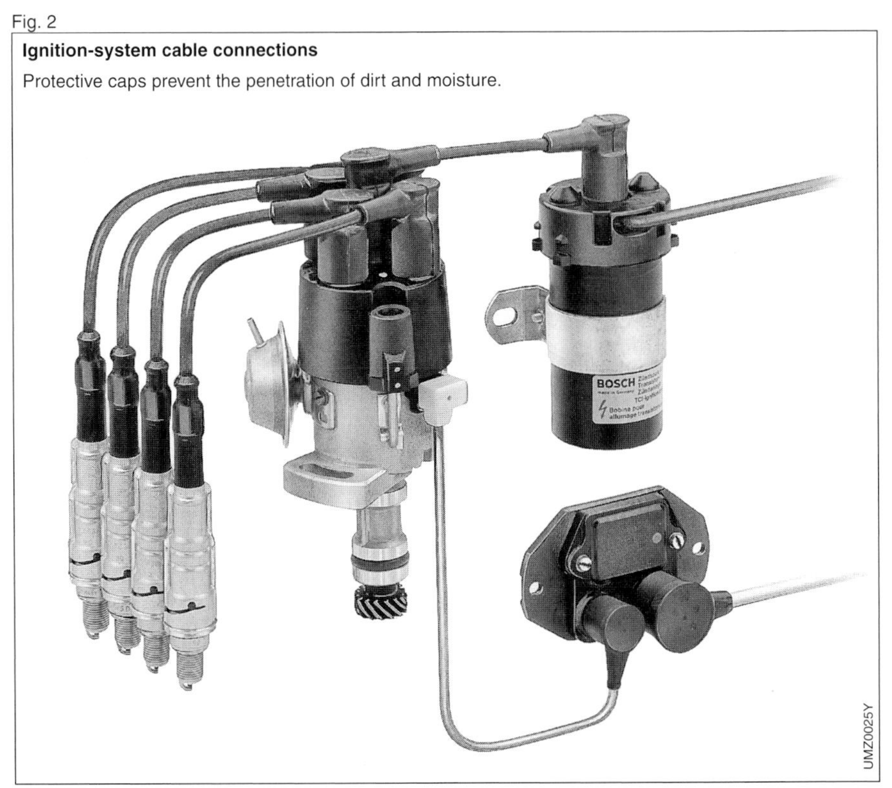

Workshop testing techniques

Bosch customer service

Customer service quality is also a measure for product quality. The car driver has more than 10,000 Bosch Service Agents at his disposal in 125 countries all over the world. These workshops are neutral and not tied to any particular make of vehicle. Even in sparsely populated and remote areas of Africa and South America, the driver can rely upon getting help quickly. Help which is based upon the same quality standards as in Germany, and which is backed of course by the identical guarantees which apply to Bosch customer-service work all over the world.
The data and performance specs for the Bosch systems and equipment are precisely matched to the engine and the vehicle. In order that these can be checked in the workshop, Bosch develops the appropriate measurement techniques, test equipment, and special tools, and equips its Service Agents accordingly.

Testing techniques for engine and ignition

Apart from the air/fuel mixture preparation, perfect ignition is of vital importance for efficient engine operation. As with all complex systems, a wide variety of disturbances and malfunctions can occur which can have a negative effect on engine function, and therefore not only impair handling and driveability, but also the composition of the exhaust gases.

Detecting malfunctions

On modern automotive systems, the engine-management ECU monitors the system's connection paths, actuators, and sensors, while at the same applying self-diagnosis to check its own functions. Malfunctions detected in the process are stored and can be retrieved at any time through the diagnosis interface. Since it

Bild 1

3D ignition display (secondary) of a 6-cylinder engine with dual-spark ignition coil at 760 min^{-1}

1 Parade display: Main sparks and back-up sparks are superimposed,
2 Positive display: Main sparks at cylinders 1, 2, and 5 (high ignition voltage in the compression stroke); back-up sparks at cylinders 4, 3, and 6 (low ignition voltage in the exhaust stroke),
3 Negative display: Main sparks at cylinders 4, 3, and 6; back-up sparks at cylinders 1, 2, and 5,
4 Firing sequence: 1-4-3-6-2-5.

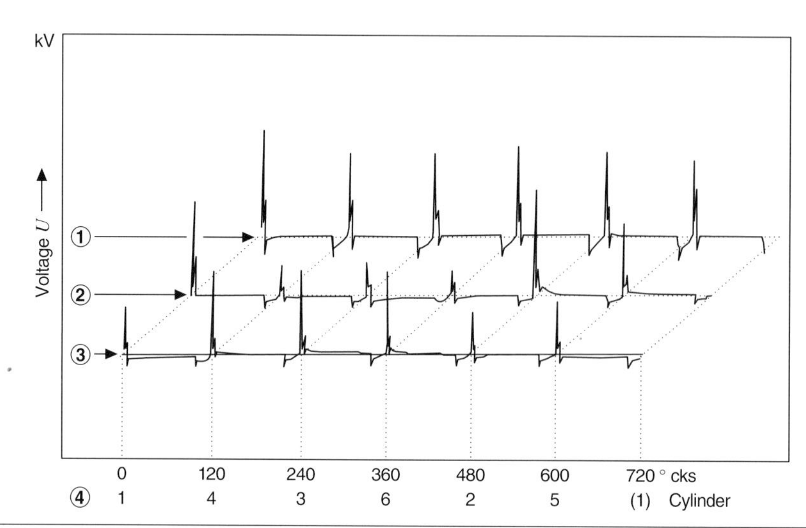

is impossible to say whether a malfunction is due to an open-circuit in the connection line or a faulty component (sensor, actuator), further tests must be carried out using an engine analyzer.

Testing with the engine analyzer

This engine analyzer is available either as a compact Pocket-Tester or in the form of a complete diagnosis system featuring a variety of functions such as exhaust-gas analysis, oscilloscope etc. When choosing an engine analyzer, one important criterium is the type of ignition system which it is to test: Conventional coil ignition (CI), distributorless semiconductor ignition, or all ignition types on engines with up to 12 cylinders.
The modern Bosch engine analyzers are suitable for all ignition systems on the market and incorporate a special menu to select the ignition system in question.
A special feature of the stationary analyzers is their ability to display simultaneously the signals from the primary and secondary circuits of several ignition coils (for instance for ignition systems with single-spark and dual-spark ignition coils).The 3D oscillogram in Fig. 1 shows the functioning of a distributorless semiconductor ignition system for a 6-cylinder engine with dual-spark ignition coils. In order to perform detailed investigations on individual cylinders, the oscillogram can be switched over to the individual-display mode. If a dual-trace oscilloscope is used, it is also possible to observe the primary and secondary-circuit oscillograms together. With its "Search" function, the Bosch Engine Analyzer FSA 560, features another distinct advantage in being able to search for irregularities in the signal waveforms of the primary and secondary sides. This is a valuable aid in high-speed trouble-shooting. Here, the signal waveform is investigated for irregularities which may have occurred during the 8 seconds immediately before the memory button was pressed. This means that in order to localise malfunctions it is possible to perform a direct comparison between the ignition voltage and spark duration at individual cylinders.

Special secondary measuring sensor

On ignition coils which are mounted directly on the spark plug, special adapters (secondary measuring sensors) are used at the ignition coil to pick-up the secondary signal. These sensors are usually in the form of a metal adapter, and depending upon the ignition coil's mechanical design, a wide variety of mounts are required, and also shielding in order to suppress interference from/with other ignition signals.
Some ignition coils are already provided with a diagnosis receptacle for inserting a standardized secondary measurement sensor. This considerably reduces the costs for the workshop.

Testing sensors and actuators

In addition to assessing the ignition system by means of the appropriate oscillograms, it is also highly important that a functional check is made of the sensors and actuators.
The individual sensors, such as knock sensor, Lambda oxygen sensor, or throttle-valve potentiometer, send signals to the electronic control unit (ECU). The actuators are responsible for the implementation of the commands sent out by the ECU. This means that these components are of vital importance for the ignition system's efficient functioning. Their signals can be checked with the engine analyzer's oscilloscope and stored for subsequent assessment or print-out.

Help in case of breakdown

Suitable engine analyzers are indispensable for mobile applications (for instance for breakdown services). These are powered from fitted batteries or directly from the system voltage of the vehicle under test.
The enhanced versions of the mobile units feature dual-trace oscilloscopes with memory function, and preselected measurement ranges which serve to simplify operation.

Spark plugs

Spark-ignition engine and externally supplied ignition

Ignition energy

High-voltage generation

Ignition in the spark-ignition engine is electrical. The electrical energy is taken from the battery. Controlled by the engine, the ignition system periodically generates high voltage. This high voltage causes a spark discharge between the electrodes of the spark plug in the combustion chamber. The energy contained in the spark ignites the compressed air-fuel mixture. The energy taken from the battery is stored in the ignition coil for the periodic generation of high voltage.

This stored energy is used at the right moment in time to generate high voltage. This high voltage is generated inductively in the ignition coil according to the transformer principle. High voltage and ignition energy are sufficient to cover the increase in required ignition voltage resulting from wear in the system (Fig. 1).

Generation of the ignition spark

If there is sufficient high voltage, the ignition spark jumps across the electrodes of the spark plug. At the instant of ignition, i.e. when the energy-storage device discharges, the voltage across the spark-plug electrodes rises very quickly until the flashover voltage (ignition voltage) is reached. As soon as the spark is discharged, the voltage across the electrode drops to the spark voltage. At the

Fig. 1

Schematic of an electronic ignition system

1 Ignition coil with ignition driver stage, **2** High-voltage distributor, **3** Spark plug, **4** ECU, **5** Coolant-temperature sensor, **6** Throttle-valve switch, **7** Engine-speed sensor, **8** Sensor wheel, **9** Battery, **10** Ignition and starting switch.

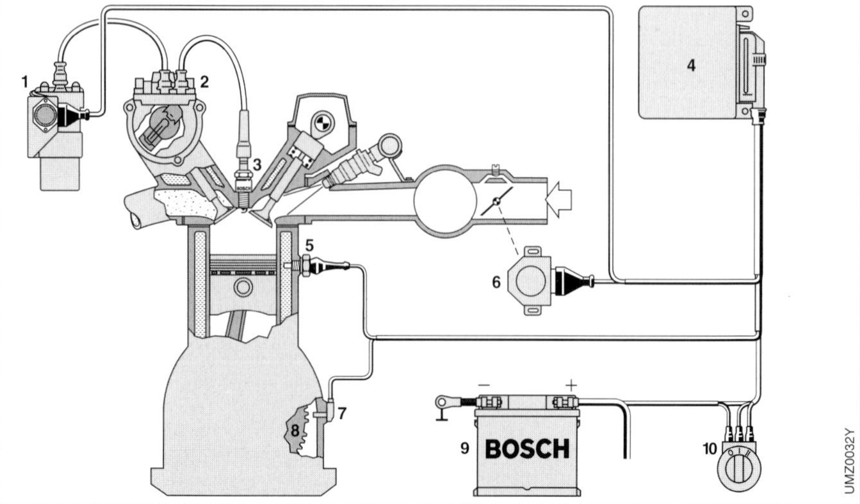

same time, a current flows in the now conductive spark gap. The air-fuel mixture is ignited during the burn time of the ignition spark (spark duration). As soon as the conditions required for discharge no longer obtain, the spark breaks off and the voltage decays to zero (Fig. 2). What has been described here applies only if the gas between the electrodes is quiescent. Higher flow velocities lead to a clear change in the spark characteristics. The spark may be extinguished and reignited in the course of the so-called "spark duration". Phenomena of this kind are known as a follow-up spark.

Spark duration

Within the "spark duration", ignitable air-fuel mixture must be reached by the spark in order to obtain reliable ignition. The "spark duration" is the length of time for which the arc burns following the initial flashover between the electrodes until the residual stored energy decays. It must be maintained long enough to ignite the combustible mixture, despite any lack of homogeneity in the mixture (inconsistencies in mixture distribution).

Ignition voltage

The ignition voltage required by the spark plug is the maximum high voltage theoretically necessary for spark discharge. The ignition voltage of the spark plug is the voltage at which the spark actually jumps across the electrodes. The high voltage causes a strong electric field between the electrodes, so that the spark gap is ionized and thus becomes conductive. The high voltage generated by the ignition system – the available ignition voltage – can exceed 30,000 V. The voltage reserve is the difference between this available ignition voltage and the maximum requirement at the spark plug. The maximum ignition voltage increases as a function of time due to the larger electrode gaps that accompany the aging process. Ignition miss occurs when this process advances to the point where the requirement exceeds the available voltage.

Voltage between spark-plug electrodes

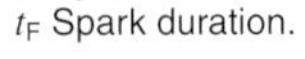

K Spark head, **S** Spark tail,
t_F Spark duration.

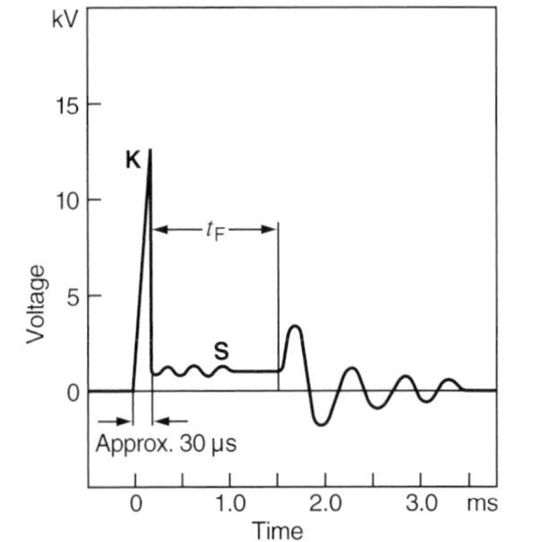

Fig. 2

Non-engine-related influences on the ignition-voltage requirement

The following factors determine the spark-plug voltage requirement:
Electrode gap: The required ignition voltage increases as a function of increasing spark plug gap.
Electrode configuration: Small electrode dimensions increase the intensity of the electrical field; this enhanced field strength can be employed to reduce the voltage requirement.
Electrode materials: Because the electrodes' electron work function varies according to conductor material, electrode materials influence the ignition voltage.
Insulator surface: If part or all of the electrical transfer between the electrodes is generated along the insulator, electrons from the insulator surface will reduce the ignition voltage.

Engine-related influences on the ignition-voltage requirement

Among the engine design factors that influence the required ignition voltage, the compression ratio and the possible use of a supercharger (boost factor) are the most significant.

Spark-plug stress factors

Spark-plug function

The function of the spark plug is to introduce the ignition energy into the combustion chamber and to initiate the combustion of the air-fuel mixture by the electrical spark between its electrodes.
In conjunction with the other components of the engine, e.g., ignition and fuel-management systems, the spark plug has a decisive effect on the operation of the spark-ignition engine. It must permit reliable cold starting, it must guarantee that there is no misfiring during acceleration, and it must withstand the engine being operated for hours on end at maximum power. These requirements apply throughout the entire service life of the spark plug.
The spark plug is positioned in the combustion chamber at the point most suitable for igniting the compressed air-fuel mixture. It must, under all operating conditions, introduce the ignition energy into the combustion chamber without developing a leak and without overheating.

Design requirements

The spark plug must be designed to withstand extreme operating conditions: The plug is exposed to both the periodic, cyclical variations within the combustion chamber and to the external climatic conditions (Figures 1 and 2).

Electrical demands
When the spark plug is operated with electronic ignition systems, voltages of up to 30,000 V can occur. It is essential that the spark plug resists insulator arcing under these conditions. The deposits resulting from the combustion process, such as soot, carbon residues, ash from fuel and oil additives, will, under certain temperature conditions, conduct electrically. However, under such conditions there must be no arcing or breakdown across the insulator even at the high ignition voltages.
The electrical resistance of the insulator must be sufficient up to 1,000 °C, with only minimal deterioration in the course of the plug's service life.

Mechanical demands
The spark plug must withstand the pressures (up to approximately 100 bar) occurring periodically in the combustion chamber, whereby it must remain fully gas-tight. In addition, high mechanical strength is required, particularly of the ceramic, which is subjected to mechanical stress when being installed, as well as being stressed in operation by the spark-plug connector and the ignition cable. The spark-plug shell must absorb the tightening forces without suffering permanent deformation.

Resistance to chemical stress
The part of the spark plug projecting into the combustion chamber can become red-hot, as well as being exposed to the high-temperature chemical processes taking place inside the chamber. Components in the fuel can form chemically-aggressive deposits on the spark plug, affecting its operating characteristics.

Resistance to thermal stress
During operation, the spark plug, in rapid succession, absorbs heat from the hot combustion gases and is then exposed to the cold air-fuel mixture which is inducted for the next cycle. The insulator must therefore maintain a high level of resistance to thermal shock.
The spark plug must also dissipate the heat it absorbs in the combustion chamber as efficiently as possible to the engine's cylinder head, and the terminal side of the spark plug should heat up as little as possible.

Fig. 1

Stress factors

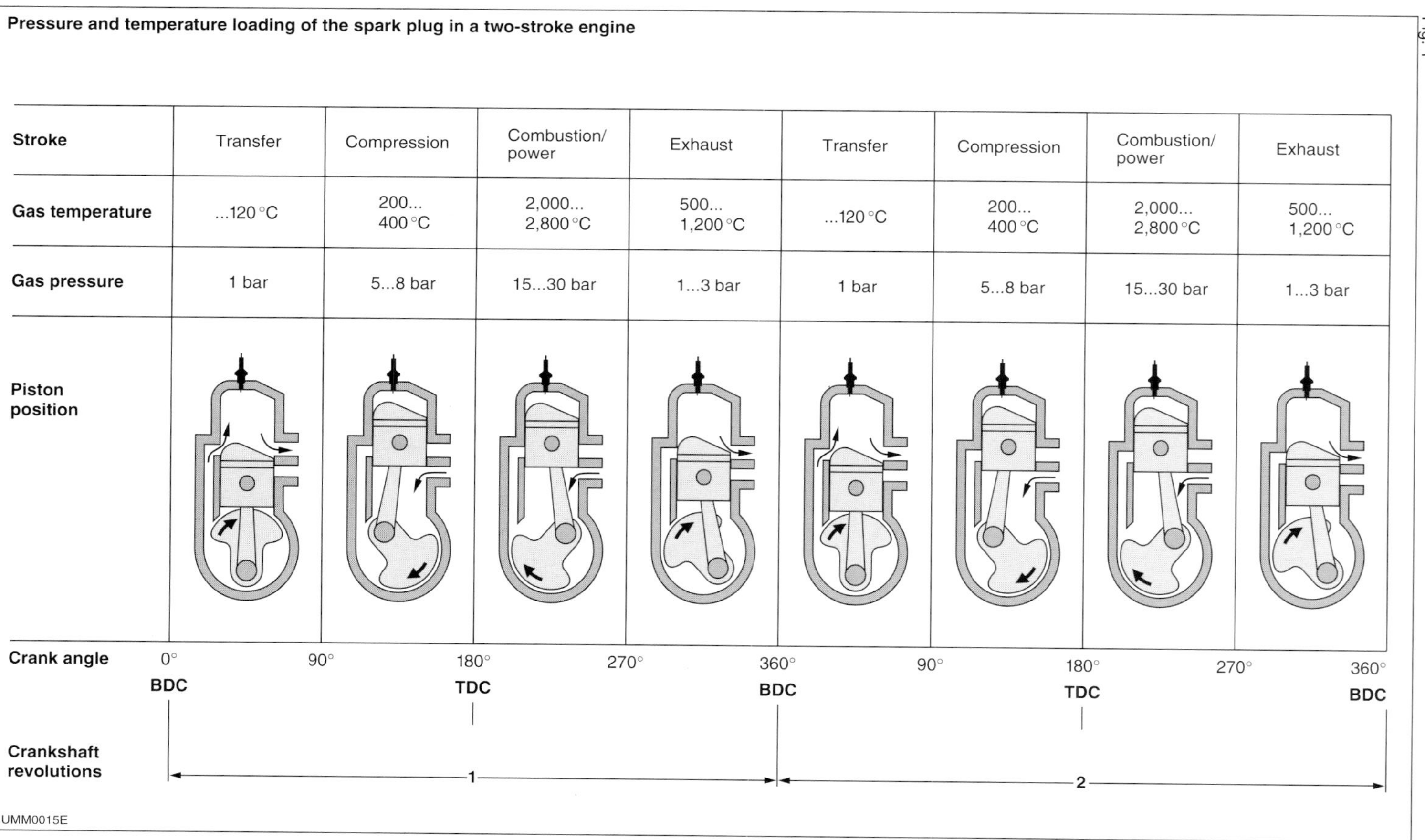

Pressure and temperature loading of the spark plug in a two-stroke engine

Stroke	Transfer	Compression	Combustion/ power	Exhaust	Transfer	Compression	Combustion/ power	Exhaust
Gas temperature	...120 °C	200... 400 °C	2,000... 2,800 °C	500... 1,200 °C	...120 °C	200... 400 °C	2,000... 2,800 °C	500... 1,200 °C
Gas pressure	1 bar	5...8 bar	15...30 bar	1...3 bar	1 bar	5...8 bar	15...30 bar	1...3 bar

Fig. 2

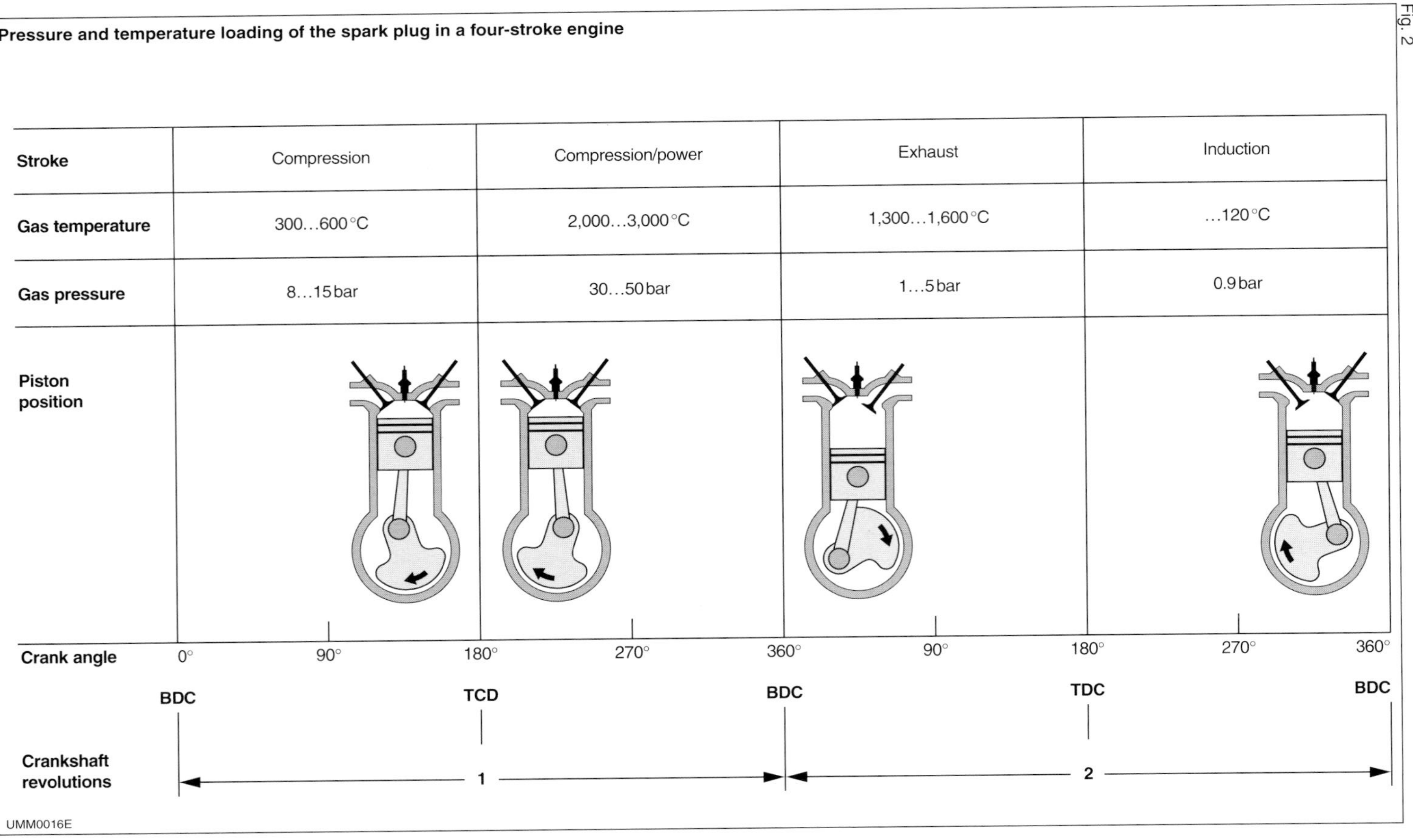
Pressure and temperature loading of the spark plug in a four-stroke engine
Stroke
Gas temperature
Gas pressure
Piston position
Crank angle
Crankshaft revolutions
Compression
300...600 °C
8...15 bar
Compression/power
2,000...3,000 °C
30...50 bar
Exhaust
1,300...1,600 °C
1...5 bar
Induction
...120 °C
0.9 bar
0°
90°
180°
270°
360°
90°
180°
270°
360°
BDC
TCD
BDC
TDC
BDC
1
2
UMM0016E

Spark-plug construction

Components

The spark plug consists of metal, ceramic and glass. These materials have different properties. Appropriate design of the spark plug makes full use of the positive properties of these materials. The terminal stud, insulator, shell and electrodes represent the most important parts of a spark plug. Center electrode and terminal stud are joined by a special conductive glass seal (Fig. 1).

Terminal stud

The steel terminal stud is melted, gastight, into the insulator with a special conductive glass seal which also serves as the electrical connection to the center electrode. On the end projecting out of the insulator, the terminal stud has a thread for attaching the spark-plug connector of the high-voltage ignition cable. In the case of connectors which conform to ISO/DIN Standards a so-called ISO/DIN terminal nut is screwed onto the thread of the terminal stud.

Insulator

The insulator is made of a special ceramic material and its function is to insulate center electrode and terminal stud from the shell. The dense micro-structure of the special ceramic ensures high resistance to electrical breakdown (puncture). The surface of the connection side of the insulator is glazed. Moisture and dirt adhere less well to this smooth glazed surface, as a result of which leakage currents are largely prevented. The insulator houses both the center electrode and the terminal stud.

The demands met in spark-plug applications for good thermal conductivity and high insulation resistance are in sharp contrast to the properties of most insulators. The material used by Bosch for the spark-plug insulator consists of aluminum oxide to which small quantities of other materials have been added. After it has been stoved and glazed, this special ceramic satisfies the demands made of the spark-plug insulator for high insulation resistance, good thermal conductivity and both mechanical and chemical strength.

Fig. 1

Construction of spark plug

1 High-voltage connector (terminal nut),
2 Al_2O_3 ceramic insulator,
3 Shell,
4 Heat-shrinkage zone,
5 Conductive glass seal,
6 Captive outer gasket,
7 Ni/Cu composite center electrode,
8 Ground electrode.

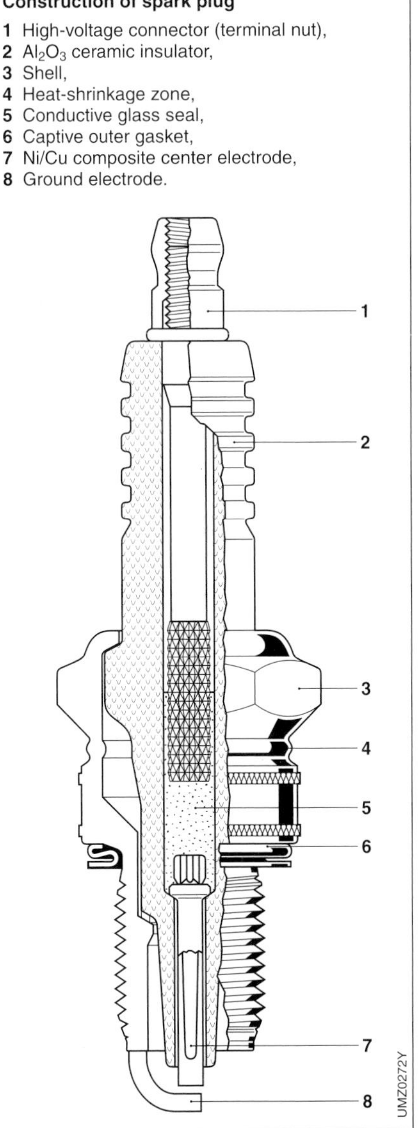

Shell

The shell is made of steel and its function is to secure the spark plug in the engine's cylinder head. The upper part of the shell has a hexagonal section to which the spark-plug wrench is applied, and the lower part is threaded. The surface of the spark-plug shell has an electroplated nickel coating to prevent corrosion, to keep the thread free and to prevent seizing, the latter point applying particularly in aluminum cylinder heads. Depending on the design of the shell, the spark plug can be provided with a seal ring (captive gasket). After the insulator has been inserted into the spark-plug shell, it is swaged and heat-shrunk in position in one operation by inductive heating under high pressure.

Electrodes

The wear on the electrodes is caused by erosion (burning away due to the ignition sparks) and by corrosion (chemical and thermal attack). These two factors cannot be treated separately as regards their effect on electrode wear. Wear increases the required ignition voltage. In addition, the electrodes must have good heat-dissipation properties.
These requirements may call for different electrode shapes and electrode materials, depending on operating conditions and application (Fig. 2).

Ground electrode

The ground electrode is welded to the shell and usually has a rectangular cross-section. Depending upon the position of the ground electrode relative to the center electrode, a distinction is made between front and side electrodes (Fig. 3). The ground electrode's service life is highly dependent upon its heat dissipation, and on both ground and center electrodes suitable composite materials are employed to improve thermal dissipation and extend effective service life (Fig. 3a). Another factor affecting service life is the ratio of thermally exposed surface to thermally conductive cross-section.
Minimal dimensions, special designs and only partial coverage of the center electrode are all employed to obtain an optimal arc pattern at the ground electrode. The external surfaces and the contours of the areas facing the center electrode can also be designed to enhance performance.

Different spark plugs feature:
- Different numbers of ground electrodes, and
- Ground electrodes of varying dimensions.

Thicker ground electrodes and multiple electrodes, therefore, can both be used to extend spark-plug service life.

Center electrode

The center electrode of the conventional spark plugs (air gap between insulator nose bore and center electrode) is

Fig. 2
Electrode arrangement
a Front electrode, **b** Side electrodes, **c** Surface-gap spark plug without ground electrode (for special applications).

a
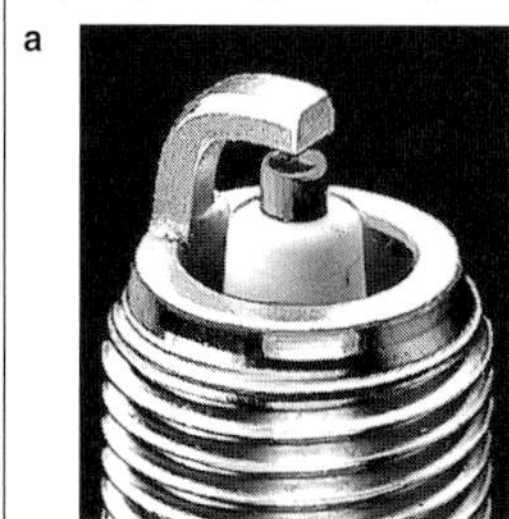
b
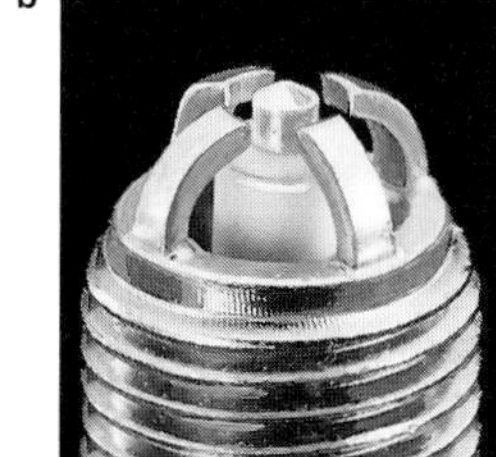
c

UMZ0216Y

melted, gas-tight, into the insulator with a special conductive seal. The electrode has a slightly smaller diameter than the bore in the insulator nose. This is necessary in order to compensate for the different internal expansion which takes place between the electrode material and the insulator ceramic. The air gap thus produced is subject to very close tolerances and is of great importance with regard to the heat range.

The cylindrical center electrode projects from the insulator nose. Center electrodes made of precious metals are smaller in diameter than the compound electrodes which have a copper core and nickel-alloy jacket.

Electrode gap

The electrode gap is the shortest distance between the center electrode and the ground electrode (Fig. 5). The smaller the electrode gap, the lower the ignition-voltage requirement.

A narrow electrode gap will reduce the voltage required to produce an arc, but the short spark can transfer only minimal energy to the mixture, and ignition miss can result. Higher voltages are required to support an arc across a larger gap. This type of gap is thus effective in transferring energy to the mixture, but the attendant reduction in ignition-voltage reserves increases the risk of ignition miss.

The electrode gap is usually about 0.7...1.2 mm (Fig. 4). The precise optimized electrode gaps for the individual engines are specified by the engine manufacturer and are given either in the owner's manual or in the Bosch spark-plug sales literature.

Electrode shape

Electrode shape affects heat dissipation, resistance to wear, the ignition-voltage requirement and the way the arc is transferred to ignite the mixture.

The electrode shape is dependent on the type of spark gap and the spark position.

Spark plugs with composite electrodes

a With front electrode, **b** With side electrodes.
1 Conductive glass seal, **2** Air gap, **3** Insulator nose, **4** Composite center electrode, **5** Composite ground electrode, **6** Ground electrodes.

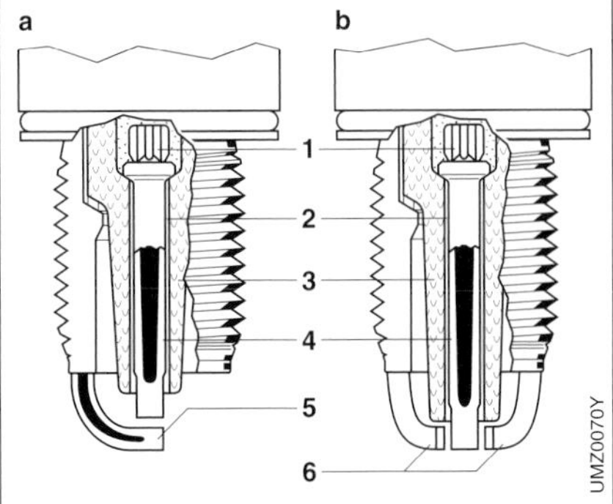

Fig. 3

Fig. 4

Relationship between electrode gap and required ignition voltage

U_0 Secondary available voltage, U_Z Required ignition voltage, ΔU Ignition voltage reserve.

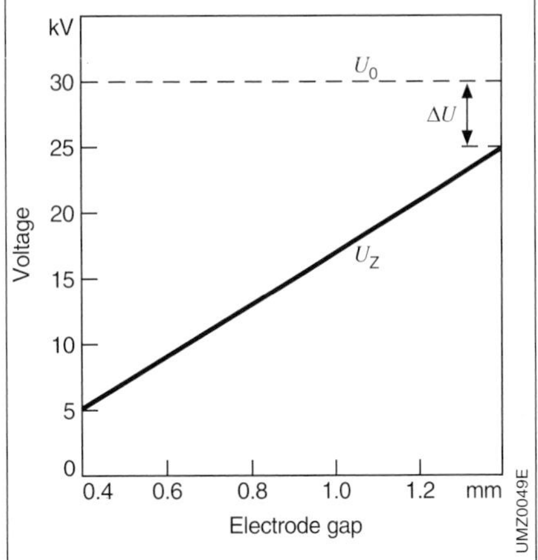

Fig. 5

Electrode gap (EA)

a Front electrode, **b** Side electrode.

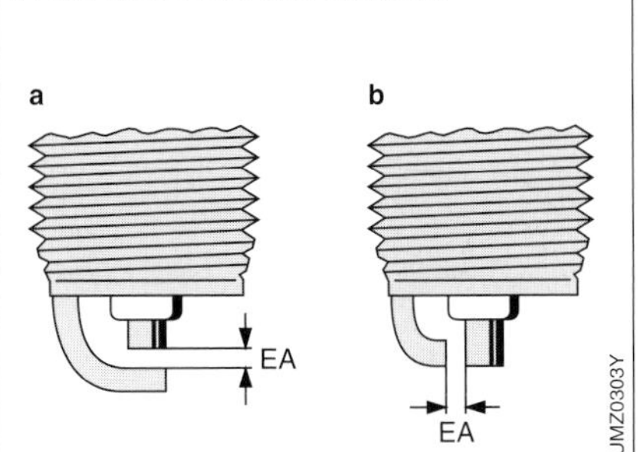

Spark gap

The mutual configuration of the electrodes defines the type of spark gap.

Air spark gap

The ignition spark arcs directly through the air-fuel mixture between the center and ground electrodes (Fig. 6a).

Surface spark gap

Under certain conditions the ignition spark first of all creeps across the surface of the insulator-nose tip and then jumps to the ground electrode across the gas-filled gap. For the same ignition voltage, the semi-surface spark is able to jump across larger gaps than the air-gap spark. The combustion conditions are improved accordingly (Fig. 6b).

Fig. 6

Spark gap

a Air spark gap,
b Surface spark gap,
c Semi-surface spark gap.

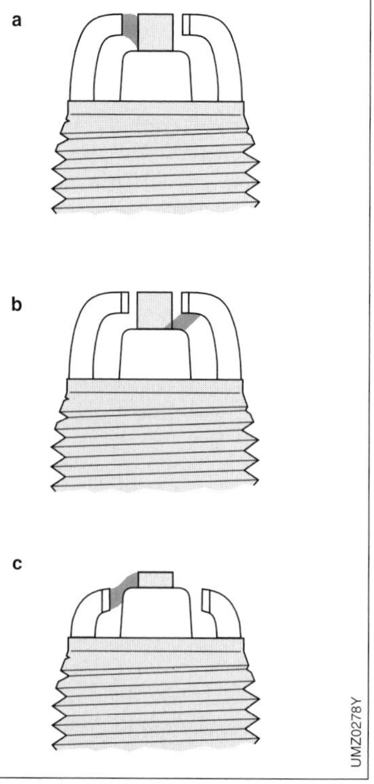

Semi-surface spark gap

The function of the semi-surface spark gap is comparable to that of the surface spark gap. Here too, the spark initially creeps from the center electrode across the surface of the insulator-tip nose and then jumps to the ground electrode across the gas-filled gap (Fig. 6c).
Here, the ground electrodes are arranged to the side of the insulator-tip ceramic so that it is impossible for an air-gap spark to be formed.

Spark position

The spark position is understood to be the position of the spark gap in the combustion chamber. The electric sparks should arc across at the point at which gas-flow conditions are particularly favorable. The spark ignites the air-fuel mixture from a position which depends upon the configuration of the electrodes and insulator and which extends a defined distance into the combustion chamber. The spark position f is referred to the end face of the spark-plug shell (Fig. 7). The normal spark position extends 3...5 mm. For special applications, spark plugs are used which have extreme spark positions. For instance, racing or special engines are equipped with spark plugs featuring a recessed spark position with the spark gap located inside the plug shell.
This considerably reduces the absorption of heat from the combustion chamber. The advantage here lies in the fact that such spark plugs do not overheat at high engine speeds.

Fig. 7

Spark position (f).

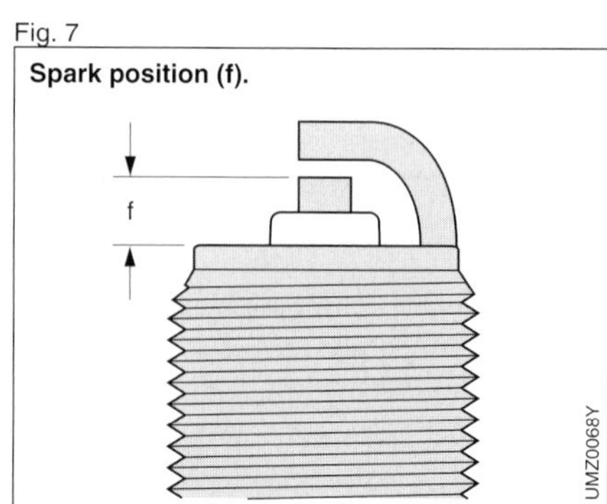

Electrode materials

Compound electrodes

The shunt sensitivity and the corrosion behavior of standard spark plugs with center electrodes made of a nickel-based alloy have been greatly improved by the development of a compound center electrode with a copper core (Fig. 8a).

Pure metals provide better thermal conductivity than alloys. At the same time though, pure metals such as nickel are more sensitive than alloys to aggressive chemical subtances in the combustion gases and to solid deposits.

For this reason, the jacket material of the compound electrode consists mainly of nickel, which is alloyed with chromium, mangangese and silicon. Each of the alloy additives has a special task to perform. Additions of manganese and silicon increase the chemical resistance, particularly against the very aggressive sulphur dioxide (the sulphur comes both from the fuel and from the lube oil).

Nickel-based alloys with silicon, aluminum and yttrium additives also improve resistance to oxidation and scaling.

The ground electrode, which must be flexible enough to allow gap adjustment, can be made of a nickel-based alloy or a composite material.

Such a structure (copper core with nickel jacket) satisfies stringent requirements for high levels of thermal conductivity and corrosion resistance in the ground electrode.

Silver center electrode

Silver displays the highest electrical and thermal conductivity of any substance. It also features extreme resistance to chemical deterioration when unleaded fuel is used. Resistance to thermal stresses can be substantially enhanced using composite particulate materials with silver as the basic substance. The characteristics described above are the reasons behind its use as an electrode material.

When solid silver is used for the center electrode, this can be reduced in diameter (Fig. 8b).

Despite the reduced diameter, the silver center electrode dissipates heat better than a comparable nickel-based electrode.

Platinum center electrode

Because platinum and platinum alloys display extreme resistance to corrosion, oxidation and melting, they are employed in the manufacture of electrodes for "Long-life" spark plugs.

Presuming the same stresses are applied, platinum electrodes used for any given application can be smaller than the equivalent nickel-based electrode (Fig. 8c).

Fig. 8

Center electrode materials

a Composite material, **b** Silver, **c** Platinum.

a

b

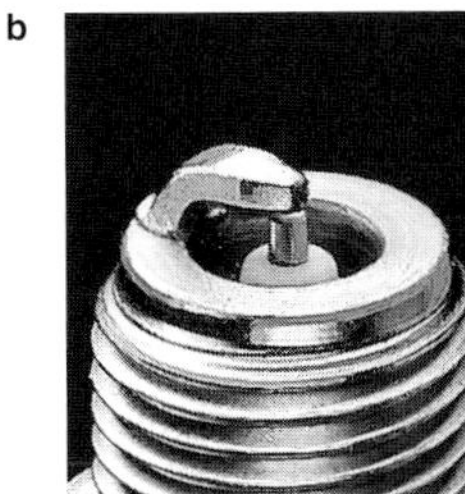

c

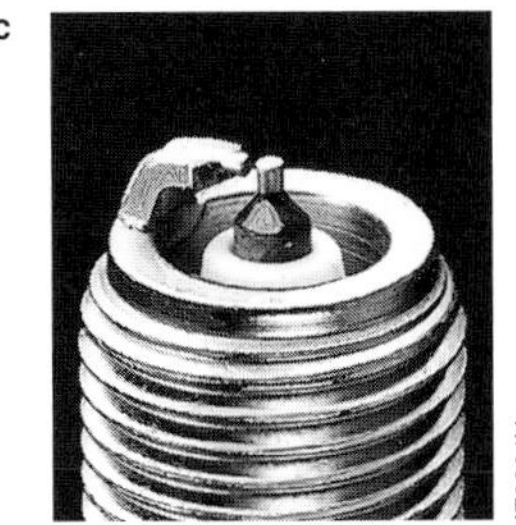

UMZ0304Y

Spark-plug heat ranges

Spark-plug operating temperatures

Operating range

When unleaded fuel is used, the parts of the insulator tip which extend into the combustion chamber should not drop below 500 °C – to ensure self-cleaning of the spark plug – and should not exceed approximately 850 °C – to prevent auto-ignition (Fig. 1).

Particulate deposits (soot) are produced in the incomplete combustion processes that characterize cold starts. Most of these deposits leave the engine along with the exhaust gases, but some remain to form a coating on both the combustion chamber and the spark plug. As these deposits form on the insulator nose, they can produce a conductive path between the center electrode and the shell. This creepage path absorbs a portion of the ignition energy, thus forming a shunt path and reducing the current available for the ignition spark. Excessive contamination can therefore prevent a spark from being generated (Fig. 2).

The deposition of combustion residues on the insulator nose is greatly dependent on its temperature and takes place mainly below about 500 °C. At higher temperatures, the carbon-containing residues on the insulator nose burn-off, as a result of which shunts cannot form, i.e. the plug "cleans" itself. The aim therefore, is to achieve an operating temperature at the insulator nose which is higher than the "self-cleaning limit" of about 500 °C, whereby the self-cleaning temperature should be reached as quickly as possible after starting.
The upper temperature limit is about 900 °C since above this temperature the air-fuel mixture can ignite prematurely on red-hot parts of the spark plug (auto-ignition). Uncontrolled ignition of this kind is highly detrimental to the engine and may cause irreparable damage within a short space of time.
For these reasons, the operating temperature of the spark plug must be kept within the above-mentioned limits.

Thermal loading capacity

The operating temperature is the equilibrium temperature which is reached

Fig. 1

Spark-plug working range

For different engine power outputs, the working range should be between 500 and 850 °C at the insulator.

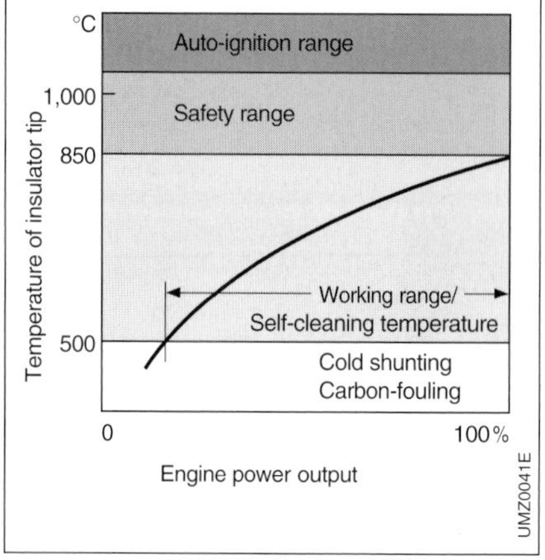

Fig. 2

Shunting due to fouled insulator nose leads to a reduced secondary available voltage

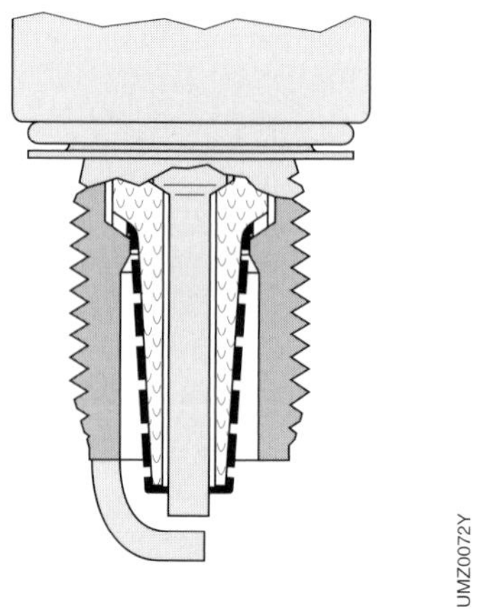

between heat absorption and heat dissipation. The spark plug is heated by the heat generated during operation in the engine's combustion chamber. The spark-plug shell has more or less the same temperature as the cylinder head, while the temperatures reached by the insulator are considerably higher.

Some (approximately 20%) of the heat absorbed by the spark plug is dissipated through the inflow of fresh mixture during the induction stroke. Most of the heat is transferred through the center electrode and insulator to the spark-plug shell and from there to the cylinder head (Fig. 3). The supply of heat to the spark plug is dependent on the engine. Engines with a high specific power output usually have higher combustion-chamber temperatures than engines with a low specific power output. The heat-absorbing properties of the spark plug must, therefore, be matched to the engine type in question.
The heat range is characteristic of the thermal loading capacity of the spark plug.

The heat range and the engine

The spark plug's heat range is an index of its capacity to withstand thermal loads. It must be adapted to the engine characteristic.
The different characteristics of automotive engines with respect to operating load, working principles, compression, engine speed, cooling, and fuel make it impossible to run all engines with a standard spark plug. The same spark plug would get very hot in one engine but would reach only a relatively low average temperature in another.
In the first case, the air-fuel mixture would ignite on the glowing parts of the spark plug projecting into the combustion chamber (auto-ignition) and, in the second case, the insulator tip would very soon be so badly fouled by combustion deposits that misfiring would occur due to shunts. In other words, one and the same spark-plug type is not suitable for all engines. To ensure that the plug runs neither too "hot" nor too "cold" in a given engine, plugs with different load capacities were developed. The so-called "heat range", which is assigned to each spark plug, is used to characterise these loading capacities. The heat range is, therefore, a yardstick for selecting the correct spark plug.

Heat ranges

Fig. 3

Thermal conduction in the spark plug

A large proportion of the heat absorbed from the combustion chamber is dissipated by thermal conduction (small proportion of cooling of approximately 20% due to flowpast of fresh A/F mixture is not taken into consideration).

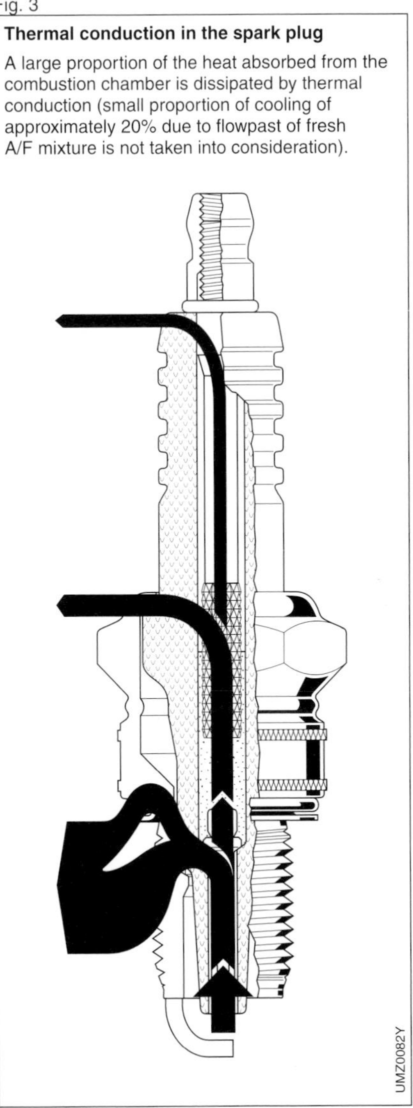

The heat range and the spark plug

Heat range and design

As the above description indicates, each spark plug must be designed to remain within a specific temperature range during operation. Therefore, in order to remain within its operating range, the spark plug for a "hot-running" engine must efficiently dissipate heat acting upon it, while the plug for a "cold-running" engine must retain the heat. Various design factors – with special emphasis on the configuration of the insulator nose – can be applied to adjust the

Fig. 4

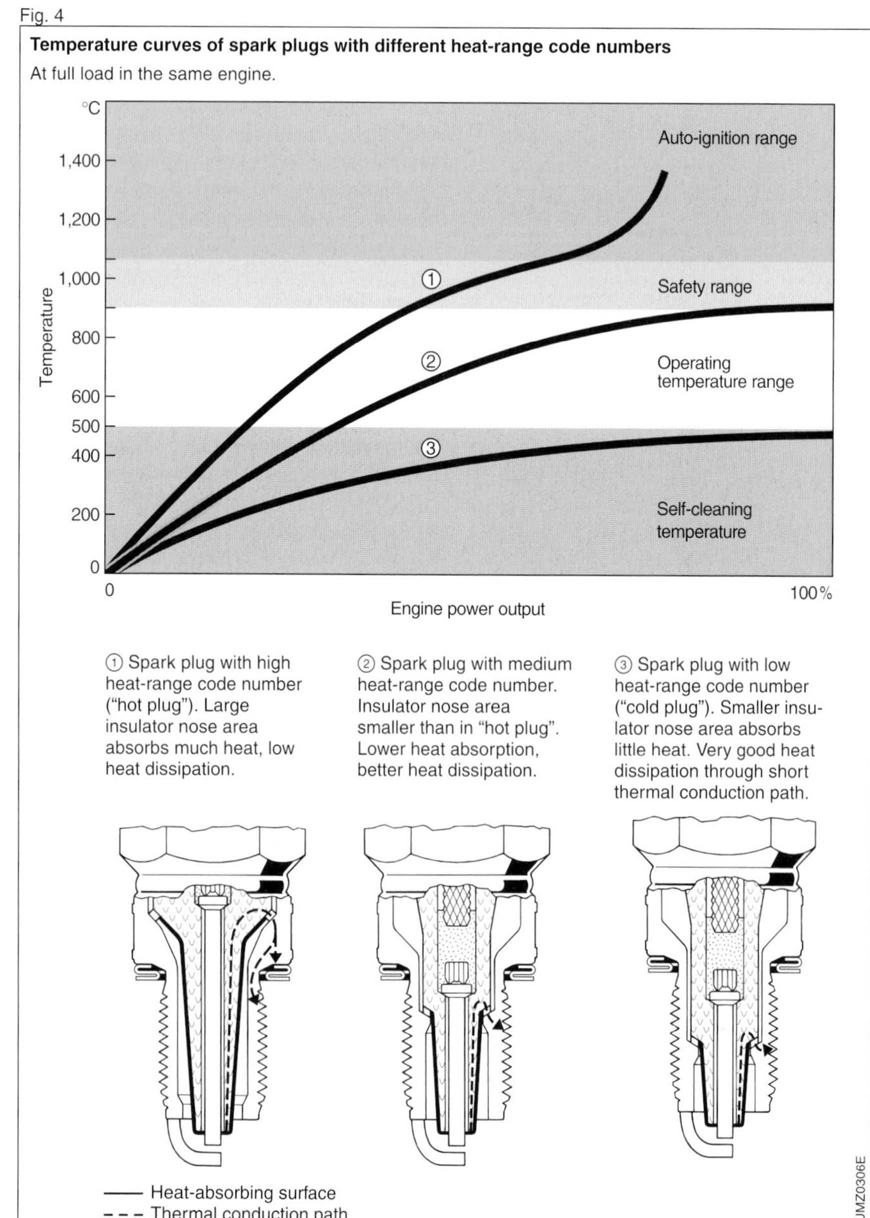

heat range for specific engines and applications.

Influence of insulator nose

Heat absorption is determined by the surface area of the insulator nose. If a large area is exposed to the combustion gases – this is achieved by having a long insulator nose – the insulator nose gets very hot. Conversely, with a short insulator nose, the area is small, and the nose remains cooler.

The heat is dissipated from the insulator nose through the center electrode and through the inner seal ring to the spark-plug shell. If the insulator nose is long, this heat transfer point formed by the seal ring is further away from the hottest point on the insulator nose than is the case with a short insulator nose. Therefore, it follows that spark plugs with a long insulator nose can absorb more heat and dissipate less heat (i.e. they are "hotter") than plugs with a short insulator nose ("cold" plug). Different lengths of insulator nose result, therefore, in different characteristics, in different heat ranges.

Heat range and heat-range code number

The heat range of a spark plug is identified by a heat-range code number.

A low heat-range code number indicates a "cold plug" with low heat absorption through its short insulator nose. A high heat-range code number applies to a "hot plug" with high heat absorption through its long insulator nose (Fig. 4).

Heat-range code numbers have been specified to make it easier to differentiate between spark plugs of different heat ranges and to determine the right plug for the engine in question. The heat-range code number is a part of the spark-plug type designation. Low code numbers (e.g., 2...4) signify "cold" plugs. High code numbers (e.g., 7...10) signify "hot" plugs.

Spark-plug selection

Temperature-measuring spark plugs

Bosch works together with the engine manufacturers in determining the best spark-plug design for a given application. The temperature-measuring spark plug provides initial information on the correct choice of spark plug. With a thermocouple in the center electrode of a spark plug, it is possible to record the temperatures in the individual cylinders as a function of engine speed and load.

This expedient represents a simple means of identifying the hottest cylinder and ensuring that the correct spark plug design is selected (Fig. 1).

Fig. 1

Temperature-measuring spark plug

1 Insulator, **2** Sheathed thermocouple, **3** Center electrode, **4** Measuring point.

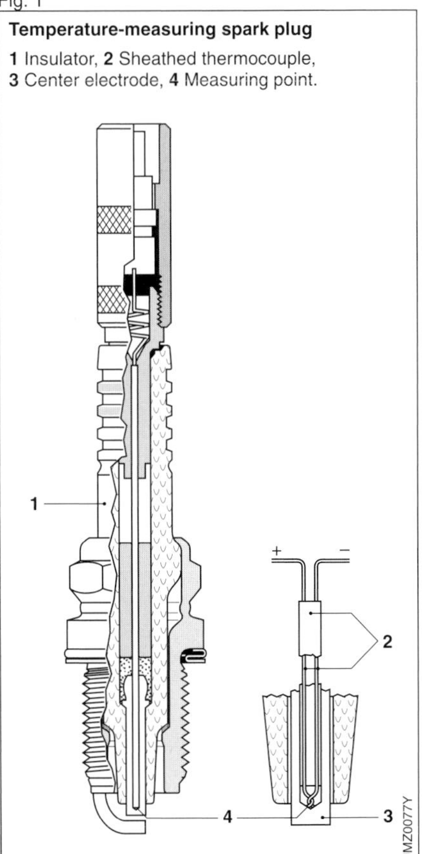

Ionic-current measuring method

With the ionic-current measuring method, the combustion process is used to specify the heat range. The ionization effect of flames makes it possible, by measuring the conductivity in the spark gap, to evaluate the time curve of the initiation combustion of the air-fuel mixture. The curve shows characteristic changes referred to the thermal loading of a spark plug (Fig. 2).
The advantage of this method over a single temperature measurement in the combustion chamber lies in the possibility of detecting auto-ignition, which depends not only on the temperature but also upon the individual engine and spark-plug design parameters.
In order to select the heat ranges of spark plugs, terms and definitions relating to the uncontrolled ignition of air-fuel mixtures have been laid down in accordance with an international agreement (ISO 2542–1972, Fig. 3):
Auto-ignition is taken to mean ignition which is independent of the ignition spark.
If ignition takes place prior to the electrical ignition point, this is pre-ignition.
If it takes place after the ignition point, this is called post-ignition. Post-ignition is non-critical as regards operation of the engine; conversely, pre-ignition may lead to serious damage. The spark plug must be selected in such a way that there is no pre-ignition.
The ionic-current measuring method makes it possible to select the heat range of a spark plug for any engine and also to determine the heat range of a given spark plug. In addition, the Bosch ionic-current measuring method makes it possible, by suppressing the ignition spark at certain intervals, to trace post-ignition and its percentage share in relation to the suppression rate as the combustion-chamber temperature rises (by advancing the ignition-timing Fig. 4).
A change in the ionic-current trace on the screen permits the precise determination, even without suppressing the electric ignition spark, of the transition from post-ignition to the beginning of pre-

Fig. 2

Diagram of ionic-current measurement

1 From ignition distributor, **2** Ionic-current adapter, **2a** Break-over diode (BOD), **3** Spark plug, **4** Ionic-current device, **5** Oscilloscope.

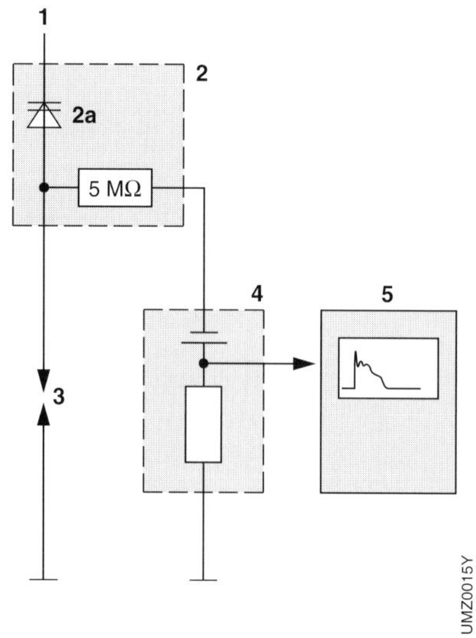

Fig. 3

Definitions with regard to heat-range selection

AI Auto-ignition, TDC Top dead center, PRI Pre-ignition, POI Post-ignition, HRR Heat-range reserve in °crankshaft, IP Ignition point in °crankshaft before TDC, α_z Ignition angle.

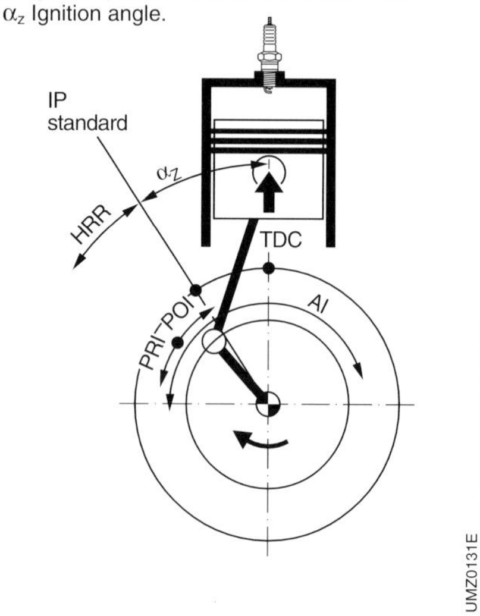

ignition. This makes the measuring method an additional aid in evaluating individual design parameters with regard to their tendency to produce auto-ignition under maximum loading.
The practical procedure is now explained with reference to an example: ionic-current measurement on three Bosch spark plugs with different heat ranges, under full-load operation, $n = 5{,}000\ \text{min}^{-1}$, ignition timing factory-set.

Ionic current testing
(example):

Spark plug series with heat-range code number	% POI[1])	PRI[2])
WR9DC	100	yes
WR8DC	50	no
WR7DC	0	no

[1]) POI = post-ignition
[2]) PRI = pre-ignition

The spark plug with the highest heat-range code number shows 100% post-ignition even with the factory-set ignition timing, i.e. each time the spark is suppressed, the compressed mixture is still ignited by the excessive heat at the insulator. Isolated cases of pre-ignition can also occur. Even the intermediate-range spark plug produces ignition every second time the current is suppressed – the safety reserves are still inadequate.
The spark plug with the lowest heat range promotes neither pre- nor post-ignition – the temperature of the insulator nose remains low enough to remove the danger of auto-ignition. This would be the recommended spark plug in this example.
Additionally, pre-ignition occurs only using a spark plug which is two heat ranges hotter. The heat range safety margin is therefore considered to be adequate.

The above remarks illustrate that spark plugs cannot simply be selected from a catalog and installed.

Characteristic ionic-current oscilloscope patterns

a Normal operating conditions,
b Suppressed ignition without post-ignition,
c Suppressed ignition with post-ignition,
d Pre-ignition.

ZZP = Ignition point

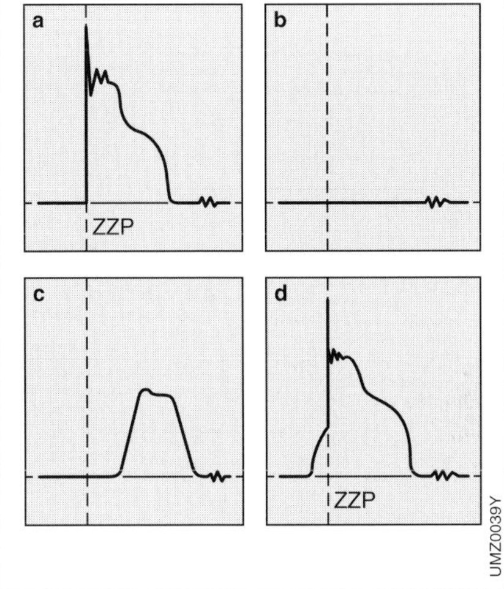

Fig. 4

Spark plug specifications are generally determined in a process that includes close cooperation between engine and spark-plug manufacturers. The recommendations of the vehicle manufacturer and the recommendations – including alternate heat ranges for different regions – contained in the Bosch sales catalogues should always be observed. Application measurements on spark plugs are best performed on the engine test stand or chassis dynamometer.
Since the engine must be operated at full load over lengthy periods of time in order to establish the hottest operating point, it is not possible to conduct tests on public roads.

Operating behavior of spark plug

Changes during operation

During operation, the spark plug is subjected both to wear and to fouling and must, therefore, be replaced at regular intervals.
In the course of its service life, the spark plug undergoes changes which increase the required ignition voltage. When the required voltage reaches a value which can no longer be compensated for by the voltage reserve, the result is misfiring.
Furthermore, the operation of the spark plug can also be adversely affected by the aging of the engine.
Deposits can lead to heat tracking, with attendant negative effects on the ignition flame. The immediate results assume the form of ignition miss, leading to a substantial increase in exhaust emissions followed by damage to the catalytic converter.

Engine-related influences

As the engine ages, leakage may occur which leads to a higher content of oil in the combustion chamber. This results in heavy deposits of soot, ash and oil carbon on the spark plug and may lead to shunts. The result is misfiring.

Electrode wear

Characteristics

Electrode wear is the removal of material from the electrodes.
A visible sign of electrode wear is the increase in the electrode gap during the plug's service life. Electrode wear can be minimized through careful selection of shapes, materials and gap configuration (surface-air gap) for the electrode.

Two processes contribute to electrode wear:
- Spark erosion, and
- Corrosion in the combustion chamber.

Spark erosion and corrosion

The discharge of electrical sparks leads to an increase in the temperature of the electrodes. In conjunction with the aggressive combustion gases, there is clear wear at high temperatures. Melted-open, microscopically small sur-

Fig. 1

Wear at center and ground electrodes

a Front-electrode spark plug, **b** Side-electrode spark plug.
1 Center electrode, **2** Ground electrode.

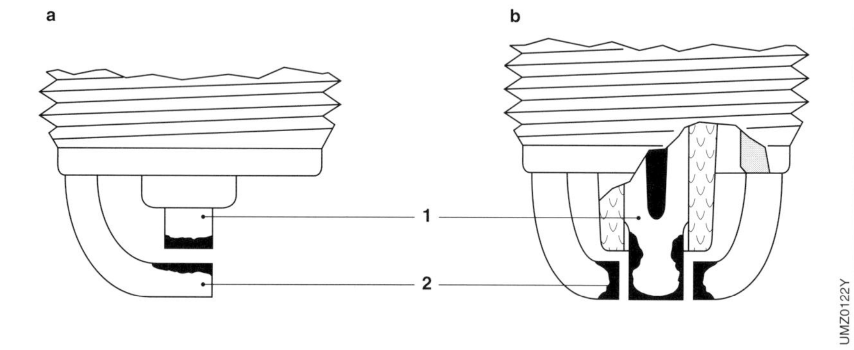

face areas are oxidized or react with other constituents in the combustion gases. The result is a removal of metal, which can be seen in the rounding of edges and also in the widening of the electrode gap (Fig. 1).

Heat-resistant materials – such as the noble metal platinum – can be used to minimize electrode wear.

Abnormal operating conditions

Abnormal operating conditions can irreparably damage the engine and the spark plug. These include:
- Auto-ignition in case of pre-ignition,
- Knocking, and
- High oil consumption (formation of ash and carbon residue).

Engine and spark plug may be damaged also by an incorrectly tuned ignition system, the use of spark plugs with the wrong heat range for the engine, or the use of unsuitable fuels.

Auto-ignition

Full-throttle operation can generate localized hot spots and cause auto-ignition at the following locations:
- At the tip of the spark plug's insulator nose,
- On the exhaust valve,
- On protruding sections of the head gasket, and
- On loose deposits (ash and carbon residue).

Auto-ignition is an uncontrolled ignition process in which the temperatures in the combustion chamber can rise to such an extent as to cause serious damage to the engine and the spark plug.

Knocking

Knocking is uncontrolled combustion with a very steep rise in pressure. It is caused by the spontaneous ignition of portions of the mixture which have not yet been reached by the advancing flame front triggered by the ignition spark. Combustion takes place considerably faster than normal (gentle) combustion. Pressure oscillations occur, with high peak pressures and high frequencies which are superimposed on the normal pressure curve (Fig. 2).
As the high-pressure waves hit the walls of the combustion chamber, their impact produces a metallic knocking sound.
The engine is subjected to severe mechanical stressing as a result of the high-pressure waves. The following engine components are endangered in particular:
- Cylinder head,
- Spark plugs,
- Valves, and
- Pistons.

Failure to recognize and deal with knocking will inevitably lead to serious engine damage. The damage is similar to that associated with the cavitation damage that occurs when supersonic flow patterns are generated. On the spark plug, pitting on the ground electrode's surface is the first sign of knocking.

Fig. 2
Pressure in the cylinder
1 With normal combustion,
2 With knocking.

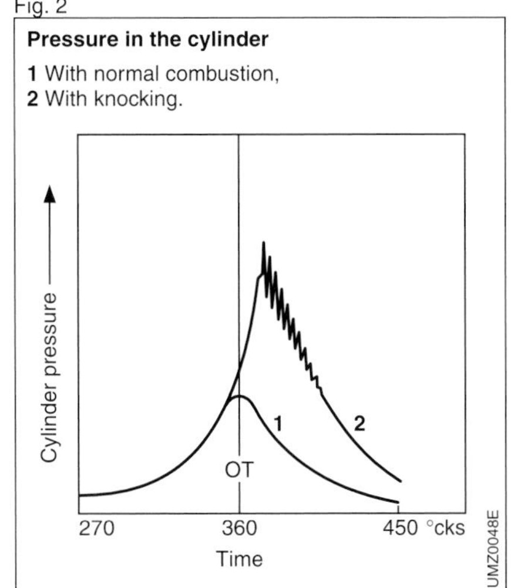

Designs

Applications

In accordance with the wide range of application areas, there are a large variety of different spark-plug designs on the market with more than 1,400 specific variants for

- Passenger cars and commercial vehicles,
- Two-wheeled vehicles,
- Boats and ships,
- Agricultural machinery and construction equipment,
- Motor saws,
- Garden-care equipment etc.

Sealing seat

Depending upon the type of engine concerned, the spark plug is sealed off against the cylinder head either by a flat-type seat or a conical-type seat (Fig. 1).
A "captive" gasket attached to the spark-plug shell is used as the sealing element in the case of the flat seat. It is specially shaped, and provides a permanent elastic seal when the spark plug is installed according to instructions.
The conical seat on the other hand does without a gasket and seals off the spark plug from the cylinder head by means of a conical surface on the spark-plug shell.

Fig. 1

Flat seat using gasket (left) and conical seat without gasket (right)

This engages directly with an appropriately shaped surface on the cylinder head.

SUPER spark plug

The majority of the Bosch spark-plug program is comprised of Bosch SUPER spark plugs (Fig. 2). For practically every vehicle there is a suitable variant available which, with its special heat range, has been precisely adapted to the engine in question.
The SUPER spark plug's main features are:

- A compound center electrode composed of a copper core surrounded by a nickel-chrome alloy jacket,
- An electrode gap which has been set at the factory for the particular engine.

The center electrode's copper core serves to efficiently conduct the heat away so that there is no danger of the electrode

Fig. 2

The Bosch SUPER spark plug

1 Compound center electrode with copper core.

overheating. The nickel-chrome alloy protects the copper core against corrosion and ensures a high level of resistance to wear caused by spark erosion.

A number of different spark gaps are formed depending upon the ground-electrode geometry and configuration. In this respect, the SUPER spark plugs can be sub-divided into three categories:
- Spark plugs in which only air-gap sparks are generated,
- Spark plugs in which only surface-gap sparks are generated, and
- Spark plugs in which both the above spark types can be generated (semi-surface gap sparks).

The last two versions are becoming more and more common since on the one hand the ignition spark can select the optimum path from the center electrode to the ground electrode, and on the other, larger spark gaps can be employed. These facts lead to an increase in ignition reliability.

SUPER 4 spark plug

The Bosch SUPER 4 spark plug (Figs. 3 and 4) is the most modern development of already existing spark plugs. It differs from conventional spark plugs in having
- Four symmetrically arranged ground electrodes,
- A silver-plated chrome-nickel-alloy center electrode with a copper core, and
- An electrode gap set for the spark plug's complete service life.

Fig. 3

The Bosch SUPER 4 spark plug

Function and principle of operation

Spark gap

In principle, the spark from the spark plug with four ground electrodes ignites the air/fuel mixture in the same manner as the spark from the plug with 2 ground electrodes. The spark is either an air-gap spark or a semi-surface gap spark. With the four ground electrodes of the SUPER 4, this means that eight different spark gaps are possible. Normally, it is purely a matter of chance which of these spark gaps is chosen (Fig. 5). The sparks are distributed uniformly around the insulator nose.

Fig. 4

Construction of the Bosch SUPER 4 spark plug

1 Terminal screw, **2** Insulator, **3** Spark-plug shell, **4** Conductive glass seal, **5** Center electrode, **6** Ground electrode.

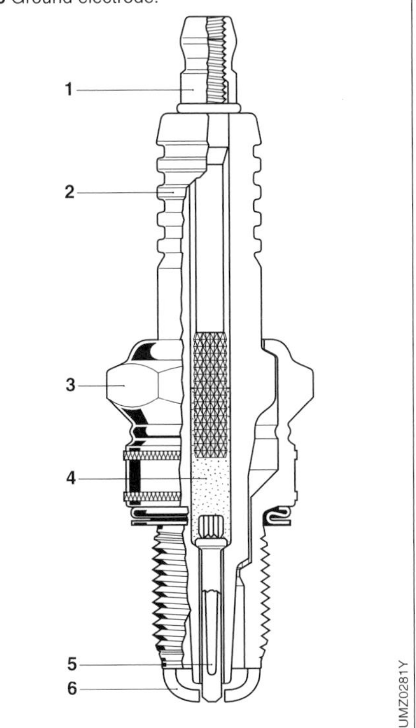

If though the insulator nose is contaminated at a given point (e.g. with soot), the spark creeps over this contamination point and jumps to the nearest ground electrode (Fig. 6). In this case, the spark burns away the contamination at the same time.

Uniform electrode wear

Since the probability of spark discharge is the same for all the electrodes, ground-electrode wear is distributed evenly between the four electrodes. The ohmic resistance of the glass seal reduces the burn-off and contributes to decreasing the electrode wear.

Heat range

The silver-plated center electrode efficiently dissipates the heat. This lowers the danger of auto-ignition due to overheating, as well as extending the safe operation range to cover higher temperatures. As a result of the surface-gap principle, the self-cleaning process starts at lower temperatures. The SUPER 4 spark plug therefore covers at least two conventional-plug heat ranges. This means that a relatively small number of SUPER 4 spark-plug types suffice for a large number of vehicle types (including those equipped with conventional air-gap spark plugs), and can be installed when a spark-plug change takes place during servicing work.

Fig. 5

Possible spark gaps

1 Air-gap spark, **2** Semi-surface gap spark.

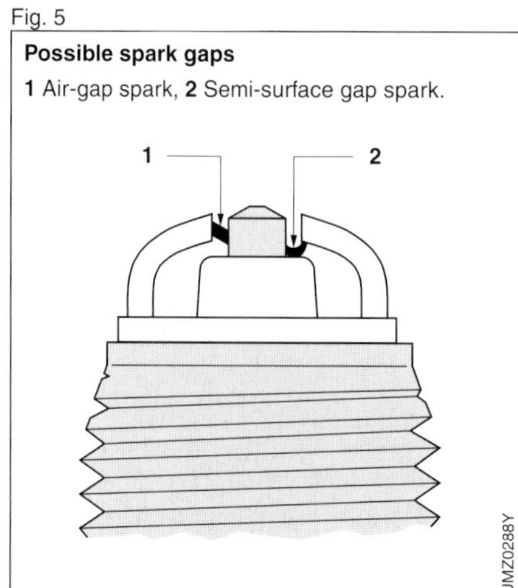

Spark-plug efficiency

The fact that the SUPER 4 spark plug's ground electrodes are so thin means that the ignition spark loses only a fraction of the energy that is lost with conventional spark plugs. And since there is therefore as much as 40% more ignition energy available, the spark-plug efficiency increases accordingly (Fig. 7).

Ignition probability

The more the excess-air factor increases (lean A/F mixture, $\lambda > 1$), the less probable it is that the A/F mixture will ignite reliably [1]). Laboratory tests have proved that with the SUPER 4 spark plug the A/F mixture can be ignited reliably up to

Fig. 6

Spark gap with contaminated insulator nose

1 Contamination at the insulator nose.

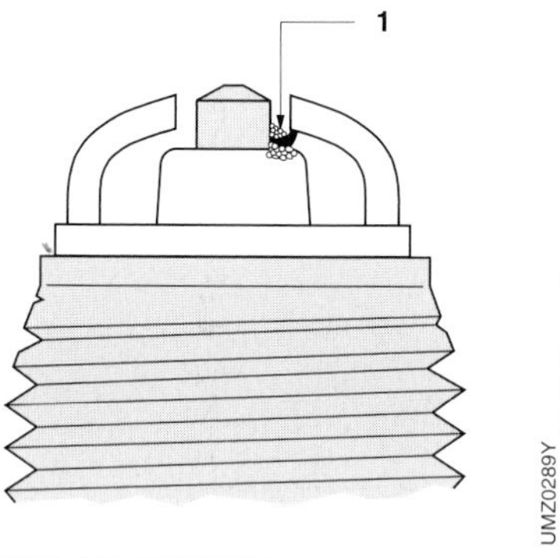

UMZ0289Y

Fig. 7

Spark-plug efficiency

1 Conventional spark plug,
2 Bosch Super 4 spark plug.

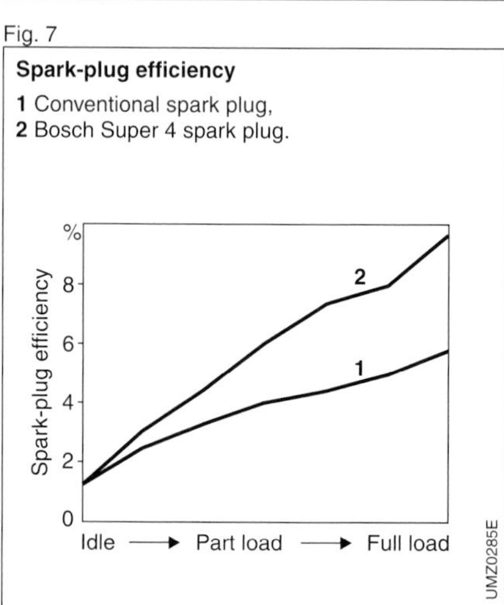

λ = 1.55, whereas with standard spark plugs in this range more than 50% of the ignition sparks fail to ignite the A/F mixture (Fig. 8).

Advantages

Compared to conventional spark plugs, the new Bosch SUPER 4 spark plug features the following improvements:

- Improved ignition reliability due to the possibility of eight spark gaps,
- Self-cleaning surface-spark technology, and
- Extended heat range.

It was thus possible to reduce the variety of SUPER 4 spark-plug types. At present, in Europe, 15 different types suffice to cover the majority of requirements. With conventional spark plugs on the other hand almost 80 different types are needed. The SUPER 4 spark plugs are also suitable for use in older vehicles so that these are also able to benefit from state-of-the-art spark-plug technology.

[1]) The access-air factor or A/F ratio λ (Lambda) defines to what extent the actually available air/fuel mixture deviates from the A/F mixture which is theoretically necessary for complete combustion:

$$\lambda = \frac{\text{Mass of air supplied}}{\text{Theoretical requirement for stoichiometric combustion}}$$

Fig. 8

Influence of the A/F mixture on the probability of ignition

1 Conventional spark plug,
2 Bosch Super 4 spark plug.

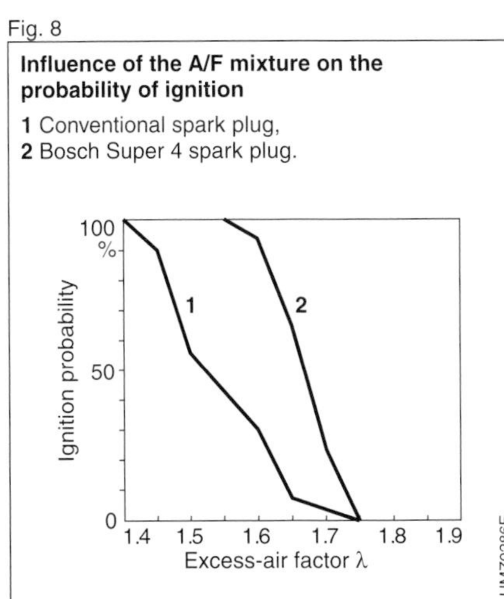

Cold-start behavior

The SUPER 4 spark plug's improved low-temperature and self-cleaning characteristics mean that it can withstand as many as 3 times more cold starts (that is, starting without the engine having been warmed up) than a conventional spark plug. Only then does it start to show the normal signs of wear.

Environmental compatibility and catalytic-converter protection

The amount of unburnt fuel, and with it the attendant HC emissions, are reduced as a result of the improved cold-start characteristics and the high level of ignition reliability. This also leads to extending the catalytic converter's service life.

Improved acceleration

Under actual driving conditions, A/F mixture lean-off takes place particularly when the vehicle is strongly accelerated. The Bosch SUPER 4 spark plug with its increased ignition reliability makes ignition miss a thing of the past and thus ensures continuous smooth acceleration.
In tests, acceleration figures from 30 to 120 km/h in 3rd and 4th gear were obtained which in each case were 0.4 s better than those with conventional spark plugs. As a result, the acceleration distance (Fig. 9) reduces by five meters and the safety for driver and passengers when overtaking increases accordingly.

Fig. 9

Comparison of acceleration in 3rd and 4th gear, from 30 to 120 km/h,

■ Conventional spark plug,
□ Bosch SUPER 4 spark plug.

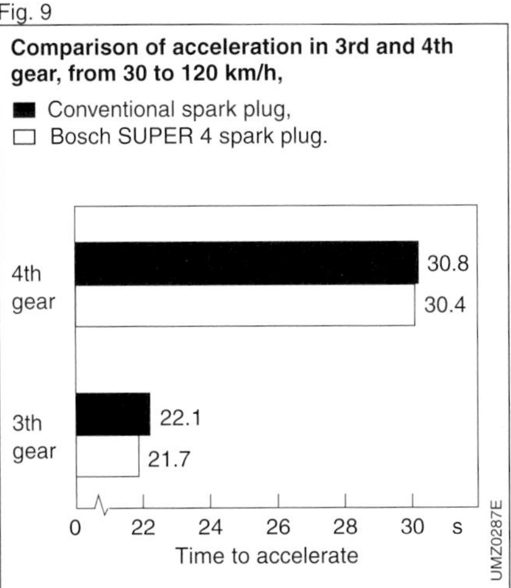

Platin plus 4 spark plug

Design

The Bosch Platin plus 4 (Fig. 10) is the latest development of the spark plugs dealt with on the preceding pages. In comparison with conventional spark plugs, the design of this surface-gap spark plug differs as follows:

- Four symmetrically arranged double-bent ground electrodes,
- One thin platinum center electrode,
- A special-alloy contact pin of improved shape,
- A newly developed ceramic insulator element featuring high electric-strength, and
- Functional improvements to the insulator-nose shape.

Method of operation

Spark gap

In principle, the ignition spark is generated in the same way as with other spark plugs: The high applied voltage leads to field peaks at the center electrode which result in flashover at the point with the maximum field strength. In contrast to spark plugs with only an air spark gap, the Platin plus 4 spark first of all travels over the surface of the insulator nose and then arcs across the air gap to the ground electrode (surface spark gap).

Uniform electrode wear

When a spark forms an arc it chooses the path of least resistance and selects the ground electrode which is nearest to its point of origin. On the Platin plus 4 though, wear is evenly distributed between four electrodes during the engine's useful life. This means that the interval between spark-plug changes is increased. The center electrode's platinum pin is resistant to erosion, and together with the improved materials used for the ground electrodes, contributes to the spark plug having a very long service life. The ohmic resistance incorporated in the conductive glass seal reduces the capacitive discharge with the result that spark erosion is also decreased. As can be seen from Fig. 11, the fact that following an 800 hour engine run on an engine test bench (corresponds to more than 100,000 km on the road) there is far less electrode wear on the Platin plus 4 plug than on a conventional air-gap spark plug means that the increase in required voltage during this period is also far lower. In addition, Figs. 12 and 13 show the spark-plug "faces" for a new Platin plus 4 plug and for one which has already operated in the engine for 800 hours. The low level of electrode wear following this endurance run is immediately apparent.

Fig. 10

Design of the Bosch Platin plus 4 spark plug

1 Terminal stud,
2 Insulator,
3 Shell,
4 Heat-shrinkage zone,
5 Gasket,
6 Conductive glass seal,
7 Contact pin,
8 Platinum pin (center electrode),
9 Ground electrode (only 2 of the 4 electrodes are shown).

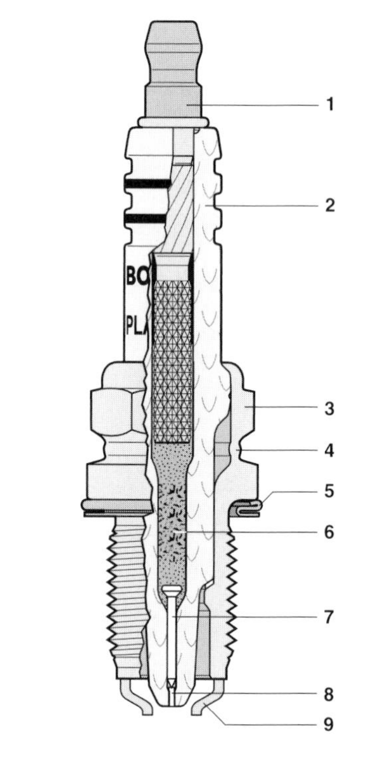

Ignition reliability

The Platin plus 4 spark plug features a very wide electrode gap (EA) of 1.6 mm. This leads to the spark plug having outstanding ignition reliability. In addition, the 4 ground electrodes are located so favorably with respect to the engine's combustion chambers that the ignition spark is not "masked" at all, and is therefore not impeded in its efficient ignition of the A/F mixture.

Cold-start characteristics

The Platin plus 4 spark plug's good cold-start characteristics are the result of the surface-gap principle employed. Since they continuously burn-off the soot on the insulator surface, surface-gap sparks are a guarantee for good self-cleaning. This means that the Platin plus 4 plug has better cold-starting characteristics than an air-gap spark plug.

Advantages

The Platin plus 4 spark plug is characterized by a number of outstanding features which are of particular importance with respect to long-term operation:

- The intervals between changes are extended to 100,000 km (60,000 miles) due to the very long useful life of the electrodes and the ceramic,
- The large number of cold engine starts which are possible,
- The excellent ignition characteristics, which lead to decisively more sophisticated running of the engine.
- Smooth development of power during the acceleration phase.

Fig. 11

Increase of required voltage along with engine running time

1 Spark plug with air gap (EA = 0.7 mm),
2 Platin plus 4 spark plug with surface spark (EA = 1.6 mm)

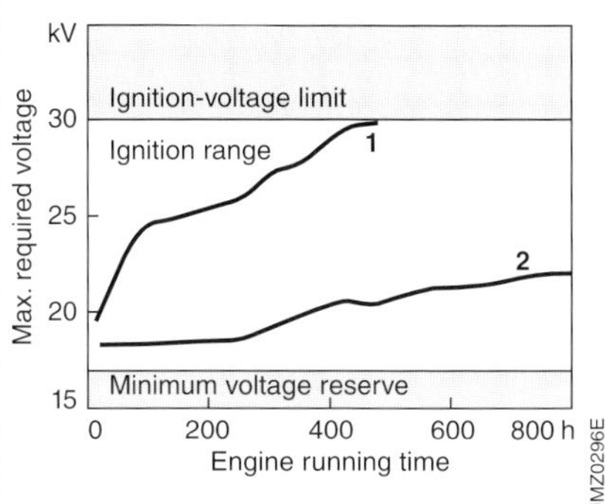

Fig. 12

"Face" of a new Platin plus 4

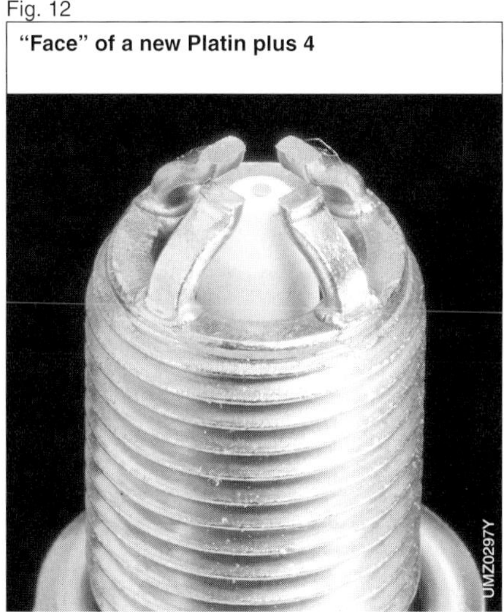

Fig. 13

"Face" of a Platin plus 4 run for 800 hours

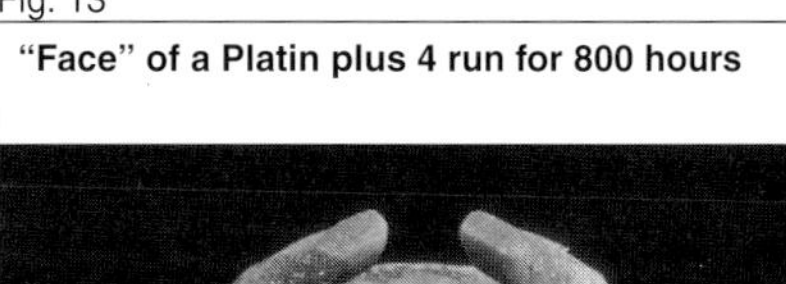

Special spark plugs

Applications

Special spark plugs are available for special requirements. Such spark plugs differ in their construction to the extent dictated by the operating and installation conditions at the engine in question.

Spark plugs for racing and competition

Due to their being continuously run at wide-open throttle, the engines used in racing and competition vehicles are subjected to extreme thermal stresses. The spark plugs used in such applications usually have noble-metal electrodes (silver, platinum) and a short insulator nose. On the one hand such spark plugs absorb very little heat due to their short insulator nose, while on the other hand the heat they do absorb is dissipated very quickly by the center electrode (Fig.14).

Spark plugs with resistor

The transmission of interference pulses to the ignition cables, and with them the interference radiation, can be reduced by a resistor in the line to the spark gap. In addition, the low current flowing during the spark plug's arcing phase reduces the electrode erosion. The resistor is formed by a special conductive glass seal between the center electrode and the terminal stud, the resistance value being achieved by the inclusion of appropriate additives in the glass.

Fully shielded spark plugs

In case of very high demands being made upon interference suppression (two-way radio, car telephone), it may be necessary to shield the spark plug.

In fully shielded spark plugs the insulator is encased in a metal sleeve. The connection is inside the insulator. The shielded ignition cable is fastened to the sleeve by means of a union nut. Fully shielded spark plugs are watertight (Fig. 15).

Fig. 14

Spark plug for racing and competition

1 Silver center electrode,
2 Short insulator nose.

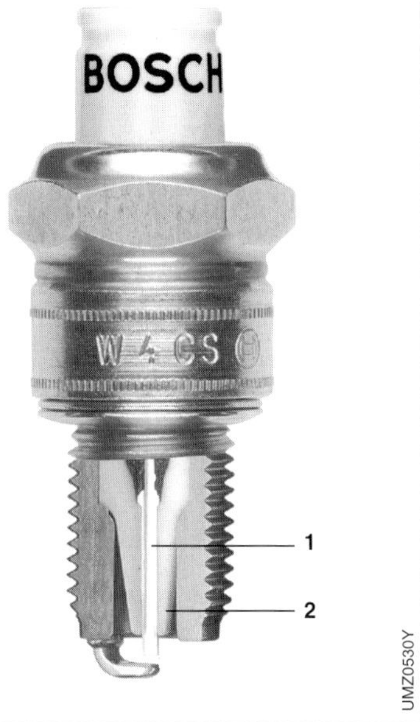

Fig. 15

Fully shielded spark plug

1 Special conductive glass seal (interference-suppression resistor),
2 Ignition-cable connection,
3 Shielding sleeve.

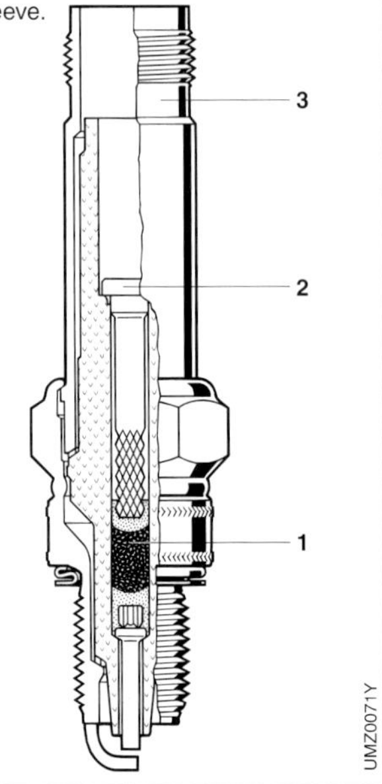

Designation codes

The individual spark-plug specifications are contained in the designation code (Fig. 16) which includes all the spark plug's essential characteristics – except its electrode gap. This is indicated on the package. The appropriate spark plug for individual engine applications is defined or recommended by the engine manufacturer and Bosch.

Fig. 16

Designation codes for Bosch spark plugs (dimensions in mm).

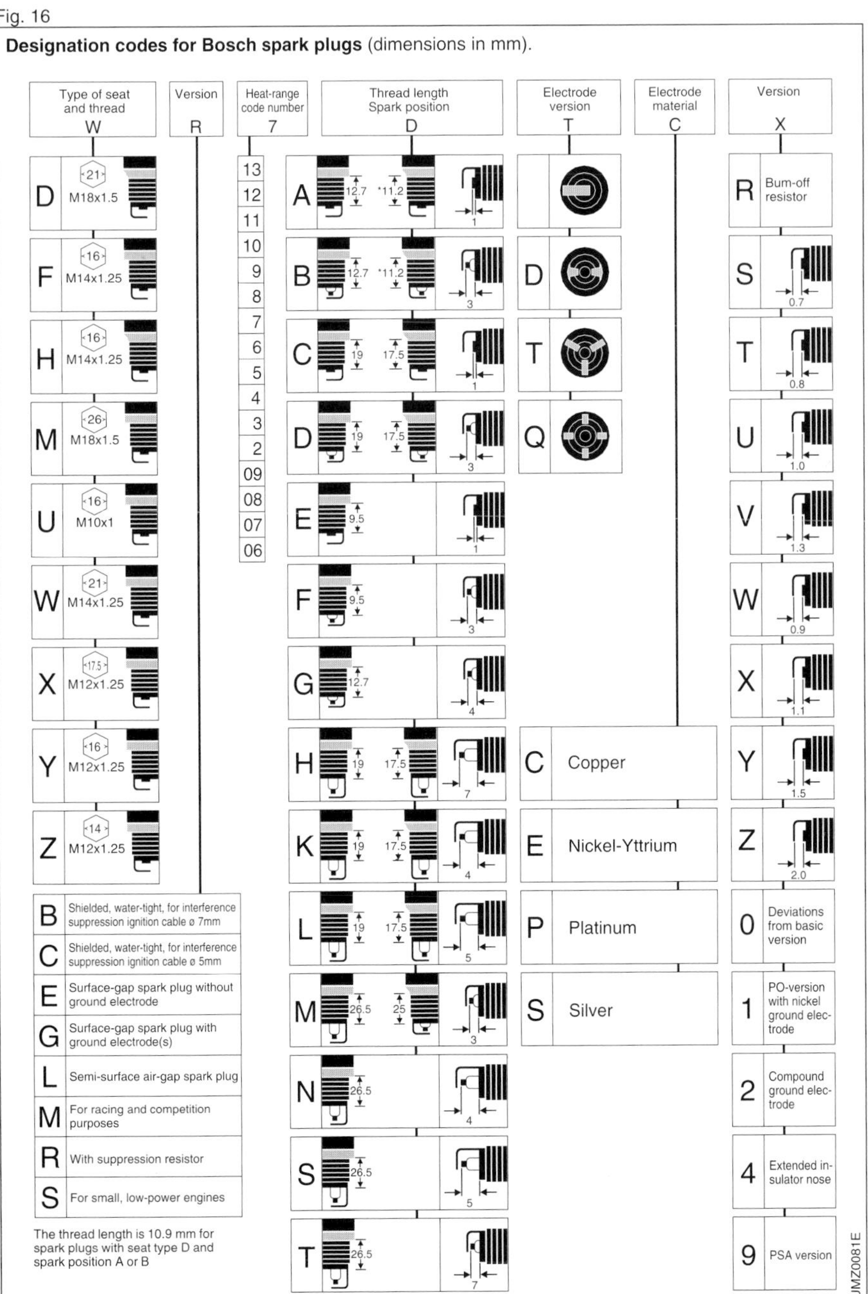

In practice

Fitting the spark plug

If the correct type is selected and correctly fitted, the spark plug is a reliable part of the ignition system.
Gap adjustments are recommended for front-electrode plugs only. The gap should not be adjusted on side-electrode spark plugs.

Removal

When removing the spark plug, first of all unscrew by a few turns. Then, using compressed air or a brush, clean the surrounding area in the cylinder head so that no dirt gets into the thread of the cylinder head or into the combustion chamber when the spark plug is removed. Only then unscrew the spark plug completely.
If the spark plug is very tight, unscrew it only a little, so as not to damage the thread in the cylinder head. Then drip oil or penetrating oil onto the threads, screw the spark plug back in again and attempt after a few minutes to unscrew it completely.

Installation

When installing the spark plug in the engine, note the following:

- The contact faces on the spark plug and on the engine must be clean.
- Bosch spark plugs are treated with an anti-corrosion oil, so no supplementary lubricants are required. Seizing up is impossible because the threads are nickel-plated.

Spark plugs should, if possible, be tightened to the torque given in Table 1 using a torque wrench. When the spark plug is tightened, the tightening torque is transmitted from the hexagon section to the seat and the thread.
This means that the insulator may come loose if the spark-plug shell is warped due to excessive tightening torque or through tilting of the spark-plug wrench. This completely destroys the spark plug's thermal properties; such a plug can cause engine damage. Therefore, the tightening torque must not exceed a given value.
The tightening torques apply to spark plugs as new, i.e. to lightly oiled spark plugs.
In practice, spark plugs are often fitted without a torque wrench. Consequently, they are usually overtightened. Bosch therefore recommends the following rule-of-thumb procedure:
1. Screw the spark plug by hand into the clean thread as far as it will go.

Table 1

Tightening torques

Spark-plug seat	Thread	Cylinder-head material	
		Cast iron Tightening torque (N · m)	Light alloy Tightening torque (N · m)
Spark-plug with flat seat	M 10x1	10...15	10...15
	M 12x1.25	15... 25	15...25
	M 14x1.25	20... 40	20...30
	M 18x1.5	30... 45	20...35
Spark-plug with conical seat	M 14x1.25	20...25	15...25
	M 18x1.5	20...30	15...23

Then apply the spark-plug wrench. One of three procedures is now used:

- Spark plugs with captive gasket (flat seal): New plugs: Using the wrench, turn the new spark plugs until first resistance is felt and then turn the wrench 90 degrees (Fig. 1),
- Used spark plugs with captive gasket: Turn used spark plugs 30 degrees after first resistance is felt,
- Spark plugs with conical (tapered) seat: Turn the spark plug 15 degrees after first resistance is felt (Fig. 2).

2. When tightening or loosening the spark plug, the wrench should not be held at an angle; the insulator will otherwise be broken off or pushed to the side, and the spark plug destroyed.

3. In the case of box wrenches with a loose tommy bar, the hole for the tommy bar must be above the spark plug so that the tommy bar can be pushed through both holes in the box wrench. If the holes are lower and the tommy bar is inserted only through one hole, the spark plug will be damaged.

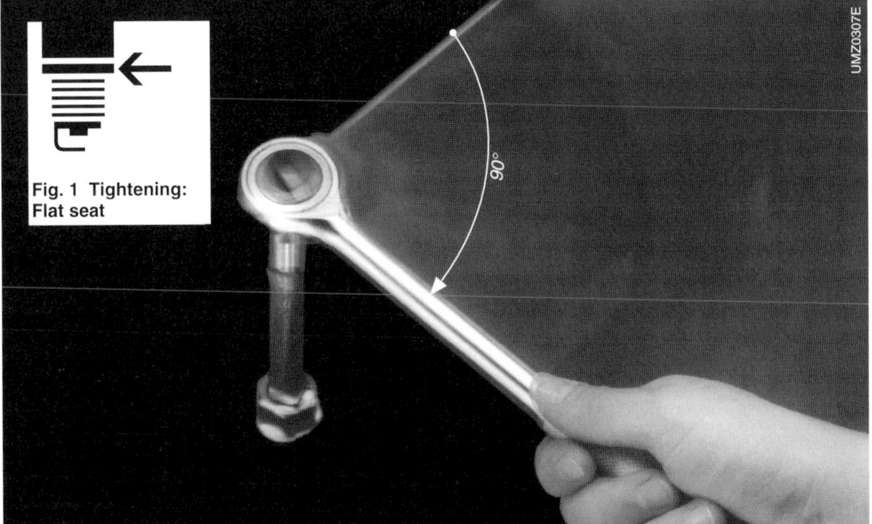

Fig. 1 Tightening: Flat seat

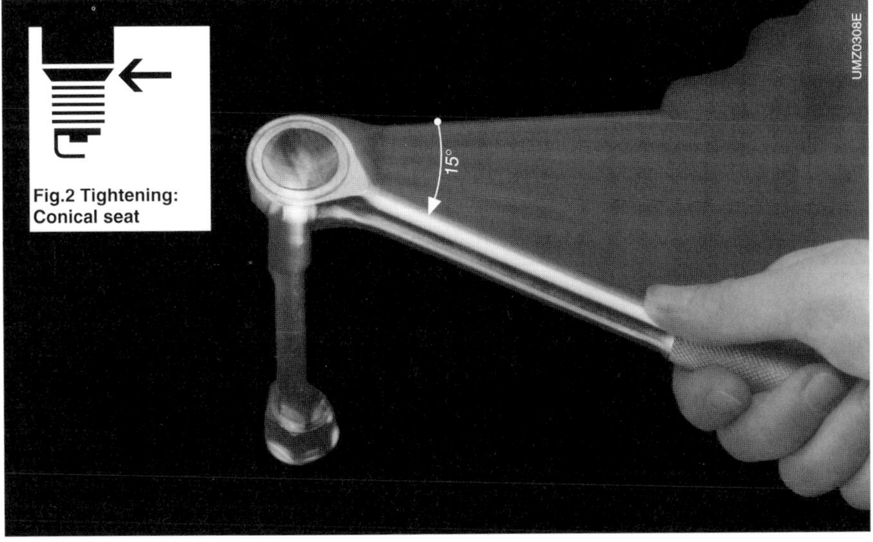

Fig.2 Tightening: Conical seat

Mistakes and their consequences

Basically, a given engine type should be fitted only with the spark plugs specified by the engine manufacturer or with those recommended by Bosch. To rule out incorrect selection from the outset, the motorist should consult the Bosch Service specialist. The desired information is also provided in sales literature, such as catalogs, product stands with information boards or application summaries. The use of unsuitable types of spark plugs can lead to serious engine damage. The most frequently encountered mistakes are:
- Incorrect heat range,
- Incorrect thread length, and
- Modifications/damage to seal face.

Incorrect heat-range code number
The heat-range code number must, under all circumstances, conform to the spark plug specifications of the motor-vehicle manufacturer or to the Bosch spark-plug recommendations.
Spark plugs with the wrong heat range for the engine can produce auto-ignition.

Incorrect thread length
The length of the thread on the spark plug must be the same as the length of the thread in the cylinder head.
If the spark-plug thread is too long, the spark plug projects too far into the combustion chamber.
Consequences:
- Possible damage to the piston,
- Coking-up of the spark-plug thread may make it impossible to remove the spark plug,
- Spark plug overheating.

If the thread is too short, the spark plug does not project far enough into the combustion chamber.
Consequences:
- Poor ignition of the mixture,
- Spark plug does not reach its self-cleaning temperature,
- The lower threads in the cylinder head become coked up.

Modifications to the seat
On spark plugs with a conical seat, it is not permissible to use a washer or a seal ring. On flat-seat spark plugs, it is permissible to use only the captive gasket fitted to the spark plug. It must not be removed or replaced by a washer.
Without the gasket, the spark plug projects too far into the combustion chamber. This impairs efficient heat transfer between spark-plug shell and cylinder head, and the plug does not seal properly.
If an additional seal ring is used, the spark plug is not seated deep enough in the thread hole, and the transfer of heat from the spark-plug shell to the cylinder head is also impaired.

Spark-plug faces

Reading the spark plugs provides valuable information on spark-plug and engine operating conditions.
The appearance of the electrodes and insulator of the spark plug – the "spark-plug face" – provides information on spark-plug operation, mixture and the combustion process within the engine (Figures 3 to 5).

Reading the spark plugs is thus an important engine-diagnosis procedure. For an accurate reading, observe the following procedural guidelines: The vehicle must be operated under standard conditions before an attempt is made to read the spark plugs. If the reading is preceded by an extended period at idle, carbon deposits on the plug will render accurate analysis impossible; this problem is more serious immediately after a cold start. The vehicle should be driven about 10 km, at a variety of moderate loads and engine speeds. Extended idling prior to shutdown is to be avoided.

Fig. 3

Spark-plug faces. Part 1

① **Normal condition**

Insulator nose grayish-white or grayish-yellow to brown. Engine is in order. Heat range of plug correct.

Mixture setting and ignition timing are correct, no misfiring, cold-starting device functioning.

No deposits from fuel additives containing lead or from alloying constituents in the engine oil. No overheating.

② **Sooted – carbon-fouled**

Insulator nose, electrodes and spark-plug shell covered with velvet-like, dull black soot deposits.

Cause: incorrect mixture setting (carburetor, fuel injection): mixture too rich, air filter very dirty, automatic choke not in order or manual choke pulled too long, mainly short-distance driving, spark plug too cold, heat-range code number too low.

Effects: Misfiring, difficult cold-starting.

Remedy: Adjust A/F mixture and choke device, check air filter.

③ **Oil-fouled**

Insulator nose, electrodes and spark-plug shell covered with shiny soot or carbon residues.

Cause: too much oil in combustion chamber. Oil level too high, badly worn piston rings, cylinders and valve guides. In two-stroke engines, too much oil in mixture.

Effects: Misfiring, difficult starting.

Remedy: Overhaul engine, adjust oil/fuel ratio (2-stroke engines), fit new spark plugs.

④ **Lead fouling**

Insulator nose covered in places with brown/yellow glazing which can have a greenish color.

Cause: Lead additives in fuel. Glazing results from high engine loading after extended part-load operation.

Effects: At high loads, the glazing becomes conductive and causes misfiring.

Remedy: Fit new spark plugs since cleaning the old ones is pointless.

Fig. 4

Spark-plug faces. Part 2

⑤ Pronounced lead fouling

Insulator nose covered in places with thick brown/yellow glazing which can have a greenish color.

Cause: Lead additives in fuel. Glazing results from high engine loading after extended part-load operation.

Effects: At high loads, the glazing becomes conductive and causes misfiring.

Remedy: Fit new spark plugs since cleaning the old ones is pointless.

⑥ Formation of ash

Heavy ash deposits on the insulator nose resulting from oil and fuel additives, in the scavening area, and on the ground electrode. The structure of the ash is loose- to cinder-like.

Cause: Alloying constituents, particularly from engine oil can deposit this ash in the combustion chamber and on the spark-plug face.

Effects: Can lead to auto-ignition with loss of power and possible engine damage.

Remedy: Repair the engine. Fit new spark plugs. Possibly change engine-oil type.

⑦ Center electrode covered with melted deposits

Melted deposits on center electrode. Insulator tip blistered, spongy, and soft.

Cause: Overheating caused by auto-ignition. For instance due to ignition being too far advanced, combustion deposits in the combustion chamber, defective valves, defective ignition distributor, poor-quality fuel. Possibly spark-plug heat-range value too low.

Effects: Misfiring, loss of power (engine damage).

Remedy: Check the engine, ignition, and mixture-formation system. Fit new spark plugs with correct heat-range code number.

⑧ Partially melted center electrode

Center electrode has melted and ground electrode is severely damaged.

Cause: Overheating caused by auto-ignition. For instance due to ignition being too far advanced, combustion deposits in the combustion chamber, defective valves, defective ignition distributor, poor-quality fuel.

Effects: Misfiring, loss of power (engine damage). Insulator-nose fracture possible due to overheated center electrode.

Remedy: Check the engine, ignition, and mixture-formation system. Fit new spark plugs.

Fig. 5

Spark-plug faces. Part 3

⑨ **Partially melted electrodes.**

Cauliflower-like appearance of the electrodes. Possible deposit of materials not originating from the spark plug.

Cause: Overheating caused by auto-ignition. For instance due to ignition being too far advanced, combustion deposits in the combustion chamber, defective valves, defective ignition distributor, poor-quality fuel.

Effects: Power loss becomes noticeable before total failure occurs (engine damage).

Remedy: Check engine and mixture-formation system. Fit new spark plugs

⑩ **Heavy wear on center electrode.**

Cause: Spark-plug exchange interval has been exceeded.

Effects: Misfiring, particularly during acceleration (ignition voltage no longer sufficient for the large electrode gap). Poor starting.

Remedy: Fit new spark plugs.

⑪ **Heavy wear on ground electrode.**

Cause: Aggressive fuel and oil additives. Unfavorable flow conditions in combustion chamber, possibly as a result of combustion deposits. Engine knock. Overheating has not taken place.

Effects: Misfiring, particularly during acceleration (ignition voltage no longer sufficient for the large electrode gap). Poor starting.

Remedy: Fit new spark plugs.

⑫ **Insulator-nose fracture.**

Cause: Mechanical damage (spark plug has been dropped, or bad handling has put pressure on the center electrode). In exceptional cases, deposits between the insulator nose and the center electrode, as well as center-electrode corrosion, can cause the insulator nose to fracture (this applies particularly for excessively long periods of use).

Effects: Misfiring, spark arcs-over at a point which is inaccessible for the fresh charge of A/F mixture.

Remedy: Fit new spark plugs.

M-Motronic engine management

The M-Motronic system

System overview

M-Motronic combines all the electronic systems for engine control in a single control unit (ECU) which, in turn, governs the actuating systems used to control the spark-ignition engine. Engine-mounted monitoring devices (sensors) gather the required operating data and relay the information to input circuits for:
- Ignition (on/off),
- Camshaft position,
- Vehicle speed,
- Gear selection,
- Transmission control,
- Air conditioner, etc.

Monitored analog data include:
- Battery voltage,
- Engine temperature,
- Intake-air temperature,
- Air quantity,
- Throttle-valve angle,
- Lambda oxygen sensor,
- Knock sensor, etc.
  as well as
- Engine speed.

Input circuits located within the ECU convert these data for subsequent operations in the microprocessor. The microprocessor, in turn, uses these data to determine the engine's momentary operating conditions; this information serves as the basis for the ECU's command signals, which are amplified by power-output stages before being transmitted to the final-control elements used to control the engine. This system combines fuel injection, highest-quality mixture preparation and the correct ignition timing to provide mutual support over the entire range of operating conditions encountered in the spark-ignition engine.

M-Motronic versions

The descriptions and illustrations on the following pages refer to a typical M-Motronic configuration (Figure 1). Other M-Motronic systems are available to meet the special demands arising from specific national regulations as well as the requirements of the individual automobile manufacturers.

Basic functions
Control of the ignition and fuel-injection processes is (independent of version) at the core of the M-Motronic system.

Auxiliary functions
Additional open and closed-loop control functions – required in response to legislation aimed at reductions in exhaust emissions and fuel consumption – supplement the basic M-Motronic functions while making it possible to monitor all components exercising an influence on the composition of the exhaust gases.
These include:
- Idle-speed control,
- Lambda oxygen control,
- Control of the evaporative emissions control system,
- Knock control,
- Exhaust-gas recirculation (EGR) for reducing NO_X emissions, and
- Control of the secondary-air injection to reduce HC emissions.

The system can also be expanded to meet special demands from automobile manufacturers by including the following:
- Open-loop turbocharger control as well as control of variable-tract intake manifolds for increased engine power output,

- Camshaft control for achieving reductions in exhaust emissions and fuel consumption as well as enhanced output, and
- Knock control along with engine and vehicle-speed governing functions, to protect engine and vehicle.

The acquisition and processing of the measured information is described in the chapters dealing with acquisition and processing of the operating data.

Vehicle management

M-Motronic supports the control units in other vehicle systems. It can, for instance, operate together with the automatic transmission's control unit to provide reductions in torque during shifting, thereby lessening transmission wear. M-Motronic can also work together with the ABS control unit to provide traction control (TCS) for enhanced vehicle safety.

The schematic system illustration below shows a maximal-configuration M-Motronic system. This type of system can be employed to satisfy

- The stringent emissions limits, and
- The requirement for an integral, onboard diagnosis (OBD) system for California vehicles from 1993 onward.

Fig. 1

System diagram of M-Motronic M5 with integrated diagnostics (OBD)

1 Carbon canister, **2** Shutoff valve, **3** Canister-purge valve, **4** Fuel-pressure regulator, **5** Injector, **6** Pressure actuator, **7** Ignition coil, **8** Phase sensor, **9** Secondary-air pump, **10** Secondary-air valve, **11** Air-mass meter, **12** Control unit (ECU), **13** Throttle-valve sensor, **14** Idle actuator, **15** Air-temperature sensor, **16** EGR valve, **17** Fuel filter, **18** Knock sensor, **19** Engine-speed sensor, **20** Engine-temperature sensor, **21** Lambda oxygen sensor, **22** Diagnosis interface, **23** Diagnosis lamp, **24** Pressure differential sensor, **25** Electric fuel pump.

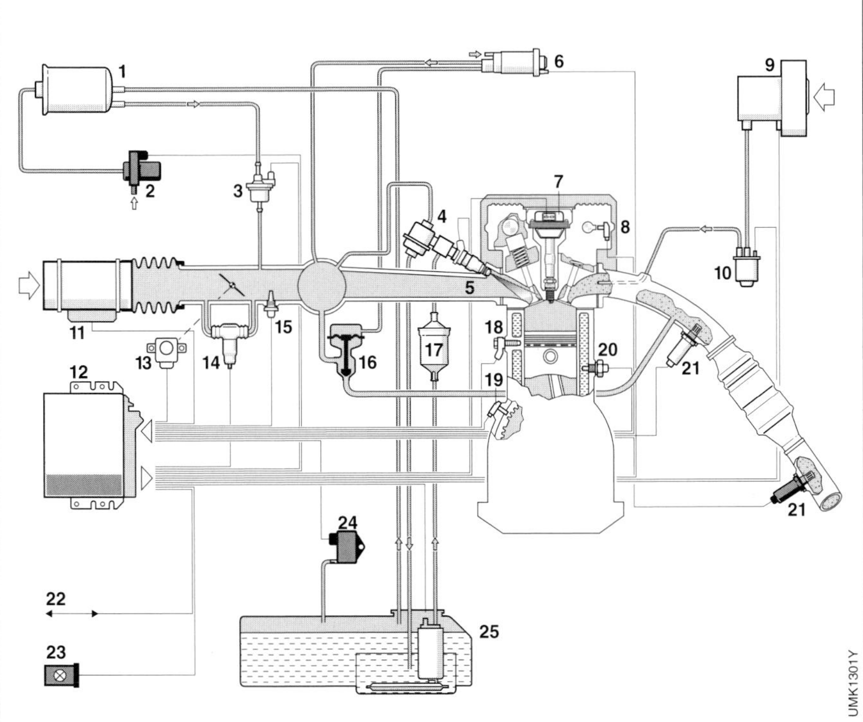

Fuel system

Fuel supply

Fuel-supply system

The fuel-supply system must be capable of providing the engine with the required quantity of fuel under all operating conditions. An electric pump draws the fuel through a filter while extracting it from the tank for delivery to the fuel-distribution rail with its electromagnetic injectors. The injectors spray the fuel into the engine's intake tract in precisely metered quantities. The excess fuel flows through the pressure regulator and back to the fuel tank (Figure 1).

The pressure regulator generally employs the pressure within the intake manifold as its reference. This characteristic pressure works in combination with the constant flow through the fuel rail (cooling effect) to prevent vapor bubbles from forming in the fuel. The resulting pressure differential at the injector usually remains constant in the 300 kPa range.
Where necessary, the fuel-supply system can also be designed to incorporate pressure attenuators to reduce pulsation in the fuel line.

Fig. 1

Fuel supply system

1 Electric fuel pump (in-tank),
2 Fuel filter,
3 Fuel rail,
4 Injector,
5 Fuel-pressure regulator.

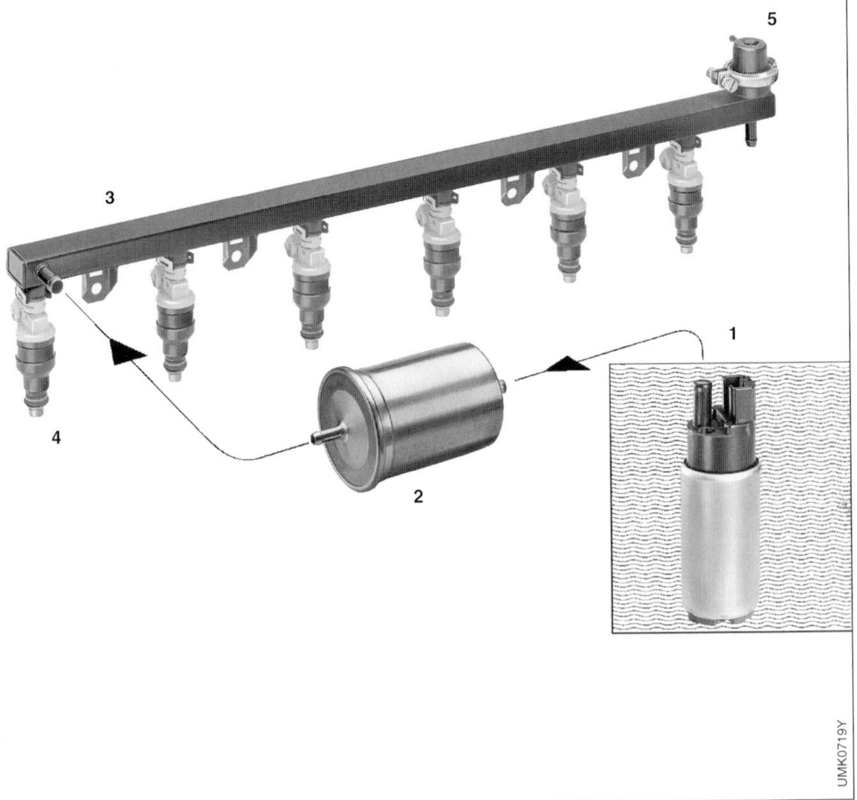

Electric fuel pump

Function

The electric fuel pump supplies a continuous flow of fuel from the tank. It can be installed either within the fuel tank itself ("in-tank") or at an external location in the fuel line ("in-line").

The in-tank pumps currently in general use (Figures 2 and 3) are integrated within the fuel tank's installation assembly along with the level sensor and a swirl plate to remove vapor bubbles from the fuel return line. When an in-line pump is used, hot delivery problems can be solved by using a supplementary in-tank booster pump to supply fuel from the tank at low pressure. To ensure that the delivery pressure in the system is maintained at the required level, the maximum supply capacity is always greater than the system's theoretical maximum requirement.

The fuel pump is activated by the engine-management ECU. A safety circuit interrupts fuel delivery when the engine is stationary with the ignition on.

Fig. 2

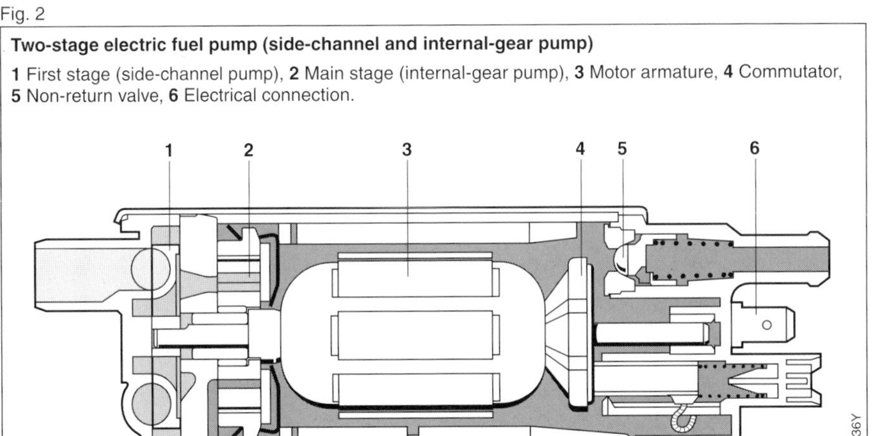

Two-stage electric fuel pump (side-channel and internal-gear pump)

1 First stage (side-channel pump), **2** Main stage (internal-gear pump), **3** Motor armature, **4** Commutator, **5** Non-return valve, **6** Electrical connection.

Fig. 3

Two-stage electric fuel pump (side-channel and peripheral pump)

1 Suction cover with supply connection, **2** Impeller, **3** First stage (side-channel pump), **4** Main stage (peripheral pump), **5** Pump housing, **6** Motor armature, **7** Non-return valve, **8** End cover with pressure connection.

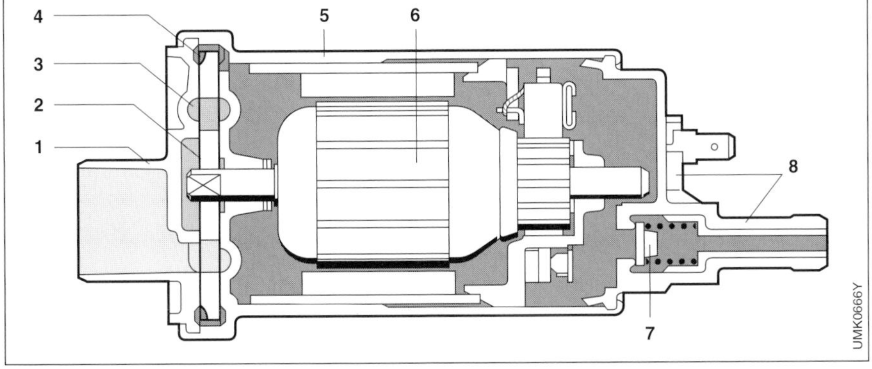

Design

The electric fuel pump consists of the following elements:
- Pump assembly,
- Electric motor and end cover.

The electric motor and pump assembly are located in a common housing, where they are immersed in circulating fuel. This arrangement provides effective cooling for the electric motor. Because no oxygen is present, it is impossible for an ignitable mixture to form; there is no danger of an explosion. The end cover contains the electrical connections, the non-return valve and the hydraulic connection for the pressure side. The non-return valve maintains system pressure for a period of time after the unit is shut down to prevent vapor bubbles from forming. Interference-suppression devices can also be included in the end-cover assembly.

Design variations

Various design principles are employed to satisfy individual system demands (Figure 4).

Positive-displacement pumps

Roller-cell (RZP) and internal-gear pumps (IZP) are both classified as positive-displacement designs. Both types of pump operate by using variable-sized, circulating chambers to expose a supply orifice and draw in fuel as their volume expands. Once the maximum volume is reached, the supply orifice closes and the discharge orifice opens. The fuel is now forced out as the effective volume in the chamber decreases. The chambers on the roller-cell pump are formed by rollers circulating in a rotor plate. A combination of centrifugal force and fuel pressure forces them outward against the eccentric roller path. The eccentricity between rotor plate and roller path provides the constant increase and decrease in chamber volume.

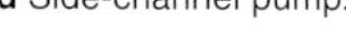
Fig. 4

Principles of operation

a Roller-cell pump,
b Peripheral pump,
c Internal-gear pump,
d Side-channel pump.

a

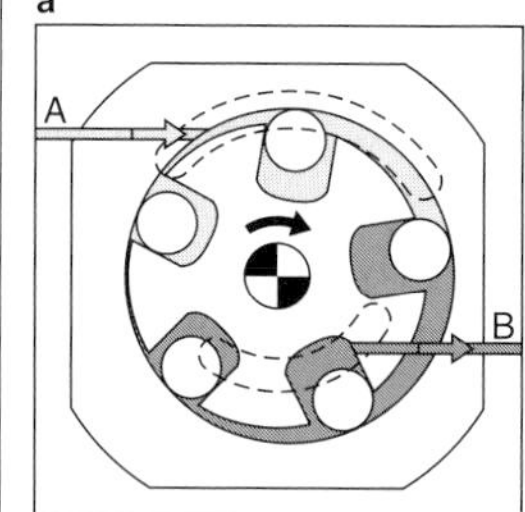

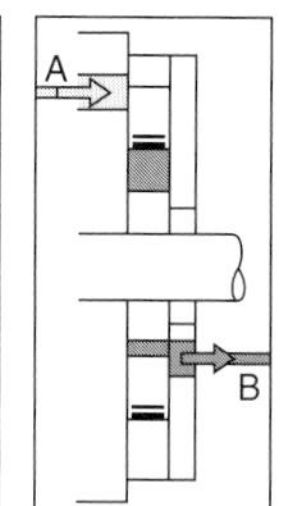

b

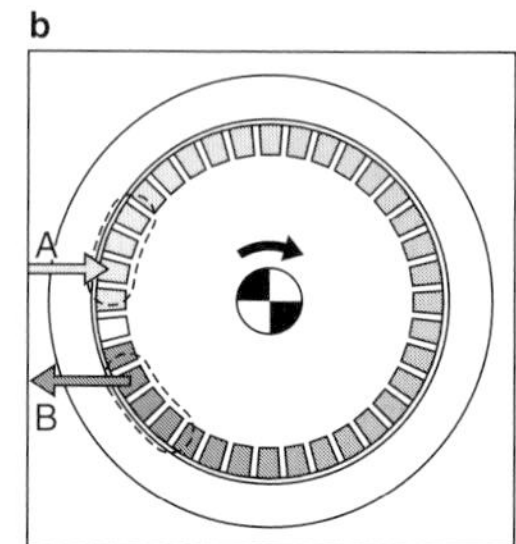

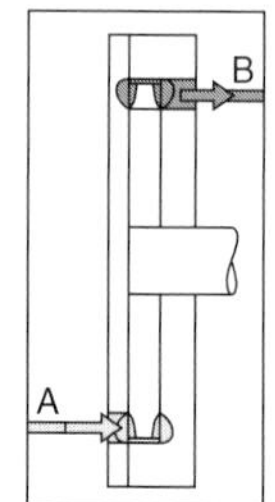

c

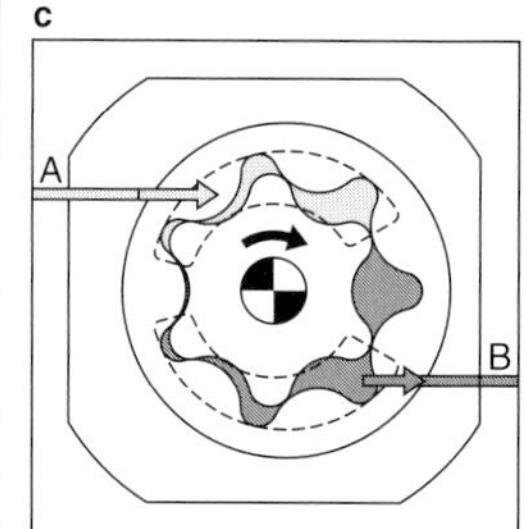

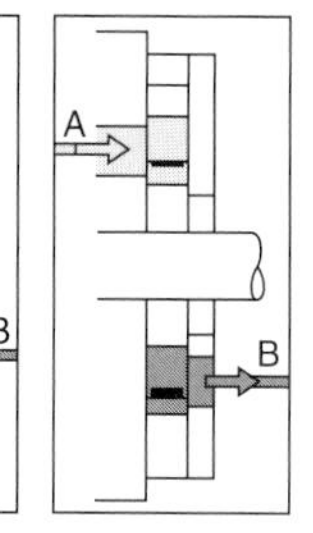

d

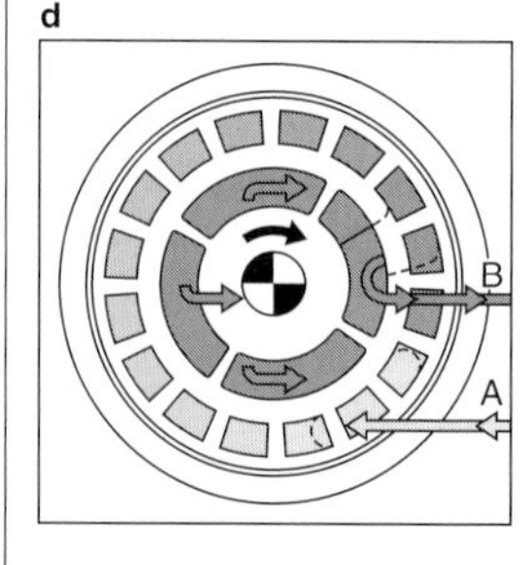

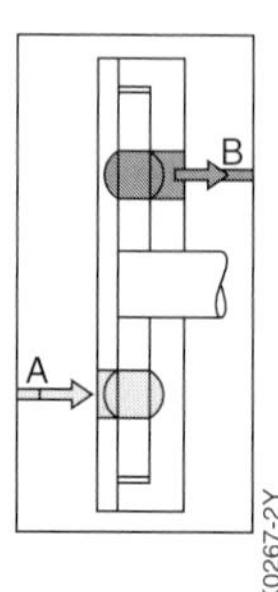

UMK0267-2Y

The internal-gear pump consists of an internal drive gear that moves against the surface of an eccentrically-mounted ring gear; the ring gear is equipped with one tooth more than the drive gear. As the mutually-sealed tooth flanks turn, variable chambers are formed between them. Roller-cell pumps can be used to obtain fuel pressures in excess of 650 kPa. Internal-gear pumps can supply up to 400 kPa, a figure adequate for virtually all Motronic applications.

Hydrokinetic flow pumps

The peripheral and side-channel pumps are both classified as hydrokinetic flow pumps. In these pumps an impeller accelerates the fuel particles before discharging them into the tract where they generate pressure via pulse exchange. The peripheral pump differs from its side-channel counterpart in its larger number of impeller blades and the shape of the impellers, as well as in the positions of the side channels, which – unlike those of the side-channel unit – are located on the circumference or periphery. Although peripheral pumps are only capable of generating maximum fuel pressures in the 400 kPa range, they do supply a continuous, virtually pulseless flow of fuel. This makes them particularly attractive for use in those vehicular applications where limiting noise is a major priority. Side-channel pumps can only produce pressures of up to 30 kPa. One important use for this type of unit is as a booster pump in systems with in-line main pumps; another major application is as the primary stage in a two-stage in-tank pump of the kind installed in vehicles susceptible to hot starting problems and/or with single-point fuel injection.

Fig. 5

Fuel filter

1 Paper element,
2 Strainer,
3 Support plate.

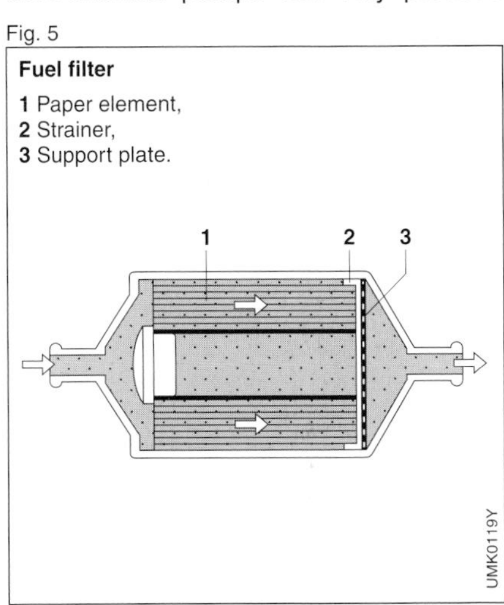

Fuel filter

Contaminants in the fuel can impair the operation of both pressure regulator and injectors. A filter is therefore installed downstream from the electric fuel pump. This fuel filter contains a paper element featuring a mean pore diameter of 10 µm. A backplate retains it in its housing. The replacement intervals are determined by the filter's volume and contamination levels in the fuel (Figure 5).

Fuel rail

The fuel flows through the fuel rail where it is evenly distributed to all injectors. The injectors are mounted on the fuel rail, which also usually includes a fuel-pressure regulator. A pressure attenuator may also be present. The dimensions of the fuel rail are selected to inhibit the local fluctuations in fuel pressure that could otherwise be triggered as the injectors run through their operating cycles. This prevents the injection quantities from reacting to changes in load and engine speed. Depending upon the particular vehicle type and its special requirements, the fuel rail can be made of steel, aluminum or plastic. It may also include an integral test valve, which can be used to bleed pressure for servicing as well as for test purposes.

Fuel-pressure regulator

Injection quantity should be determined exclusively by injection duration. Thus the difference between the fuel pressure in the distribution rail and the pressure in the intake tract must remain constant. A means is thus required for adjusting the fuel pressure to reflect variations in the load-sensitive manifold pressure. The fuel-presure regulator regulates the amount of fuel returning to the tank to maintain a constant pressure drop across the injectors. The pressure regulator is generally positioned at the far end of the fuel rail to avoid impairing the flow within the rail. However, it can also be mounted in the fuel-return line.

The fuel-pressure regulator is designed as a diaphragm-controlled overflow pressure regulator (Figure 6). A rubber-fiber diaphragm divides the pressure regulator into two sections: fuel chamber and pressure chamber. The spring presses against a valve holder integrated within the diaphragm. This force causes a flexibly mounted valve plate to push against a valve seat. When the pressure exerted against the diaphragm by the fuel exceeds that of the spring, the valve opens and allows fuel to flow directly back to the tank until the diaphragm assembly returns to a state of equilibrium, with equal pressure exerted on both of its sides. A pneumatic line is provided between the spring chamber and the intake manifold downstream from the throttle valve, allowing the chamber to respond to changes in manifold vacuum. Thus the pressures at the diaphragm correspond to those at the injectors. As a result, the pressure drop at the injectors remains constant, as it is determined solely by the spring force and surface area of the diaphragm.

Fuel-pressure attenuator

The injectors' operating cycles and the periodic discharge of fuel that characterize the positive-displacement fuel pump both induce fluctuations in fuel pressure. Under unfavorable circumstances, the mountings for the electric fuel pump, fuel lines and fuel rail can transmit these vibrations to the vehicle's body. Noise from this source can be prevented using specially designed mounting elements and fuel-pressure attenuators. The layout of the fuel-pressure attenuator (Figure 7) is similar to that of the pressure regulator. In both cases a spring-loaded diaphragm separates the fuel from the air space. The spring force is calculated to lift the diaphragm from its seat as soon as the fuel reaches operating pressure. This provides a variable fuel chamber which can accept fuel to ease pressure peaks and then release it again when pressure falls. The spring chamber can be fitted with a manifold-vacuum line to stay within the optimum operating range in the face of fluctuations in the fuel's absolute pressure. The pressure attenuator also shares the pressure regulator's installation flexibility, as it can also be mounted in the rail or in the fuel-return line.

Fig. 6

Fuel-pressure regulator

1 Intake-manifold connection, **2** Spring, **3** Valve holder, **4** Diaphragm, **5** Valve, **6** Fuel supply, **7** Fuel return.

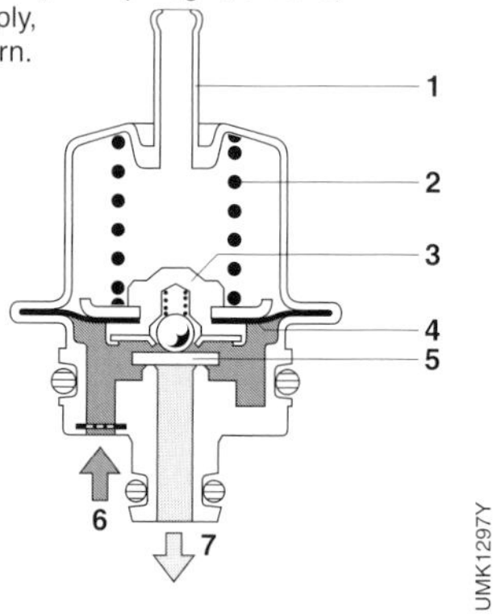

Fig. 7

Fuel-pressure attenuator

1 Spring, **2** Spring plate, **3** Diaphragm, **4** Fuel supply, **5** Fuel return.

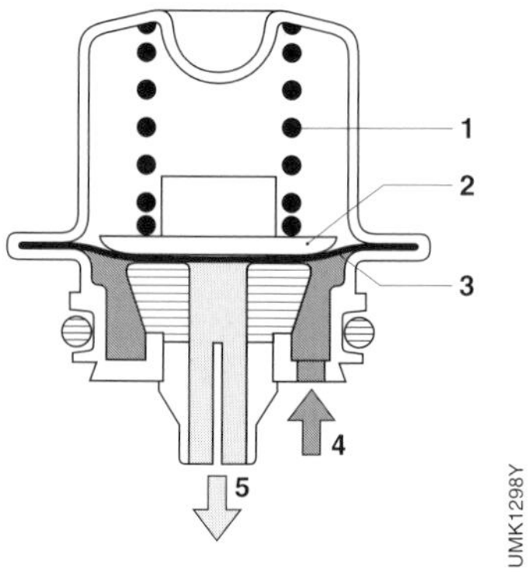

Fuel injection

Uncompromising demands for smooth running and low emissions in automobiles have made it necessary to provide thorough and precise mixture formation for every single work cycle. The fuel mass must be injected in quantities that are precisely metered to match the amount of intake air; in today's applications exact injection timing is acquiring increasing significance. For this reason, every cylinder is assigned an electromagnetic injector. The injector sprays the fuel – in precise quantities at a point in time determined by the ECU – directly toward the cylinder intake valve(s). Thus condensation along the walls of the intake tract of the kind that leads to deviations from the desired Lambda value is largely avoided. Because the engine's intake manifold conducts only combustion air, its geometry can be optimized to meet the engine's dynamic gas-flow requirements.

Electromagnetic injector

The electromagnetic injector contains a solenoid armature mounted on a valve needle (Figures 8 and 9), and travels through precise motions within the valve body. When the unit is at rest, a coil spring presses the valve needle against the seat to seal off the flow of fuel through

Fig. 8

Injector (top-feed)

1 Filter strainer in fuel supply, **2** Electrical connection, **3** Solenoid winding, **4** Valve housing, **5** Armature, **6** Valve body, **7** Valve needle.

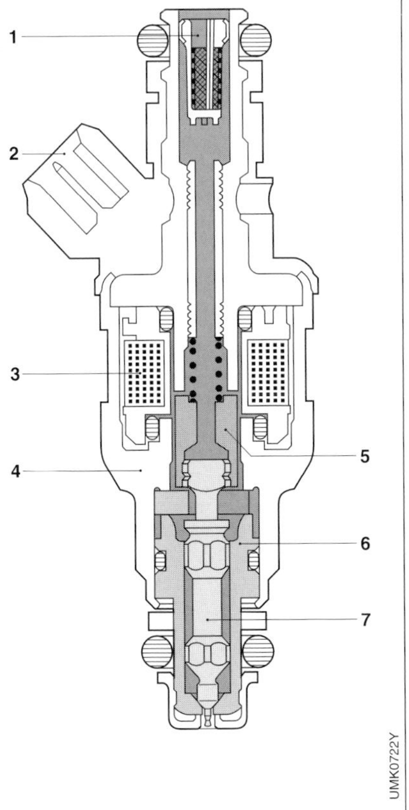

Fig. 9

Injector (bottom-feed)

1 Electrical connection, **2** Filter screen in fuel supply, **3** Solenoid winding, **4** Valve housing, **5** Armature, **6** Valve body, **7** Valve needle.

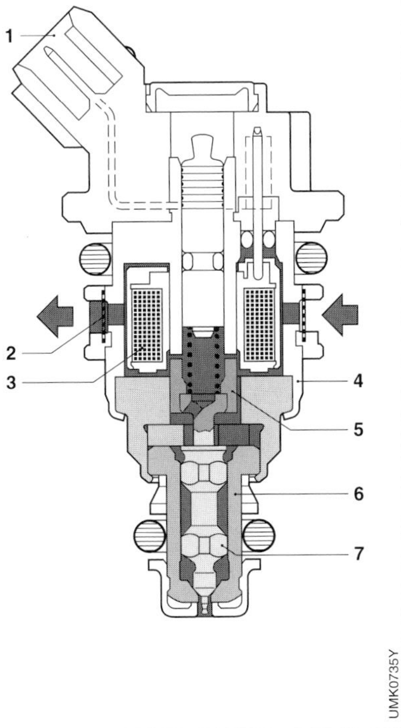

the outlet orifice and into the intake manifold. When the control transmits an activation current to the solenoid winding in the valve housing, the solenoid armature rises between 60 and 100 µm, lifting the valve needle in the process; the fuel can now flow through the calibrated orifice. The response times lie between 1.5 ... 18 ms at a control frequency of 3...125 Hz, depending upon the type of injection and the momentary engine speed and load conditions.

Different injector designs are employed to meet varying requirements:

Top-feed injector

Fuel enters the top-feed injector from above and flows through its vertical axis. This unit is installed in a specially-formed opening in the fuel rail. Sealing is provided by an upper seal ring, while a clip holds the unit in place. The lower end, which also has a seal ring, extends into the engine's intake manifold (Figure 8).

Bottom-feed injector

The "bottom-feed" injector is integrated within the fuel-rail assembly, where it is constantly immersed in flowing fuel. The fuel supply enters the unit from the side ("bottom-feed"). The fuel rail itself is mounted directly on the intake manifold.

Fig. 10

Injectors (bottom-feed) integrated in fuel rail

1 Fuel supply, **2** Injector, **3** Electrical connection, **4** Contact rail, **5** Pressure regulator, **6** Fuel return.

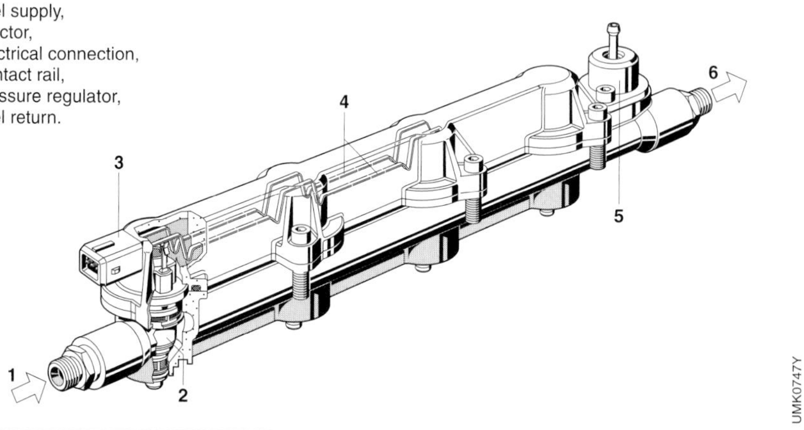

Fig. 11

Metering layouts and fuel preparation

1 Ring-gap metering, **2** Single-orifice metering, **3** Multi-orifice metering, **4** Multi-orifice metering with dual-stream injector.

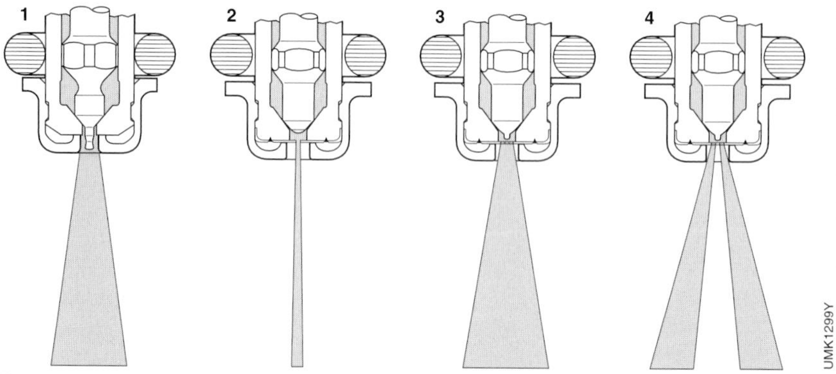

The injector is retained in the fuel rail by either a clip, or a cover on the rail which can also house the electrical connections. Two seal rings prevent the fuel from escaping. This type of modular design provides several advantages; these include good starting and driving response with hot engines as well as low installation height (Figures 9 and 10).

Mixture formation

A variety of different fuel-metering arrangements are employed to satisfy the demand for the effective fuel atomization necessary for ensuring maximum homogeneity in the air-fuel mixture while simultaneously holding intake-tract condensation to a minimum. The injector's discharge orifice is specially calibrated to fulfill these requirements in the respective applications (Figure 11). On units with ring-gap metering, a section of the valve needle (pintle) extends through the valve body. The resulting ring gap forms the calibrated fuel-discharge orifice. The lower end of the pintle features a machined breakaway edge where the fuel atomizes before emerging in a tapered pattern.

On injectors with single-orifice metering, the pintle is replaced by a thin injection-orifice disk with a calibrated opening. Virtually none of the thin jet of fuel lands on the walls of the intake tract. However, fuel atomization is limited. Injectors featuring multi-orifice metering are fitted with an injection-orifice disk of the kind used in the single-orifice units, the difference being that the multi-orifice disk contains numerous calibrated openings. These are arranged to provide a tapered spray pattern similar to that achieved with annular-orifice metering devices, and provide comparable fuel atomization. The orifices can also be designed to provide two or more spray patterns. This makes it possible to achieve optimum fuel distribution via separate injection into the individual inlet runners on multi-valve engines. Meanwhile, air-shrouded injectors can provide even better mixture formation.

Combustion air traveling at the speed of sound is extracted from the intake manifold at a location upstream from the throttle valve; it then proceeds through a calibrated opening located directly on the injection-orifice disk. The interaction between fuel and air molecules provides thorough atomization. To allow air to be drawn in through the opening, a partial vacuum referred to atmospheric pressure is required in the intake manifold. The air-shrouded design is thus most effective during part-throttle operation (Figure 12).

Fig. 12

Air-shrouded injector

1 Air supply, **2** Fuel supply.

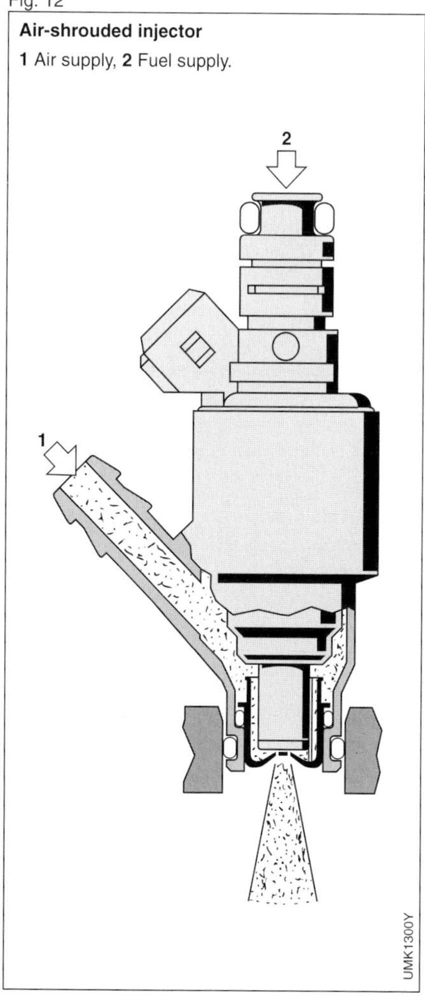

High-voltage ignition circuit

The high-voltage ignition circuit generates the high-tension current required for ignition and then delivers it to the correct spark plug at precisely the right instant.

The Motronic high-voltage circuit can be constructed according to any one of several design options:
- High-voltage circuit with one ignition coil, one power-output stage and a high-tension distributor (rotating voltage distribution).
- High-voltage circuit featuring one single-spark coil and one power-output stage per cylinder (stationary or electronic voltage distribution).
- High-voltage circuit with one dual-spark coil and one power-output stage for every two cylinders (stationary or electronic voltage distribution).

Ignition coil

Function

The ignition coil stores the ignition energy and produces the high voltages required to generate a spark when the ignition is triggered.

Design and operation

The ignition coil operates according to the laws of induction. The unit consists of two magnetically coupled copper coils (primary and secondary windings). The energy stored in the primary winding's magnetic field is transmitted to the secondary side. The transformation ratio for current and voltage is a function of the ratio of the numbers of coils contained in the respective windings of the primary and secondary circuits (Figure 1).

Modern ignition coils consist of single plates, combined to form a closed ferrous circuit, and a plastic housing. Within the housing the primary winding is wound on a bobbin mounted directly on the core. Further outward is the secondary winding; in the interests of enhanced arcing resistance, this assumes the shape of a disk or chamber winding. The housing, meanwhile, is filled with epoxy resin to provide effective insulation between the two windings, and between the windings and the core. The specific design configuration is tailored to match the individual application.

Ignition driver stage

Function and operation

Ignition driver stages featuring multi-stage power transistors control the flow of primary current through the coil, replacing the breaker points found on earlier ignition systems.
In addition, the ignition output stage is also charged with limiting both primary current and primary voltage. The primary voltage is restricted to prevent excessive increases in the supply of secondary voltage, which could damage components in the high-tension circuit. Restrictions on primary current hold the ignition system's energy output to a specified level. Power-output stages can be either internal (forming an integral part of Motronic) or external (located outside the Motronic unit).

Fig. 1

Ignition coils (diagram)

Rotating distribution: **a** Single-spark coil.
Stationary distribution: **b** Single-spark coil, **c** Dual-spark coil.

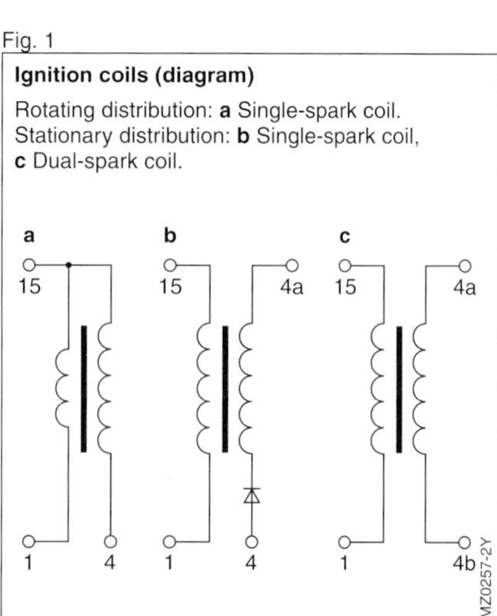

High-voltage generation
The M-Motronic ECU activates the ignition driver stage during the calculated dwell period. It is within this period that the primary current within the coil rises to its specified intensity. The level of the primary current and the primary inductance in the ignition coil determine the level of energy stored in the ignition system. When the firing point arrives, the ignition driver stage interrrupts the flow of current. The flux in the magnetic field induces secondary voltage in the coil's secondary winding.
The potential secondary voltage (secondary voltage supply) depends upon a number of factors. These include the amount of energy stored in the ignition system, the capacity of the windings and the coil's transformation ratio, as well as the secondary load and the restrictions on primary voltage determined by the ignition system's driver stage. The supply of secondary voltage must always lie above the level of voltage required to generate a spark at the spark plug (ignition voltage requirement) The spark energy must be adequate for initiating combustion in the mixture, even if subsidiary sparking occurs.

When the primary current is switched, an undesired voltage of approx. 1...2 kV is induced in the secondary winding (switch-on voltage); its polarity is the opposite of that of the high-tension voltage. It is essential that arcing at the spark plug (switch-on spark) be avoided. Systems with rotating distribution use a distributor spark discharge gap for effective suppression of this phenomenon. On systems with stationary spark-distribution and single-spark coils a special diode is incorporated in the high-voltage circuit to perform the same function. With stationary spark-distribution and dual-spark coils, the high arcing voltage encountered when two spark plugs are connected in series suppresses the switch-on spark, without supplementary measures being necessary.

Voltage distribution

Rotating voltage distribution
On conventional ignition systems, the high-tension voltage generated by the coil is transmitted to the correct cylinder by a mechanical distributor. Because M-Motronic uses electronics to regulate the distributor's auxiliary functions (formerly mechanical advance adjustment as a function of load and engine speed), the distributor can be simplified. The high-voltage distributor's individual components are:
- The insulated cover,
- The rotor with suppression resistor,
- The distributor cap with discharge terminals, and
- The interference-suppression shield.

The distributor's rotor is mounted directly on the camshaft.
Reliable high-voltage distribution can only be guaranteed within a certain dwell-angle range which is inversely proportional to the number of cylinders. Centrifugal rotor adjustment provides adequate range extension on 6-cylinder engines, but 8-cylinder systems usually need two 4-cylinder units.

Stationary voltage distribution
Distributorless, stationary or electronic voltage distribution is available in two alternative versions:

System with single-spark ignition coils
Each cylinder is equipped with a coil and an output stage, which the M-Motronic unit triggers in the appropriate firing order. Because distributor losses can no longer occur, it is possible to make the coils especially small. The preferred installation position is directly above the spark plug. Stationary distribution with single-spark ignition coils is universally suited for use with any number of cylinders. There are no restrictions on the adjustment range for ignition advance. It must be noted that the unit does require additional synchronization; this is provided by a camshaft sensor (Figure 2).

System with dual-spark ignition coils

A single coil and one ignition output stage are assigned to two cylinders. Each end of the secondary winding is connected to one of the spark plugs. The cylinders are selected so that the compression stroke on one coincides with the exhaust stroke of the other. When the ignition fires, a spark arcs at both spark plugs. Because it is important to ensure that the spark produced during the exhaust stroke ignites neither residual gases nor fresh intake gas, there is a small restriction on the potential range of ignition advance angles.
This system does not require a synchronization sensor at the camshaft (Figure 3).

Connectors and interference suppressors

High-voltage cables

The high-voltage generated at the coil must be transmitted to the spark plug. This function is discharged by h.v. strength copper wires embedded in synthetic insulation material, with specially designed plug connectors for joining the high-tension components mounted on their ends. Every high-voltage lead represents a capacitive load for the ignition system, and reduces the supply of secondary voltage accordingly; thus the cables should be kept as short as possible.

Interference resistors, interference supression

The pulse-shaped high-tension discharge characteristic of every arc at the spark plug also represents a source of radio interference. The current peaks associated with discharge are limited by suppression resistors in the high-voltage circuit. To hold radiation of interference emanating from the high-voltage circuit to a minimum, the suppression resistors should be installed as close to the actual source as possible. Partial or complete encapsulation of the ignition system can supply additional reductions in interference. The resistors for interference suppression are generally installed in the spark-plug connectors and the plug connections, while rotating distributors can also include a resistor at the rotor. Spark plugs with integral interference-suppression resistors are also available. It must, however, be remembered that

Fig. 2

Single-spark ignition coil

1 External low-voltage terminal, **2** Multiplate iron core, **3** Primary winding, **4** Secondary winding, **5** Internal high-voltage terminal with spring-loaded contact, 6 Spark plug.

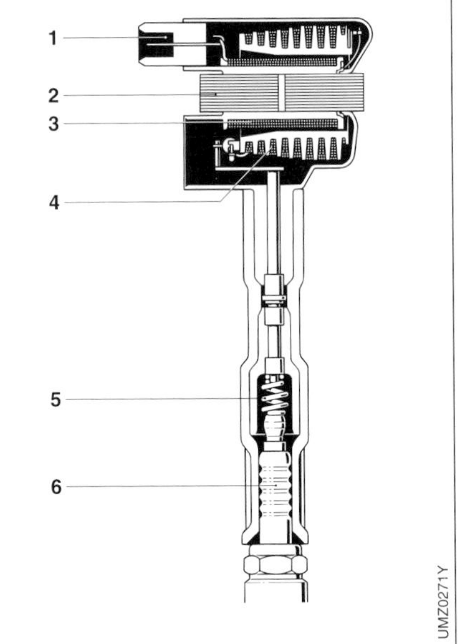

Fig. 3

Dual-spark ignition coil (stationary voltage distribution)

1 Low-voltage terminal, **2** Iron core, **3** Primary winding, **4** Secondary winding, **5** High-voltage connections.

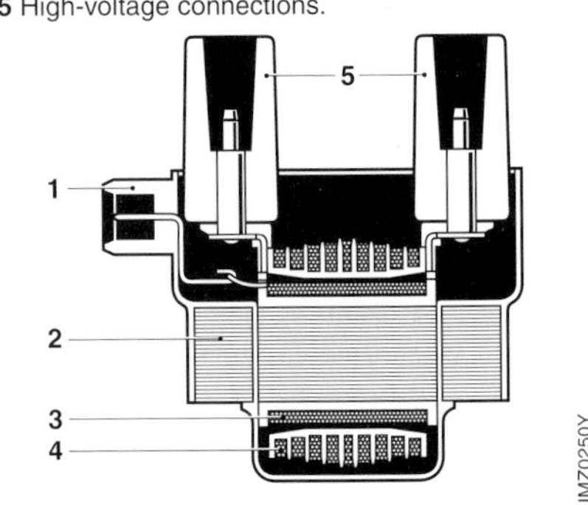

increases in the resistance in the secondary circuit will be accompanied by additional losses within the ignition circuit, the ultimate result being a reduction in the ignition energy available at the spark plug.

Fig. 4

Spark plug

1 Terminal nut,
2 Al_2O_3 ceramic insulator,
3 Case,
4 Heat-shrinkage zone,
5 Conductive glass,
6 Seal,
7 Compound center electrode Ni/Cu,
8 Ground electrode.

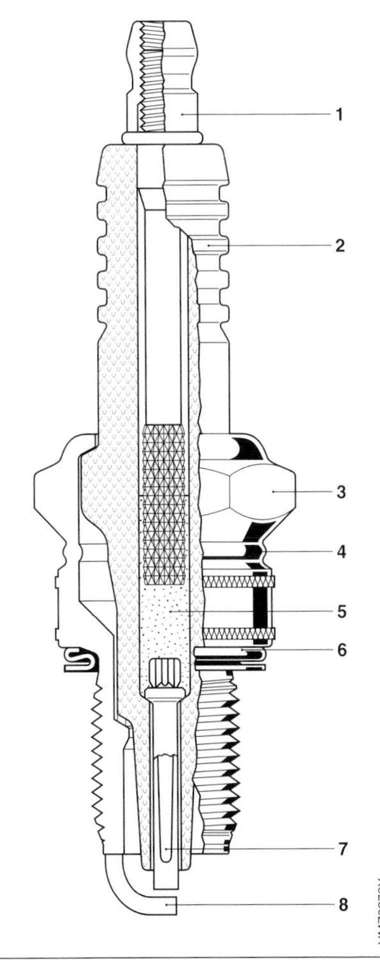

Spark plug

The spark plug generates the spark for igniting the air-fuel mixture in the combustion chamber. It is a ceramic-insulated, gas-tight, high-voltage conductor leading into the combustion chamber. Once the sparking voltage is reached, the path between the center electrode and the ground electrode becomes conductive and converts the remaining energy from the coil into a spark.

The voltage level required for ignition depends upon the electrode gap, electrode geometry, combustion-chamber pressure and the A/F ratio at the firing point.
The spark plug's electrodes are subject to wear in normal engine operation. This wear results in higher voltage requirements. The ignition system must be capable of providing enough secondary voltage to ensure that adequate ignition voltage remains available under all operating conditions for the life of the spark plug.

Operating-data acquisition

Engine load

One of the most important variables used for determining injection quantity and ignition advance angle is the engine's load state (load monitoring).

The various Motronic systems employ the following load sensors to monitor engine load:

- Air-flow sensor,
- Hot-wire air-mass meter,
- Hot-film air-mass meter,
- Intake manifold pressure sensor, and
- Throttle-valve sensor.

In the Motronic systems the throttle-valve sensor generally assumes the function of a secondary load sensor, supplementing one of the primary load sensors listed above. It is also employed as a primary load sensor in some isolated cases.

Air-flow sensor

The air-flow sensor is located between the air filter and the throttle valve, where it monitors the volumetric flow rate [m^3/h] of the air being drawn into the engine. The force of the air stream acts against the constant return force of a spring, and the air flap's deflection angle is monitored via potentiometer. The potentiometer voltage is transmitted to the ECU for comparison with the potentiometer's initial supply voltage. The resulting voltage ratio serves as an index of the induction air's volumetric flow rate. The ECU ensures accuracy by compensating for the effects of potentiometer aging and temperature when processing the resistance (Figure 1).

In order to prevent pulsation in the intake air stream from setting up oscillations in the air-flow sensor flap, the system also includes a counterflap and a "damping volume." The air-flow sensor is equipped with a temperature sensor. This transmits a temperature-sensitive resistance value to the control unit, allowing it to compensate for variations in air density arising from changes in the temperature of the intake air.

The air-flow sensor is still a component in many of the M-Motronic and L-Jetronic systems currently in production. The load sensors described in the following section are preferably installed, and will replace the flap-controlled air-flow sensor in future systems.

Air-mass meters

The hot-wire and hot-film air-mass meters are both "thermal" load sensors. They are installed between the air filter and the throttle valve, where they monitor the mass flow [kg/h] of the air being drawn into the engine. The two meters operate according to a common principle.

Fig. 1

Air-flow sensor in intake system

1 Throttle valve, **2** Air-flow sensor, **3** Intake air temperature signal to ECU, **4** ECU, **5** Air-flow sensor signal to ECU, **6** Air filter. Q_L Intake air quantity, α Deflection angle.

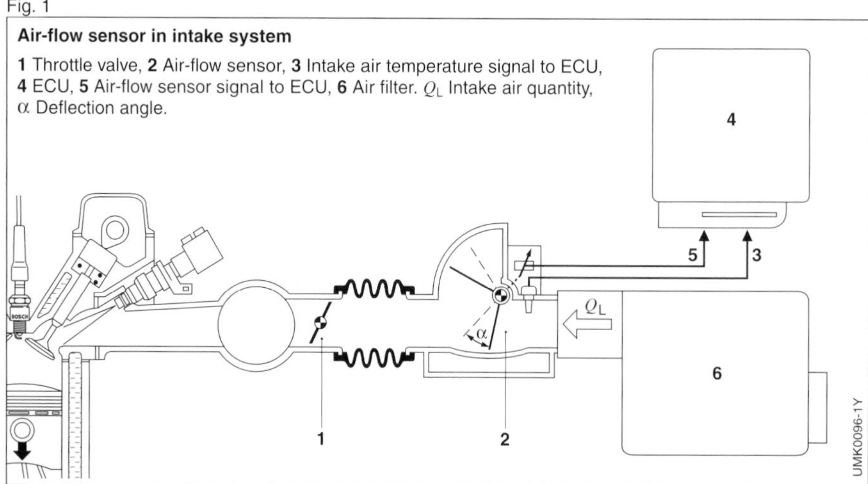

An electrically heated element is mounted in the intake-air stream, where it is cooled by the flow of incoming air. A control circuit modulates the flow of heating current to maintain the temperature differential between the heated wire (or film) and the intake air at a constant level. The amount of heating current required to maintain the temperature thus provides an index for the mass air flow. This concept automatically compensates for variations in air density, as this is one of the factors that determines the amount of warmth that the surrounding air absorbs from the heated element.

Fig. 2

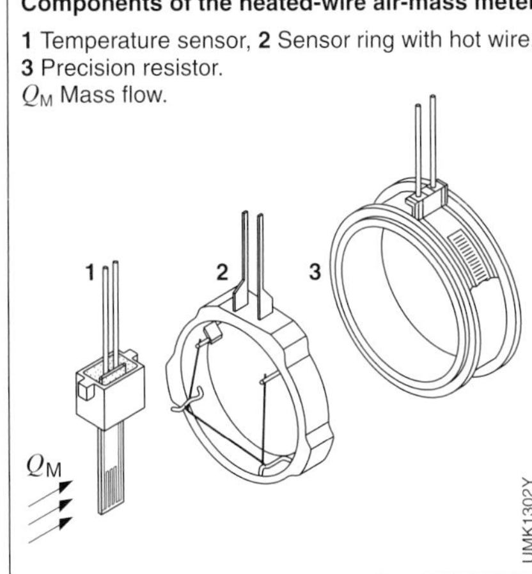

Components of the heated-wire air-mass meter

1 Temperature sensor, **2** Sensor ring with hot wire, **3** Precision resistor.
Q_M Mass flow.

Fig. 3

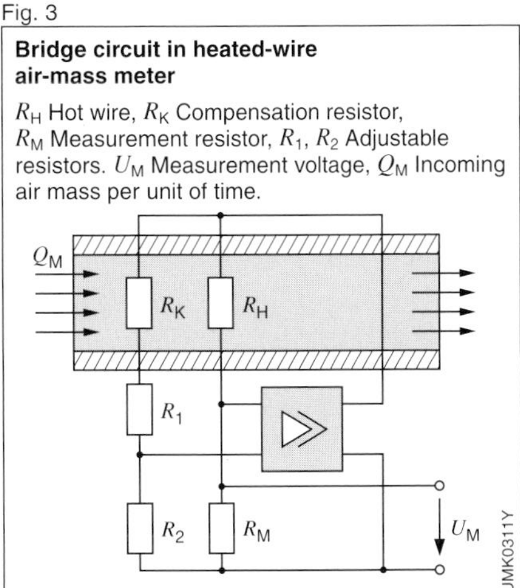

Bridge circuit in heated-wire air-mass meter

R_H Hot wire, R_K Compensation resistor, R_M Measurement resistor, R_1, R_2 Adjustable resistors. U_M Measurement voltage, Q_M Incoming air mass per unit of time.

Hot-wire air-mass meter

The heated element on the hot-wire air-mass meter is a platinum wire only 70 μm in diameter. A temperature sensor is integrated within the hot-wire air-mass meter to provide compensation data for intake-air temperature. The main components in the control circuit are a bridge circuit and an amplifier. The heated wire and the intake-air temperature sensor both act as temperature-sensitive resistors within the bridge (Figures 2 through 4). The heating current generates a voltage signal, proportional to the mass air flow, at a precision resistor. This is the signal transmitted to the ECU.

Fig. 4

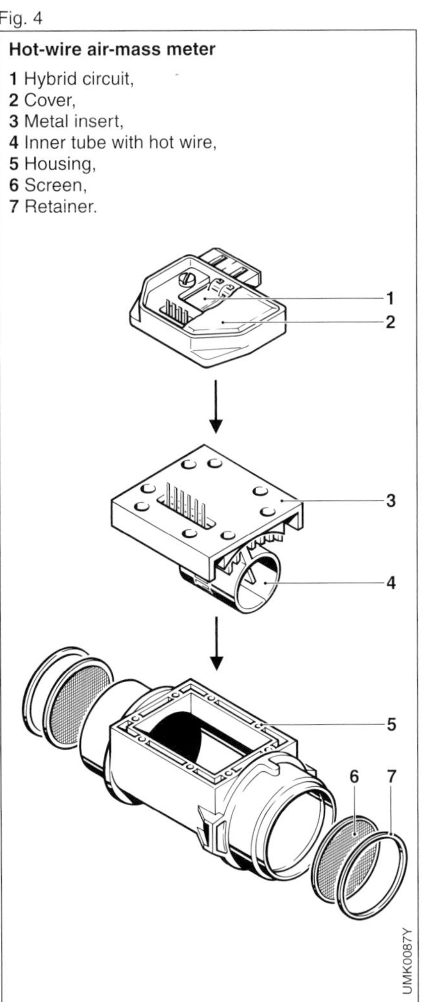

Hot-wire air-mass meter

1 Hybrid circuit,
2 Cover,
3 Metal insert,
4 Inner tube with hot wire,
5 Housing,
6 Screen,
7 Retainer.

To prevent the "drift" that could result from contaminant deposits on the platinum wire, the wire is heated up to "burn-off" temperature for one second after the engine is switched off. This process vaporizes and/or splits off the deposits and cleans the wire.

Hot-film air-mass meter

The heated element on the hot-film air-mass meter is a platinum film resistor (heater). It is located on a ceramic plate together with the other elements in the bridge circuit. The temperature of the heater is monitored by a temperature-sensitive resistor (flow sensor) also included in the bridge.

The separation of heater and flow sensor facilitates design of the control circuitry. Saw cuts are employed to ensure thermal decoupling between the heating element and the intake-air temperature sensor.

The complete control circuitry is located on a single layer. The voltage at the heater provides the index for the mass air flow. The hot-film air-mass meter's electronic circuitry then converts the voltage to a level suitable for processing in the ECU (Figures 5 through 7).

This device does not need a burn-off process to maintain its measuring precision over an extended period. In recognition of the fact that most deposits collect on the sensor element's leading edge, the essential thermal-transfer elements are located downstream on the ceramic layer. The sensor element is also designed to ensure that deposits will not influence the flow pattern around the sensor.

Intake-manifold pressure sensor

A pneumatic passage connects the intake-manifold to this pressure sensor, which monitors the absolute pressure [kPa] within the intake manifold.

The unit can be constructed as an installation component for the ECU or as a remote sensor for mounting on or near the intake manifold. A hose connects the installation unit to the manifold.

Fig. 5

Hot-film air-mass meter

a Housing, **b** Hot-film sensor (installed in center of housing).
1 Heat sink, **2** Intermediate component, **3** Power chip, **4** Hybrid circuit, 5 Sensor element.

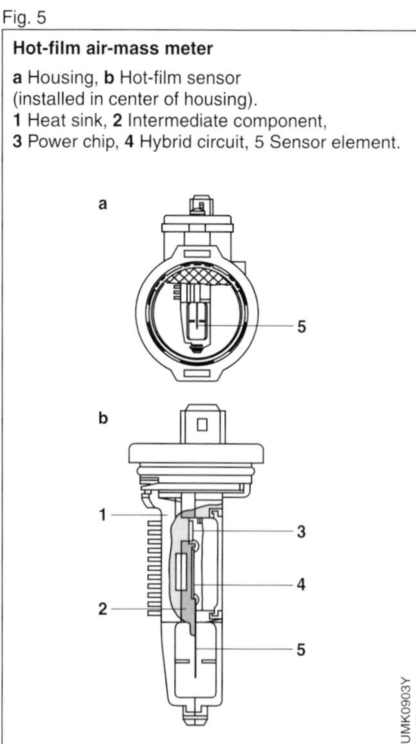

Fig. 6

Hot-film sensor element

1 Ceramic substrate, **2** Sawcut.
R_K Temperature-compensation sensor,
R_1 Bridge resistor, R_H Heater resistor,
R_S Sensor resistor.

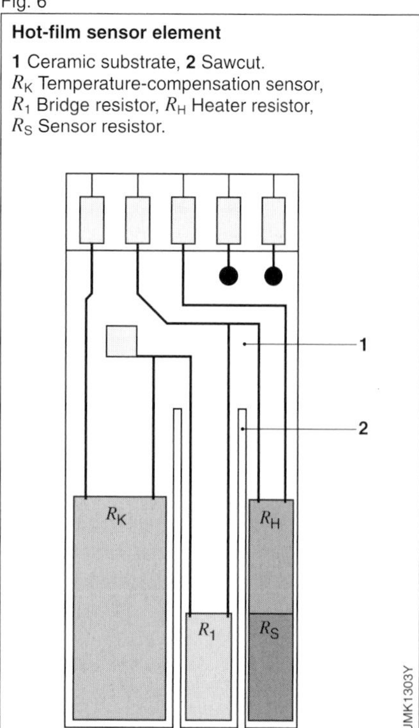

The sensor is divided into a pressure cell with two sensor elements and a chamber for the evaluation circuitry. Sensor elements and evaluation circuitry are located on a single ceramic layer (Figure 8).

The sensor element consists of a bell-shaped thick-layer diaphragm enclosing a reference volume with a specific internal pressure. The diaphragm's deflection is determined by the pressure in the intake manifold.

A series of piezo-resistive resistor elements is arranged on the diagram; the conductivity of these elements varies in response to changes in mechanical tension. These resistors are incorporated in a bridge circuit in such a manner that any deflection at the diaphragm will lead to a change in the bridge balance. The bridge voltage thus provides an indication of intake-manifold pressure (Figure 9).

The evaluation circuit amplifies the bridge voltage, compensates for temperature effects and linearizes the pressure response curve. The output signal from the evaluation circuit is transmitted to the ECU.

Throttle-valve sensor

This sensor provides a secondary load signal based on the angle of the throttle valve. The applications for this secondary load signal include providing information for dynamic functions, load-range recognition (idle, full or part-throttle), and serving as a backup signal in the event of main-sensor failure.

The throttle-valve sensor is attached to the throttle-valve assembly where it shares a common shaft with the throttle valve. A potentiometer evaluates the throttle valve's deflection angle and transmits a voltage ratio to the ECU via a resistance circuit (Figures 10 and 11).

Circuit of hot-film air-mass meter

R_K Temperature-compensation sensor, R_H Heater resistor, R_1, R_2, R_3 Bridge resistors, U_M Measurement voltage, I_H Heater current, t_L Air temperature, Q_M incoming air mass per unit of time.

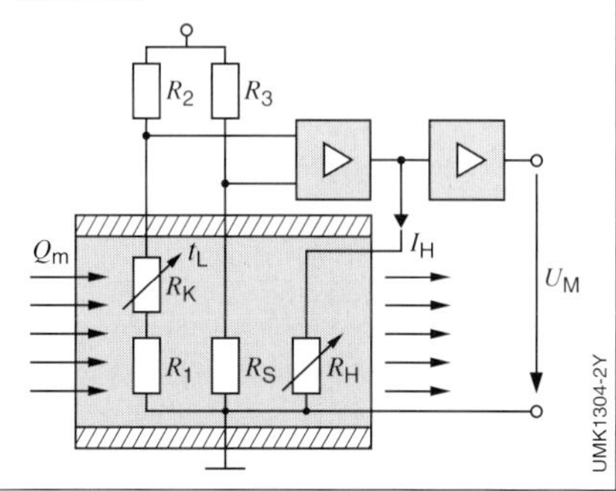

Fig. 7

Fig. 8

Pressure sensor (for installation in ECU)

1 Pressure connection, **2** Pressure cell with sensor elements, **3** Sealing edge, **4** Evaluation circuitry, **5** Thick-film hybrid.

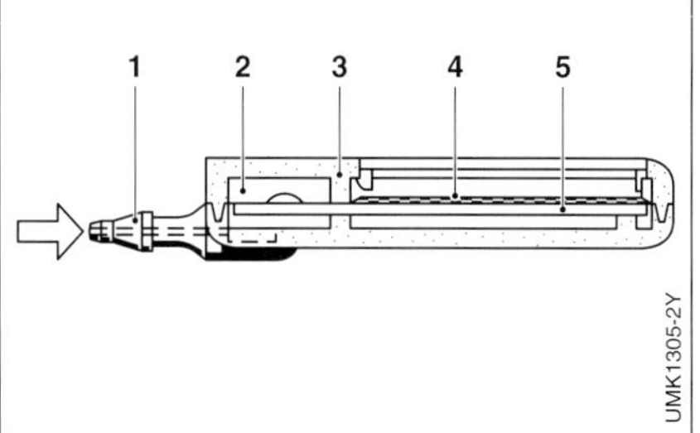

Fig. 9

Thick-film diaphragm in pressure sensor

1 Piezoresistive resistors, **2** Base diaphragm, **3** Reference pressure chamber, **4** Ceramic substrate.
p Pressure.

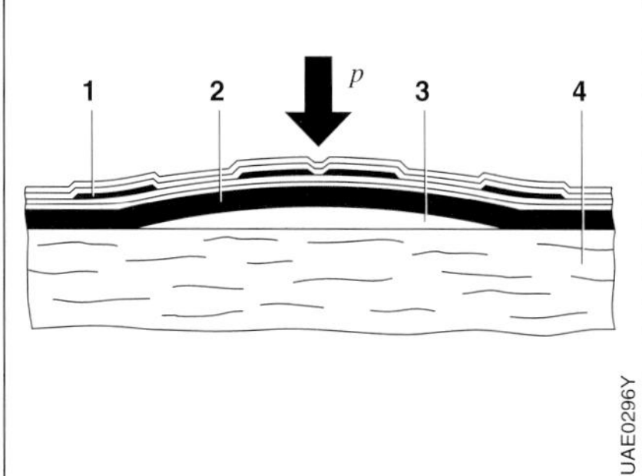

More exacting precision is required when the throttle-valve sensor is used as the primary load sensor. This higher level of precision is obtained by using a throttle-valve sensor incorporating two potentiometers (two angle ranges) as well as improved suspension.
The control unit determines the mass of the intake air by monitoring throttle-valve angle and engine speed. Data from temperature sensors allows the unit to respond to variations in the air mass due to temperature change.

Fig. 10

Throttle-valve sensor

1 Throttle-valve shaft, **2** Resistor track 1, **3** Resistor track 2, **4** Wiper arm with wiper, **5** Electrical connection.

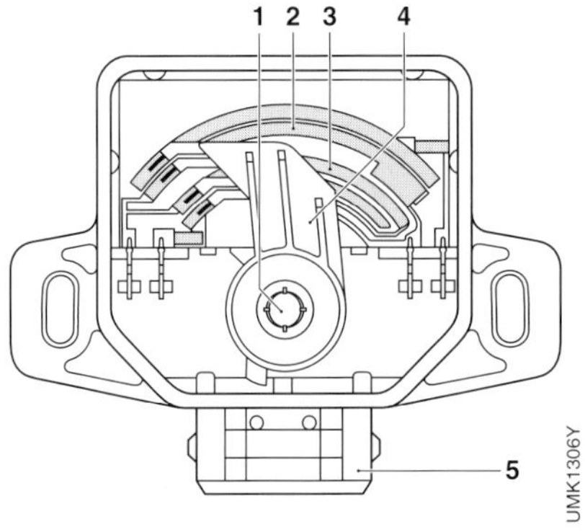

UMK1306Y

Fig. 11

Throttle-valve sensor circuitry

U_M Measurement voltage, R_1, R_2 Resistor tracks 1 and 2, R_3, R_4, R_5 adjustment resistors.
1 Throttle valve.

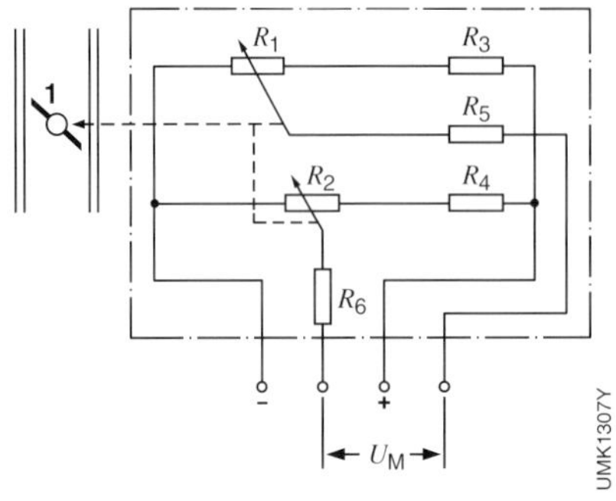

UMK1307Y

Engine speed, crankshaft and camshaft positions

Engine speed and crankshaft position

The degree of piston travel within the cylinder is employed as a measured variable for determining the firing point. The pistons in all cylinders are connected to the crankshaft via the connecting rods. A sensor at the crankshaft thus provides the information on the locations of the pistons in the cylinders.
The speed at which the crankshaft changes its position is the engine speed, defined in the number of crankshaft revolutions per minute (rpm). This also represents another important Motronic input variable for the Motronic unit, and is calculated from the crankshaft position signal. Although the signal from the crankshaft sensor basically indicates crankshaft position, which is then converted to an engine-speed signal in the ECU, the device has come to be known as the engine-speed, or rpm sensor.

Generating the crankshaft position signal

Installed on the crankshaft is a ferromagnetic ring gear with a theoretical capacity of 60 teeth, whereby two teeth are missing on the gear in question (tooth gap). An inductive speed sensor registers the 58-tooth sequence. This sensor consists of a permanent magnet and a soft-iron core with a copper winding (Figure 12).

The magnetic flux field at the sensor responds as the teeth on the sensor gear pass by, generating AC voltage (Figure 13). The amplitude of this AC voltage decreases as the interval between sensor and sensor gear increases, and rises in response to higher engine speeds. Sufficient amplitude is already available at a minimal engine speed (20 min^{-1}). Pole and tooth geometry must be matched. The evaluation circuit in the ECU converts the sinus voltage with its highly varying amplitudes into square-wave voltage with a constant amplitude.

Calculating the crankshaft position

The flanks of the square-wave voltage are transmitted to the computer via an interrupt input. A gap in the tooth pattern is registered at those points where the flank interval is twice as large as in the previous and subsequent periods. The tooth gap corresponds to a specific crankshaft position for cylinder no. 1. The computer synchronizes the crankshaft position according to this point in time. It then counts 3 degrees further for every subsequent positive or negative tooth flank. The ignition signals, however, must be transmitted in smaller stages. The duration between two tooth flanks is thus further divided by four. The time unit thus derived can be multiplied by two, three or four and added to a tooth flank for the ignition advance angle (allowing steps of 0.75 degrees).

Calculating segment duration and engine speed from the engine-speed sensor signal

The relatonship between the cylinders of the four-stroke engine is such that two crankshaft rotations (720 degrees) elapse between the start of each new cycle at cylinder no. 1.

This interval is the mean ignition interval, and is referred to as the segment time T_S. When the interval is distributed equally, the result is:

Interval	Degrees	Teeth
2 cylinders	360	60
3 cylinders	240	40
4 cylinders	180	30
5 cylinders	144	24
6 cylinders	120	20
8 cylinders	90	15
12 cylinders	60	10

Fig. 12

Engine-speed sensor

1 Permanent magnet, **2** Housing, **3** Engine housing, **4** Soft-iron core, **5** Winding, **6** Ring gear with reference point (tooth gap).

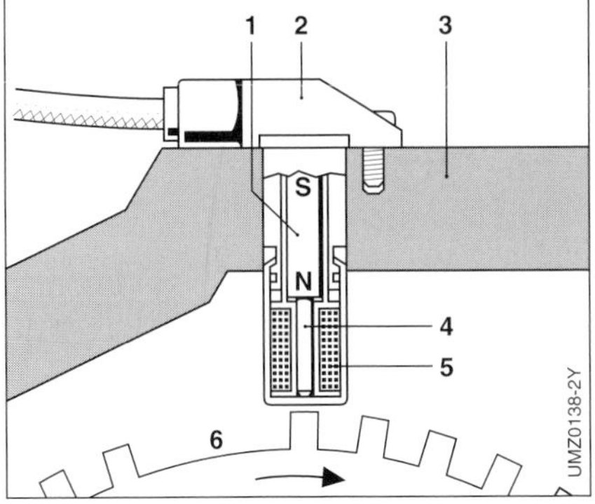

Fig. 13

Signal patterns for ignition, crankshaft and camshaft

a Ignition-coil secondary voltage, **b** Crankshaft rpm sensor signal, **c** Camshaft Hall-sensor signal.
1 Close, 2 Ignition.

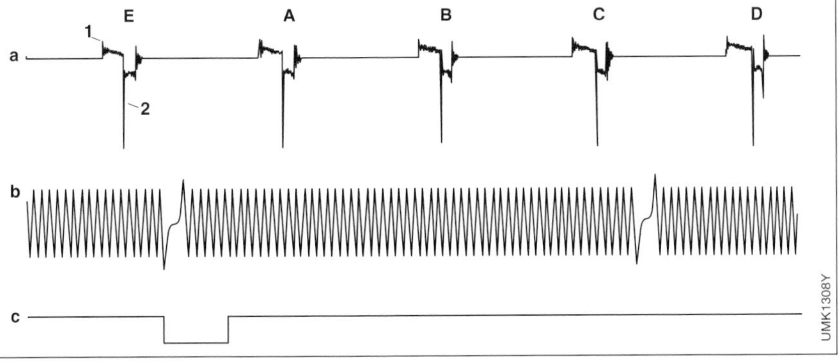

Ignition, injection and the engine speed derived from the segment time are recalculated for every new interval. The figure for rotational speed describes the mean crankshaft rpm within the segment time and is proportional to its reciprocal.

Camshaft position

The camshaft controls the engine's intake and exhaust valves while operating at half the rotating speed of the crankshaft. When a piston travels to top dead center, the camshaft uses the settings of the intake and exhaust valves to determine whether the cylinder is in the compression phase to be followed by ignition, or in the exhaust phase. This information cannot be derived from the crankshaft position.

If the ignition is equipped with a high-voltage distributor with a direct mechanical link to the camshaft, the rotor will point to the correct cylinder automatically: the ECU does not require supplementary information on the position of the camshaft. In contrast, Motronic systems featuring stationary voltage distribution and single-spark ignition coils require additional information, as the ECU must be able to decide which ignition coil and spark plug are due to be triggered. To do so, it must be informed as to the camshaft's position.

The position of the camshaft must also be monitored in those systems where separate injection timing is used for each individual cylinder, as is the case with sequential injection (SEFI).

Hall-sensor signal

Camshaft position is usually monitored with a Hall sensor. The monitoring device itself consists of a Hall element with a semiconductor wafer through which current flows. This element is controlled by a trigger wheel that turns together with the camshaft. It consists of a ferromagnetic material and generates voltage at right angles to the direction of the current as it passes the Hall element (Figure 13).

Calculating camshaft position

As the Hall voltage lies in the millivolt range, it is processed within the sensor before being transmitted to the ECU in the form of a switching signal. In the simplest case, the computer responds to trigger-wheel gaps by checking to see whether Hall voltage is present and whether or not cylinder no. 1 is on its power stroke.

Special trigger-wheel designs make it possible to use the camshaft signal as a backup for emergency operation in case of crankshaft (engine-speed) sensor failure. However, the resolution provided by the camshaft signal is too imprecise to allow it to be employed as a permanent replacement for the crankshaft speed sensor.

Mixture composition

Excess-air factor λ

The lambda oxygen sensor monitors the excess-air factor λ. Lambda defines the number for the mixture's A/F ratio. The catalytic converter functions best at $\lambda = 1$.

Lambda oxygen sensor

The Lambda oxygen sensor's outer electrode extends into the exhaust stream, while the inner electrode is exposed to the surrounding air (Figure 14).

The essential constituent of the oxygen sensor is a special-ceramic body featuring gas-permeable platinum electrodes on its surface. Sensor operation is based on the ceramic material's porosity, which allows oxygen in the air to diffuse (solid electrolyte). The ceramic material becomes conductive at higher temperatures. Voltage is generated at the electrodes when different oxygen levels are present on the respective sides. A stoichiometric air/fuel ratio of $\lambda = 1$ produces a characteristic jump (jump function) in the response curve (Figure 16).

The oxygen sensor's voltage and internal resistance are both sensitive to temperature. Reliable control operation is possible with exhaust-gas temperatures exceeding 350 °C (unheated sensor), or 200 °C (heated sensor).

Heated Lambda oxygen sensor

The design of the heated Lambda oxygen sensor is largely the same as that of the unheated version (Figure 15). A ceramic heater element warms the sensor's active ceramic layer from the inside, ensuring that its ceramic material remains hot enough for operation – even at low exhaust-gas temperatures. The heated sensor is protected by a tube with a restricted flow opening to prevent the sensor's ceramics from being cooled by low-temperature exhaust gases.

The heated sensor reduces the waiting period between engine start and effective closed-loop control. It also provides reliable control with lower-temperature exhaust gases (e.g., at idle). Heated sensors offer shorter reaction times for reduced closed-loop response intervals. This type of sensor also offers greater latitude in selecting the installation position.

Fig. 14

Lambda oxygen sensor location in exhaust pipe

1 Special ceramic coat, **2** Electrodes, **3** Contact, **4** Housing contact, **5** Exhaust pipe, **6** Ceramic shield (porous), **7** Exhaust gas, **8** Air.

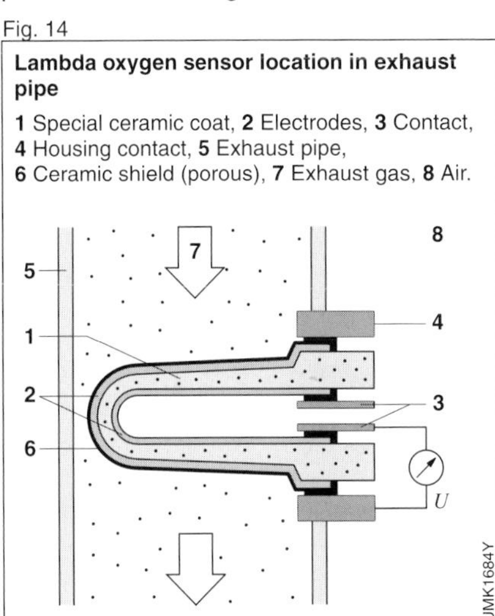

Fig. 16

Lambda oxygen sensor voltage curve at 600 °C operating temperature

a Rich mixture (air deficiency), **b** Lean mixture (excess air).

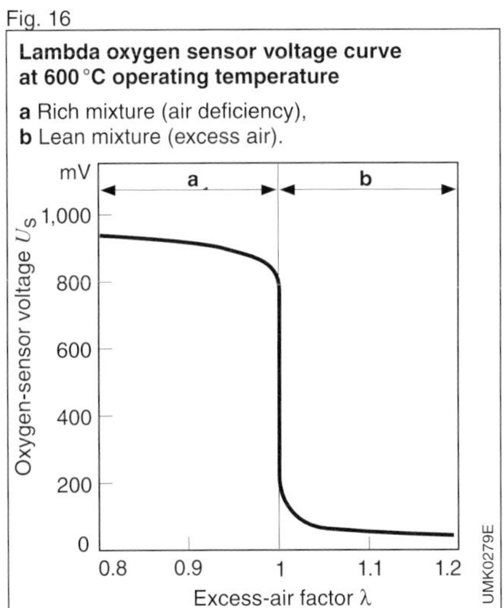

Fig. 15

Heated Lambda oxygen sensor

1 Probe housing, **2** Ceramic shield tube, **3** Electrical connections, **4** Shield tube with slits, **5** Active ceramic sensor layer, **6** Contact, **7** Shield, **8** Heating element, **9** Clamp connections for heater element, **10** Disc spring.

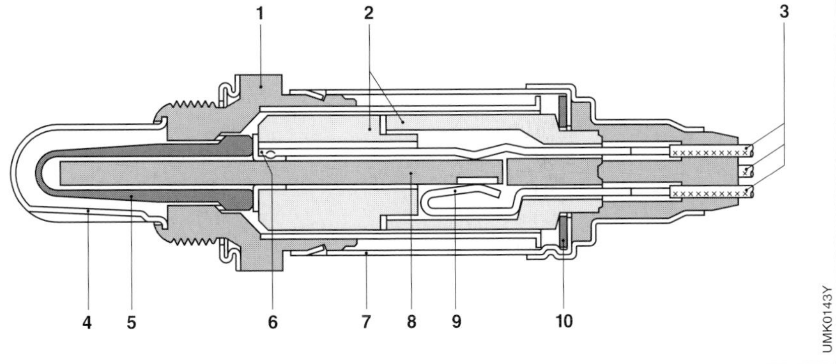

UMK0143Y

Combustion knock

Under certain conditions combustion in the spark-ignition engine can degenerate into an abnormal process characterized by a typical "knocking" or "pinging" sound. This phenomenon is an undesirable combustion process known as "knocking", which limits the engine's output and specific efficiency levels. It occurs when fresh mixture preignites in spontaneous combustion before being reached by the expanding flame front.

Normally initiated combustion and the piston's compressive force lead to the pressure and temperature peaks that produce self-ignition in the end gas (remaining unburned mixture). Flame velocities in excess of 2,000 m/s can occur, as compared to speeds of roughly 30 m/s for normal combustion. This abrupt combustion process produces substantial local pressure increases in the end gas. The resulting pressure wave propagates until stopped by impact with the cylinder walls representing the outer extremity of the combustion chamber.

Chronic preignition is accompanied by pressure waves and increased thermal stresses at the cylinder-head gasket, piston and in the vicinity of the valves. All of these factors can lead to mechanical damage.

Fig. 17

Knock sensor

1 Seismic mass, **2** Cast mass, **3** Piezoelectric ceramic, **4** Contacts, **5** Electrical connection.

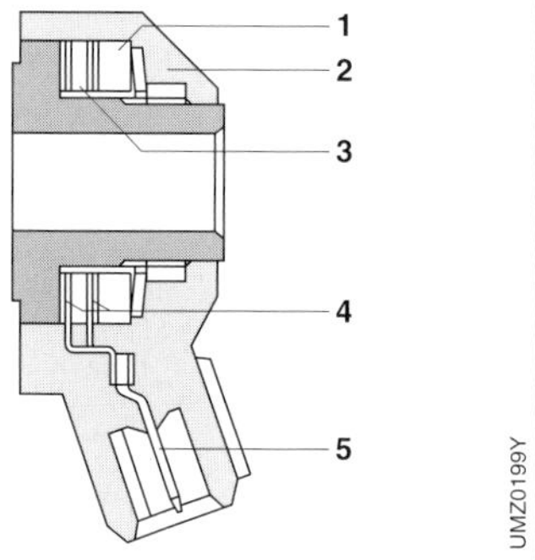

1 2 3 4 5

UMZ0199Y

Fig. 19

"Listening" positions for the knock sensors

1 The knock sensor is located between the second and third cylinders. **2** If two knock sensors are fitted, these are located between the first and second, and the third and fourth cylinders.

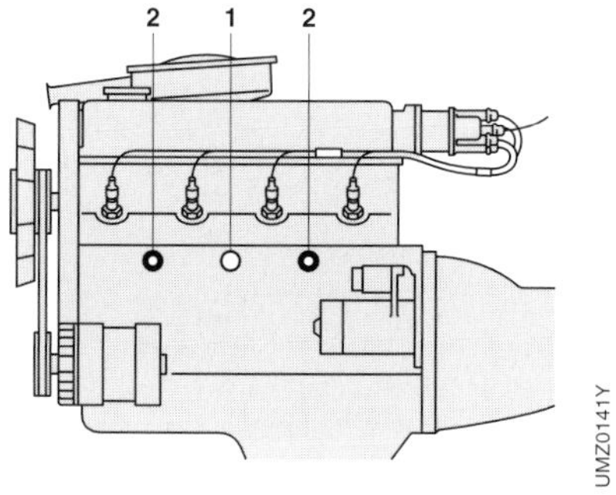

2 1 2

UMZ0141Y

Fig. 18

Knock-sensor signals

The knock sensor supplies a signal **c** corresponding to the pressure pattern **a** in the cylinder. The filtered pressure signal is shown in **b**.

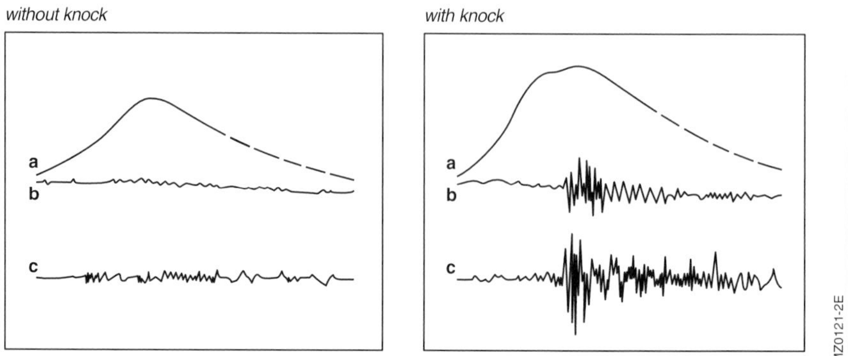

without knock

a
b
c

with knock

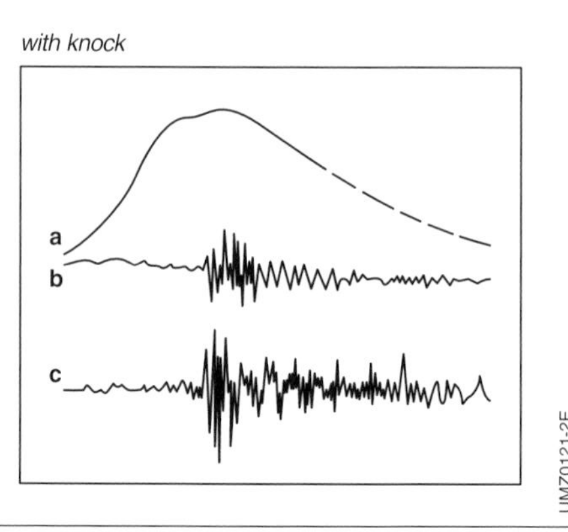

The characteristic vibration patterns generated by combustion knock can be monitored by knock sensors for conversion into electrical signals, which are then transmitted to the Motronic ECU (Figures 17 and 18). Both the number and positions of the knock sensors must be carefully selected. Reliable knock recognition must be guaranteed for all cylinders and under all engine operating conditions, with special emphasis on high loads and engine speeds. As a general rule, 4-cylinder engines are equipped with one, 5 and 6-cylinder engines with two, and 8 and 12-cylinder engines with two or more knock sensors (Figure 19).

Fig. 20

Engine-temperature sensor

1 Electrical connection, **2** Housing, **3** NTC resistor.

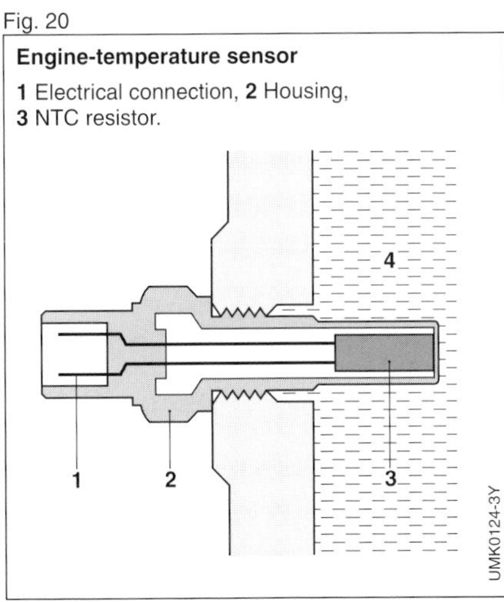

Fig. 21

Temperature-sensor response curve (NTC)

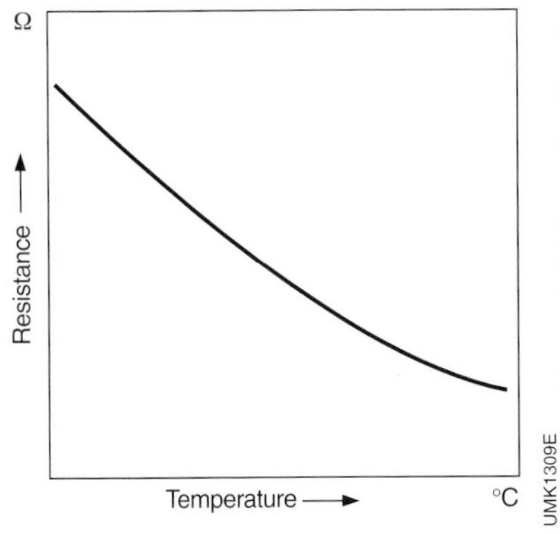

Ω

Resistance →

Temperature →

°C

UMK1309E

Engine and intake-air temperatures

The engine-temperature sensor incorporates a temperature-sensitive resistor which extends into the coolant circuit whose temperature it monitors. A sensor in the intake tract registers the intake-air temperature in the same fashion (Figure 20).

The resistor is of the negative temperature coefficient type (NTC, see Figure 21) and forms part of a voltage-divider circuit operating with a 5 V supply. An analog-digital converter monitors the resistor's voltage drop, which provides an index of the temperature. Compensation for the non-linear relationship between voltage and temperature is provided by a table stored in the computer's memory; the table matches each voltage reading with a corresponding temperature.

Battery voltage

The electromagnetic injector's opening and closing times are affected by the battery's voltage. Should voltage fluctuations occur in the vehicle's electrical system, the ECU will prevent response delays by adjusting the duration of the injection process. At low battery voltages the ignition circuit's dwell times must be extended to provide the coil with the opportunity to accumulate sufficient ignition energy.

Operating-data processing

Processing load signals

Monitored variables

The ECU uses the signals for load and engine speed to calculate a load signal corresponding to the air mass inducted into the engine during each stroke. This load signal serves as the basis for calculations of injection duration and for addressing the programmed response curves for ignition advance angle (Figure 1).

Monitoring air mass

Hot-wire or hot-film air-mass meters measure the air mass directly, producing a signal that is suitable for use as a parameter in load-signal calculations. When an air-flow meter is used, density correction is also required before air mass and load signal can be determined.
In special cases monitoring errors due to heavy pulsation in the intake manifold also receive compensation in the form of a pulsation correction.

Monitoring pressure

Pressure-monitoring systems (using a pressure sensor to determine load) differ from air-mass monitoring systems in that no direct formulas are available for defining the relationship between intake pressure and the air-mass intake. The ECU therefore calculates the load signal with the aid of corrections stored in a program map.
Subsequent compensation is provided for changes in temperature and residual gas relative to the initial state.

Fig. 1

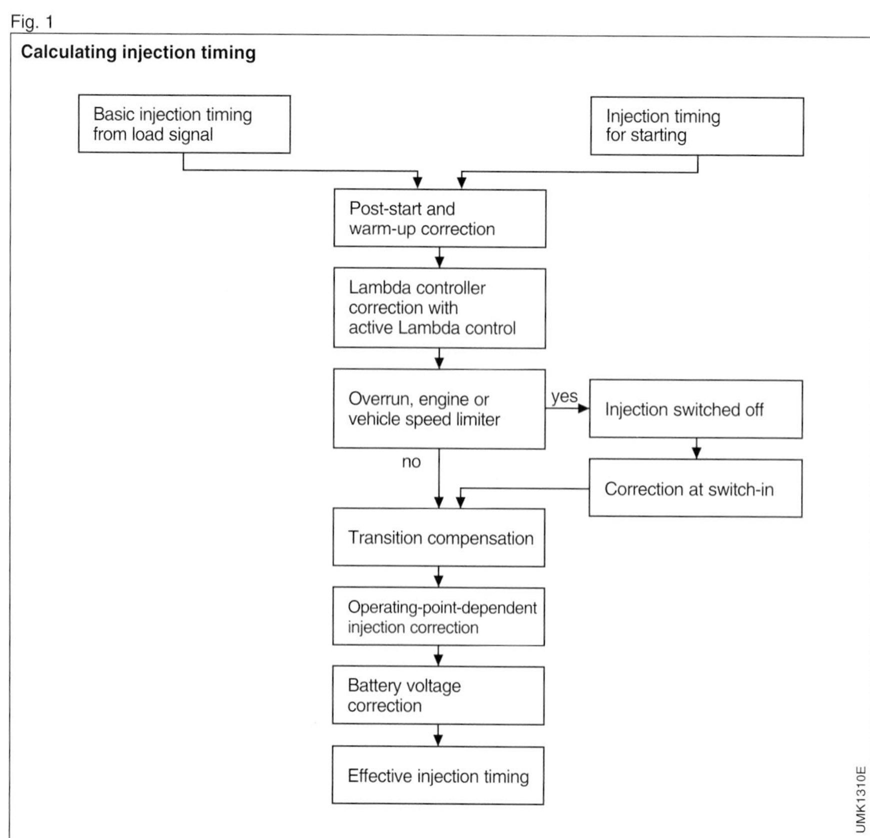

Measuring throttle-valve angle

When a throttle-valve sensor is used, the ECU determines the load signal with reference to engine speed and throttle-valve angle. Compensation for variations in air density are based on readings for temperature and ambient pressure.

Calculating injection timing

Base injection timing

The base injection timing is calculated directly from the load signal and from the injector constants, and defines the relationship between the duration of the activation signal and the flow quantity at the injector. This constant thus varies according to injector design. When the injection duration is multiplied by the injector constant the result will be a fuel mass corresponding to a particular air mass for each stroke.

The base setting is selected for an excess-air factor of $\lambda = 1$.

This remains valid for as long as the pressure differential between fuel and intake manifold stays constant. When it varies, a λ correction map compensates for this influence on injection times.

Meanwhile, a battery-voltage corrector compensates for the effects that fluctuations in battery voltage have on the injectors' opening and closing times.

Effective injection time

The effective injection time results when the correction factors are included in the calculations. The correction factors are determined in corresponding special functions and provide adjustment data for varying engine operating ranges and conditions. The correction factors are used both individually and in combinations according to applicable parameters.

Fig. 2

Comparison of injection types

a Simultaneous injection, **b** Group injection, **c** Sequential injection.

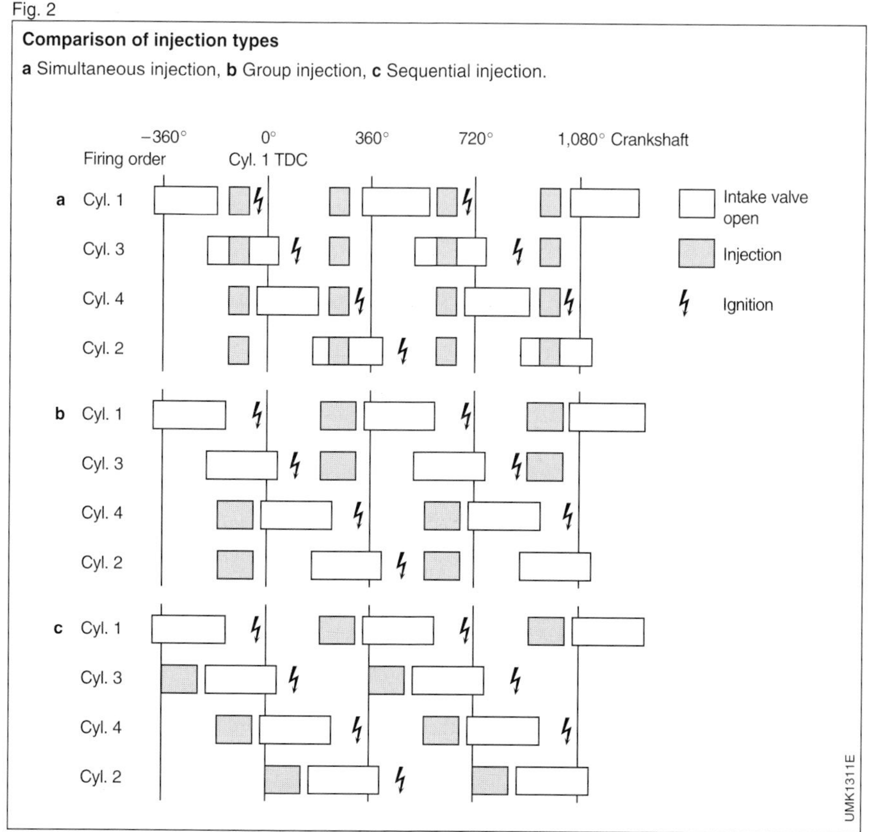

The process for calculating the injection time is illustrated in Figure 1. The individual operating ranges and conditions will be explained in more detail in the following chapters.
Once cylinder filling drops below a certain level, the mixture will cease to ignite. Restricting the injection time thus prevents the formation of unburned hydrocarbons in the exhaust gas.
For starting, the injection time is calculated separately using criteria independent of the calculated load signal.

Injection mode
In addition to the injection time, the injection mode is yet another important parameter for fuel economy and exhaust emissions. The range of options depends upon the type of injection system (Figure 2):
- Simultaneous injection,
- Group injection, or
- Sequential injection.

Simultaneous injection
With simultaneous injection, the injection process is triggered twice per cycle at a specific point in time at all injectors, that is, once for each camshaft revolution, or once for every two crankshaft revolutions. The injection mode is static.

Group injection
Group injection combines the injectors in two groups, with each group being triggered once per cycle. The time interval between the two triggering points is equal to one crankshaft rotation period. This arrangement makes it possible to use the engine operating point as the essential criterion in selecting the injection mode while also preventing undesirable spray through the open intake valves throughout a wide range in the program map.

Sequential injection
This type of injection provides the highest degree of design latitude. Here, the injection processes from the individual injectors take place independently of each other at the same point in the cycle referred to the respective cylinder. There are no restrictions on injection timing, which can be freely adapted to correspond with the optimization criteria.

Comparison
Group and sequential injection require a wider injector variation range (range extending from the minimum quantity at idle to the maximum under full throttle) than simultaneous injection.

Controlling dwell angle

The dwell angle varies the ignition coil's current-flow period according to engine speed and battery voltage. The dwell angle is selected to ensure availability of the required primary current at the end of the current flow time throughout the widest possible range of operating conditions.
The dwell angle is based on the ignition coil's charge time, which, in turn, depends upon battery voltage (Figure 3). A supplementary dynamic reserve makes it possible to supply the required current even during sudden shifts to high engine speeds.
The charge time is restricted in the upper rpm range to maintain an adequate arcing time at the spark plug.

Fig. 3
Primary current pattern at various system voltages

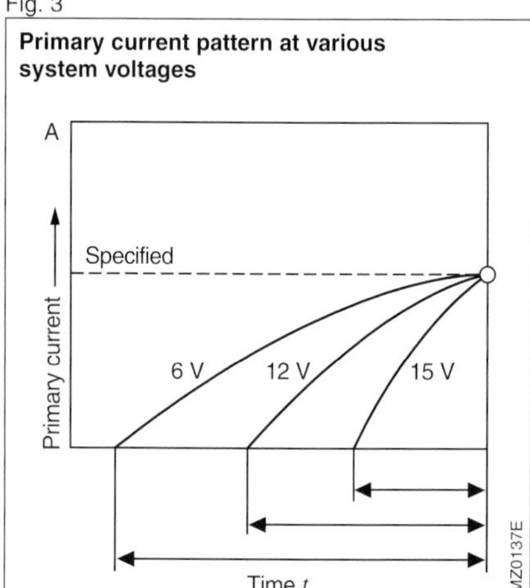

Controlling ignition advance angle

A program map containing the basic ignition timing for various engine loads and speeds is stored in the memory of the M-Motronic ECU. This ignition advance angle is optimized for minimal fuel consumption and exhaust emissions.
Data for engine and intake-air temperature (monitored via sensors) provide the basis for corrections to compensate for temperature variations. The unit can supply additional corrections and/or revert to other program maps to adapt to all operating conditions. Thus the mutual effects of torque, emissions, fuel consumption, preignition tendency and drivability can all be taken into account. Special ignition-angle correction factors are active during operation with secondary air injection or exhaust-gas recirculation (EGR) as well as in dynamic vehicle operation (e.g., when accelerating).
The various operating ranges (idle, part throttle, full throttle, start and warm-up) continue to be taken into account. Figure 4 shows a flow chart describing ignition-angle processing.

Operating-data processing

Fig. 4

Calculating ignition timing

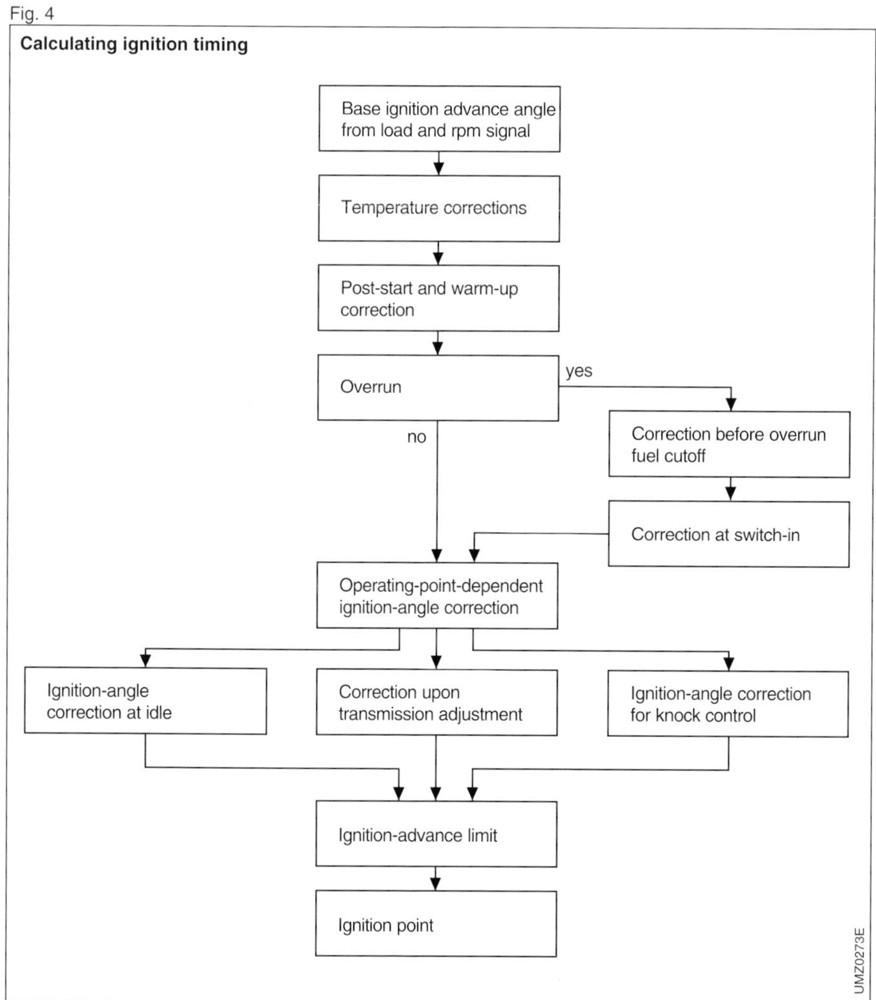

Operating conditions

Start

Special calculations are employed to determine the injection quantity for the duration of the starting procedure.
In addition, special injection timing is used for the initial injection pulses. The injection quantity is augmented in accordance with engine temperature to promote formation of a fuel film on the walls of the intake manifold, thereby compensating for the engine's higher fuel requirements as it runs up to speed. As soon as the engine starts to turn over, the quantity of supplementary fuel is reduced and then cancelled once the engine starts to run.
Ignition advance angle is also specially adjusted for starting. The adjustment occurs with reference to engine temperature and engine speed.

Post-start phase

The post-start phase is characterized by further reductions in the supplementary injection quantity. The reductions are based on engine temperature and the elapsed time since the end of the starting process. The ignition advance angle is also adjusted to correspond to the revised fuel quantities and the different operating conditions. The post-start phase terminates with a smooth transition to the warm-up phase.

Warm-up phase

Different strategies can be employed for the warm-up phase, depending upon engine and emissions-control design. The decisive criteria are drivability, emissions and improved fuel economy. A lean warm-up combined with retarded ignition timing raises the temperature of the exhaust gas. Another way to obtain high exhaust temperatures is to employ a rich warm-up mixture together with secondary-air injection. Here air is injected into the exhaust system downstream from the exhaust valves for a brief period after the engine starts. A secondary-air pump can provide this additional air. When the temperature is high enough, this excess air supports oxidation of HC and CO in the exhaust system while simultaneously generating the desired high exhaust temperatures (Figure 1).
Both of these measures help the catalytic converter to begin effective operation sooner.
The effects of the adjustments to ignition angle and injection timing can be supplemented by higher idle speeds. These are provided by a specially designed air injection unit, and also result in shorter warm-up times at the catalytic converter.

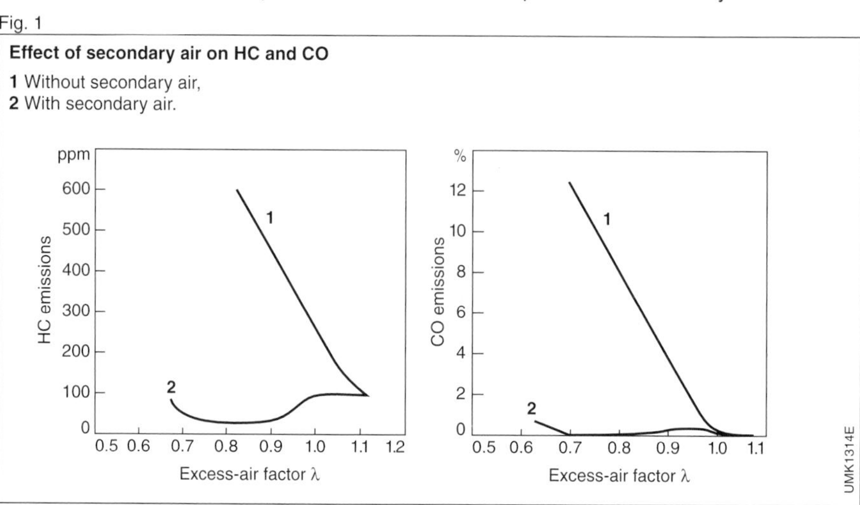

Fig. 1
Effect of secondary air on HC and CO
1 Without secondary air,
2 With secondary air.

Once the converter reaches operating temperature the injection is governed to λ = 1. This is accompanied by a corresponding adjustment in ignition angle.

Transition compensation

Acceleration/deceleration

A portion of the fuel sprayed into the intake manifold does not reach the cylinder in time for the next combustion process. Instead, it forms a condensation layer along the walls of the intake tract. The actual quantity of fuel stored in this film increases radically in response to higher loads and extended injection times.

A portion of the fuel injected when the throttle valve opens is used for this film. A corresponding amount of supplementary fuel must thus be injected to compensate and prevent the mixture from leaning out. Because the additional fuel retained in the wall film is released as the load factor drops, the injection time must be reduced by a corresponding amount during deceleration.

Figure 2 shows the resulting curve for injection times.

Overrun fuel cutoff/ renewed fuel flow

When the throttle closes, the injection is switched off in the interests of reduced fuel consumption and lower exhaust emissions.

Fig. 2

Transition injection timing

1 Injection timing from load signal,
2 Effective injection timing, **3** Additional fuel,
4 Fuel reduction, **5** Throttle-valve angle α_{DK}.

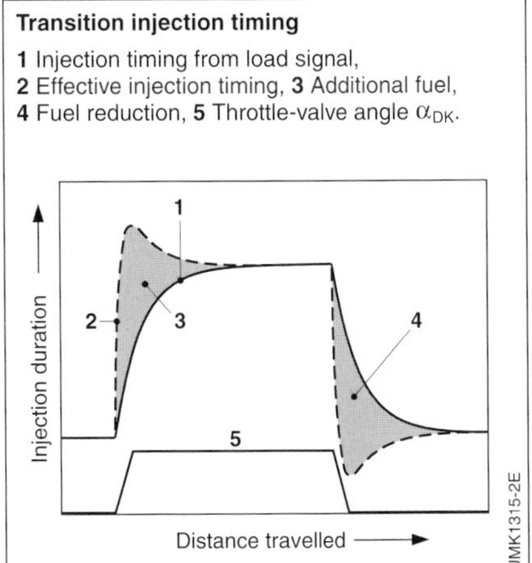

The injection stop is preceded by a reduction in the ignition advance to attenuate torque jump during the transition to trailing throttle.

The injection starts again once a specific reactivation speed – located above idle speed – is reached. Various activation speeds are stored in the ECU. These vary according to different parameters such as engine temperature and rpm dynamics, and are calculated to prevent excessive drops in engine speed, regardless of operating conditions.

When the injection resumes, it sprays in supplementary fuel to rebuild the fuel wall layer. The ignition advance angle is also adjusted to provide a smooth torque increase.

Closed-loop idle-speed control

Idle

The fuel consumption at idle is largely determined by the engine's efficiency and by the idle speed. A substantial proportion of the fuel consumed by vehicles in heavy urban traffic is actually burned at idle. The idle speed should thus be as low as possible. At the same time, the idle should never fall so far that rough running or stalling occur, even under additional loads such as electrical equipment, air conditioner, automatic transmission in gear, power steering, etc.

Idle-speed control

The idle-speed control must maintain a balance between torque generation and engine load in order to ensure a constant idle speed. The load on the idling engine is a combination of numerous elements, including internal friction within the engine's crankshaft and valvetrain assemblies as well as the ancillary drives (for instance, for the water pump). The idle-speed control compensates for this internal friction, which, in turn, changes over the life of the engine. These loads are also extremely sensitive to temperature variations.
In addition to these internal sources of friction, there are also external factors such as the load from the air conditioner that was mentioned above. The load from these external factors is subject to substantial variations as ancillary devices are switched on and off. Modern engines with small flywheel weights and large-volume intake manifolds are especially sensitive to these load fluctuations.

Input variables

In addition to the signal from the engine-speed sensor, the idle-speed control circuit also requires information on throttle-valve angle in order to recognize the idle state (foot off accelerator pedal). Engine temperature is also monitored to allow advance compensation for the effects of temperature. An air mass is specified with reference to engine temperature and the target idle-speed; this idle speed is then corrected in closed-loop operation. Where present, the input signals from the air conditioner and automatic transmission also facilitate the correction process and provide supplementary support for closed-loop idle-speed control.

Fig. 3

Bypass actuator with hose connections

Fig. 4

Manifold-mounted bypass actuator

Actuator adjustments

Physically, for idle-speed control there are three possibilities for adjustment intervention:

Air control

The proven control procedure is to regulate air flow by means of a bypass around the throttle valve, or to adjust the throttle valve itself using either a variable throttle stop or a direct actuator unit as found in "Electronic Throttle Control (ETC)."

On bypass actuators designed for hose connection, the bypass around the throttle valve consists of air hoses and an actuator (Figure 3). More modern are bypass actuators for direct installation; this type of bypass-air regulation device is flange-mounted directly on the throttle-valve assembly.

Figure 4: Example of a single-winding rotary actuator for direct installation.

One disadvantage associated with bypass actuators is that they add to the throttle-valve's own leakage air. Once the engine is well run-in, the combined air flowing through the throttle valve and the bypass actuator could well exceed the air quantity that the engine needs at idle. At this point effective idle regulation is no longer possible. This liability disappears when adjustments to the throttle valve itself are employed to regulate the air flow. The idle throttle device uses an electric motor and gear drive to vary the position of the throttle valve's idle stop (Figure 5). Delays in idle response are encountered when adjustments to air flow are used in systems with large-volume intake manifolds.

Adjustments to ignition advance angle

The second (and faster reacting) option is to adjust the ignition advance angle. Systems with an rpm-sensitive ignition advance angle react to sinking engine speeds by increasing the ignition advance to provide a boost in torque.

Mixture composition

Strict emissions-control regulations and a limited range of practical possibilities relegate the mixture-adjustment option to virtual insignificance.

Fig. 5

Throttle-valve assembly with integral idle actuator

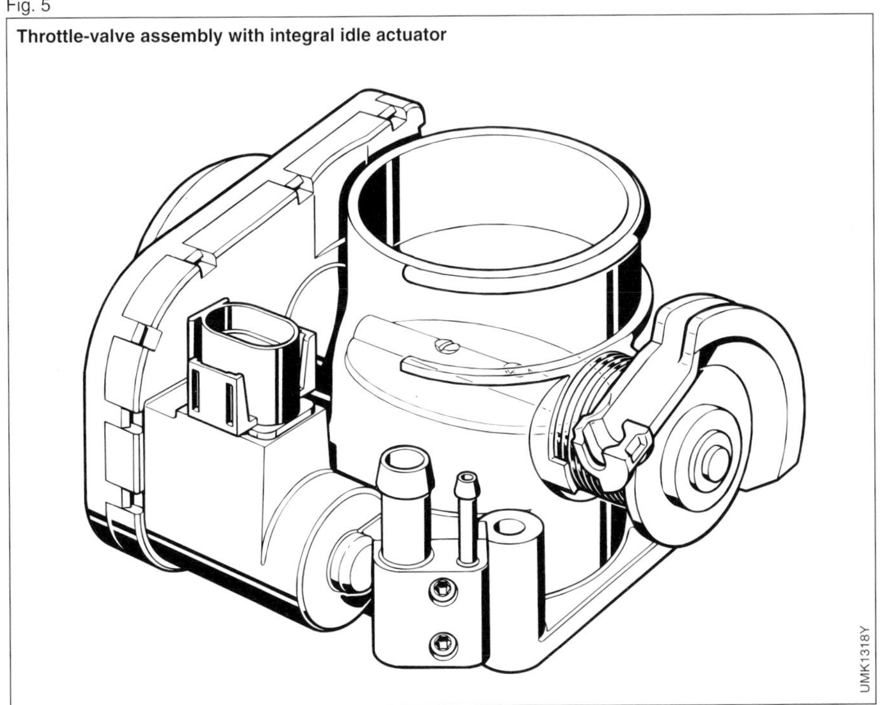

Lambda closed-loop control

Post-treatment of exhaust gases in a three-way catalytic converter is an effective means for reducing concentrations of harmful exhaust emissions. The converter transforms the three pollutants CO, HC and NO_X into H_2O, CO_2 and N_2.

Control range

The range available for simultaneous conversion of all three of the above components is extremely narrow, and is termed the "lambda window" (λ = 0.99...1). This means that closed-loop Lambda control is essential.

A Lambda oxygen sensor is installed in the exhaust system upstream from the catalytic converter, where it monitors the exhaust-gas oxygen content.

Lean mixtures ($\lambda > 1$) produce a sensor voltage of approx. 100 mV, while a rich mixture ($\lambda < 1$) generates approx. 800 mV. At $\lambda = 1$ the sensor voltage jumps abruptly from one level to the other (Figure 6).

The ECU uses the signal from the air-mass meter and the monitored engine speed to generate an injection signal. At the same time, it also produces a supplementary Lambda-control factor from the Lambda-sensor signal for use in correcting the injection time.

Fig. 6

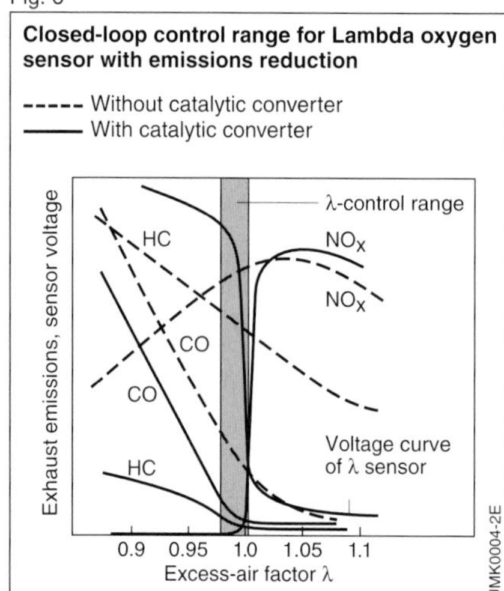

Closed-loop control range for Lambda oxygen sensor with emissions reduction

Operation

The Lambda oxygen sensor must be operational before the Lambda closed-loop control circuit can function. An auxiliary evaluation circuit monitors this factor on a continuing basis. A cold oxygen sensor or damaged circuitry (short or open circuits) will generate implausible voltage signals which are rejected by the ECU. The heated Lambda sensors used in most systems are ready for operation after only 30 seconds.

Cold engines require a richer mixture ($\lambda < 1$) to idle smoothly. For this reason the Lambda closed-loop control circuit can only be activated once a set temperature threshold has been passed.

Once the lambda control is activated, the ECU uses a comparator to convert the signal from the sensor into a binary signal.

The controller reacts to the transmitted signal ($\lambda > 1$ = mixture too lean, or $\lambda < 1$ = mixture too rich), by modifying the control variables (with an initial jump followed by a ramp progression). The injection time is adjusted (lengthened or shortened), and the control factor reacts to the continuing data transfer by settling into a constant oscillation (Figure 7). The duration of the oscillation periods is determined by the flow times of the gas, while the "ramp climb" maintains largely constant amplitudes within the load

Fig. 7

Diagram of Lambda closed-loop control circuit

1 Air-mass meter, **2** Engine, **3a** Lambda oxygen sensor 1, **3b** Lambda oxygen sensor 2 (as required), **4** Catalytic converter, **5** Injectors, **6** ECU. U_S Oxygen-sensor voltage, U_V Injector control voltage, V_E Injected fuel quantity.

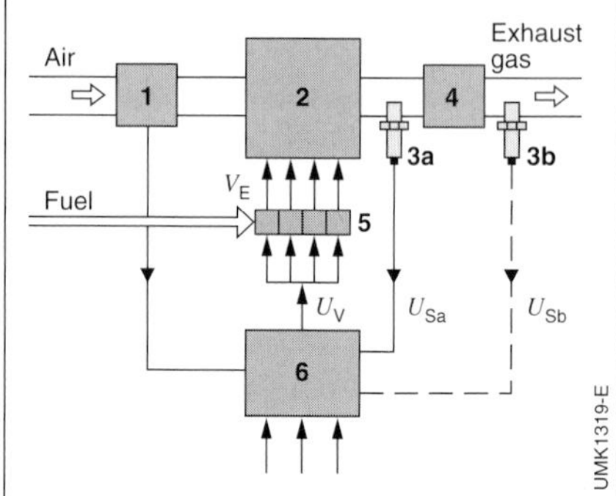

speed range despite variations in the travel time of the gas.

Lambda shift

The optimum conversion range and the voltage jump at the oxygen sensor do not coincide precisely. An asymmetrical control oscillation pattern can be used to shift the mixture into the optimal range ($\lambda = 1$). The asymmetry is obtained either by delaying the switch in the control factor after the voltage jump (from lean to rich) at the oxygen sensor or by providing an asymmetrical jump. This is the case when the voltage jump at the oxygen sensor during the transition from lean to rich is different from that produced at the change from rich to lean.

Adapting the pilot settings to the Lambda closed-loop control

Lambda closed-loop control corrects the subsequent injection process on the basis of the previous measurement at the oxygen sensor, whereby an unavoidable time lag results from the gas travel times. For this reason, the approach to a new operating point is accompanied by deviations from $\lambda = 1$ due to incorrect pilot control, a condition alleviated once the closed-loop control picks up the new cycle. Thus a special pilot control mechanism is needed to maintain compliance with emissions limits. The pilot-control is defined during adaptation to the engine and the Lambda response curve is stored in the ROM. However, revisions may become necessary due to the drift that can occur during the life of the vehicle. Among the drift factors are variations in the density and quality of the fuel. When the Lambda controller consistently repeats the same corrections in a certain load and engine-speed range, the pilot-control adaption mechanism recognizes this state. It corrects the pilot control for this range and records the correction in a RAM memory chip (with an uninterrupted current supply). The corrected pilot control is thus ready to respond immediately at the next start, assuming duty until the Lambda closed-loop control becomes operative.

Power interruptions in the current to the permanent memory are also recognized; the adaptation process then commences again with neutral default values providing the initial basis for operation.

Dual-sensor Lambda closed-loop control

Installing the Lambda oxygen sensor downstream of the catalytic converter provides better protection against contamination from the exhaust gas (here, "downstream" = on the tailpipe side). This type of backup sensor can provide a second control signal to augment the one from the main sensor upstream of the converter (here, "upstream" = on the engine side). The second signal is superimposed on the first to provide stable mixture composition over an extended period (Figure 7).
The superimposed control modifies the asymmetry in the constant oscillation pattern that is associated with control concepts based on the oxygen sensor mounted upstream from the converter; this compensates for the lambda shift. A Lambda closed-loop control strategy based solely on the downstream-mounted oxygen sensor would feature excessive response delays due to the gas travel times.

Evaporative-emissions control systems

Origins of fuel vapors

The fuel in the fuel tank is heated by:
- Heat radiation from outside sources, and
- The excess fuel from the system return line which was heated during its passage through the engine compartment.

The result is the HC emissions that are usually emitted from the fuel tank in the form of vapor.

Limiting HC emissions

Evaporative emissions are subject to legal limits.

Evaporative-emissions control systems restrict these emissions. These systems are equipped with an activated charcoal filter (the so-called carbon canister) located at the end of the fuel tank's vent line. The activated charcoal in the canister binds the fuel vapors and allows only air to escape into the atmosphere, as well as serving as a pressure-release device. In order to ensure that the charcoal can continually regenerate, an additional line leads from the carbon canister to the intake manifold. Vacuum is produced in this line when the engine runs, causing a stream of atmospheric air to flow through the charcoal on its way to the manifold. The fuel vapors stored in the activated charcoal are entrained by the air stream and conducted to the engine for combustion. A so-called canister-purge valve in the line to the intake manifold meters the flow of this regeneration or "cleansing" air (Figure 8).

Regeneration flow

The regeneration flow is an air-fuel mixture of necessarily indeterminate composition, since fresh air as well as air containing substantial concentrations of gasoline vapor can come from the carbon canister. The regeneration flow thus represents a major interference factor for the Lambda control system. A regeneration flow representing 1 % of the intake air and consisting solely of fresh air will

Fig. 8

Evaporative-emissions control system

1 Line from fuel tank to carbon canister,
2 Carbon canister,
3 Fresh air,
4 Canister-purge valve,
5 Line to intake manifold,
6 Throttle valve.

Δp Difference between manifold pressure p_s and atmospheric pressure p_u.

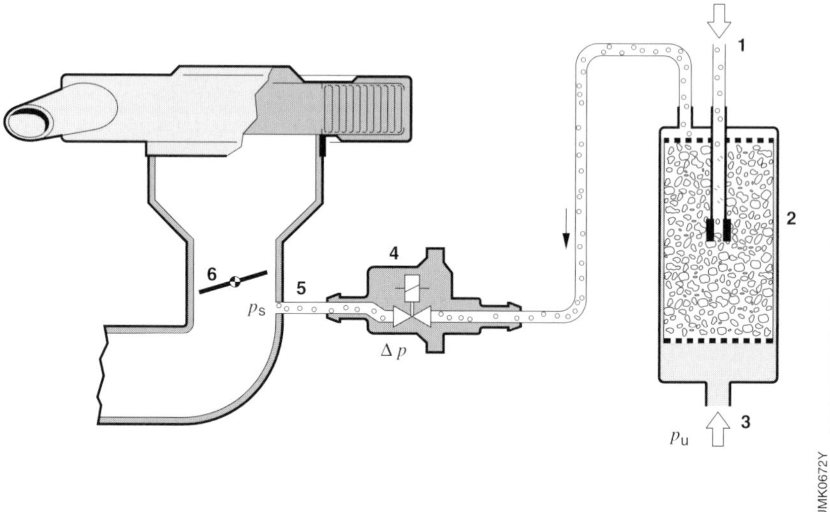

lean out the intake mixture by 1 %. A flow with a substantial gasoline component can enrichen the mixture by something in the order of 30 %, due to the effects on the A/F ratio λ of fuel vapor with a stoichiometric factor of 14.7. In addition, the specific density of fuel vapor is twice that of air.

Canister-purge valve

The canister-purge valve's control mechanism must ensure adequate air-purging of the carbon canister while holding Lambda deviations to a minimum (Figure 9).

ECU control operations

The canister-purge valve closes at regular intervals in order to allow the mixture adaptation process to function without being interfered with by tank ventilation.
The canister-purge valve opens in a "ramp-shaped" pattern. The ECU "memorizes" the resulting Lambda control deviations as mixture corrections from the fuel-regeneration system. The system is designed to operate with up to 40 % of the total fuel coming from the regeneration flow.
With the Lambda control system inactive, only small regeneration quantities are accepted, as there would be no control mechanism capable of compensating for the mixture deviations that would occur.
The valve closes immediately in the overrun fuel cutoff mode to prevent unburned gasoline vapors from entering the catalytic converter.

Fig. 9

Canister-purge valve

1 Hose connection, **2** Non-return valve, **3** Leaf spring, **4** Sealing element, **5** Solenoid armature, **6** Sealing seat, **7** Solenoid coil.

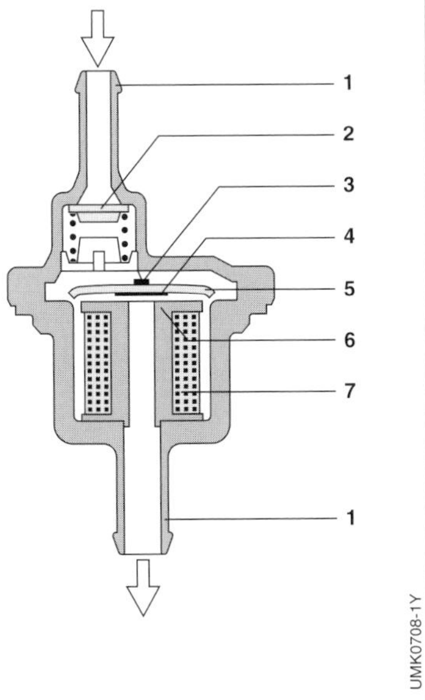

Knock control

Electronic control of the ignition timing allows extremely precise adjustments of the ignition-advance angle based on engine rpm, temperature and load factor. Nevertheless, a substantial safety margin to the knock limit must be maintained. This margin is necessary in order to ensure that no cylinder will reach or go beyond its preignition limit, even when susceptibility is increased by risk factors such as engine tolerances, aging, environmental conditions and fuel quality. The resulting engine design, with its fixed safety margin, is characterized by lower compression and reduced ignition advance. The ultimate results are sacrifices in fuel economy and torque.
These liabilities can be avoided by using a knock sensor, whereby experience has shown that the compression ratio can be raised with accompanying improvements in both fuel economy and torque. With this device, it is no longer necessary to select the default ignition advance angle with the most knock-sensitive conditions in mind. Instead, a best-case scenario can be used (e.g., engine compression at lower tolerance limit, optimum fuel quality, cylinder with minimum preignition tendency). This makes it possible to operate each individual cylinder at the preignition limit for optimum efficiency in virtually all operating ranges for the life of the vehicle.
The essential prerequisite allowing this kind of ignition-angle program is reliable recognition of all preignition beyond a specified intensity, extending to every

cylinder and throughout the engine's entire operating range. The knock sensors are solid-body sonic detectors installed at one or several suitable points on the engine. Here they detect the characteristic oscillation patterns that accompany knock and transform them into electrical signals before transmitting them to the Motronic ECU for processing. This ECU employs a special processing algorithm to detect incipient preignition in any of the combustion cycles in the respective cylinders. When this condition is recognized, the ignition advance angle is reduced by a programmed increment. When the knock danger subsides, the ignition for the affected cylinder is then slowly advanced back to the default setting. The knock recognition and knock-control algorithms are designed to prevent the kind of preignition that results in engine damage as well as audible knock (Fig. 10).

Adaptation

Real-world engine operation results in the individual cylinders having different knock limits, and therefore different ignition points. In order to adapt the default values to reflect the respective knock limits under varying operating conditions, individual ignition-retard increments are stored for each cylinder.
The data are stored in the non-volatile memory programs in the permanent RAM for engine speed and load factor. This allows the engine to be operated at optimum efficiency under all operating conditions without any danger of audible combustion knock, even during abrupt changes in load and rpm.
The engine can even be approved for operation with fuels having a low anti-knock quality. Generally, the engine is adjusted for use with premium-grade gasoline. Operation with regular-grade gasoline can also be approved.

Knock control on turbocharged engines

Systems combining boost pressure and knock control are especially effective on engines with exhaust-gas turbochargers.

Closed-loop knock control

Control algorithm with active ignition intervention on 4-cylinder engine.
$K_{1...3}$ Knock at cylinders 1...3 (no knock at cylinder 4)
a Ignition retard, **b** Step width for advance, **c** Advance.

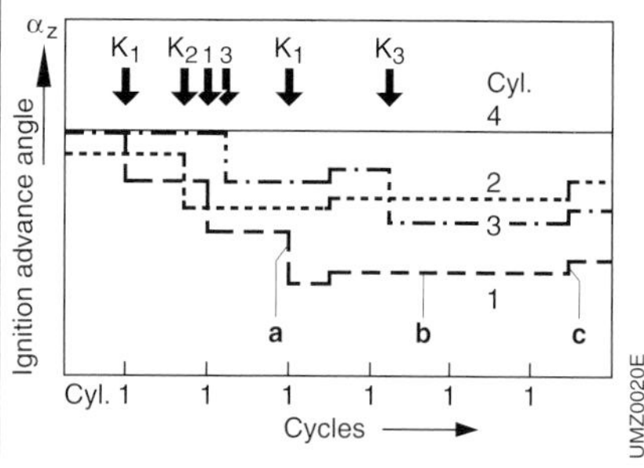

Fig. 10

The initial response to ignition knock is to reduce the timing advance angle. Further action to reduce the pre-ignition tendency in the form of a reduction in boost pressure is initiated only once the ignition-retard limit – which varies according to the temperature of the exhaust gas – has been reached. This makes it possible to maintain exhaust-gas temperatures within acceptable limits while operating the turbocharged engine at the knock limit for optimum efficiency.

Boost-pressure control

Turbocharger boost

The exhaust-gas turbocharger has prevailed in the face of competition from other supercharging methods, such as pressure-wave and mechnical supercharging. Turbochargers make it possible to achieve high torque and output from small-displacement, high-efficiency powerplants. Because a turbocharged engine can be smaller than its naturally-aspirated counterpart producing the same amount of power, it boasts a higher power-to-weight ratio.
Automotive industry research has demonstrated that compared to a standard naturally aspirated engine, a small-displacement turbocharged engine with electronic boost control can provide improvements in fuel economy equal to those achieved with a prechamber diesel. The

main components of the exhaust-gas turbocharger are the compressor and the exhaust-gas impeller mounted on the other side of the same shaft. The exhaust-gas turbocharger transforms a portion of the energy in the exhaust gas into the rotation energy used to power the compressor. This, in turn, draws in fresh air and compresses it before blowing it through the intercooler, throttle valve, intake manifold and into the engine.

Actuators for exhaust-gas turbochargers

Passenger-car engines must be capable of generating substantial torque at low engine speeds. This is the reason why the turbocharger housing is dimensioned to operate most efficiently with lower mass exhaust-gas flow rates (for example, full load at n = 2,000 min^{-1}). To prevent the turbocharger from overloading the engine at higher exhaust-gas mass flow rates, a bypass mechanism (waste gate) must be included in the housing to divert a portion of the flow around the turbine and into the downstream exhaust system. The bypass valve generally assumes the form of a flap in the turbine housing. Less common is a disk valve installed parallel to the turbine in a separate housing. Variable turbine geometry has still not been used for spark-ignition engines, but could also be combined with boost-pressure control (Figure 11).

Fig. 11

Actuator for electronic boost-pressure control

1 Pulse valve,
p_2 Boost pressure,
p_D Pressure in diaphragm unit,
TVM Triggering signal from ECU to pulse valve,
V_T Flow volume through turbine,
V_{WG} Flow volume through waste gate.

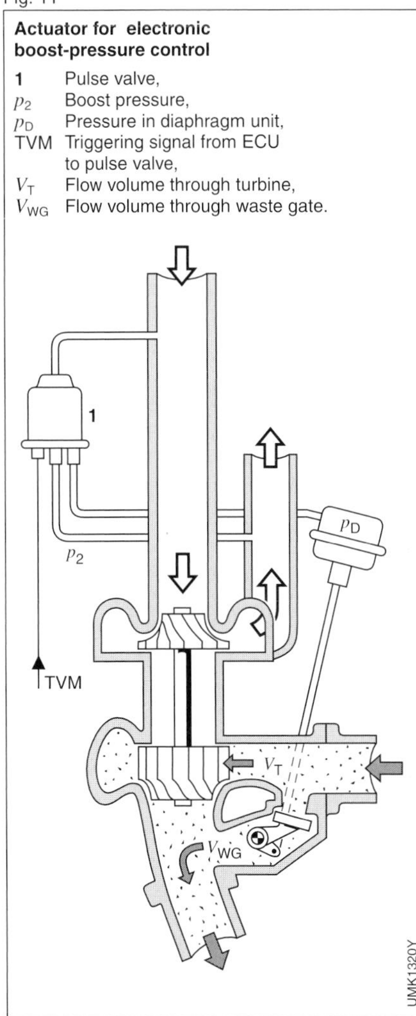

Boost-pressure (closed-loop) control

Pneumatic-mechanical closed-loop control systems use a turbocharger actuator directly exposed to the boost pressure at the turbocharger outlet. This layout provides only limited latitude for tailoring the progression of the torque curve through the engine speed range. There is only a full-load limit. It is not possible to provide control compensation for tolerances at full-load boost. At part load the closed bypass valve reduces efficiency. Acceleration from low engine speeds can be accompanied by delayed turbocharger reaction (turbo lag).

These disadvantages can be avoided with electronic boost-pressure control (Fig. 11). The specific fuel consumption can be reduced in some part-throttle operating ranges. The system operates by opening the bypass valve, with the following results:

- Residual work on the part of the engine and turbine output are reduced,
- Pressure and temperature at the compressor outlet are lowered, and
- The pressure differential at the throttle valve is lowered.

A linear relationship between torque curve and throttle-valve angle is also obtained, with improved sensitivity at the accelerator pedal. In order to provide the improvements listed above, the exhaust-gas turbocharger and the actuator must be perfectly adapted for use in the individual engine. The affected elements in the actuator are:

- The electropneumatic pulse valve,
- The effective diaphragm surface, stroke and spring in the diaphragm unit, and
- The diameter of the valve disk/flap at the wastegate.

Depending upon the load sensor, the setpoints stored in the M-Motronic ECU with electronic boost-pressure control are for pressure, air quantity or air mass. The setpoints for various engine speeds and throttle-valve angles are stored in a program map.

Actuators within the closed-loop control circuit adjust the monitored actual value to coincide with the value prescribed for the particular operating conditions. The calculated value is transmitted through the controller output in the form of a signal (pulse-width modulated) to the pulse valve. Within the actuator this signal modifies the control pressure and the stroke to change the effective opening at the bypass valve.

The temperature of the exhaust gases between the turbocharged engine and the turbine should never be allowed to exceed certain limits. This is why Bosch only uses boost-pressure control in conjunction with knock control. Knock control is the only means of allowing the engine to run with the maximum potential ignition advance throughout its service life. Running the engine at the optimal ignition advance angle for the specific operating conditions results in extremely low exhaust-gas temperatures.

Additional adjustments to boost pressure and/or mixture can be used to lower the temperature of the exhaust gas even further.

Limiting engine and vehicle speed

Extremely high engine speeds can lead to destruction of the engine (valve train, pistons). Engine-speed limiting prevents the maximum approved engine speed (redline) from being exceeded.
M-Motronic provides the option of restricting engine and vehicle speed by phasing out the injection.

When the maximum engine speed n_0 (or the maximum vehicle speed) is exceeded, the unit responds by suppressing the injection signals. This limits the speed of the engine (or vehicle).

The injection resumes normal operation once the speed falls below a narrow threshold.
This process is repeated at rapid intervals within an engine-speed tolerance range located around the prescribed maximum.

A reduction in operating response and smoothness calls the driver's attention to the engine speed and provides the motivation for an appropriate response.
Figure 12 illustrates the engine-speed curve's reaction to the engine-speed limitation.

Fig. 12

Limiting maximum engine speed n_0 by suppressing injection pulses
a Fuel-cutoff range.

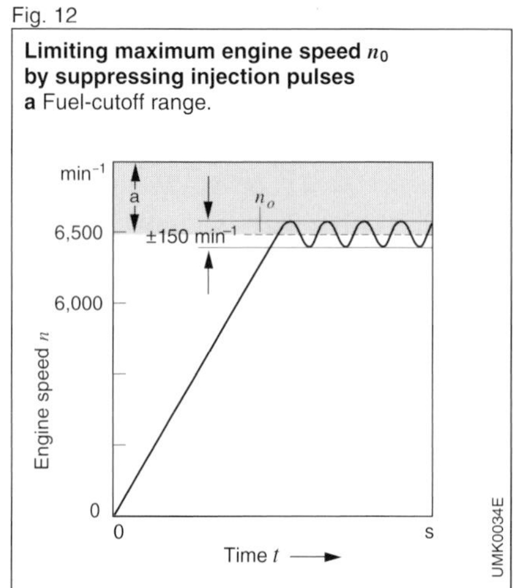

Exhaust-gas recirculation (EGR)

During valve overlap a certain amount of residual gas is returned from the combustion chamber to the intake manifold. This recirculated gas is then entrained along with the fresh air on the next intake stroke.

The actual amount of recirculated gas is determined by the valve overlap, and is thus a fixed variable relative to the various points on the operating curve.

The proportion of recirculated gas can be adjusted using either "external" exhaust-gas recirculation (EGR) with a M-Motronic-controlled exhaust-gas recirculation valve (Fig. 13), or with variable camshaft timing.

Up to a certain point, at least, larger amounts of recirculated gas can exercise a positive effect on energy conversion and thus on fuel consumption. An increase in recirculated gas also leads to lower combustion-chamber temperatures with corresponding reductions in the NO_x emission.

At the same time, once a certain point is passed a higher residual-gas component will lead to incomplete combustion. The results here are higher emissions of unburned hydrocarbons, increased fuel consumption and rough idling (Figure 14).

Effect of residual exhaust-gas recirculation on fuel consumption and emissions

1 Excess-air factor λ (residual exhaust-gas component RG = constant),
2 Recirculated gas component RG (λ = constant).

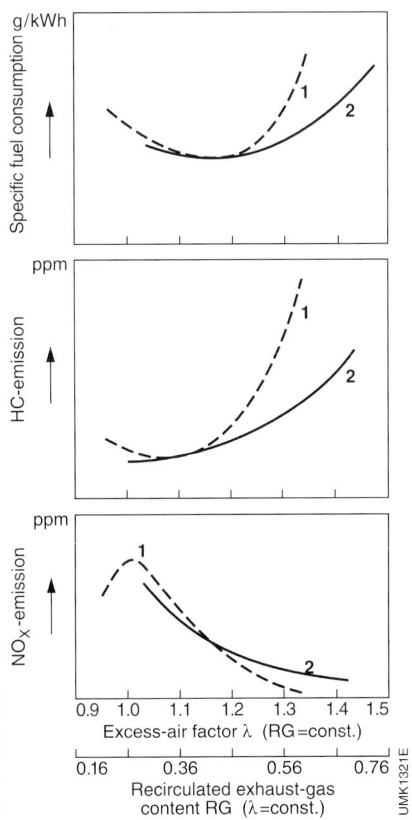

Fig. 14

Fig. 13

Exhaust-gas recirculation (example).

1 Exhaust-gas recirculation (EGR),
2 Electropneumatic converter,
3 EGR valve,
4 ECU,
5 Air-mass meter.
n Engine speed.

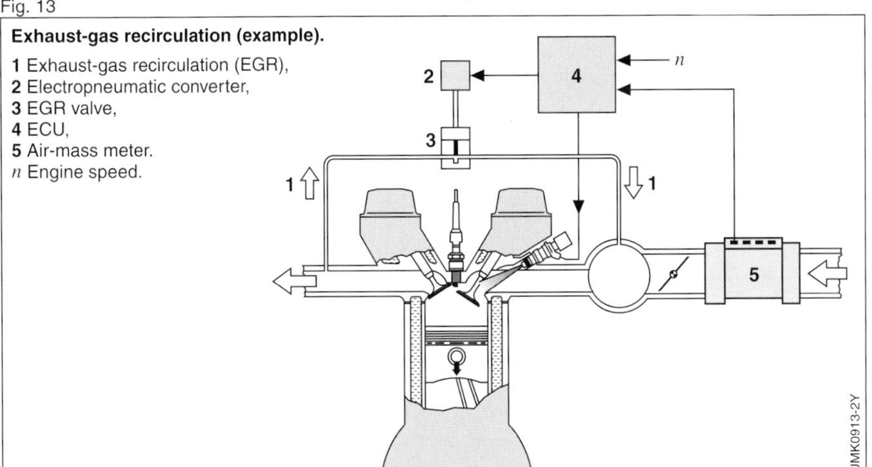

Camshaft timing

Camshaft timing can influence the spark-ignition engine in a variety of ways:
- Higher torque and output, lower emissions and fuel consumption,
- Control of the mixture composition, and,
- Graduated or infinitely-variable intake and exhaust adjustment.

"Intake closes" timing plays a decisive role in determining the amount of cylinder charge for a given engine speed. When the intake valve closes early, maximum air will be inducted at low rpm, while longer intake durations shift the maximum toward higher engine speed ranges.

The phase in which the valves overlap (at "intake opens" and "exhaust closes") determines the amount of internal residual-gas recirculation.

Longer "valve open" durations due to advanced intake-valve timing will raise the proportion of recirculated gas, as they increase the mass of the gas returned to the intake manifold for reinduction. This reduces the mass of fresh air being drawn in at a given throttle-valve opening; at any given load point the throttle valve must open further to compensate. This "dethrottling" effect reduces the gas-exchange circuit to improve efficiency and lower fuel consumption.
The proportion of recirculated gas sinks when the intake cycle is shifted toward "retard." This provides improvements in fuel economy and operating smoothness.

Variable camshaft angle

Hydraulic or electric actuators turn the camshaft by increments corresponding to specified engine speeds or operating points (this system requires that at least one intake and one exhaust camshaft be located in the cylinder head). This varies the timing for "intake/exhaust opens" and/or "intake/exhaust closes" (Figure 15).
As an example, if the actuators turn the intake camshaft to delay "intake opens/closes" at idle or at high rpm, the result will be a reduction in the proportion of recirculated gases at idle, or enhanced cylinder charging at higher engine speeds.
When the intake camshaft is turned toward earlier "intake opens/closes" at low or moderate engine speeds or in certain load ranges, the result is a higher maximum air charge to the cylinder.
This would also lead to a larger proportion of recirculated gas in the part-throttle range, with corresponding effects on fuel consumption and exhaust emissions.

Fig. 15

Intake camshaft rotation

1 Retarded, **2** Normal, **3** Advanced.

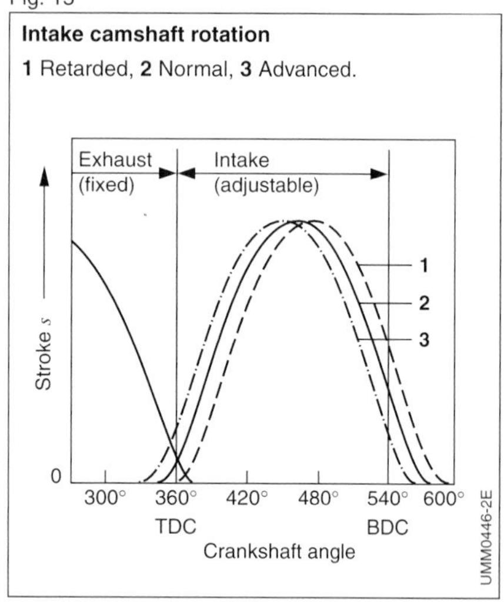

Fig. 16

Variable camshaft timing

1 Standard, **2** Auxiliary lobes.

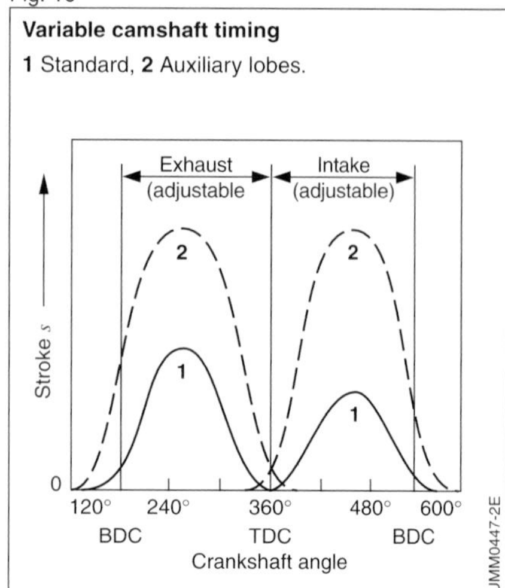

Camshaft lobe control

Systems with camshaft lobe control modify the valve timing by alternately activating cam lobes with two different shapes. The first lobe supplies optimum valve timing and lift for intake and exhaust valves during lower and mid-range operation. A second cam lobe provides longer valve-opening times and lift, and becomes operational when the rocker arm to which it is connected is locked onto the standard rocker arm in response to engine speed (Figure 16).

An optimal but complicated process is infinitely-variable valve timing and lift adjustment. This concept employs cam lobes with 3-dimensional geometry and sliding camshafts to provide maximum latitude in engine design (Figure 17).

Operating conditions

Fig. 17

Infinitely-variable valve timing and lift adjustment

a minimum,
b maximum lift.

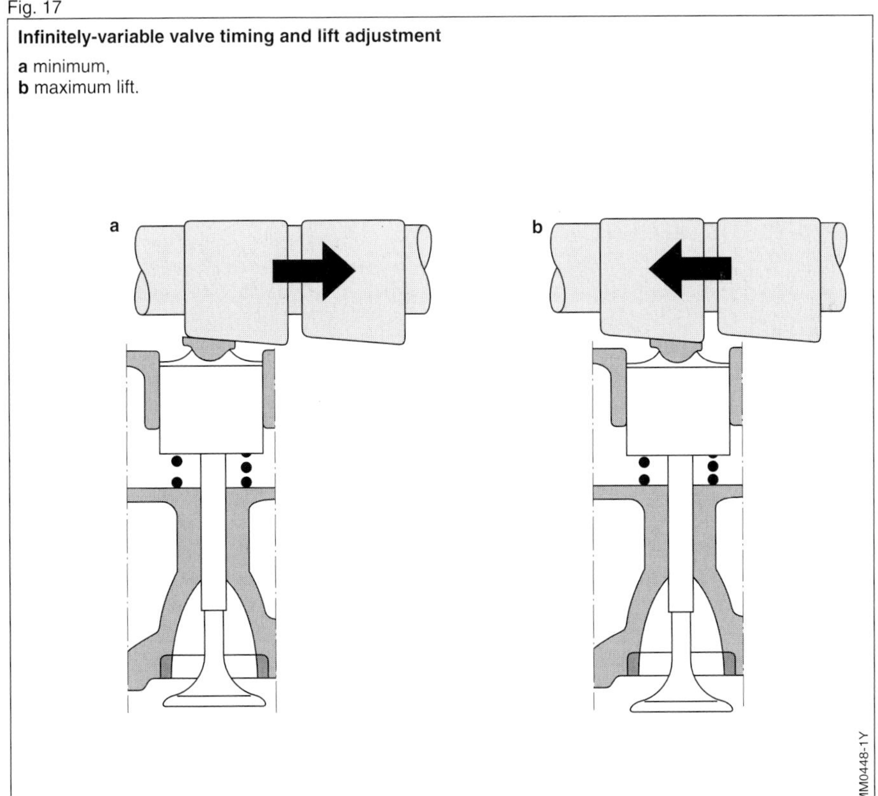

Variable-geometry intake manifold

The tandem objectives of engine design are maximum torque at low engine speeds and high output at the rated maximum. The engine's torque curve is proportional to the mass of the intake air as a function of engine speed.

One effective means of influencing torque is to provide the intake manifold with the appropriate geometrical configuration. The simplest method for providing intake boost is to exploit the dynamics of the incoming air. To ensure balanced distribution of the air-fuel mixture, intake manifolds for carburetor or single-point (Mono-Jetronic) injection systems need short intake runners with minimal variations in lengths.

The intake runners for multipoint systems transport only air; the injectors discharge the fuel. This arrangement offers a wider range of options in intake-manifold design. The standard manifold for a multipoint injection system consists of individual curved runners and a plenum chamber with throttle valve.

General principles:
- Short curved runners allow high maximum output accompanied by sacrifices in torque at lower engine speeds; whereby long runners provide an inverse response pattern.
- Large-volume plenum chambers can provide resonance effects in certain engine-speed ranges, leading to improved cylinder filling. They are also subject to potential faults in dynamic response; these assume the form of mixture variations under rapid load change.

Variable intake-manifold geometry can be used to obtain an almost ideal torque curve. Geometry can be changed for instance as a function of engine load, engine speed, and throttle-valve setting. Various changes are possible:

Fig. 18

Resonance "supercharging"

a Configuration, **b** Curve of air input.
1 Resonator runner, **2** Resonance chamber,
3 Cylinders, **4** With resonance boost,
5 Normal intake manifold.

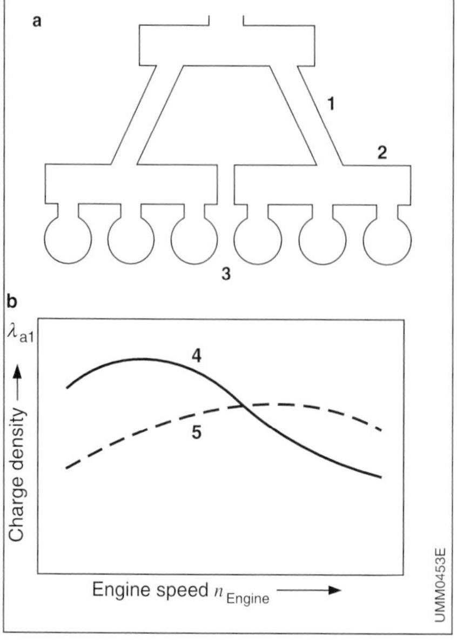

Fig. 19

Variable-geometry intake systems

Switchable: **a** two-stage, **b** three-stage.
A, B Cylinder groups; **1, 2** Flaps, engine speed determines opening time.

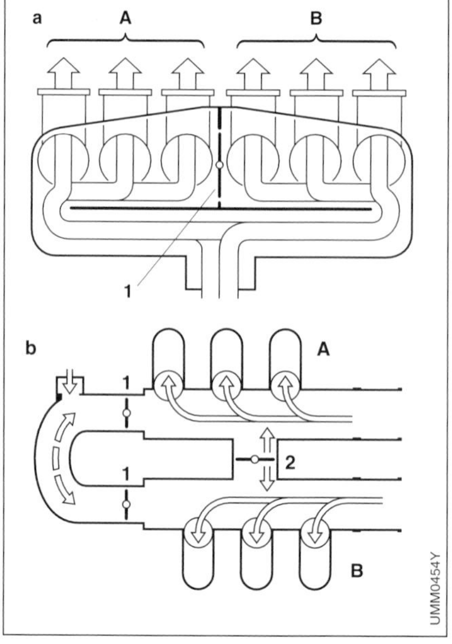

- Runner-length adjustment,
- Change-over between runner lengths or runner diameters,
- In case of multiple runners, the selective cutoff of an individual runner for each cylinder,
- Changeover between different plenum-chamber volumes.

Intake oscillation boost

Each cylinder has an individual, fixed-length intake runner, usually connected to a plenum chamber. The energy balance is defined by a process in which the induction force from the piston is converted into kinetic energy in the gas column upstream from the intake valve. This kinetic energy then serves to compress the fresh charge.

Resonance boost

Resonance boost systems use short runners to connect groups of cylinders with equal ignition intervals to resonance chambers. These, in turn, are connected via resonance tubes to the atmosphere or a plenum chamber, allowing them to act as Helmholtz resonators (Figure 18).

Variable-geometry intake systems

Both types of dynamic supercharging augment the achievable cylinder charge, especially in the lower engine-speed range.

Variable-response intake systems use devices such as flaps to separate and connect system areas assigned to various groups of cylinders (Figure 19).
Variable-length intake runners operate with the first resonance chamber at low rpm. The length of the runner changes as engine speed increases, at which point a second resonance chamber opens (Figure 20).
Figure 21 shows the effects of variable intake-runner geometry on the brake mean effective pressure (bmep), which is used as an index for cylinder charging.

Fig. 20

Infinitely-variable length intake system

1 Fixed housing,
2 Rotating drum (air distributor),
3 Drum air-supply opening,
4 Air-supply opening for intake runners,
5 Seal (e.g., leaf spring),
6 Intake runners,
7 Intake valve,
8 Intake air stream.

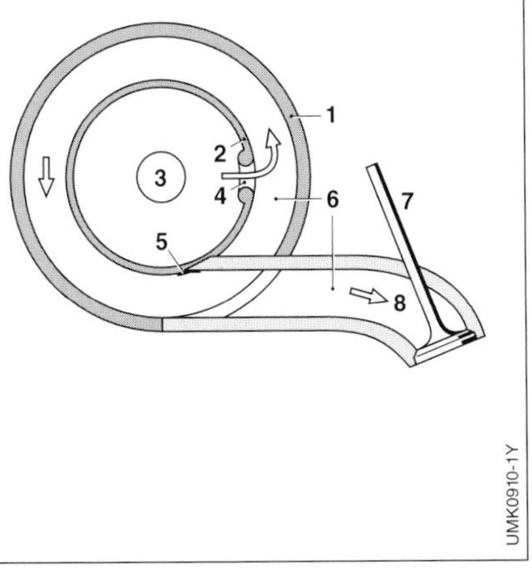

UMK0910-1Y

Fig. 21

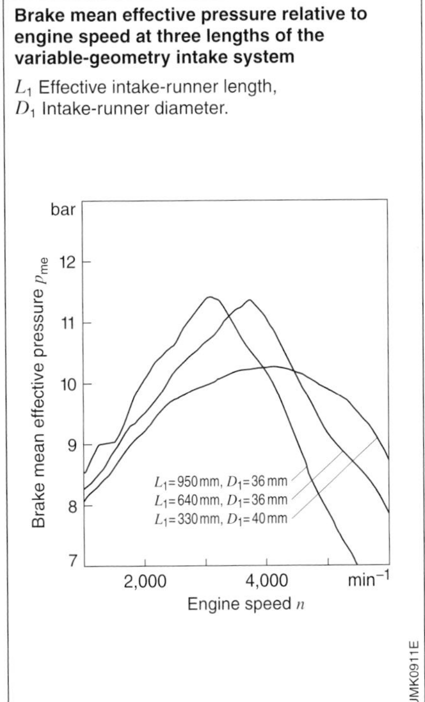

Brake mean effective pressure relative to engine speed at three lengths of the variable-geometry intake system

L_1 Effective intake-runner length,
D_1 Intake-runner diameter.

Integrated diagnosis

Diagnostic procedure

An "on-board diagnosis (OBD) system" is standard equipment with M-Motronic. This integral diagnostic unit monitors ECU commands and system responses. It also checks the individual sensor signals for plausibility. This test procedure is carried out constantly during normal vehicle operation.

The ECU stores recognized errors together with the operating conditions under which they occurred. When the vehicle is serviced, a tester can be used to read out and display the stored errors through a standardized interface. The information facilitates trouble-shooting for the service personnel.

In response to the demands of the California Air Resources Board, (CARB), diagnosis procedures extending far beyond those in earlier tests have been developed. All components whose failure could cause a substantial increase in harmful emissions must be monitored.

Diagnosis areas

Air-mass meter

The process for monitoring the air-mass meter provides an example of the M-Motronic system's self-diagnosis function. While the injection duration is being determined on the basis of intake-air mass, a supplementary comparison injection time is calculated from throttle-valve angle and engine speed. If the ECU discovers an excessive variance between the two, its initial response is to store a record of the error. As vehicle operation continues, plausibility checks determine which of the sensors is defective. The control unit does not store the corre-sponding error code until it has unequivocally determined which sensor is at fault.

Combustion miss

Combustion miss, as can result from factors such as worn spark plugs or faulty electrical contacts, allows unburned mixture to enter the catalytic converter. This mixture can destroy the converter, and represents an extra load on the environment. Because even isolated combustion failures result in higher emissions, the system must be able to recognize them. Figure 1 shows the effects of combustion miss on emissions of hydrocarbons (HC), carbon monoxide (CO) and oxides of nitrogen (NO_X).

Many potential methods of detecting combustion miss were tested, and monitoring the running response of the crankshaft proved to be the most suitable. Combustion miss is accompanied by a shortfall in torque equal to the increment which would normally have been generated in the cycle where the error occured. The result is a reduction in rotation speed. At high speeds and low loads the interval from ignition to ignition (period duration) is extended by only 0.2 %. This means that the rotation must be monitored with extreme precision, while extensive computations are also required to distinguish combustion miss from other interference factors.

Catalytic converter

Yet another diagnostic function monitors the efficiency of the catalytic converter. For this purpose the Lambda oxygen sensor upstream of the catalytic converter is supplemented by a second downstream oxygen sensor. A correctly-operating converter will store oxygen, thus attenuating the Lambda control oscillations. As the catalytic converter ages, this response deteriorates until finally the signal pattern from the upstream sensor approaches that received from the downstream sensor. A comparison of the signals from the oxygen sensors thus provides the basis for determining the catalytic converter's condition. A warning lamp alerts the driver in the event of a defect.

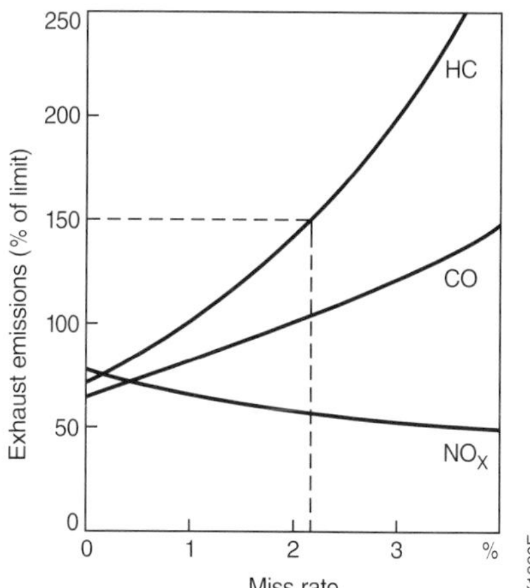

Fig. 1

Fig. 2

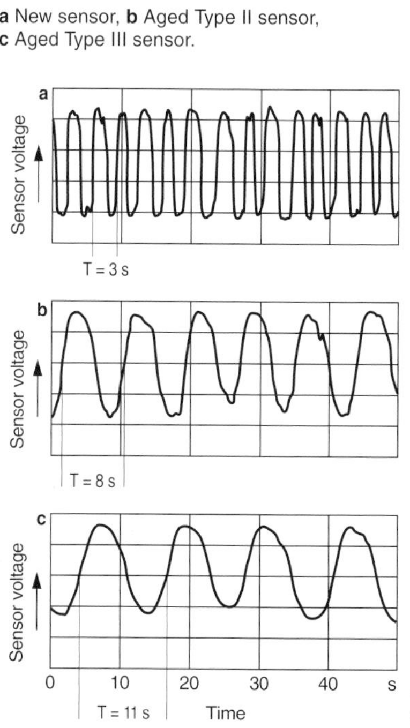

Lambda oxygen sensor

A stoichiometric air-fuel mixture must be maintained if the catalytic converter is to perform to its full potential. This is taken care of by the signals from the Lambda oxygen sensors. The fact that two oxygen sensors are fitted in each exhaust tract makes it possible to use the sensor downstream from the converter to check for control variations in the upstream unit. A Lambda oxygen sensor that has been exposed to excessive heat for a considerable period of time may react more slowly to changes in the air-fuel mixture. This increases the period duration for the Lambda control's two-state controller (Figure 2). A diagnosis function monitors this control frequency and informs the driver of excessive delays in sensor response via warning lamp.

The Lambda oxygen sensors' heating resistance is checked by measuring current and voltage. The M-Motronic ECU controls the heater resistance element directly, with no relay in between, to allow this test to be performed. The sensor signal is subjected continuously to plausibility checks, and the system responds to implausible signals by denying access to other functions depending upon the Lambda control. The appropriate error code is also stored in the fault memory.

Fuel supply

When the air-fuel mixture deviates from stoichiometric for extended periods of time, this condition is taken into account together with the mixture adaptation. If the deviations exceed specific predefined limits, this indicates that a fuel-system component or a fuel-metering device has moved outside specification tolerances. An example would be a faulty pressure regulator or load sensor, whereby the error could also stem from a leak in the intake manifold or exhaust system.

Secondary-air injection

The secondary-air injection activated after cold starts must also be monitored, as its failure would also influence emissions. This can be done using the signals from the Lambda oxygen sensors when the secondary-air injection is active, or it can be activated and observed at idle using a Lambda control test function.

Exhaust-gas recirculation (EGR)

Various options are available for diagnosing the exhaust-gas recirculation system, whereby two are in general use. With the first option, a sensor monitors the temperature increase at the location where the hot exhaust gases return to the intake manifold while the EGR is operating.

With the second procedure, the exhaust-gas recirculation valve (EGR valve) is opened all the way with the vehicle on trailing throttle (overrun fuel cutoff). The exhaust gases flowing into the intake manifold cause its internal pressure to rise. A pressure sensor measures and evaluates the increase in manifold pressure.

Tank system

Emissions emanating from the exhaust system are not the only source of environmental concern; fuel vapors from the fuel tank are also a problem. In the near term the legal requirements will be limited to a relatively simple check of canister-purge valve operation. A means of recognizing leaks in the evaporative-emissions storage system will be required at a later date. Figure 3 illustrates the basic principle employed for this diagnosis. A cutoff valve closes off the storage system.
Then, preferably with the engine idling, the canister-purge valve is opened and the intake-manifold pressure spreads through the system. An in-tank pressure sensor monitors the pressure build-up to determine whether leaks are present.

Fig. 3

Vacuum test to detect leaks in evaporative-emissions control system

1 Intake manifold,
2 Canister-purge valve,
3 Shutoff valve,
4 Fuel tank,
5 Pressure differential sensor,
6 Safety valve.

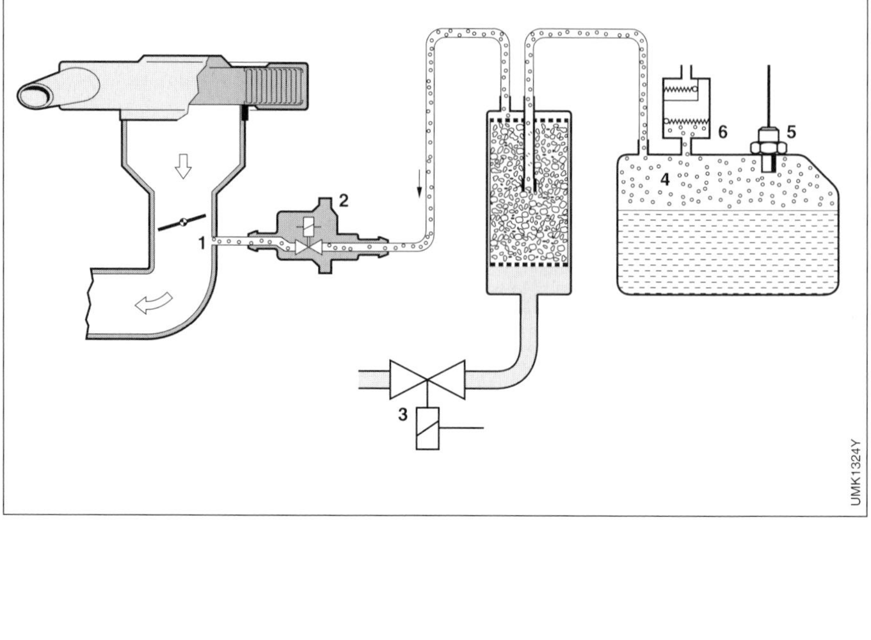

Other monitoring devices

The main emphasis of the new statutes applies to the engine-management system, but other systems (for instance, automatic transmission) are also monitored. These report any faults to the engine-management ECU, which then assumes responsibility for triggering the diagnosis lamp. Greater system complexity and ever more stringent environmental regulations are making diagnosis increasingly important.

Emergency running mode (limp-home)

In the interval between the initial occurrence of a fault and vehicle service, default settings and emergency-running functions assume responsibility for the ignition and air-fuel mixture. This allows the vehicle to continue operating, albeit with sacrifices in driving comfort. The ECU responds to errors that have been recognized in an input circuit by replacing the missing information or reverting to a default value.
When an output unit fails, specific backup measures corresponding to the individual problem are implemented. Thus the ECU reacts to a defect in the ignition circuit by switching off the fuel injection at the affected cylinder in order to prevent damage to the catalytic converter.

When the vehicle is serviced, the Bosch engine tester can be used to read out and display the faults and errors detected during operation (Figure 4).

Fig. 4
Bosch engine tester

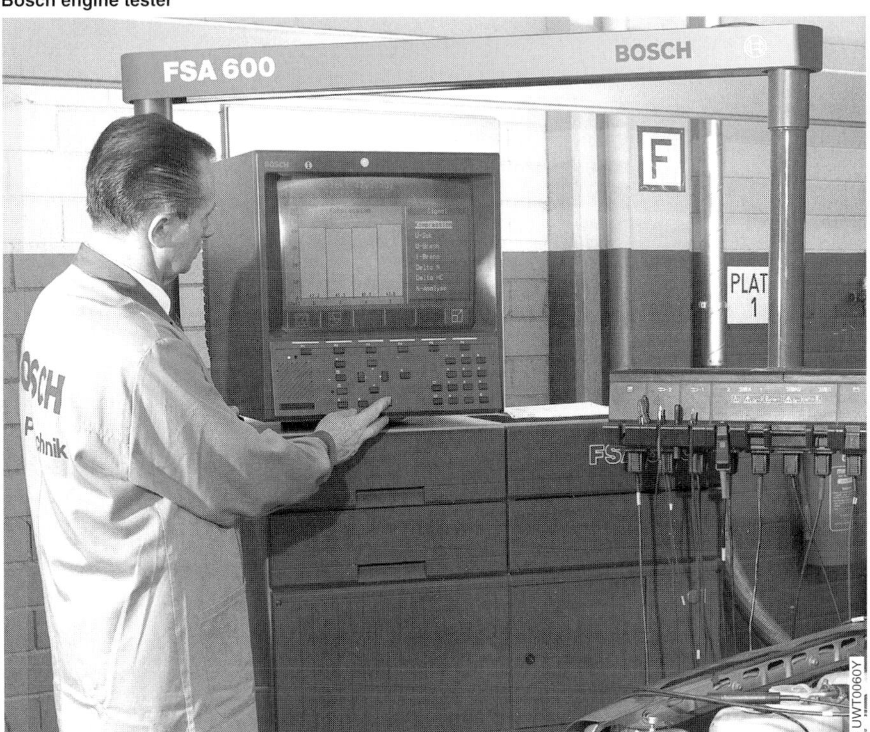

ECU

Function

The ECU is the "computer and control center" for the engine-management system. It employs stored functions and algorithms (processing programs) to process the input signals transmitted by the sensors. These signals serve as the basis for calculating the control signals to the actuators (e.g., ignition coil, injectors) which it manages directly via power output stages (Figure 1).

Physical design

The ECU is a metal housing containing a printed-circuit board with electronic componentry.
A multiple-terminal plug connector provides the link between ECU and sensors, actuators and power supply. Depending upon the specific ECU and the number of system functions, the plug can be of 35-, 55- or 88-pole design.

The amplifiers and power output components for direct actuator control are installed on heat sinks in the ECU. Efficient heat transfer to the bodywork is necessary due to the the amount of heat that these components produce.

Environmental conditions

The ECU must withstand temperature extremes, moisture and mechanical loads with absolutely no impairment of operation. Resistance to electromagnetic interference, and the ability to suppress radiation of high-frequency static, must also be of a high order.

The ECU must be capable of errorless signal processing within an operating range extending from –30 °C to +60 °C, at battery voltages that range from 6 V (during starting) to 15 V.

Power supply

A voltage regulator provides the ECU with the constant 5 V operating voltage needed for the digital circuitry.

Signal inputs

Various processes are employed to transmit the input signals to the ECU. The signals are conducted through protective circuits, while signal converters and amplifiers may also be present. The microprocessor can process these signals directly.

An analog/digital converter (A/D) within the microprocessor transforms analog signals (for instance, information on intake-air quantity, temperature of engine and intake air, battery voltage, Lambda oxygen sensor) into digital form.
The signal from the inductive sensor with information on engine speed and crankshaft reference point is processed in a special circuit to suppress interference pulses.

Signal processing

The input signals are processed by the microprocessor within the ECU. In order to function, this microprocessor must be equipped with a signal-processing program stored in a non-volatile memory (ROM or EPROM). This memory also contains the specific individual performance curves and program maps used for engine control.
Due to the large number of engine and vehicle variations, some ECU's are equipped with a special version-code feature. This allows the manufacturer or service technician to feed supplementary program data into the program maps stored in the EPROM, making it possible to provide the operating characteristics desired for the particular version. Other types of ECU are designed to allow complete data banks to be programmed into the EPROM at the end of production (end-of-line programming). This reduces the number of individual ECU configurations required by the manufacturer.
A read/write memory component (RAM) is needed for storing calculated values and adaptation factors as well as any system errors that may be detected (diagnosis). This RAM requires an uninterrupted power supply to function properly. This memory chip will lose all

data if the vehicle battery is disconnected. The ECU must then recalculate the adaptation factors after the battery is reconnected. To prevent this, some units therefore use an EEPROM instead of a RAM to store these required variables.

Transmitting the signal

The output stages triggered by the microprocessor supply sufficient power for direct control of the actuators. These output stages are protected against short circuits to ground, irregularities in battery voltage and electrical overload that could destroy them.

At several output stages, the OBD diagnosis function recognizes errors and reacts by deactivating (where necessary) the defective output. The error entry is stored in the RAM. The service technician can then read out the error using a tester connected to the serial interface. Another protective circuit operates independently of the ECU to switch off the electric fuel pump when the engine-speed signal falls below a certain level. When some ECU's are switched off at the ignition/steering lock (Terminal 15, or "ignition off"), a holding circuit holds the main relay open until program processing can be completed.

Fig. 1

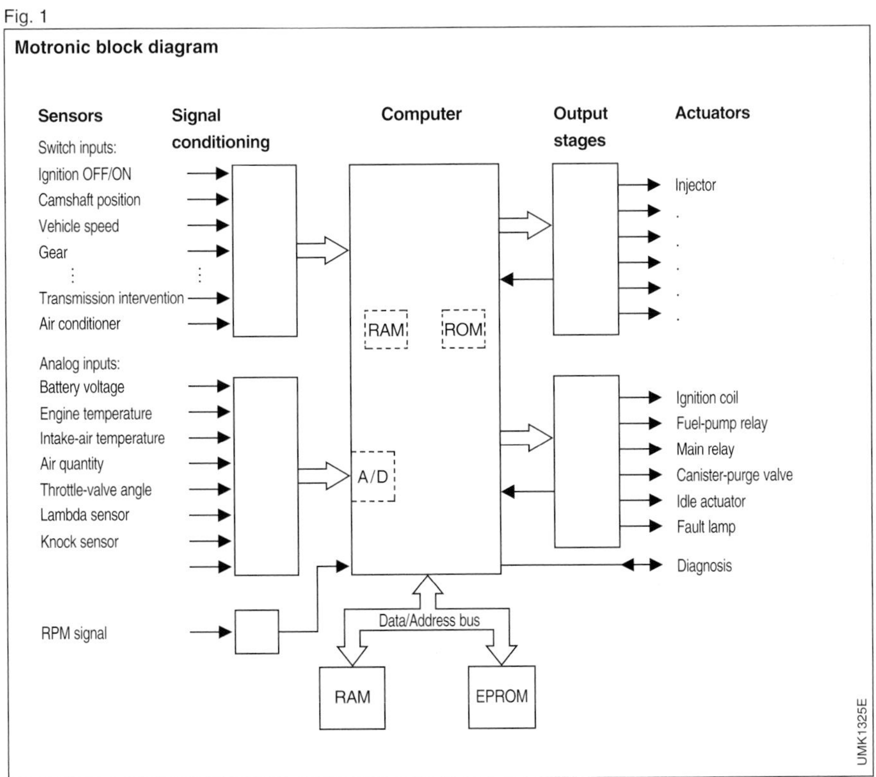

Interfaces to other systems

System overview

Increased application of electronic control systems in vehicles in areas such as
- Transmission control,
- Electronic throttle control (EMS, E-Gas, drive-by-wire),
- Electronic engine management (Motronic),
- Antilock braking system (ABS),
- Traction control system (TCS),
- On-board computer, etc.,

has made it necessary to combine the respective ECU's in networks. Data communications between control systems reduce the number of sensors and allow better exploitation of the individual system potentials.
The interfaces can be divided into two categories:
- Conventional interfaces, with binary signals (switch inputs), pulse-duty factors (pulse-width-modulated signals),
- Serial data transmission, e.g., Controller Area Network (CAN).

Conventional interfaces

In conventional automotive data-communications systems, each signal is assigned to a single line. Binary signals can only be transmitted as one of the two conditions "1" or "0" (binary code), for instance, air-conditioning compressor "ON" or "OFF".
Pulse-duty factors (potentiometer) can be employed to relay more detailed data, such as throttle-valve aperture.
The increasing levels of data exchange between the various electronic components in the vehicle means that conventional interfaces are no longer capable of providing satisfactory performance. The complexity of today's wiring harnesses is already difficult to manage, while the requirements for data communications between ECU's are on the rise (Figure 1).
These problems can be solved by using CAN (Controller Area Network), a bus system (bus bar) specially designed for automotive use.
Provided that the ECU's are equipped with a serial CAN interface, CAN can be used to relay the signals from the sources listed above.

Serial data transmission (CAN)

There are three basic applications for CAN in motor vehicles:
- To connect ECU's,
- Bodywork and convenience electronics (Multiplex),
- Mobile communications.

The following is limited to a description of communications between ECU's.

ECU networking

Here electronic systems such as M-Motronic, electronic transmission control, etc. are combined within a single network. Typical transmission times lie between approx. 125 kBit/s and 1MBit/s. These times must be high enough to maintain the required real-time response. One of the advantages that serial data transmission enjoys over conventional interfaces (e.g., pulse-duty factor, switching and analog signals) is the high speeds achieved without high loads on the central processing units (CPU's) in ECU's.

Fig. 1

Conventional data communications

GS Transmission control, ETC Electronic throttle control, ABS Antilock braking system,
TCS Traction control system,
MSR Engine-drag torque control.

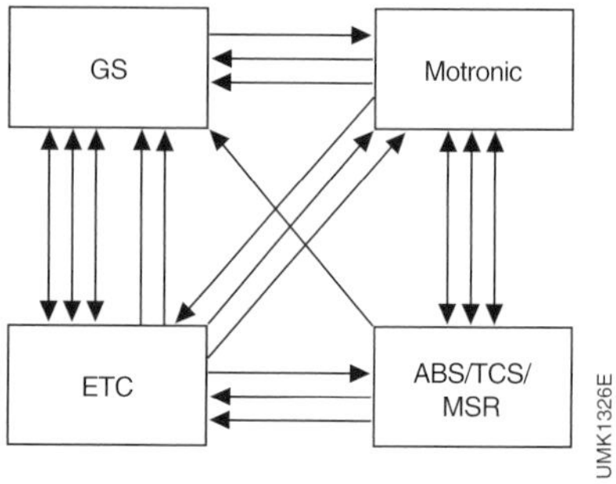

Bus configuration

CAN works on the "multiple master" principle. This concept combines several equal-priority ECU's in a linear bus structure (Figure 2). The advantage of this structure is the fact that failure of one subscriber will not affect access for the others. The probability of total failure is thus much lower than with other logical arrangements (such as loop or star structures). With loop or star architecture, failure in one of the subscribers or the central ECU will result in total system failure.

Content-keyed addressing

The CAN bus system addresses the data according to content. Each message is assigned a permanent 11-bit identifier. This identifier indicates the contents of the message (e.g., engine speed). Each station processes only those data whose identifiers are stored in its acceptance list (acceptance check). This means that CAN does not need station addresses to transmit data, and the junctions do not need to administer system configuration.

Bus arbitration

Each station can begin transmitting its highest priority message as soon as the bus is free. If several stations start transmitting simultaneously, the resulting bus-access conflict is resolved using a "wired-and" arbitration arrangement. This arrangement assigns first access to the message with the highest priority, with no loss of either time or data bits. When a station loses the arbitration, it automatically reverts to standby status and repeats its transmission attempt as soon as the bus indicates that it is no longer occupied.

Fig. 2

Linear bus structure

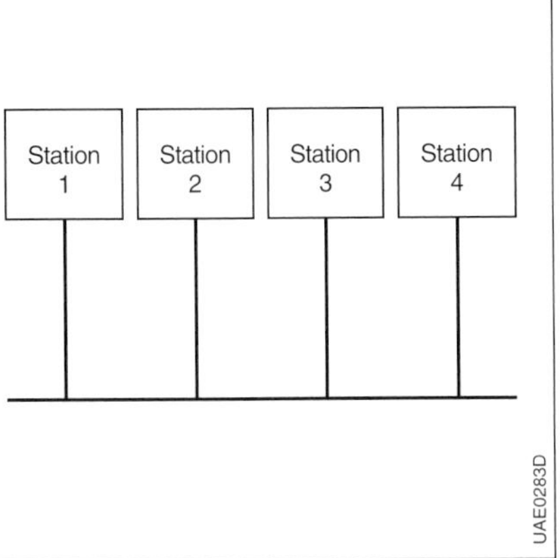

Station 1 | Station 2 | Station 3 | Station 4

UAE0283D

Message format

A data frame of less than 130 bits in length is created for transmissions to the bus. This ensures that the queue time until the next – possibly extremely urgent – data transmission is held to a minimum. The data frames consist of seven consecutive fields.

Standardization

The International Standards Organisation (ISO) has recognized CAN as a standard for use in automotive applications with data streams of over 125 kBit/s, along with two additional protocols for data rates up to 125 kBit/s.

ME-Motronic engine management

The overall Motronic system

System overview

The Motronic system contains all of the actuators (servo units, final-control elements) required for intervening in the spark-ignition engine management, while monitoring devices (sensors) register current operating data for engine and vehicle. These sensor signals are then processed in the input circuitry of a central electronic control unit (ECU) before being transferred to the ECU microprocessor (function calculator). The information provided (Figs. 1 and 2) includes data on:
- Accelerator-pedal travel,
- Engine speed,
- Cylinder charge factor (air mass),
- Engine and intake-air temperatures,
- Mixture composition, and
- Vehicle speed.

The microprocessor employs these data as the basis for quantifying driver demand, and responds by calculating the engine torque required for compliance with the driver's wishes. Meanwhile, the driver or a transmission-shift control function selects the conversion ratio needed to help define engine speed.
The microprocessor generates the required actuator signals as the first stage in setting the stipulated operating status. These signals are then amplified in the driver circuit and transmitted to the actuators responsible for engine management. By ensuring provision of the required cylinder charge together with the corresponding injected fuel quantity, and the correct ignition timing, the system furnishes optimal mixture formation and combustion.

ME version

The following descriptions focus on a typical version of ME-Motronic. Within the type designation, the letter "M" stands for the classical Motronic function of coordinated control for injection and ignition, while "E" indicates integration of the ETC electronic throttle control.

Basic functions

Motronic's primary function is to implement the engine's operational status in line with the driver's demands. The system's microprocessor responds to this demand by translating the accelerator-pedal travel into a specified engine output. When converting the required engine-output figure to the parameters for actually controlling engine output, that is
- The density of the cylinders' air charges,
- The mass of the injected fuel, and
- The ignition timing,

the system takes into account the extensive range of current operating data as monitored by its sensors:

Auxiliary functions

ME-Motronic complements these basic functions with a wide spectrum of supplementary open and closed-loop control functions, including:
- Idle-speed control,
- Lambda closed-loop air-fuel mixture control,
- Control of the evaporative-emissions control system,
- Exhaust-gas recirculation (EGR) for reductions in NO_X emissions,
- Control of the secondary-air injection to reduce HC emissions, and
- Cruise control.

These secondary functions have been rendered essential by a combination of

factors. While these include legal mandates for reduced exhaust emissions and a continued demand for further enhancements of fuel economy, they also embrace higher expectations now directed toward safety and driving comfort.
The system can also be expanded to incorporate the following supplements:
- Turbocharger and intake-manifold geometry-control functions (→ to enhance power output),
- Camshaft control for engines with variable valve timing (→ to enhance power output while simultaneously reducing both fuel consumption and exhaust emissions), and
- Knock control, and engine-speed control and vehicle-speed control (→ to protect engine and vehicle).

Torque-based control concept

The prime objective behind this torque-control strategy is to correlate this large and highly variegated range of objectives. This is the only way to allow flexible selection of individual functions for integration in the individual Motronic versions according to engine and vehicle type.

Torque coordination

Most of the above auxiliary open and closed-loop control functions exercise a feedback effect on engine torque. This frequently leads to the simultaneous appearance of mutually conflicting demands. In a torque-based system, all of these functions reflect the driver's behavior in that they demand a specific engine torque. ME-Motronic's flexible-response torque-based control system can prioritize these mutually antagonistic requirements and implement the most important ones. This is the advantage of the torque-based structure. All functions submit individual and independent requests for torque.

Vehicle management

The CAN (Controller Area Network) bus system allows Motronic to maintain communications with the various control units governing other systems in the vehicle. One example of this cooperation is the way Motronic operates with the automatic transmission's ECU to implement torque reductions during gear changes, thus reducing wear on the transmission. If TCS (traction control system) is installed, its ECU responds to wheelslip by transmitting the corresponding data to the Motronic unit, which then reduces engine torque.
This is yet another benefit resulting from flexible torque-based control.

Diagnosis (OBD)

ME-Motronic is complemented by components designed for on-board monitoring (OBD), allowing it comply with the stringent emissions limits and the stipulations for integrated diagnosis.

Fig. 1

ME-Motronic – Schematic diagram

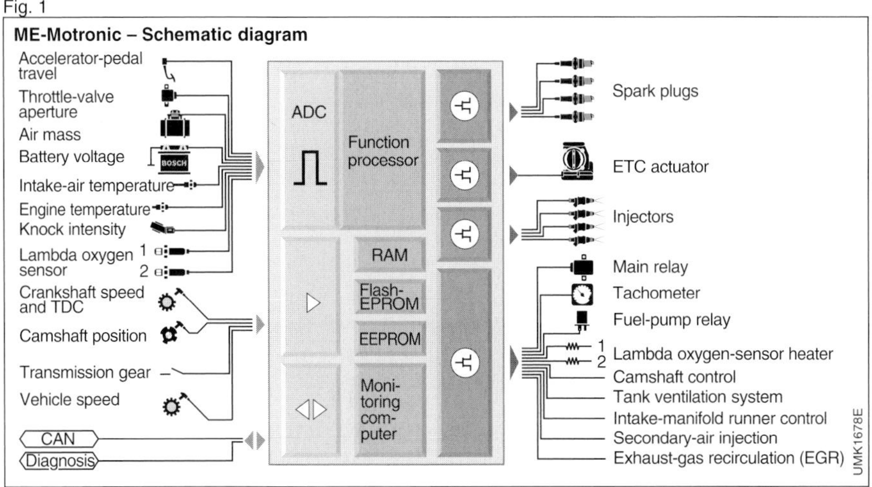

Fig. 2

ME-Motronic system diagram

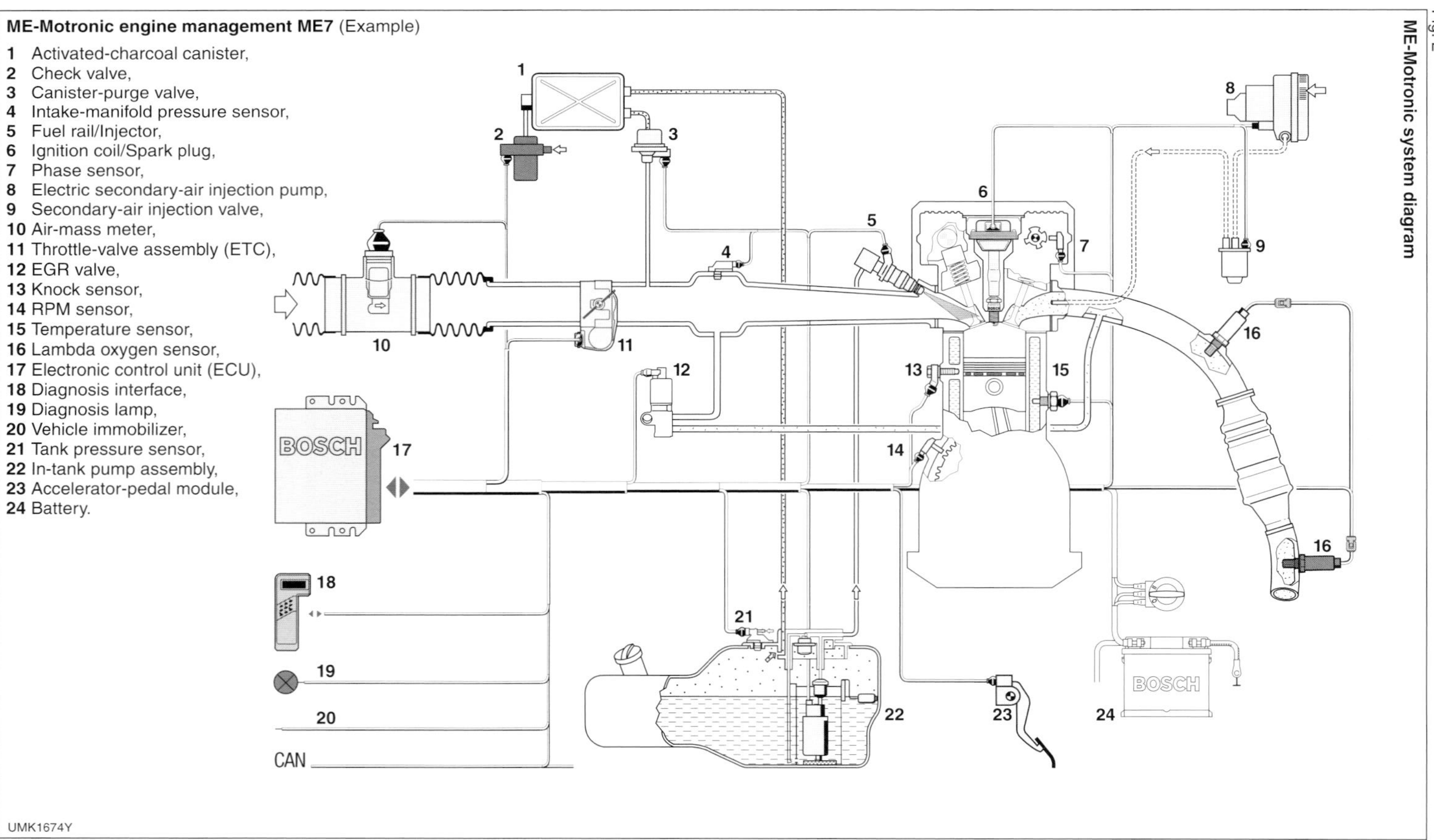

ME-Motronic engine management ME7 (Example)

1 Activated-charcoal canister,
2 Check valve,
3 Canister-purge valve,
4 Intake-manifold pressure sensor,
5 Fuel rail/Injector,
6 Ignition coil/Spark plug,
7 Phase sensor,
8 Electric secondary-air injection pump,
9 Secondary-air injection valve,
10 Air-mass meter,
11 Throttle-valve assembly (ETC),
12 EGR valve,
13 Knock sensor,
14 RPM sensor,
15 Temperature sensor,
16 Lambda oxygen sensor,
17 Electronic control unit (ECU),
18 Diagnosis interface,
19 Diagnosis lamp,
20 Vehicle immobilizer,
21 Tank pressure sensor,
22 In-tank pump assembly,
23 Accelerator-pedal module,
24 Battery.

Cylinder-charge control systems

Throttle-valve control

On spark-ignition engines with external mixture formation, the prime factor determining output force and thus power is the cylinder charge. The throttle valve controls cylinder charge by regulating the engine's induction airflow.

Conventional systems

Conventional layouts rely on mechanical linkage to control the throttle valve. A Bowden cable or linkage rod(s) translate accelerator-pedal travel into throttle-valve motion.

To compensate for the cold engine's higher levels of internal friction, a larger air mass is required and supplementary fuel must be injected. Increased air flow is also required to balance drive-power losses when ancillaries such as air-conditioning compressors are switched on. This additional air requirement can be met by an air-bypass actuator, which controls a supplementary air stream routed around the throttle valve (Figure 2). Yet another option is to use a throttle-valve actuator designed to respond to demand fluctuations by readjusting the throttle valve's minimum aperture. In both cases, the scope for electronic manipulation of airflow to meet fluctuating engine demand is limited to certain functions, such as idle control.

Principle of air control using air bypass valve

1 Idle valve (bypass valve), **2** ECU, **3** Throttle valve, **4** Bypass tract.

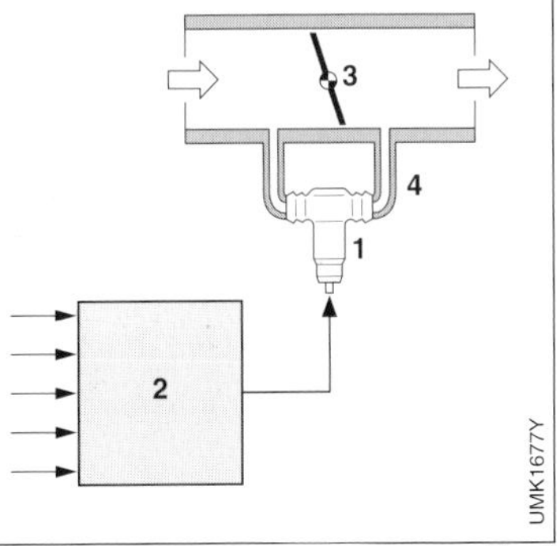

Fig. 2

Systems with ETC

In contrast, ETC (electronic throttle control) employs an ECU to control throttle-valve travel. The throttle valve forms a single unit along with the throttle-valve actuator (DC motor) and the throttle-valve angle sensor: This is the throttle-valve assembly (Figure 1).

Two mutually-opposed potentiometers monitor accelerator-pedal travel as the basis for controlling this type of throttle-

Fig. 1

ETC system

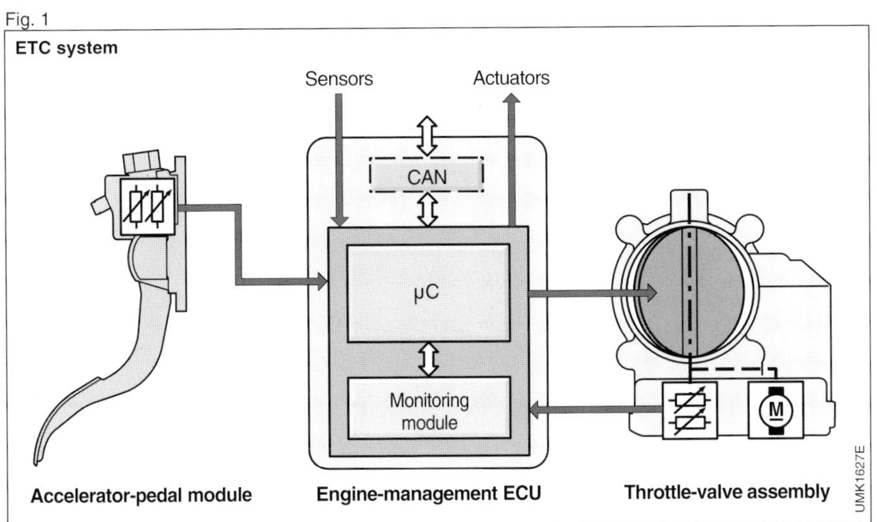

valve assembly. The ECU calculates the throttle-valve aperture that corresponds to the driver's demand, makes any adjustments needed for adaptation to the engine's current operating conditions, and then generates a corresponding trigger signal for transmission to the throttle-valve actuator. The throttle-valve travel sensor with its two mutually-opposed potentiometers permits precise and accurate response to positioning commands.
The dual-potentiometer setup at the throttle valve is complemented by dual potentiometers to monitor accelerator pedal travel; this arrangement serves as an integral part of the overall ETC monitoring function by furnishing the desired system redundancy. This subsystem continuously checks and monitors all sensors and calculations that can affect throttle-valve aperture whenever the engine is running. The system's initial response to malfunctions is to revert to operation based on redundant sensors and process data. If no redundant signal is available, the throttle valve moves into its default position.
The ME-Motronic system integrates ETC control within the same engine-management ECU used to govern the ignition, injection and numerous auxiliary functions. This renders retention of a special dedicated ECU for ETC unnecessary. Figure 3 shows the components in an ETC system.

Gas-exchange control

Although throttle-valve control represents the primary method of regulating the flow of fresh air into the engine, a number of other systems are also capable of adjusting the mass of fresh and residual gases in the cylinder:
- Variable valve timing on both the intake and exhaust sides,
- Exhaust-gas recirculation (EGR),
- Variable-geometry intake manifold (dynamic boost), and
- Exhaust-gas turbocharger.

Variable valve timing

In defining valve timing it is important to recognize that fluctuations in factors such as engine speed and throttle-valve angle induce substantial variations in the flow patterns of the gas columns streaming into and out of the cylinder. This means that when invariable (fixed) valve timing

Fig. 3

ETC components

1 DV-E5 throttle-valve assembly,
2 Engine-management ECU,
3 Accelerator-pedal module (FPM).

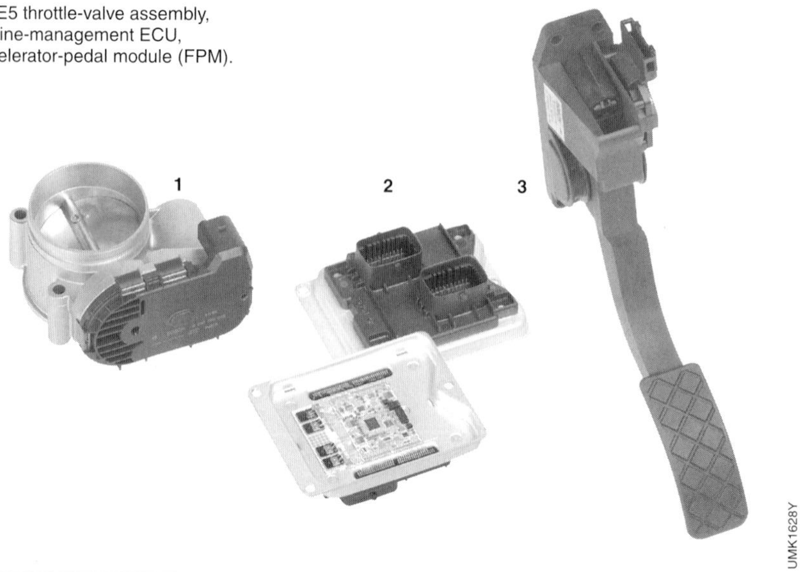

1 2 3

UMK1628Y

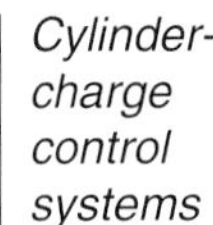

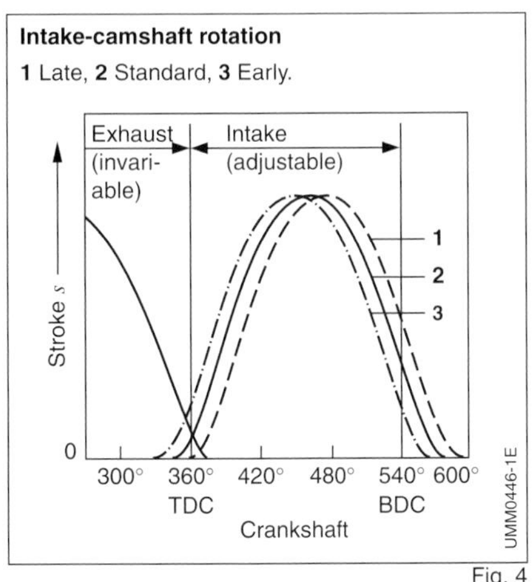

Intake-camshaft rotation
1 Late, **2** Standard, **3** Early.

Fig. 4

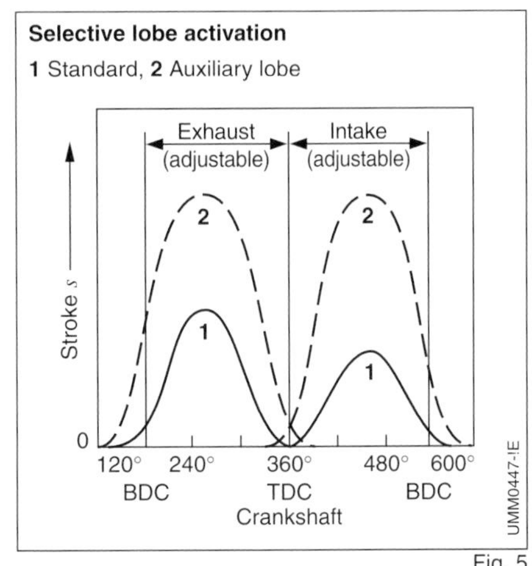

Selective lobe activation
1 Standard, **2** Auxiliary lobe

Fig. 5

is used the gas-exchange process can only be optimized for a single operating status. In contrast, variable valve timing can be employed to adapt gas flow to various engine speeds.

Camshaft adjustment

In conventional engines a chain or toothed timing belt serves as the mechanical link between the crankshaft and camshaft(s). On engines with adjustable camshaft, at least the angle of the intake-camshaft relative to the crankshaft can be varied. Nowadays, adjustment of both intake and exhaust camshafts relative to the crankshaft is being increasingly encountered. The adjustment process relies on electric or electrohydraulic actuators. Figure 4 shows how the open phase of the intake valve "shifts" relative to TDC when the intake camshaft's timing is modified. One option is to turn the camshaft to retard the "intake opens/closes" phase at idle to reduce residual gases and obtain smoother idling.
At high engine speeds the intake valve's closing point can be delayed to obtain maximum charge volumes. The same objective can be achieved at low to moderate engine speeds and/or in specific part-throttle ranges by varying the timing of the intake camshaft and shifting the entire intake cam-phase forward (advanced "intake opens/ closes").

Selective camshaft-lobe control

Systems with selective camshaft-lobe control modify valve timing by alternately activating cam lobes with two different ramp profiles.
The first lobe furnishes optimal valve timing and lift for intake and exhaust valves at the low end and in the middle of the engine's operating range.
A second cam lobe is available for increased valve lift and extended phase durations. This lobe is activated when the rocker arm to which it is connected locks onto the standard rocker arm once the engine crosses a specific speed threshold (Figure 5).
Infinitely-variable valve timing and valve-lift adjustment represents the optimum, but it is very complicated. This concept employs extended cam lobes featuring three-dimensional ramp profiles in conjunction with linear shifts in camshaft position, and grants maximum latitude for perfecting engine performance (Figure 6). This strategy can be used to obtain substantial torque increases throughout the engine's operating range.

Exhaust-gas recirculation (EGR)

Variable valve timing, as already mentioned in the section covering this subject, represents one way of influencing the mass of the residual gas remaining in the cylinder following combustion; this process is referred to as "internal" EGR.

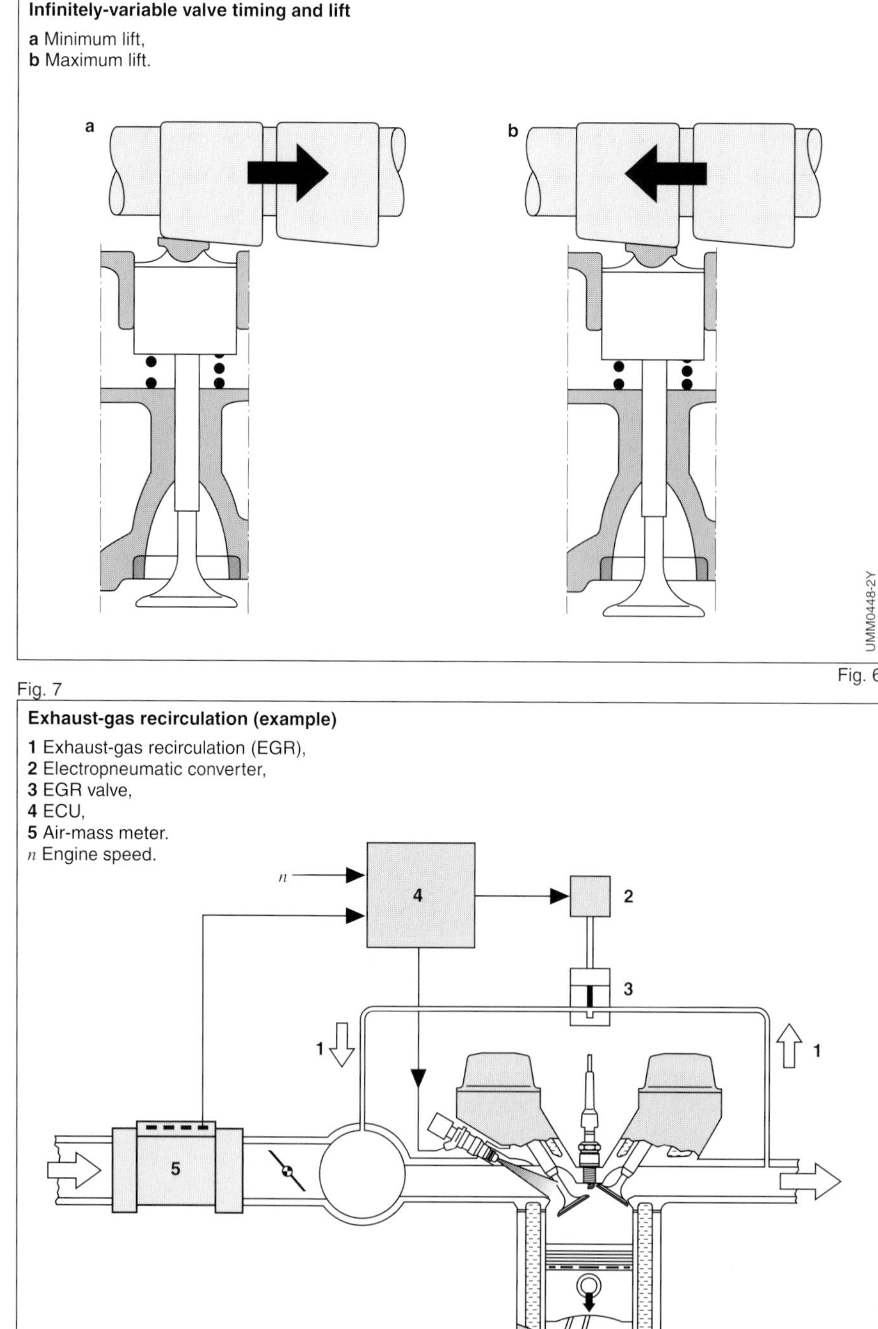
Infinitely-variable valve timing and lift
a Minimum lift,
b Maximum lift.
a
b
UMM0448-2Y
Exhaust-gas recirculation (example)
1 Exhaust-gas recirculation (EGR),
2 Electropneumatic converter,
3 EGR valve,
4 ECU,
5 Air-mass meter.
n Engine speed.
n
4
2
3
1
1
5
UMK0913-1Y

Fig. 6

Fig. 7

Another option available for varying the proportion of residual gases is to apply "external" EGR. Motronic controls this process by modifying the EGR valve's lift to reflect current engine operating conditions (Figure 7). The EGR system taps into the exhaust and, via the EGR valve, diverts a portion of the gases back into the fresh mixture. This is how the EGR valve defines the residual-gas component in the cylinder charge.
Exhaust-gas recirculation is an effective way to reduce emissions of nitrous oxides. Adding previously combusted exhaust gases to the fresh air-fuel mixture lowers peak combustion temperatures, and because generation of nitrous oxides is temperature-sensitive, NO_X emissions are reduced at the same time.
Assuming that the mass of the fresh-air charge remains constant, the overall charge will increase when exhaust gas is recirculated. This means the engine will produce the same torque at wider throttle-valve apertures (less throttling effect). The result is enhanced fuel economy.

Dynamic pressure-charging

Because maximum possible torque is proportional to fresh-gas cylinder charge, maximum torque can be raised by compressing the intake air before it enters the cylinder.
The gas-exchange process is not governed solely by valve timing; intake and exhaust-tract configuration are also important factors. Periodic pressure waves are generated inside the intake manifold during cylinder intake strokes. These pressure waves can be exploited to boost the fresh-gas charge and maximize possible torque generation.
Intake manifolds for multipoint injection systems consist of individually tuned runners and the plenum chamber with its throttle valve. Careful selection of the length and diameter of runners and of plenum chamber dimensions can be employed to exploit pressure waves in the air traveling through the intake tract. This strategy can be used to increase the density – and with it the mass – of the fresh-gas charge.

Intake wave ram effect

The pressure waves generated by the reciprocating piston propagate through the intake runners and are reflected at their ends. The idea is to adapt the length and diameter of the runners to the valve timing in such a way that a pressure peak reaches the intake valve just before it closes. This supplementary pressurization effect increases the mass of the fresh gas entering the cylinder.

Resonance pressure charging

Resonance boost systems employ short runners as links between groups of cylinders with equal ignition intervals and resonance chambers. These, in turn, are connected via resonance tubes to the atmosphere or a plenum chamber, allowing them to act as Helmholtz resonators (Figure 8). Sometimes the required plenum volumes are considerable, and the resulting storage effect can have a negative influence on dynamic response (in the form of mixture vacillation during sudden changes in load).

Fig. 8

Resonance boost

a Layout, **b** Air flow path.
1 Resonance tube, **2** Resonance chamber,
3 Cylinder, **4** With resonance boost,
5 With standard intake manifold.

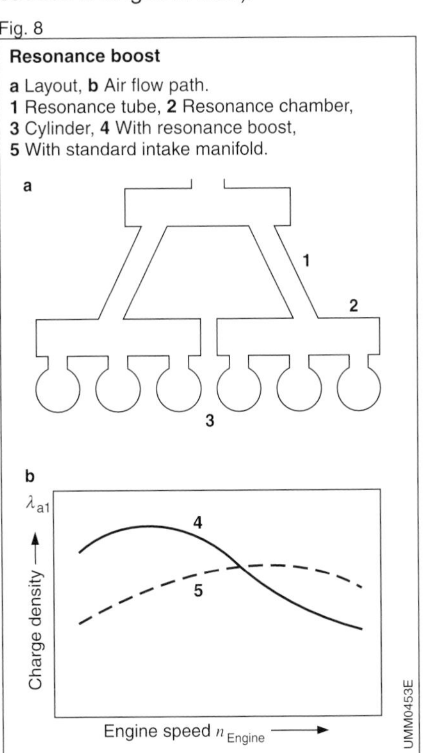

Variable-geometry intake systems

Switchable: **a** two-stage, **b** three-stage.
A, B Cylinder groups: **1, 2** Flaps,
engine speed determines opening point.

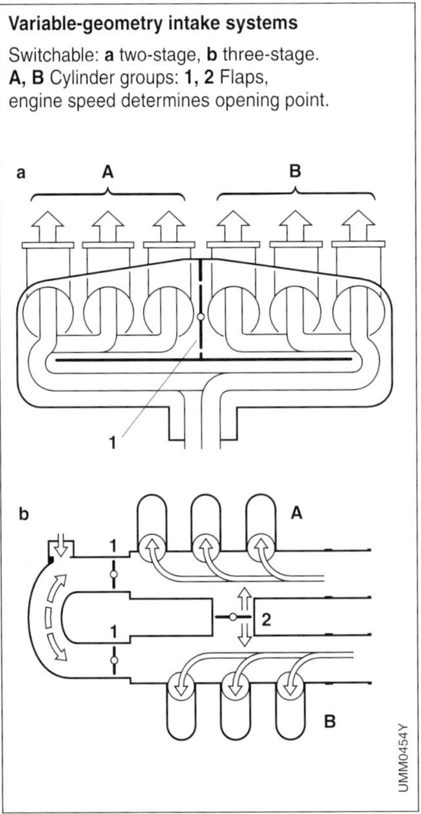

Fig. 9

Brake mean effective pressure (bmep) over engine speed for three different lengths of an infinitely-variable intake system

L_1 Effective intake-runner length,
D_1 Intake-runner diameter.

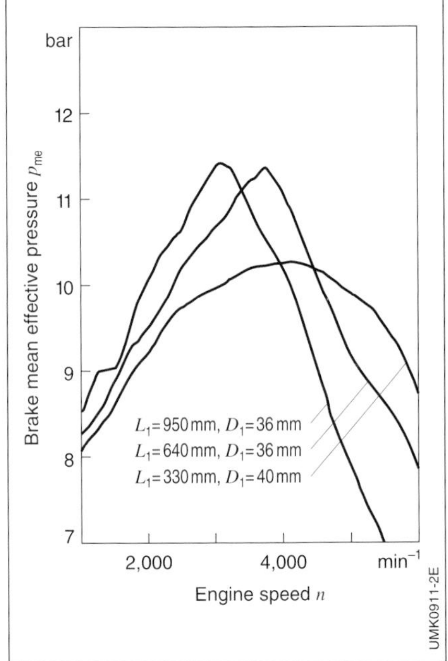

Fig. 11

Fig. 10

Infinitely-variable-length intake system

1 Fixed housing,
2 Rotating drum (air distributor),
3 Drum air-entry orifice,
4 Intake runner air-entry orifice,
5 Seal (e.g., leaf spring),
6 Intake runners,
7 Intake valve,
8 Induction air flow.

UMK0910-2Y

Variable-geometry intake manifold

Both of the dynamic charge-flow enhancement concepts described above are suitable for increasing the maximum available charge volume, especially at the low end of the engine-speed range.

However, the variable-geometry intake manifold (selective intake-tract activation) can be used to obtain a virtually ideal torque curve. This concept opens up numerous possibilities, with adaptive response to variations in such factors as engine load factor, engine speed and throttle-valve angle:

- Adjustment of runner length,
- Alternation between different runners of varying lengths and diameters,
- Selective deactivation of individual runners leading to individual cylinders in multitract systems, and
- Alternating selection of different plenum-chamber volumes.

For switching such variable-geometry air-intake systems, as engine speed varies flaps or similar devices are used to open and close the connections between the groups of cylinders (Fig. 9).

At low rpm the variable-length intake runners operate in combination with the initial resonance chamber. The length of the runners is then subjected to ongoing modification as engine speed rises before the process culminates in the opening of a second resonance chamber (Figure 10).

Figure 11 provides an index of air-flow efficiency by illustrating how variable intake-manifold geometry influences brake mean effective pressure (bmep) as a function of engine speed.

Exhaust-gas turbocharging

Yet another and even more effective option in the quest for higher induction-gas density levels is to install a supercharging device. Among the familiar techniques for furnishing forced induction on spark-ignition engines, the most widespread is turbocharging. Turbocharging makes it possible to extract high torques and powers from engines with minimal piston displacement by running them at high levels of volumetric efficiency. Compared to naturally-aspirated engines, if we presume identical power outputs, it is the turbocharged powerplant's lower weight and more compact dimensions that represent its salient advantages.

Until recently turbochargers were primarily viewed as a way to raise power-to-weight ratios, but today the focus is shifting toward achieving torque increases in the low and mid-range sectors of the engine's rev spectrum. This strategy is very often applied in conjunction with electronic boost-pressure control.

The exhaust-gas turbocharger's primary components are the impeller and the turbine, which are mounted at opposite ends of a common shaft.

The energy employed to drive the turbocharger is extracted from the engine's exhaust stream. Although, on the one hand, this process exploits energy that in naturally-aspirated engines would otherwise be wasted (owing to the inherent expansion limits imposed by the crankshaft assembly), on the other, the turbo process generates higher exhaust back-pressures as the price for impeller power.

The exhaust turbine converts a portion of the exhaust-gas' energy into rotational energy to drive the impeller, which draws in fresh air before compressing it and dispatching it through the intercooler, throttle-valve and intake manifold on its way to the engine.

Turbocharger boost-control actuator

Even at the bottom end of the rev range, passenger-car engines are expected to produce high torque. In response, turbine housings are basically designed for efficient operation with modest exhaust-gas mass flow rates, i.e., for WOT as low as $n \leq 2{,}000\ \text{min}^{-1}$.

To prevent the turbocharger from overboosting the engine when the exhaust-gas stream rises to higher mass flow rates, the unit must incorporate a bypass valve. This bypass valve, or

Fig. 12

Cross section of a Roots-type supercharger

1 Housing,
2 Rotor.

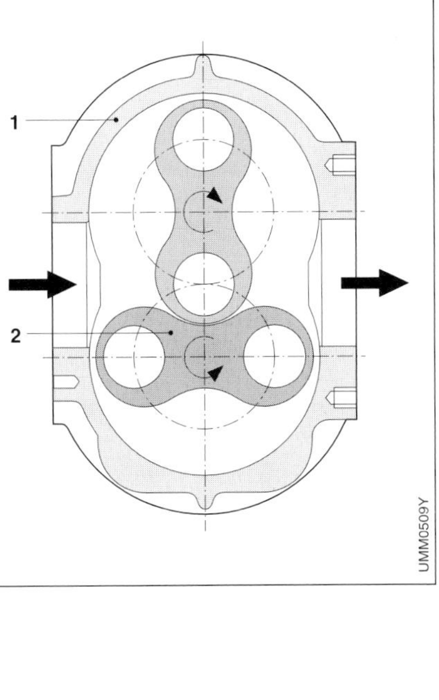

wastegate, diverts a portion of the exhaust gases in these higher ranges, routing them around the turbine and feeding the gas back into the exhaust system on the other side. This bypass valve is generally a flap valve integrated within the turbine housing, although less frequently it is in the form of a plate valve in a separate casing parallel to the turbine.

Mechanical supercharging

The motive force used to power the mechanical compressor is taken directly from the IC engine. Mechanical superchargers are available in a variety of designs, and Figure 12 shows the Roots compressor. A belt from the crankshaft usually drives the supercharger at a fixed ratio, so crankshaft and supercharger rotate at mutually invariable rates. One difference relative to the turbocharger is that the supercharger responds immediately to increases in rpm and load factor, with no lag while waiting for the impeller to accelerate. The result is higher engine torque in dynamic operation.

This advantage is relativized by the power needed to turn the compressor. This must be subtracted from the engine's effective net output, and leads to somewhat higher fuel consumption. However, Motronic can lessen the significance of this factor by controlling operation of a compressor clutch to switch off the supercharger at low rpm and under light loads.

Fig. 1

Fuel-supply system with return line

1 Fuel tank,
2 Electric fuel pump,
3 Fuel filter,
4 Fuel-pressure regulator,
5 Injector.

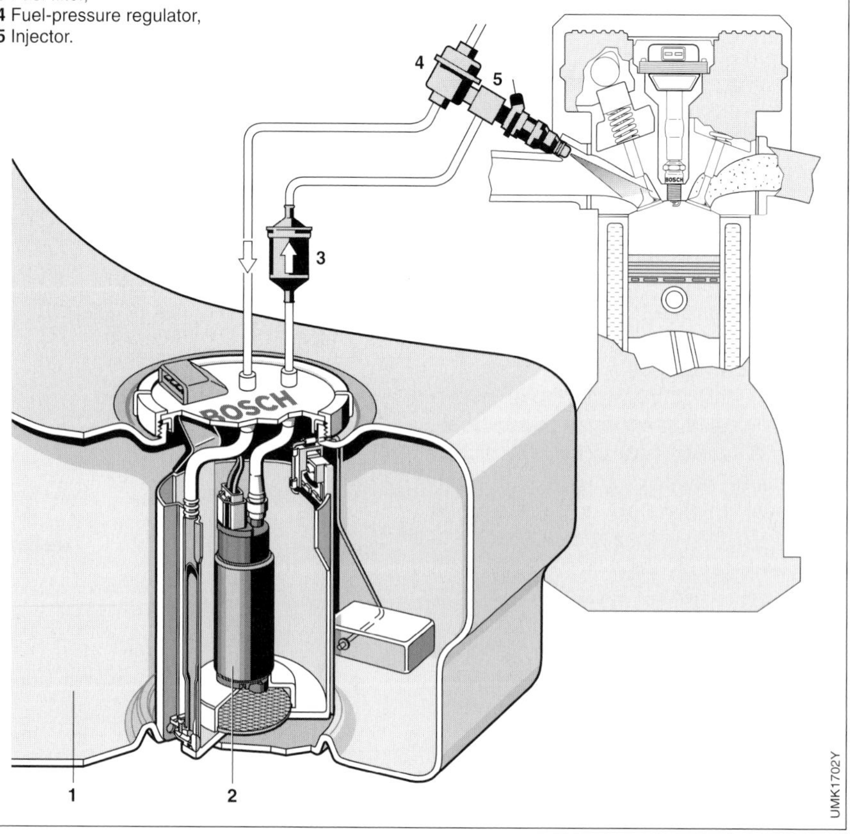

EKP 13.5 electric fuel pump

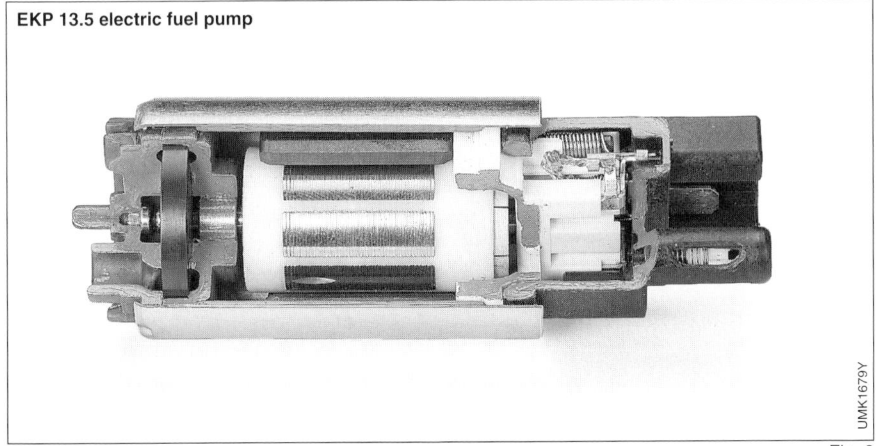

Fig. 2

Fuel system

Fuel supply and delivery

Systems with and without return lines

The fuel system's function is to ensure a consistently reliable supply of the fuel mass needed to meet the engine's requirements under any and all operating conditions. An electric pump draws the fuel from the tank and forces it through a filter for delivery to the fuel (distribution) rail with its solenoid-triggered injectors. The injectors then spray the fuel into the engine's intake manifold in precisely metered quantities. In systems equipped with a return line the excess fuel then flows through the pressure regulator and back to the fuel tank (Figure 1). Until quite recently this layout represented the state-of-the-art, but now returnless fuel-supply systems are becoming increasingly common.

Both systems rely on an electric fuel pump to provide the fuel circuit with a continuous supply of fuel from the tank. A pressure regulator, operating at a typical supply pressure of 300 kPa, maintains system pressure by controlling the flow of fuel returning to the tank. The high pressure inhibits formation of undesirable vapor bubbles in the fuel.

In the returnless system the pressure regulator is mounted immediately adjacent to the pump. This means that the fuel return line from the engine can be omitted, thus lowering production costs as well as fuel temperatures within the tank. These lower temperatures mean lower hydrocarbon emissions, and promote improved performance from the evaporative-emissions control system.

Electric fuel pump

The electric fuel pump maintains a continuous flow of fuel from the tank. It can be installed either within the tank itself (in-tank) or mounted externally in the fuel line (in-line).

The in-tank pumps currently in general use (Figure 2 shows the EKP 13.5 as a representative example) are integrated within pump assembly units along with the fuel-gauge sensor and a swirl baffle designed to remove vapor from the fuel in the return line. Hot-delivery problems in systems with in-line pumps can be solved by installing a supplementary booster pump within the tank to maintain a low-pressure fuel flow up to the main pump unit. The pump's maximum delivery capacity always exceeds the system's theoretical maximum requirement to ensure that it remains consistently capable of maintaining system pressure under all operating conditions.

The fuel pump is switched-on by the engine-management ECU. An interrupt circuit or software-based function stops fuel delivery whenever the engine is stationary with the ignition switched on.

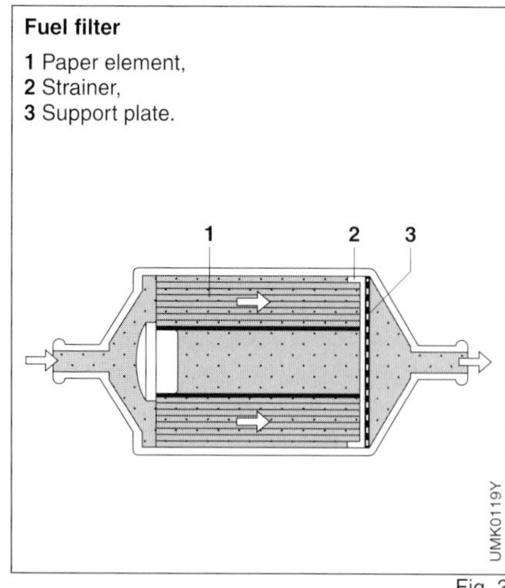

Fig. 3

Fig. 4

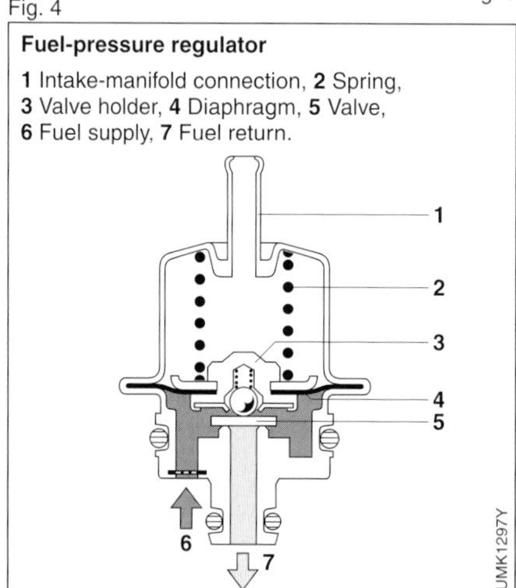

Fig. 5

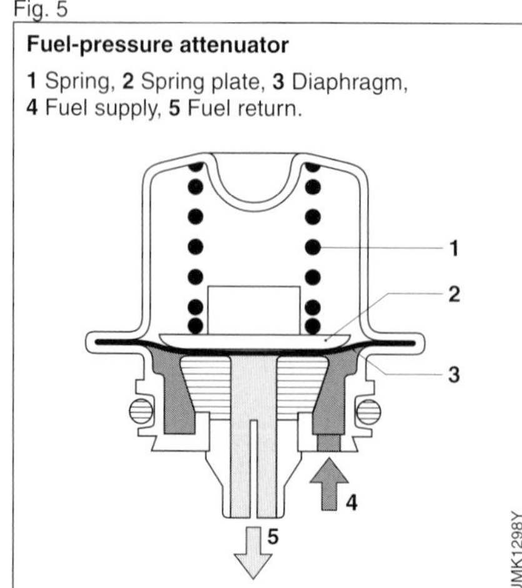

Fuel rail

The fuel then flows through the fuel rail, whence it is evenly distributed to all injectors. The injectors are mounted on the fuel rail, while rails in systems with a return line also include a fuel-pressure regulator. A pressure attenuator may also be installed. Fuel rails are carefully dimensioned to prevent local fuel-pressure variations caused by resonances which occur when the injectors open and close. This prevents irregularities in injection mass flow that might otherwise arise during load and rpm transitions. Depending upon the particular vehicle type and its special requirements, fuel rails can be manufactured in steel, aluminum or plastic. The rail may also incorporate a test valve, which can be used to bleed pressure during servicing as well as for test purposes.

Fuel filter

Fuel-borne contaminants can impair the operation of both pressure regulator and injectors. A filter is therefore installed in the line downstream from the electric fuel pump. This fuel filter contains a paper element with a mean pore diameter of 10 µm (Figure 3).

Fuel-pressure regulator

Injection quantities are determined by injection duration and the pressure differential between the fuel in the rail and the intake-manifold pressure.

Return-line systems utilise a pressure regulator to maintain the pressure differential between fuel system and intake manifold at a constant level. This pressure regulator regulates the amount of fuel returning to the tank to maintain a constant pressure drop across the injectors (Figure 4). In order to ensure that the fuel rail is efficiently flushed with fuel, the pressure regulator is generally mounted at its far end.

In systems without a return line the pressure regulator is installed within the in-tank pump assembly, whence it maintains the pressure within the fuel rail at a constant level relative to ambient pressure. As this implies, the system does not maintain a constant pressure differential between rail and manifold, so injection duration must be calculated accordingly.

Fuel-pressure attenuator

The injectors' cyclical operating phases and the periodic fuel-discharge characteristic of the positive-displacement fuel pump both induce pressure waves in the fuel system. Under unfavorable circumstances the mounts on the electric fuel pump, the fuel lines and the rail itself can transmit these vibrations to the fuel tank and to the vehicle's body. Noise stemming from these sources can be inhibited through the use of specially designed mounts and with fuel-pressure attenuators. The fuel-pressure attenuator (Figure 5) shares its general design layout with the pressure regulator, with a spring-loaded diaphragm separating the fuel from the air space in both cases.

Fuel injection

Uncompromising demands for running refinement and low emissions make it vital to provide consistently excellent air-fuel mixture quality in every engine cycle. The system must inject fuel in quantities precisely metered to reflect the induction air's mass (duration), while the start of injection is also a significant factor (timing). This is why multipoint injection systems feature a solenoid-operated injector for each individual cylinder. At the injection point specified by the ECU the injector sprays a precisely metered quantity of fuel into the area directly in front of the cylinder's intake valve(s). Thus condensation along the walls of the intake tract leading to undesired deviations from the prescribed lambda values is largely avoided. Because the engine's intake manifold conducts only combustion air, its geometry can be optimized solely on the basis of the engine's dynamic gas-flow requirements.

Solenoid injector

Design and function

The essential components of the injector are

- The valve casing with solenoid winding and electrical connection,
- Valve seat with injector-nozzle disc, and
- Reciprocating valve needle with solenoid armature.

A filter screen in the fuel supply line guards the valve against contamination, while two O-rings seal the injector against the fuel rail and the intake manifold. The spring and the force of fuel pressure against the valve seat insulate the fuel-supply system from the intake manifold for as long as the coil remains without current flow (Figure 6).
When the injector's solenoid winding is energized, the coil responds by

Fig. 6

EV6 injector design

1 O-rings,
2 Strainer,
3 Valve housing with electrical connection,
4 Coil winding,
5 Spring,
6 Valve needle with solenoid armature,
7 Valve seat with hole plate.

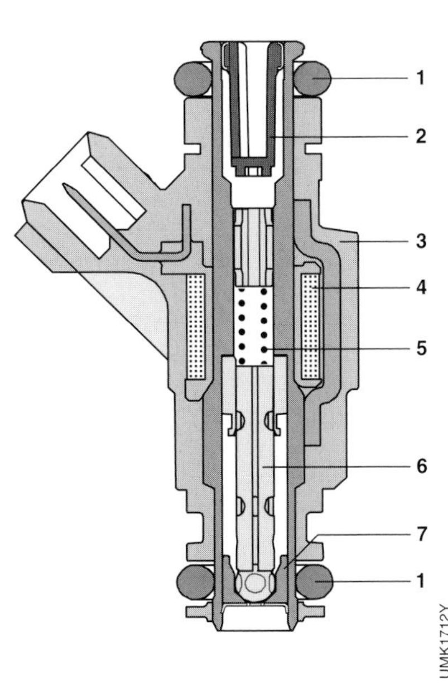

generating a magnetic field. This lifts the armature, the valve needle rises from the seat, and fuel flows through the injector. System pressure and the exit aperture defined by the orifices in the injector nozzle are the primary factors in determining the injected fuel quantity per unit of time. The valve needle closes again as soon as the trigger current ceases to flow.

Designs

Four injector configurations are employed to cover virtually all current Motronic systems (Figure 7):

– The EV1 injector has been in production since the early 70s. Thanks to ongoing development, this injector continues to satisfy all of the essential requirements encountered in modern gasoline-injection systems.

– The EV6 is based on the EV1. It features superior hot-starting for improved performance in returnless fuel-supply systems. This is important, considering the fact that the fuel supplied to the injector is hotter in a returnless system than it is in systems with fuel return. There are also a range of different EV6 installation lengths available.

– So-called "air shrouding" can be added to further improve mixture formation with the EV6 (refer to Figure 9).

– The EV12 has been developed from the EV6. For intake manifolds having difficult geometries, the injection point can be shifted forward 20 mm to obtain an ideal position.

Spray formation

The injector's spray pattern as determined by discharge geometry and angle, as well as droplet size, influence the air-fuel mixture formation process. Different spray-patterns are needed to adapt the injector for use with specific geometrical configurations in cylinder head and intake manifold. Various spray concepts are available to meet these requirements (Figure 8).

Tapered spray pattern

After being discharged through individual orifices in the nozzle plate, the emerging fuel streams converge to form a tapered spray cloud. A pintle at the bottom end of the needle and extending beyond the nozzle can also be employed to obtain a tapered form. Engines with a single

Fig. 7

View of different injectors

a EV1 injector,
b EV6 injector,
c EV12 injector.

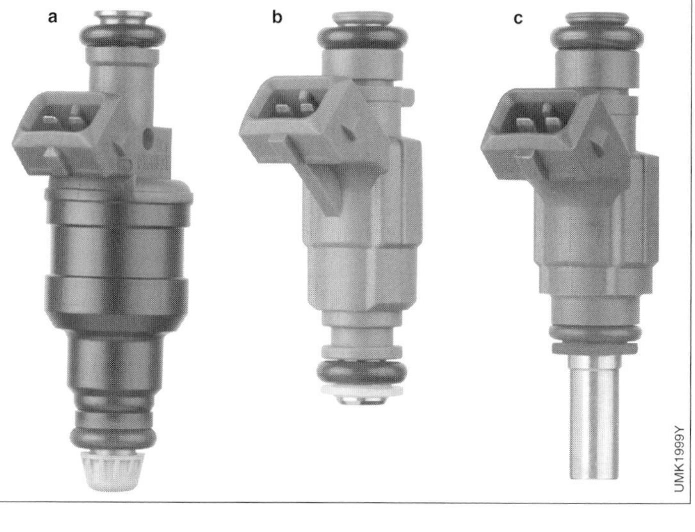

intake valve are a typical application area for tapered-spray injectors, which focus their discharge toward the opening between the intake valve and the wall of the intake manifold.

Dual-stream injector

Dual-stream patterns are employed in engines with two intake valves. The nozzle plate's holes are arranged to concentrate the emerging fuel in two spray patterns. Each of these clouds serves one intake valve.

Air shrouding

Air-shrouded valves exploit the pressure differential between intake manifold and the ambient atmosphere to enhance the quality of mixture formation. Air is conducted through a shroud in the discharge region of the nozzle plate. This air accelerates to extremely high velocities as it travels through this narrow passage before then emerging and promoting more intense atomization in the air-fuel mixture (Figure 9).

Fig. 8

Discharge patterns

a Tapered pattern,
b Dual-stream pattern.

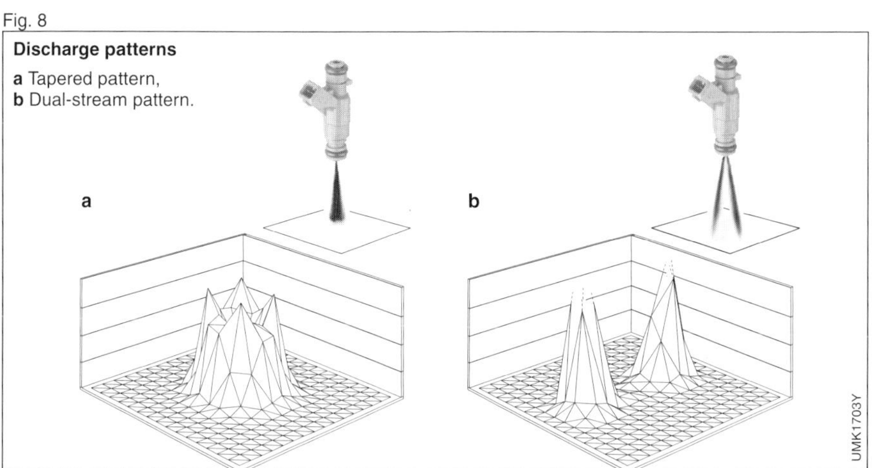

Fig. 9

EV6 with external air shroud

1 Air,
2 Fuel,
3 Air-fuel mixture.

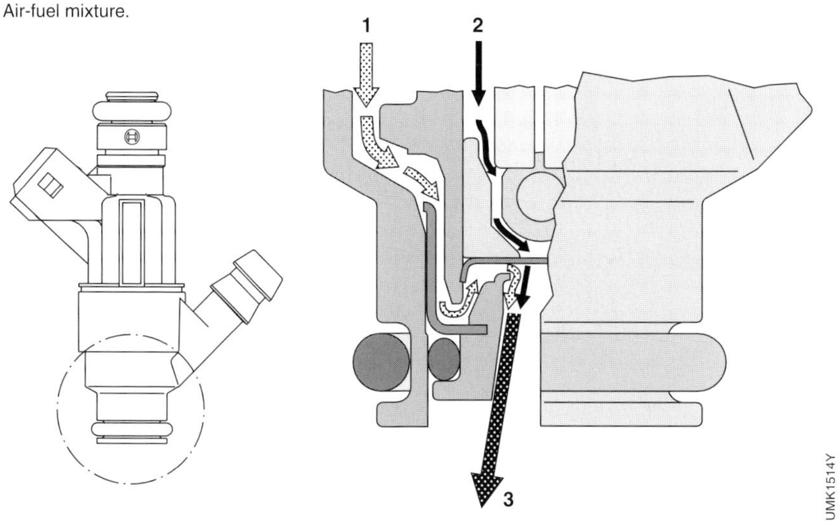

1
2
3
UMK1514Y

Operating-data acquisition

Driver demand

There is no mechanical linkage to connect the accelerator pedal with the throttle valve in engine-management systems with ETC throttle control. Instead, the accelerator pedal's position is monitored by an accelerator-pedal travel sensor for transformation into an electrical signal within the pedal module. The engine-management system interprets this signal as "driver demand."
The accelerator-pedal module is an operational unit unifying all required pedal functions and mechanical components. Although this concept renders all in-vehicle adjustment redundant, the fact that a wide variety of individual installation geometries together with very cramped conditions are frequently encountered often makes it necessary to resort to specific configurations according to vehicle type.
The accelerator-pedal travel sensor incorporates a redundancy feature (two potentiometers) to facilitate diagnosis and ensure backup operation. The sensors operate with separate reference voltages from mutually independent sources, and their signals are also processed separately within the engine ECU.

Air charge

Engines with design concepts based on manifold injection display a linear relationship between air-charge density and the force generated in the combustion process, which, in turn, corresponds to the engine's load factor. This is why air charge is more than simply a primary parameter for calculating injection quantity and ignition timing in the ME-Motronic system. In a torque-based system such as ME-Motronic, cylinder charge also serves as the basis for calculating instantaneous engine torque generation.

Using the following sensors, the ME-Motronic monitors cylinder charge with the following sensors:
- Hot-film air-mass meter (HFM),
- Intake-manifold pressure sensor (DS-S),
- Ambient-pressure sensor (DS-U),
- Boost-pressure sensor (DS-L, on turbocharged powerplants), and
- Throttle-valve sensor (DKG).

Concepts for monitoring induction charge vary according to engine, and all sensors are not present in all systems. Values for those parameters that are not directly monitored are derived from other, monitored data.

HFM5 hot-film air-mass meter

The hot-film sensor is a "thermal" air-flow monitor. Basically speaking, this meter (or sensor) is positioned somewhere between air filter and throttle valve,

Fig. 1

Sensor element in hot-film air-mass meter

1 Electrical connections, **2** Electrical terminals, **3** Evaluation electronics, **4** Air intake, **5** Sensor element, **6** Air discharge, **7** Casing.

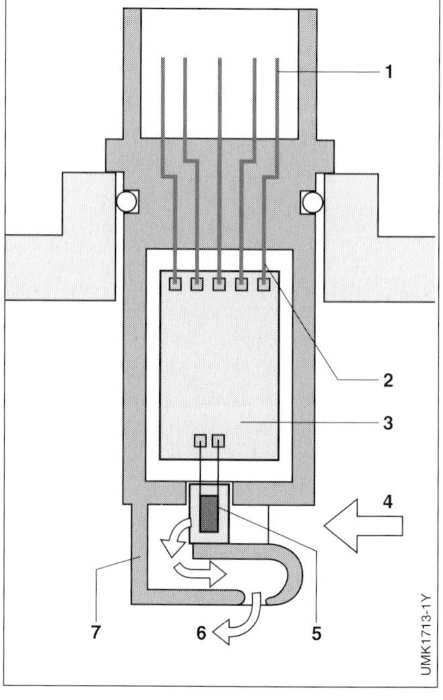

although it can for example be located as an "insert" in the air-filter housing or in a sensor tube in the air-intake tract. Figure 1 shows the meter's design.

The air-flow meter must register the engine's mass air intake (kg/h) with the utmost precision. At high load factors, in particular, there is a tendency for reverse flow pulses from the piston to generate waves in the air upstream from the throttle valve. These reverse undulations should not be allowed to detract from the air-mass meter's monitoring accuracy.

The HFM5 hot-film air-mass meter is a micromechanical device featuring a hot zone heated to a specific temperature. Temperatures drop on each side of this zone. If no air flow is present, then the descending thermal gradients in the sectors on each side of the hot zone will be identical. Air-flow leads to a more radical temperature progression curve for the induction side, in response to the cooling effect that the incoming air exerts on the sensor element. Although the air flowing past the opposite (engine) side of the sensor also has a cooling effect, the air warmed by the heater element ultimately raises relative temperatures on this side. The result is the thermal progression pattern in the illustration. Temperatures T_1 and T_2 are found at the two monitoring points M_1 and M_2, respectively, with the actual differential being determined by induction-air mass. This differential ΔT is then converted into a voltage.

This monitoring concept also makes it possible to register reverse flow, a situation reflected by T_2 being lower than T_1.

Fig. 2

Monitoring concept of hot-film air-mass meter

1 Temperature curve, no air flow,
2 Temperature curve, with air flow,
3 Sensor, **4** Heated zone, **5** Diaphragm,
6 HFM5 with measurement tube, **7** Air current.
M_1, M_2 Scan points, T_1, T_2 Temperatures,
ΔT Temperature differential generates sensor signal

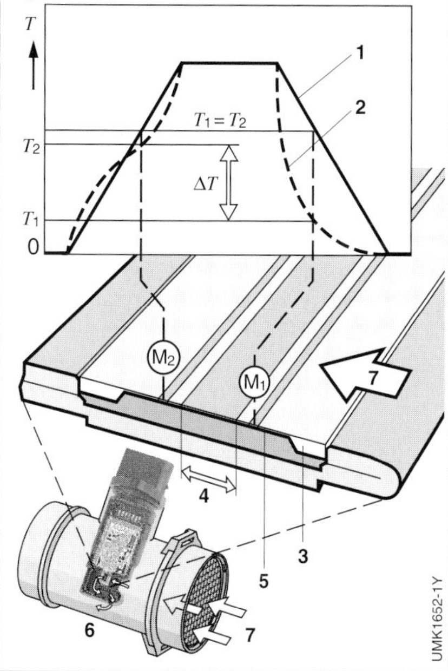

HFM5 signal voltage over air mass

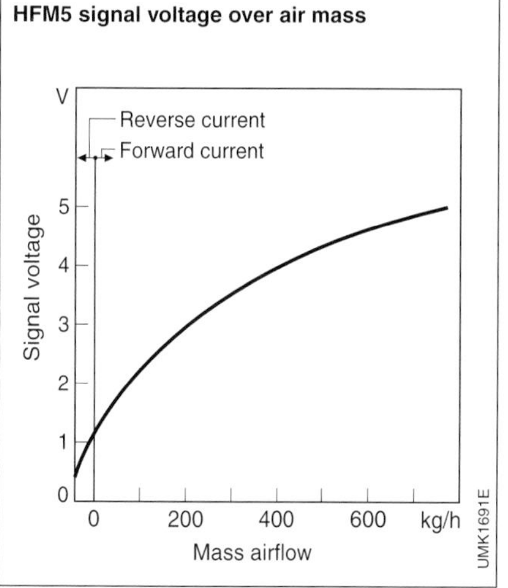

Fig. 3

The temperature differential is a direct index of induction-air mass. The relationship between air mass and the voltage generated by the processing circuit defines the sensor's characteristic curve (Figure 3), which has sectors for both forward and reverse flow. Monitoring precision is further enhanced by using a voltage provided by the Motronic ECU as a reference for the sensor signal. The characteristic curve has also been plotted to aid the Motronic's integrated diagnosis in recognizing problems such as open circuits.

A separate intake-air temperature sensor can also be integrated in the system.

Pressure sensor (for integration in ECU)

1 Pressure connection,
2 Pressure cell with sensor elements,
3 Seal rim,
4 Evaluation circuit,
5 Thick-layer hybrid (ceramic substrate).

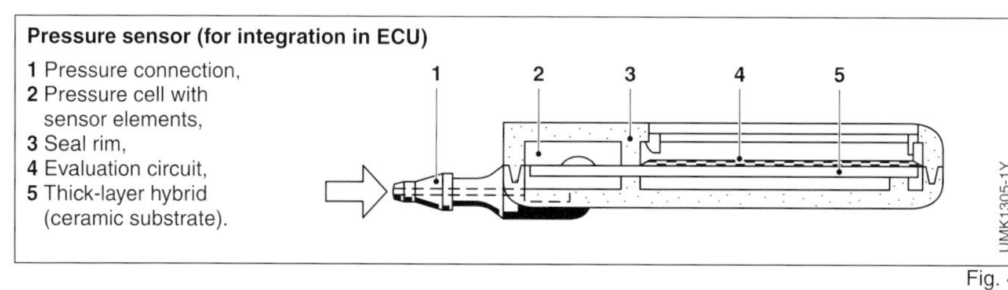

Fig. 4

The sensor tube is available in a variety of diameters designed to reflect the engine's specific airflow requirements.

Intake-manifold pressure sensor

A pneumatic link allows the pressure sensor to monitor absolute pressure levels (kPa) within the intake manifold. The sensor's monitoring range extends from 2...115 kPa (20...1,150 mbar).
This micromechanical sensor is available either for integration within the ECU or as a remote sensor for local mounting on or near the manifold.
This sensor, with its pneumatic link to the manifold, consists of a pressure cell with two sensor elements and a space for the processing circuitry, both of which are mounted on a single ceramic substrate (Figure 4). The processing circuit amplifies the electrical voltages generated at various levels as a reflection of pressure variations. It also compensates for thermal effects and converts the pressure curve to linear form. The signal from the processing circuit is transmitted to the ECU.

Ambient-pressure sensor

The ambient-pressure sensor shares its design with the manifold-pressure sensor, but is located inside the ECU. Barometric pressure readings are needed by systems that rely on throttle-valve aperture instead of an air-mass meter to monitor the incoming air supply. The precise data on ambient air density registered by this sensor serve as a significant factor in numerous diagnosis functions.

Boost-pressure sensor

In order to regulate boost pressure, this must first of all be measured. This function is taken over by a pressure sensor with a measurement range extending up to 250 kPa (2,500 mbar).

Throttle-valve sensor

ME-Motronic uses the throttle valve to control engine output. The throttle-valve sensor is employed to verify that the throttle valve responds to commands by assuming the specified angle (closed-loop position control).
The throttle-valve sensor forms an integral component within the throttle-valve assembly, with system redundancy provided by sensors incorporating two separate potentiometers operating with individual reference voltages.

Engine speed, crankshaft and camshaft angles

Engine speed and crankshaft angle

The momentary piston position (travel) within the cylinder serves the system as one parameter for defining the ignition firing point. Because each piston in each cylinder is joined to the crankshaft via a connecting rod, the crankshaft or engine-speed sensor can provide data indicating piston travel in each cylinder. The rate at which the crankshaft changes its position is the engine speed, quantified in the number of crankshaft revolutions per minute (rpm). Engine speed, which is yet another vital operating parameter for Motronic, is also calculated from crankshaft position.

Generating the crankshaft-angle signal

Installed on the crankshaft is a ferromagnetic sensor rotor with a theoretical capacity for 60 teeth, but with a 2-tooth gap, giving an actual sequence of 58 teeth. This 58-tooth sequence is scanned by an inductive rpm sensor featuring a soft-iron core and a permanent magnet (Figure 5). The sensor's magnetic field responds to the passing teeth by generating AC voltage. The amplitude of this voltage rises radically at progressively higher engine speeds. Adequate amplitudes for operation are available at minimal engine speeds, extending as low as 20 min^{-1}.

Tooth and sensor-pole geometries must be precisely matched. The processing circuit in the ECU converts the sinus-wave voltage with its highly inconsistent amplitudes into a constant-amplitude square-wave pattern.

Determining crankshaft angle

The flank data for the square-wave voltage are transmitted to the computer through an interrupt input. A gap in the tooth pattern is registered at those points where the flank interval is twice as large as in the previous and subsequent sectors. This tooth gap corresponds to a specific crankshaft angle, with cylinder no. 1 as the reference. The computer uses this point as its crankshaft synchronization reference. It then counts 6 degrees further for each subsequent negative tooth flank. Because the ignition must be triggered based on more minute increments, the interflank periodicity is divided by eight. This new unit can then be multiplied by one, two, three, etc., and appended to a flank position as the basis for determining the firing point, allowing timing adjustments in increments of 0.75 degrees.

Fig. 5

Inductive RPM sensor

1 Permanent magnet, **2** Housing, **3** Engine block, **4** Soft-iron core, **5** Winding, **6** Toothed rotor with reference point (gap).

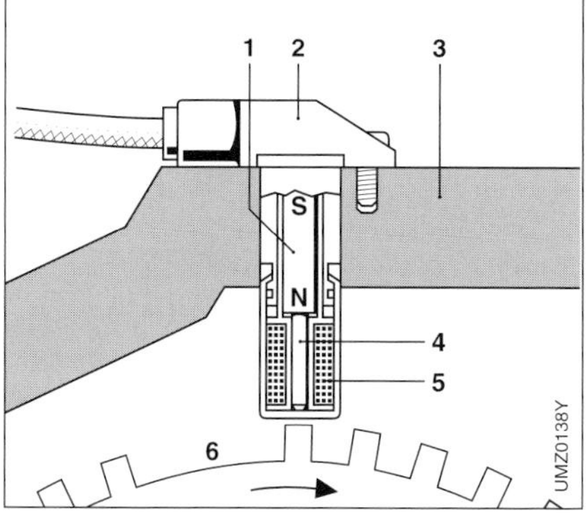

Calculating segment duration and engine speed from engine-speed sensor signal

The cylinder offset in the four-stroke engine results in two crankshaft revolutions (720 degrees) elapsing between the start of each new cycle on cylinder no. 1. This period defines the mean ignition interval, and the intermediate duration is the segment time T_S. Equal distribution of the intervals results in:

Table 1

Interval	Degrees	Teeth
2 cylinders	360	60
3 cylinders	240	40
4 cylinders	180	30
5 cylinders	144	24
6 cylinders	120	20

Ignition, injection and the engine speed derived from the signal duration are recalculated with each new interval. The figure for rotational speed describes the mean crankshaft rpm within the segment period and is proportional to its reciprocal.

Camshaft position

The camshaft controls the engine's intake and exhaust valves while rotating at half the rate of the crankshaft.

When a piston travels to top dead center, it is the positions of the intake and exhaust valves, as defined by the camshaft, that determine the cylinder's actual cycle phase. TDC can mark the end of the compression phase and the start of ignition, or, alternatively, the end of the exhaust stroke. Crankshaft data alone cannot indicate which.

Motronic systems with static (distributorless) ignition and dedicated coils – such as ME-Motronic – differ from

systems with rotating spark distribution by requiring supplementary data on cycle phase. The ECU must decide which coil and spark-plug combination are to be triggered, and it derives this information from the camshaft's position.
Data for camshaft position are also needed by systems that adapt injection timing for each cylinder individually, such as those with sequential injection.

Hall-sensor signal
Camshaft position is usually monitored with a Hall sensor. This consists of a Hall element with a semiconductor wafer through which current flows. This ferromagnetic Hall element responds to activation by a trigger wheel rotating in unison with the camshaft by generating voltage at right angles to the direction of the current passing through it.

Determining camshaft position
As the Hall voltage lies in the millivolt range, it must be processed in the sensor before being transmitted to the ECU in the form of a switching signal. The basic procedure is for the microprocessor to respond to trigger-wheel gaps by checking for Hall voltage and determining whether or not cylinder no. 1 is on its power stroke.
Special trigger-wheel designs make it possible to use the camshaft signal as an emergency back-up in the event of crankshaft sensor failure. However, the resolution provided by the camshaft signal is much too low to allow its use as a permanent replacement for the crankshaft rpm sensor.

Mixture composition

Excess-air factor λ

The lambda excess-air factor (λ) quantifies the relative masses of the air and fuel in the mixture. Optimal performance is obtained from the catalyst at $\lambda = 1$. Because the lambda, or O_2 sensor, monitors the concentration of oxygen in the exhaust gas, its signals are an index of the excess-air factor lambda (λ).

Oxygen (lambda) sensor

The oxygen sensor (lambda probe) is a "dual-threshold" unit capable of indicating both rich ($\lambda<1$) and lean ($\lambda>1$) mixtures. The radical transitions that characterize this sensor's response curve (Figure 7) facilitate mixture adjustments to achieve $\lambda=1$.
The wide-band sensor provides information on the current excess-air factor, and can also be used to maintain richer and leaner mixtures.

Two-state oxygen (lambda) sensor based on the Nernst concept
The outside of the oxygen sensor's electrode extends into the exhaust stream, while the inside remains in contact with the surrounding air (Figure 6). The essential component of the oxygen sensor is a special ceramic body featuring gas-permeable platinum electrodes on its surface. Sensor operation relies on the porous nature of the ceramic material, which allows oxygen in the air to diffuse (solid electrolyte). This ceramic becomes conductive when heated. Voltage is generated at the electrodes in response to differences in the oxygen levels on the inside and outside of the sensor. A stoichiometric air-fuel ratio of $\lambda = 1$ produces a characteristic jump (step function) in the response curve (Figure 7).

Fig. 6
Position of lambda oxygen sensor in exhaust pipe (schematic)
1 Ceramic coating, **2** Electrodes, **3** Contacts, **4** Housing contacts, **5** Exhaust pipe, **6** Ceramic shield (porous), **7** Exhaust gas, **8** Ambient air, *U* Voltage.

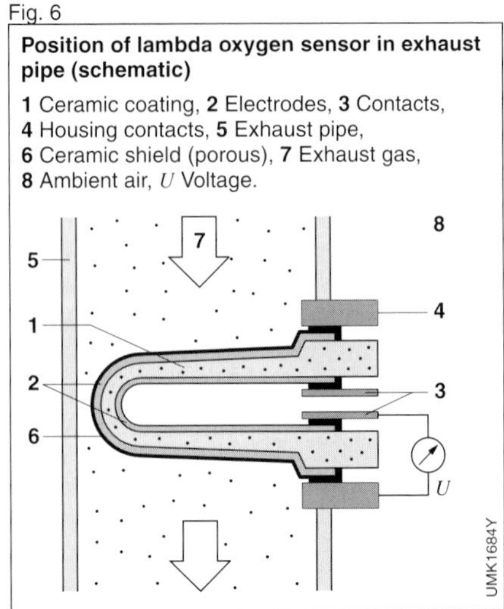

The oxygen sensor's voltage and internal resistance are both sensitive to temperature. Reliable control operation becomes possible once exhaust-gas temperatures exceed 350 °C with unheated sensors, or 200 °C with heated units.

Heated oxygen (lambda) sensor

The heated sensor reduces the waiting period between engine start and initiation of effective closed-loop control by providing reliable performance with cooler exhaust gas (e.g., at idle). Because heated sensors warm more quickly, they also have shorter reaction times, reducing lag in closed-loop operation. This type of sensor also offers greater latitude in the selection of installation positions.

A ceramic heater element warms the sensor's active ceramic layer from the center to ensure that the active material remains hot enough for operation, even while exhaust-gas temperatures remain modest.

The heated sensor is protected by a guard tube incorporating a restricted flow opening to prevent its ceramic material from being cooled through exposure to low-temperature exhaust gases.

Fig. 7

Two-state lambda oxygen sensor voltage curve at 600 °C operating temperature

a Rich mixture (air deficiency),
b Lean mixture (excess air).

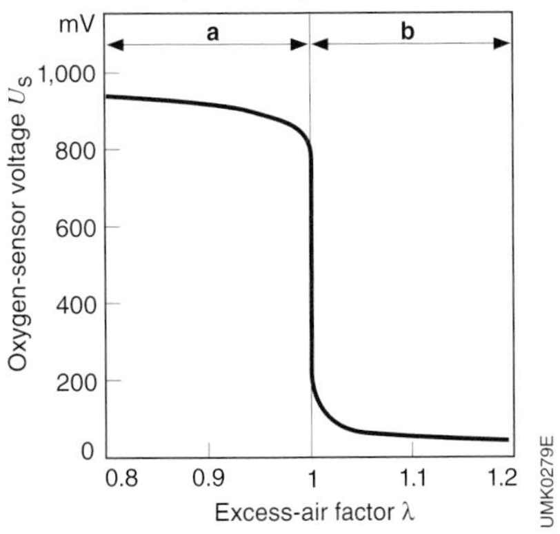

mV; Oxygen-sensor voltage U_S: 0, 200, 400, 600, 800, 1,000
Excess-air factor λ: 0.8, 0.9, 1, 1.1, 1.2
a, b
UMK0279E

Wide-band oxygen (Lambda) sensor

While the dual-threshold sensor signals rich and lean mixtures, the wide-band sensor can actually be used to measure the excess-air factor via the exhaust gas. This facility promotes improved dynamic response in the closed-loop control system, regardless of whether the momentary mixture specification is $\lambda = 1$.

The wide-band sensor expands on the principle of the Nernst cell by incorporating a second electrochemical cell, the pump cell. It is through a small slot in this pump cell that the exhaust gas enters the actual monitoring chamber (diffusion gap) in the Nernst cell. Figure 8 shows a schematic diagram of the sensor's design. This configuration contrasts with the layout used in the dual-threshold cell by maintaining a consistently stoichiometric air-fuel ratio in the chamber. Application of pumping voltage to the pump cell results in oxygen discharge when the exhaust gas is lean, and oxygen induction if it is rich. The resulting pumping current is an index of the excess-air factor in the exhaust gas. The basic progression of the curve for pumping current is portrayed in Figure 9. Lean exhaust leads to a positive pumping current, which is needed to maintain a stoichiometric composition in the atmosphere within the diffusion gap,

Fig. 8

Schematic portrayal of continuous wide-band lambda oxygen sensor and position of sensor in exhaust pipe

1 Nernst cell, **2** Reference cell, **3** Heater,
4 Diffusion gap, **5** Pump cell, **6** Exhaust pipe.

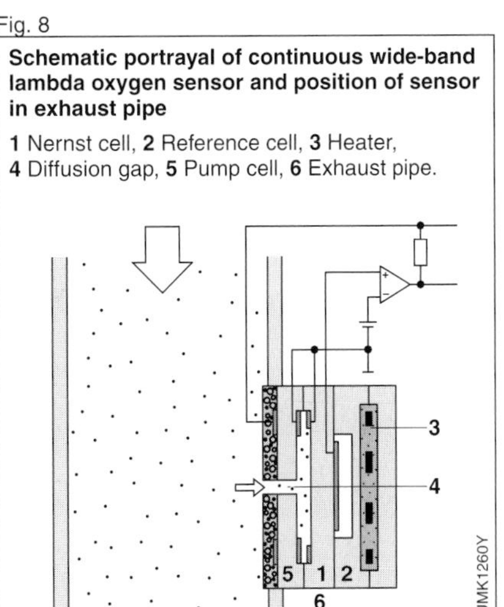

Pump current I_p in a wide-band lambda oxygen sensor over exhaust-gas air factor

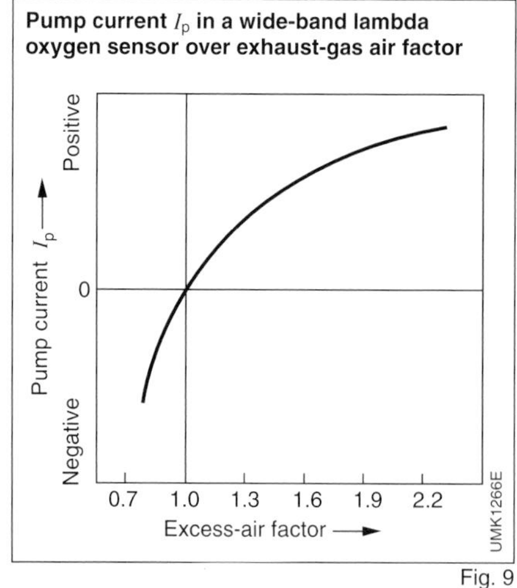

Fig. 9

while rich exhaust is indicated by a negative pumping current.

The wide-band sensor thus differs substantially from its dual-threshold counterpart. While the dual-threshold unit uses the voltage at the Nernst cell as a direct measurement signal, the wide-band sensor employs special processing and control circuitry to adjust the pumping current. The resulting current is then monitored as an index of the exhaust gas' excess-air factor. Because sensor operation is no longer dependent on the step-function response of the Nernst cell, air factor can be monitored as a continuous progression.

Combustion knock

Under certain conditions, combustion in the spark-ignition engine can degenerate into an abnormal process characterized by a typical "knocking" or "pinging" sound. This undesirable combustion phenomenon marks the outer limits of ignition timing advance, and thus, at the same time, defines the boundaries of power-generation potential and efficiency. It occurs when fresh mixture preignites in spontaneous combustion before being reached by the expanding flame front. During an otherwise normally initiated combustion event, the pressure and temperature peaks created by the piston's compressive force generate self-ignition in the end gas (remaining unburned mixture). This process can be accompanied by flame velocities in excess of 2000 m/s, as compared to speeds of roughly 30 m/s for normal combustion. This abrupt combustion leads to substantial pressure rises in the end gas, and the resulting pressure wave continues to propagate until halted by the cylinder walls.

Chronic preignition is characterized by pressure pulses and high thermal stresses acting on the cylinder-head gasket, the piston and around the valves. It can produce mechanical damage in all of these locations.

The characteristic oscillations induced by combustion knock can be monitored by knock sensors for conversion into electrical signals, which can then be transmitted to the Motronic ECU (Figures 10 and 11).

Both the number and installation positions of the knock sensors must be defined with the utmost care. Reliable knock detection must be guaranteed for all cylinders and under all operating conditions, with special emphasis on high engine speeds and load factors. As a general rule, 4-cylinder engines are equipped with one, 5 and 6-cylinder engines with two, 8 and 12-cylinder engines with two or more knock sensors.

Engine and intake-air temperatures

The engine-temperature sensor incorporates a thermally sensitive resistor which extends into the coolant circuit it monitors. Figure 12 shows the design structure of this sensor.

The electrical resistor is characterized by a Negative Temperature Coefficient (NTC), indicating that its electrical resistance is inversely proportional to temperature. Figure 13 shows the basic response curve of resistance over temperature. The NTC resistor forms part of a voltage divider circuit operating on a 5-volt power supply. The voltage from the NTC resistor varies with temperature, and an analog-digital converter registers these data as an index of thermal conditions in the coolant. Compensation for the non-linear relationship between voltage and temperature is provided by a table stored in the computer's database, which matches each reading with a corresponding temperature. The sensor in the intake tract monitors the temperature of the induction air according to the same concept.

Engine-temperature sensor

1 Electrical terminals, **2** Housing, **3** NTC resistor, **4** Coolant.

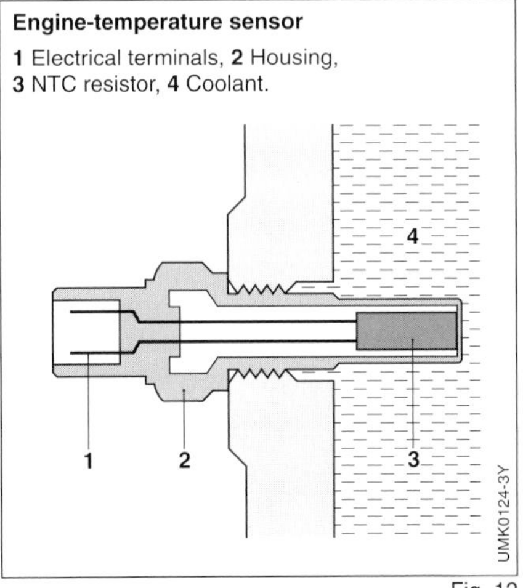

Fig. 12

Fig. 10

Knock sensor

1 Seismic mass, **2** Casting, **3** Piezoelectric ceramic layer, **4** Contact paths, **5** Electrical terminals.

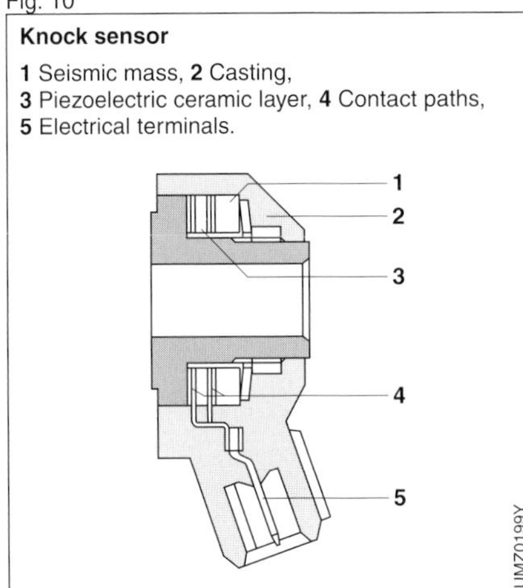

Fig. 13

Temperature sensor (NTC) response curve

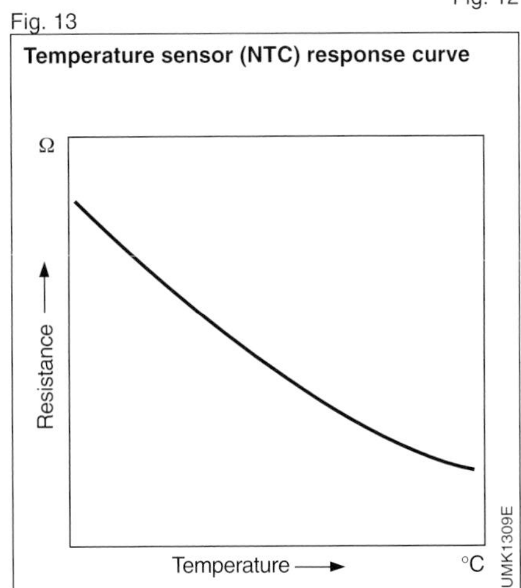

Fig. 11

Knock-sensor signals

The knock sensor transmits signal c which reflects pressure-wave pattern **a** in the cylinder. The filtered pressure signal is portrayed in **b**.

without knock

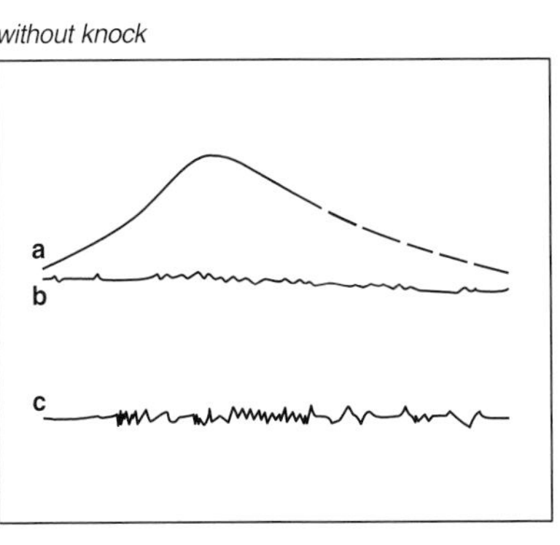

with knock

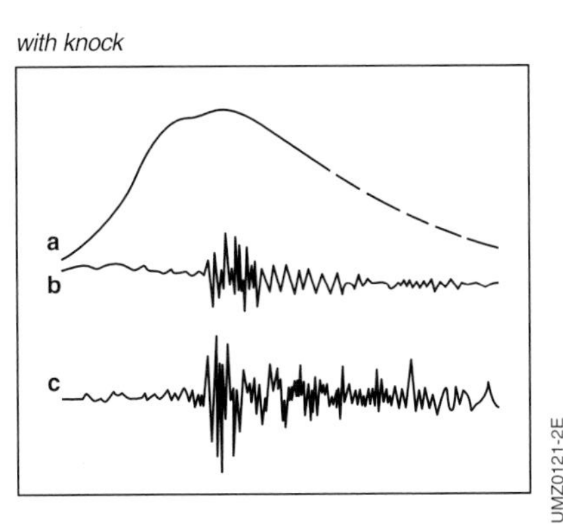

Operating-data processing

Torque-led control concept

Purpose

The engine-management system's primary assignment is to convert driver demand into engine power and torque. The driver needs this engine power to overcome running resistance during steady-state operation as well as for accelerating the vehicle.

On today's spark-ignition engines this entails calculating the air-charge requirement along with the corresponding injection quantity and optimal ignition timing. Once these parameters have been defined, the system can proceed to active control of the actuators that regulate them (throttle-valve assembly, injectors, ignition coils).

In addition to governing cylinder charge, injection and timing, engine-management systems have also assumed control of a number of auxiliary functions, many of which also consume engine power.

A distinguishing feature of ME-Motronic is its torque-led control concept. Numerous subcomponents within the overall Motronic system (idle control, rpm governor, etc.) join the systems controlling the drivetrain (e.g., TCS, transmission-shift control) and general vehicular functions (such as air-conditioner operation) in relaying requests for adaptation of current engine output to the basic Motronic system. To cite one example, the air-conditioner control system requests an increase in engine output prior to engaging the a/c compressor clutch.

Earlier practice entailed direct implementation of such commands at the control-parameter level (cylinder charge, fuel mass and ignition timing) on an uncoordinated, individual basis. ME-Motronic goes a step further by prioritizing and coordinating individual demands before using the available control parameters to implement the resulting specified torque (Figure 1). This coordinated control strategy makes it possible to obtain optimal emissions and fuel consumption from the engine at every operational coordinate.

An essential element within the torque-based control concept is the ETC electronic accelerator pedal which permits the throttle-valve control mechanism to advance beyond merely reflecting pedal inputs. Formerly the driver used the accelerator pedal to activate a mechanical linkage and determine throttle-valve aperture. The driver

Fig. 1

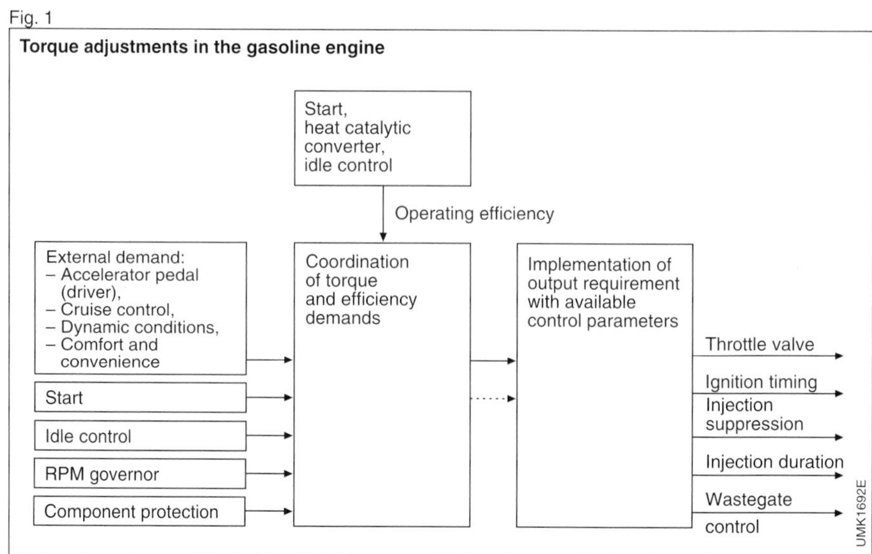

exercised direct control over the cylinder charge, and the engine-management system's options for initiating separate adjustments were limited to activation of a bypass routed around the throttle valve.

Calculating specified torque

The basic parameter underlying ME-Motronic's torque-led control concept is the internal torque produced during combustion. This is the physical force produced by gas pressure during the compression and power strokes. The engine's actual net torque production is obtained by subtracting such factors as friction, pumping (gas-transfer) losses and drive power for ancillary equipment (water pump, alternator, etc.) from this internal force.

The ultimate goal of the torque-led control system is to select precisely the engine control parameters needed for accurate response to driver demand while simultaneously compensating for all losses and supplementary requirements. Because Motronic "knows" the optimal specifications for charge density, injection duration and ignition timing for any desired torque, it can consistently maintain optimal emissions and fuel economy.

Adjustment of actual torque

ME-Motronic's torque coordinator has two potential control paths to choose from when regulating internal torque generation (Figure 2). One path, furnishing gradual reaction, is controlled by triggering the throttle valve (ETC), while the rapid-response path relies on manipulation of ignition timing and/or deactivating the injection at individual cylinders. The slower path, also known as the charge-control path, is responsible for static operation. The charge requirement calculated for a given torque generation determines the cylinder charge, which is then provided by the throttle valve. The rapid-response (ignition timing) path can react very quickly to dynamic variations in torque generation.

Calculating cylinder charge

The air mass within the cylinder following closure of the intake valve is the air charge. There is also a "relative (air) charge" which is independent of piston displacement. It is defined as the ratio of the current charge to a charge obtained under specified standard conditions (p_0 = 1,013 hPa, T_0 = 273 K).

The relative charge must be known in order to calculate injected fuel quantity.

Fig. 2

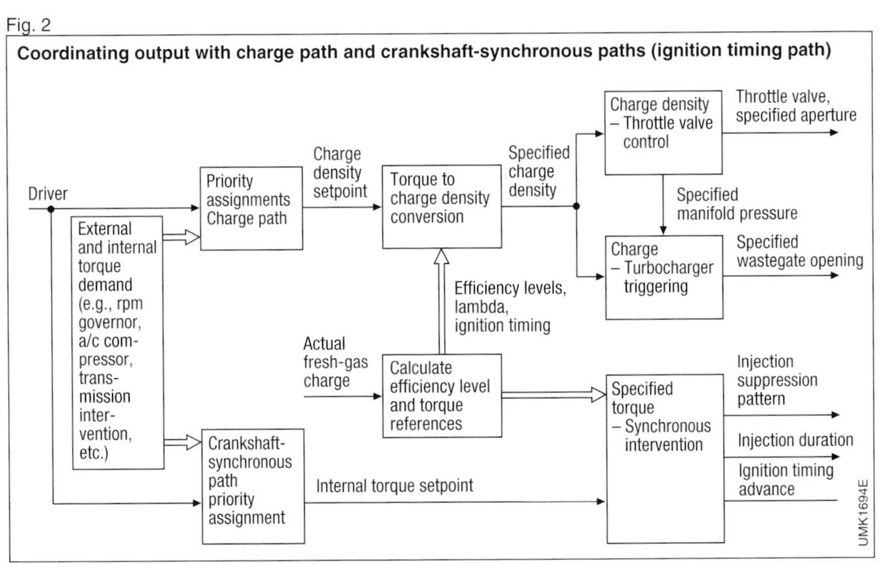

On today's spark-ignition engines it is also the primary parameter for influencing engine output, which is why it is incorporated in the torque structure as a control parameter. Because no means for directly monitoring charge density is available, it must be calculated from available sensor signals with the aid of a simulation model. The requirements for the charge model are:

- Precise determination of charge density under all operating conditions (dynamic, selective-flow intake manifold, variable valve timing, etc.),
- Accurate response to exhaust-gas components in systems with variable-rate EGR (controlled external or internal EGR),
- Calculation of the control command parameter for "throttle-valve aperture" corresponding to any given charge-density requirement.

Intake-manifold simulation model

The actual mass of the air within the cylinder is relevant for fuel metering and torque calculations. In the absence of a method to directly monitor cylinder charge, it is calculated using an intake-manifold simulation model. Depending on the system's induction-charge sensor (air-mass meter or intake-manifold pressure sensor), the raw data for this model is either monitored directly or simulated.

Induction air mass

Here, the decisive parameter is the mass of the incoming air. While charge density can be calculated directly from induction air mass during static engine operation, abrupt variations in throttle-valve aperture result in a time lag. This stems from the fact that (for example) the first response to an opening throttle valve is for the intake manifold to fill with air. A disparity exists between the air mass actually entering the combustion chamber and the air monitored by a sensor such as the hot-wire air-mass meter (HFM) for the duration of this lag period. It is only after pressure levels in the manifold start to rise that more air can flow into the combustion chamber.

Intake-manifold pressure

These considerations elevate manifold pressure to the status of a primary factor; the relationship between the relative cylinder charge – the relevant factor – and the intake-manifold pressure can be portrayed using a linear equation (Figure 3).

The linear equation's offset is defined by the partial pressure emanating from internal residual gases, making it a function of valve overlap, rpm and ambient barometric pressure, while the gradient is determined by engine speed, valve overlap and combustion-chamber temperature.

Other flow into the manifold

A supplementary mass flow, over and above the air entering through the throttle valve, results from activation of such systems as the evaporative-emissions control. The regeneration flow required by this system can be varied with the aid of a tank vent valve (purge valve). With manifold pressure as a known quantity, it is possible to calculate the regeneration flow for use in the intake-manifold model simulation process.

Monitoring charge density with the HFM

When an HFM is fitted this directly measures the air mass entering the intake manifold. This process entails

Fig. 3

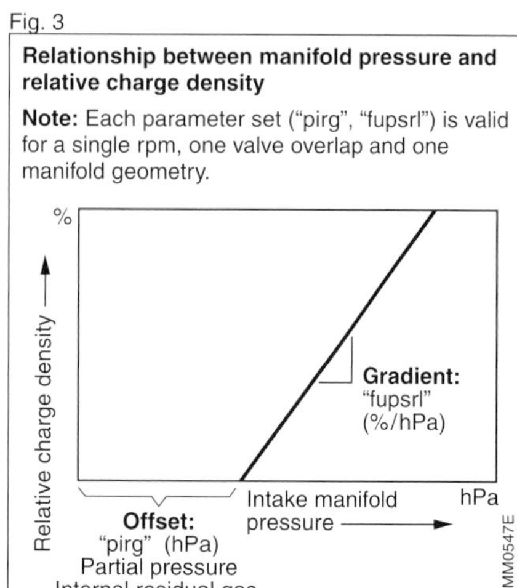

multiplying the mean mass airflow monitored during an intake stroke (segment) by the intake stroke's duration for conversion into a relative charge density. The other parameters required for intake-manifold model simulation (such as intake-air temperature) are either monitored directly or calculated in the modeling process (in this case intake-manifold pressure, but also secondary parameters such as combustion-chamber temperature).

Monitoring charge density with a manifold-pressure sensor

If an intake-manifold pressure sensor is present to act as a "primary charge-density sensor" then manifold pressure can be monitored directly. The system calculates the mass of the air entering the intake manifold based on manifold pressure.

Cylinder charge control

The density of the charge entering the cylinder is also controlled with the intake-manifold model, exploiting the fact that gas flow through valves (in this case the throttle valve) can be formulated as an equation. The main factors are the entry pressure immediately in front of the throttle valve, the pressure drop, the temperature and the effective aperture; all parameters calculated with the intake-manifold model. Other parameters relevant for specific throttle valves (such as friction losses in the air current) must be quantified using test-stand measurements.

Now the intake-manifold model can be "turned around" to calculate a throttle-valve aperture from the desired cylinder charge density (which has been calculated by ME-Motronic's torque-led control facility). This aperture is transmitted to the throttle-valve actuator's position controller as a command value.

Calculating injection timing

Calculating injection duration

Cylinder-charge density can be used as the basis for calculating the fuel mass required to obtain a stoichiometric air-fuel ratio. The injector constant, which varies according to injector design, can then be incorporated into the calculations to produce the injection duration.

Injection duration is also affected by the differential between the fuel's supply pressure and injection counterpressure. The standard fuel supply pressure is 300 kPa (3 bar). This pressure can be maintained using any of a variety of reference sources. Fuel-supply systems with return lines maintain constant supply pressures relative to manifold pressure. This strategy ensures that the pressure differential across the injectors remains constant in the face of changing manifold pressures, so that roughly consistent flow rates result. Returnless fuel systems rely on a different concept, maintaining their 300 kPa supply pressure relative to ambient pressure. Fluctuations in the pressure within the intake manifold produce variations in the differential between its own pressure and that of the fuel supply. A compensation function corrects this potential error source.

As the injectors open and close they induce pressure waves in the fuel-supply system. This leads to flow-rate inconsistencies when the injector is opened. An adaptation factor correlated with engine speed and injection duration is used to compensate.

The opening duration calculated up to this point will be valid if we assume that the injector has already opened and is discharging fuel at a constant flow rate, but the injector's opening time must also be considered in real-world operation. This opening duration displays significant variations depending on the voltage being supplied by the battery. There may be substantial lag before the valve opens completely, especially in the starting phase or when the battery is partially discharged. A supplementary injection duration based on battery voltage is added to the base duration to compensate for this effect.

Excessively short injection durations would lend disproportionate influence to the valve opening and closing times. This is why a minimum injection duration is defined to guarantee precise fuel metering. This minimal duration is less than the injection period required for minimum potential cylinder charging.

Injection timing
Optimal combustion depends on correct injection timing as well as precise metering. The fuel is usually injected into the intake manifold while the intake valve is still closed. Termination of the injection period is defined by something known as the injection advance, which is indicated in crankshaft degrees, and uses intake-valve closure as a reference. The injection duration can then be correlated with engine speed to obtain a point for initiating injection defined as an angle. Current operating conditions are also reflected in the calculations to define the injection advance angle.
ME-Motronic triggers an individual injector for each cylinder, making it possible to preposition a separate fuel charge for each cylinder (sequential injection). This option is not available with systems that rely on only one injection valve (single-point injection) or simultaneous activation of several injectors at once (group injection).

Calculating the ignition angle

The "reference ignition angle" is calculated based on the engine's current steady-state operating status. Its essential determinants are instantaneous cylinder charge, engine speed, and mixture composition (as indicated by the excess-air factor λ). The ignition angle is corrected to compensate for the particular operating conditions encountered during starting and in the warm-up phase. A simplified representation of the "reference ignition angle" in ME-Motronic would define it as the earliest potential ignition angle under any given operating conditions. Under standard operating conditions, with the engine warmed to its normal running temperature, this angle is defined by a minimum interval separating it from the knock threshold.
This reference ignition angle can then be further retarded by the knock control (to avoid combustion knock) and the crankshaft-synchronized torque-guidance output (to reduce torque).
The reference ignition angle is combined with the correction factors listed above to produce the so-called "basic ignition angle".
The actual ignition angle specified by the system reflects the addition of a supplementary correction factor designed to compensate for phase error in the engine-speed sensor.

Calculating the dwell angle

The purpose of the ignition system is to supply enough energy to ensure complete combustion of the air-fuel mixture at precisely the right instant. Energy availability is essentially defined by the dwell period for charging the primary circuit; the end of this period usually coincides with the firing point. The ECU specifies a dwell angle corresponding to the ignition coil's charge requirement. It activates the coil's primary current at the start of the dwell period and then interrupts it to initiate ignition at the firing point. This is how ME-Motronic controls distributorless ignition systems (DLI).
The system refers to a program map to determine the dwell angles for specific engine speeds and battery voltages, while its final output also includes a temperature correction.
The start of the dwell period is defined by the difference between the end of dwell and the dwell angle. The dwell angle is calculated from the dwell period using a time/angle conversion equation. The end of the dwell period is defined to coincide with the firing point (ignition timing).
The system basically has two options for defining the start and the end of the dwell period:
– As an angle,
– As a period of time.

When defined as an angle, the segment time is used to convert the dwell period into an angle. During dynamic variations in engine speed this produces a timing error, as the segment times used for calculating angle position are already outdated. Positive dynamic engine-speed changes (acceleration) lead to attenuated dwell periods, while negative dynamics (deceleration) produce extended dwell angles. Compensation for the dwell period reductions that accompany acceleration is provided by an injection advance, which must always be added to the basic duration. This dynamic injection advance declines as engine speed increases. In contrast, pronounced dynamic changes at low rpm can retard the dwell timing to such an extent that the dwell period becomes too brief for recharging the coil. The response is to transmit the dwell period termination point as a time function at low rpm. This ensures generation of adequate ignition energy regardless of dynamic fluctuations.

Operating conditions

The various engine operating conditions are primarily distinguished by variations in torque generation and engine speed. Figure 1 shows the different ranges.
Conditions characterized by high rates of dynamic change in rpm and load facor are especially significant, as they place special demands on the mixture-formation system (i.e., condensation and vaporization of the fuel film on the manifold walls). Other important elements are starting phase, and the subsequent transition phase that continues until engine and exhaust system warm to their normal operating temperatures.

Starting

Special calculations are employed to regulate charge control, injection and ignition timing for the duration of the starting process. At the outset of this process the air within the intake manifold is stationary, and the manifold's internal pressure reflects that of the surrounding atmosphere. This rules out using the manifold simulation model for throttle-valve control. Instead, throttle-valve position is specified as a fixed parameter based on starting temperature.
In an analogous process special "injection timing" is specified for the initial injection pulses.
The injected fuel quantity is augmented in accordance with engine temperature to promote formation of a fuel film on the walls of the intake manifold and the cylinders, thereby compensating for the engine's higher fuel requirement as it runs up to speed.
The system immediately starts to fade out the start-enrichment as soon as the engine turns over, until it is completely cancelled upon termination of the starting phase (600...700 min^{-1}).
The ignition angle is also adapted to the starting process. It is adjusted as a function of engine temperature, intake-air temperature, and engine speed.

Cylinder recognition

Before the first ignition spark can be generated, the system must reliably identify which cylinder is currently on its compression stroke. Ignition during the intake stroke could initiate backfiring through the manifold, leading to component damage.

Fig. 1

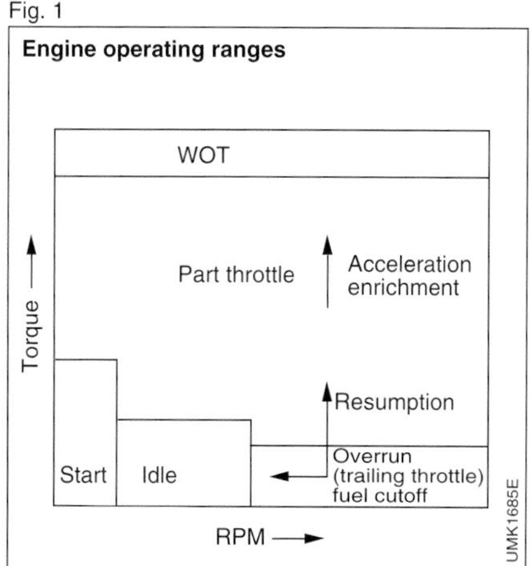

Signal patterns for ignition, crankshaft and camshaft on a 6-cylinder engine with standard sensor rotor

a Ignition-coil secondary voltage, **b** Crankshaft rpm-sensor signal, **c** Hall-sensor signal (standard sensor rotor) at camshaft.
1 Close, **2** Ignition.
A Ignition at Cylinder No. 1, **B** Ignition at Cylinder No. 5, **C** Ignition at Cylinder No. 3,
D Ignition at Cylinder No. 6, **E** Ignition at Cylinder No. 4.

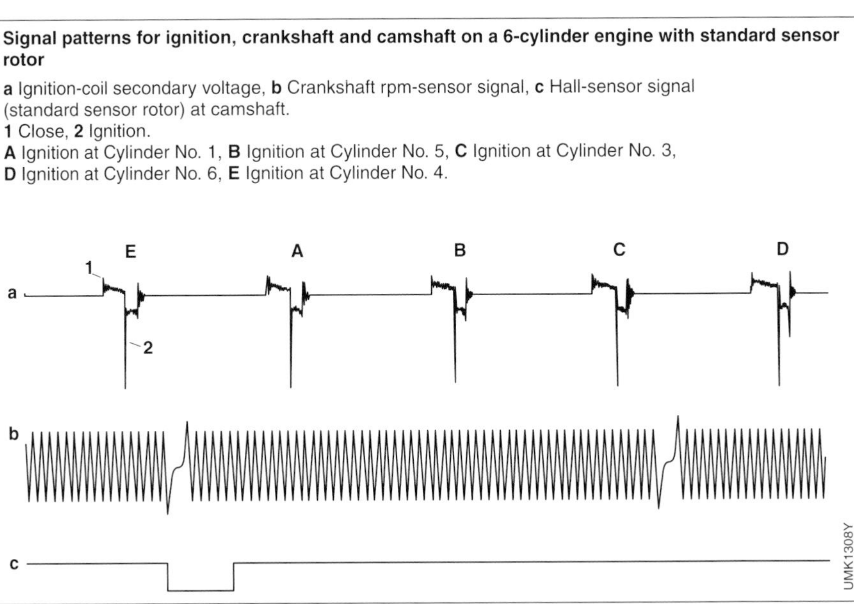

Fig. 2

The system determines cylinder phases by correlating the respective signals from crankshaft and camshaft sensors. The camshaft sensor rotor features at least one segment. The signal progression for one camshaft rotation is shown in Figure 2, which also shows the signal pattern generated by the crankshaft sensor as it scans through all 58 rotor teeth on each rotation. When the gap in the signal from this sensor (Figure 2, Curve b) coincides with the depression in curve c, this indicates that cylinder no. 1 is on its compression stroke, meaning that it is next in line to fire (Ignition A in Figure 2). The first ignition spark in the start phase cannot be triggered until the level of the camshaft signal at the first gap in the crankshaft-sensor signal has been evaluated; this is how the system determines which cylinder is next in the firing order (Figure 2: Cyl. 1 or 6).

Fig. 3

Rapid-start sensor rotor

Rapid start

The "rapid-start" mode reduces the time that elapses between starter engagement and cylinder recognition. This shortens overall starting times for enhanced convenience while also reducing the loads placed on starter and battery.

The rapid start relies on equipment such as a special "rapid-start" sensor rotor mounted on the camshaft (Figure 3). This sensor rotor generates a unique flank profile (Figure 4), enabling the system to recognize cylinder phases – and thus transmit the first ignition spark – before the first gap in the crankshaft signal is reached.

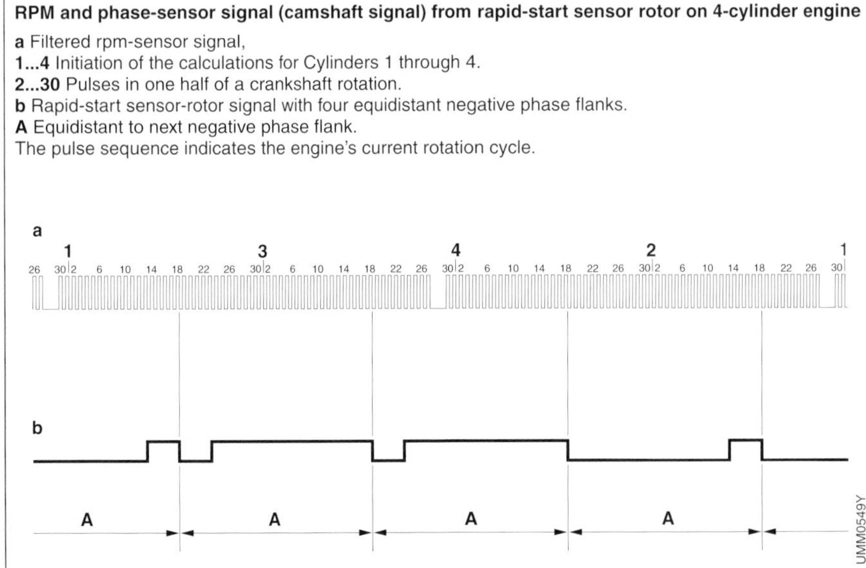

RPM and phase-sensor signal (camshaft signal) from rapid-start sensor rotor on 4-cylinder engine
a Filtered rpm-sensor signal,
1...4 Initiation of the calculations for Cylinders 1 through 4.
2...30 Pulses in one half of a crankshaft rotation.
b Rapid-start sensor-rotor signal with four equidistant negative phase flanks.
A Equidistant to next negative phase flank.
The pulse sequence indicates the engine's current rotation cycle.

Fig. 4

Post-start phase

The post-start phase (immediately following termination of the starting phase) is marked by further reductions in the charge densities and injected-fuel quantities employed for starting. System response in this phase is defined by the rise in engine temperature and the period that has elapsed since the starting phase ended.
Ignition angles are also adjusted to correspond to the revised injected-fuel quantities and different operating status. The post-start phase trails off in a smooth transition to the warm-up phase.

Warm-up and catalytic-converter heating

After starting at low engine temperatures, cylinder charge, injection and ignition are all adjusted to compensate for the engine's greater torque requirements; this process continues up to a suitable temperature threshold.
The prime concern in this phase is rapid warming of the catalytic converter, as quick transition to catalytic-converter operation permits drastic reductions in exhaust emissions. The strategy employs a portion of the exhaust gas for "cat-converter heating" during this phase, while accepting the resulting sacrifices in engine efficiency.
There are basically two concepts:
- Secondary air injection into a rich mixture with retarded ignition timing, and
- Lean warm-up with extremely retarded (late) ignition timing.

Both concepts entail using retarded ignition timing to operate the engine at a low level of efficiency. The initial results are higher exhaust-gas temperatures and reduced torque generation. The torque-based control automatically compensates for this loss by prescribing higher cylinder-charge densities. This produces a larger quantity of hot exhaust gas for use in heating the catalytic converter with minimal delay. The catalytic converter's rapid warm-up and the consequent early onset of operation furnish a substantial reduction in exhaust emissions.

Lean warm-up
The combination of lean warm-up with the extremely retarded ignition point leads to the post-oxidization of the

Effects of secondary-air injection on CO and HC emissions (6-cylinder engine, 2.8 litre, 145 kW).
1 Without secondary-air injection, **2** With secondary-air injection.
v Vehicle velocity.

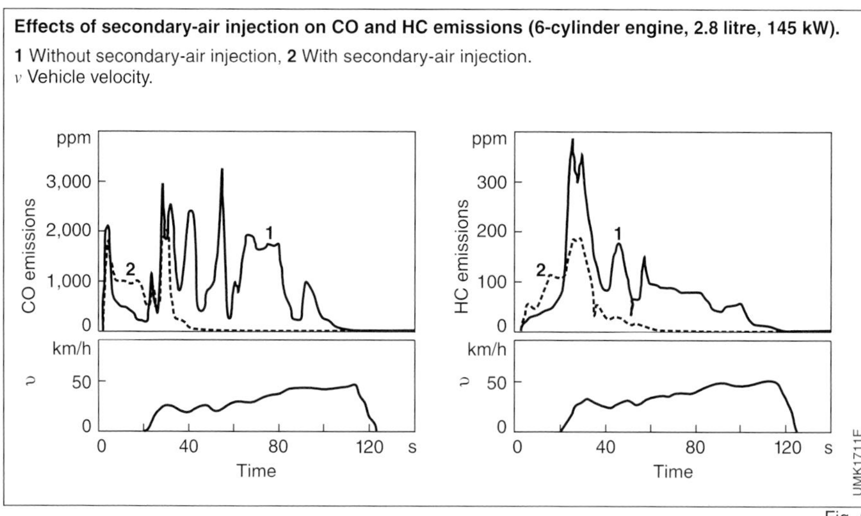

Fig. 5

unburnt hydrocarbons which result from inefficient combustion.

The term "lean warm-up" stems from the use of a slightly lean base mixture to supply the oxygen required to support this oxidation process.

Although this concept's advantage is the freedom to dispense with supplementary components, limits on its potential for heat generation mean that the catalytic converter must be installed close to the engine to minimize thermal losses.

Secondary-air injection

This concept expands on the low-efficiency strategy by operating the engine on high levels of excess fuel ($\lambda < 0.6$) to increase the carbon monoxide (CO) and hydrocarbon (HC) content of the exhaust gas. Fresh ("secondary" air which is not involved in the internal combustion process) is then injected directly downstream of the exhaust valves to support oxidation of CO and HC. This produces heat energy, which then flows to the catalytic converter, enabling it to reach operating temperature with minimal delay.

An electric vacuum pump draws the required secondary air from within the air-filter housing or through a special coarse filter. Injection into the exhaust system is then regulated by a deactivation valve and a check valve designed to prevent hot exhaust gases from flowing back into the secondary-air injection system. ME-Motronic triggers the secondary-air pump and air valve at the indicated intervals. A wide-band Lambda oxygen sensor facilitates precise diagnosis of the secondary-air pump.

This process produces enough heat for use with catalytic converters situated further away from the engine. Figure 5 shows the respective curves for hydrocarbon and carbon-monoxide emissions in the initial seconds of the emissions test with and without secondary-air injection.

Idle

The engine generates no torque at idle; the power generated in the combustion process being needed to sustain engine operation and to drive the ancillary devices. Under these conditions, the torque that the engine needs to remain in operation combines with the idle speed to define fuel consumption. Because a substantial portion of the fuel consumed by vehicles in heavy stop-and-go traffic is actually burned in this kind of use, it pays to hold friction losses during idling at the lowest possible level. This translates into specifying low idle speeds.

ME-Motronic's closed-loop idle control reliably maintains a stable idle at the defined level regardless of variations in operating conditions. These variations can stem from from factors such as fluctuating current draw in the electrical system, air-conditioner compressors, gear engagement on automatic-transmission vehicles, active power steering, etc.

WOT (full load)

At Wide-Open Throttle (WOT) there are no throttling losses, and the engine produces the maximum potential power available at any given rpm.

Transition response

Acceleration/deceleration

A portion of the fuel discharged into the intake manifold does not reach the cylinder in time for the subsequent combustion process. Instead, it forms a condensation layer along the walls of the intake manifold. The actual quantity of fuel stored in this film rises radically in response to higher load factors and extended injection durations.

A portion of the fuel injected when the throttle valve opens is absorbed for this film. As a result, a corresponding quantity of supplementary fuel must be injected to compensate and prevent the mixture from going lean under acceleration. Because the additional fuel retained in the wall film is released again once the load factor drops, injection durations must also be reduced by a corresponding increment during deceleration.

Figure 6 shows the corresponding curves for injection duration.

Fig. 6

Transitional injection duration

1 Injection signal from charge-density calculations, **2** Corrected injection duration, **3** Supplementary injected fuel quantity, **4** Reduction of injected fuel quantity, **5** Throttle valve angle α_{TV}.

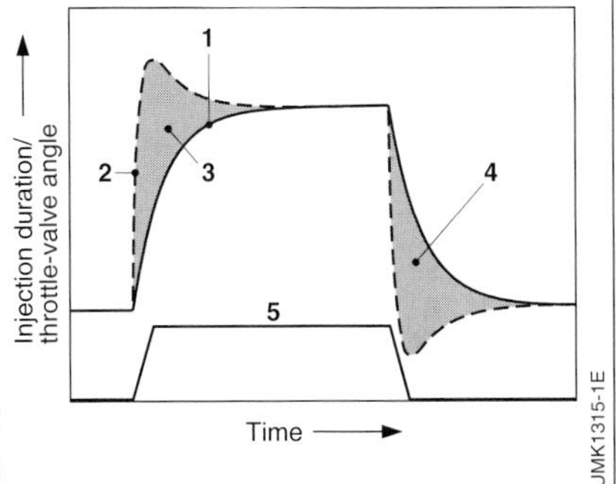

Overrun fuel cutoff/renewed fuel flow

Overrun, or trailing throttle, indicates a condition in which the power being provided by the engine at the flywheel is negative. Under these conditions the engine's friction and gas-flow losses can be exploited to slow the vehicle. The engine can continue to run with or without active fuel injection.

For passive, injectionless trailing-throttle operation the injection is deactivated to reduce fuel consumption and exhaust emissions. ME-Motronic's torque-based control can regulate suppression of the fuel-injection pulses to prevent radical torque jumps during the transition to trailing throttle by relying on gradual instead of abrupt reductions in specified output.

Injection resumes once rpm falls to a specified reactivation speed located at a point above idle. Actually, the ECU is programmed with a range of reactivation speeds. These vary to reflect changes in parameters such as engine temperature and dynamic variations in engine speed, and are calculated to prevent the rpm from falling below the defined minimum threshold.

Once injection resumes, the system starts by using the initial injection pulses to discharge supplementary fuel and rebuild the wall fuel layer. When fuel injection is resumed, the slow, controlled increase of engine torque by the torque-based control ensures that torque buildup is smooth (gentle transition).

Closed-loop idle-speed control

Purpose

The engine does not furnish torque at the flywheel during idling. To ensure consistent idling at the lowest possible level, the closed-loop idle-speed control system must maintain a balance between torque generation and the engine's "power consumption."
Power generation is needed at idle in order to satisfy load requirements from a number of quarters. These include internal friction at the engine's crankshaft and valve-train assemblies, as well as such ancillary equipment as the water pump.
The engine's internal friction losses are subject to substantial variation in response to temperature fluctuations, while friction also changes, albeit at a much slower rate, over the course of the engine's service life.
The load imposed by external factors (such as the a/c compressor) also fluctuates through a wide range as ancillaries are switched on and off. Modern engines are especially sensitive to these variations, owing to their lower reciprocating and flywheel masses as well as higher intake-manifold (storage) volumes.

Operating concept

ME-Motronic's torque-based concept relies on closed-loop idle-speed control to quantify the output needed to maintain the desired idle speed under any operating conditions. This output rises as engine speed decreases, and drops as it increases.
The system responds to recognition of new "interference factors" such as activation of the a/c compressor or engagement of a drive range in an automatic transmission by requesting more torque. Torque demand must also be increased at low engine temperatures to compensate for higher internal friction losses and/or maintain a higher idle speed.
The sum of all these output demands is relayed to the torque coordinator, which then proceeds to calculate the corresponding charge density, mixture composition and ignition timing.

Lambda closed-loop control

Post-treatment of exhaust gases in a 3-way catalytic converter represents an effective means of reducing concentrations of harmful exhaust pollutants. The converter can reduce hydrocarbons (HC), carbon monoxide (CO) and oxides of nitrogen (NO_X) by 98 % and more (Figure 1), converting them to water (H_2O), carbon dioxide (CO_2) and nitrogen (N_2). This level of efficiency is contingent on engine operation within a very narrow scatter range surrounding the stoichiometric air-fuel ratio of λ = 1.

Fig. 1

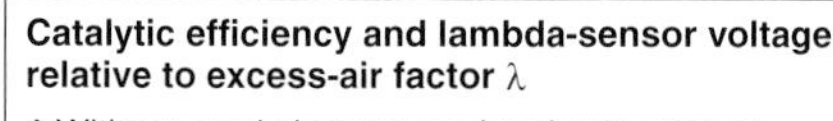

Catalytic efficiency and lambda-sensor voltage relative to excess-air factor λ

1 Without catalytic post-combustion treatment,
2 With catalytic post-combustion treatment,
3 Voltage curve with 2-threshold λ sensor.

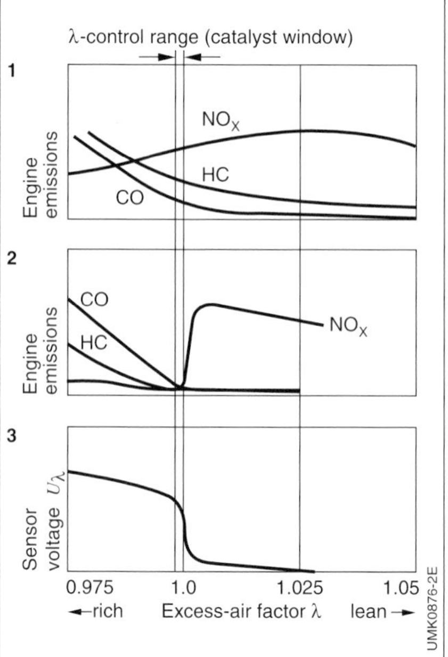

Two-state lambda closed-loop control

Control range

The “lambda window,” corresponding to the range available for effective simultaneous processing of all three “classical” exhaust-gas components, is extremely restricted. Closed-loop lambda control is needed to maintain operation within this window (λ = 0.99...1).

The two-state oxygen sensor monitors the exhaust stream's oxygen content from a position on the engine side of the catalytic converter. Lean mixtures ($\lambda > 1$) induce sensor voltages of approx. 100 mV, while rich mixtures ($\lambda < 1$) generate roughly 800 mV. At $\lambda = 1$ the sensor voltage suddenly jumps from one level to the other.

ME-Motronic includes this signal from the oxygen sensor in its calculations of injection duration. Figure 2 is a schematic portrayal of the circuit's configuration.

Fig. 2

Lambda closed-loop control: Operational schematic

1 Air-mass meter,
2 Engine,
3a Lambda oxygen sensor 1,
3b Lambda oxygen sensor 2 (with two-state control only),
4 Catalytic converter,
5 Injectors,
6 ECU.
U_S Sensor voltage,
U_V Injector control voltage.

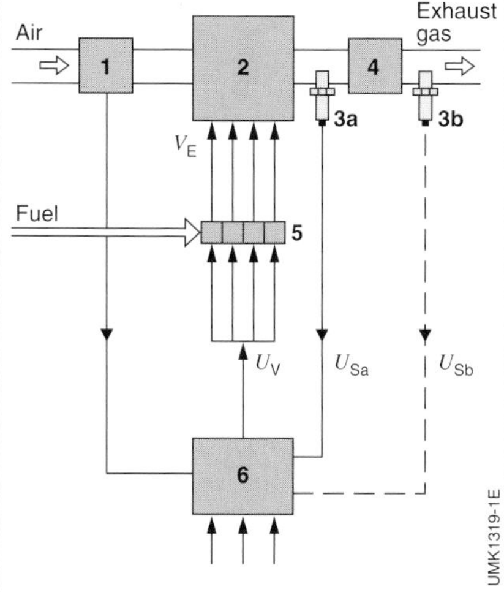

Operation

A closed-loop lambda control system can only function in tandem with a fully operational oxygen sensor. An auxiliary processing circuit monitors the sensor on a continuing basis.

A cold oxygen sensor or damaged circuitry (short or open circuits) will lead to implausible voltage signals, which the ECU will reject. Depending upon individual configuration and installation position, the heated oxygen sensors found today in most systems can assume operation after only 15 to 30 seconds.

Cold engines require a richer mixture ($\lambda < 1$) to run smoothly. This is why the closed-loop lambda control circuit is only released for active intervention once a defined temperature threshold has been reached.

Once the lambda control assumes operation, the ECU uses a comparator to convert the sensor signal into binary form.

The lambda closed-loop control reacts to incoming signals ($\lambda > 1$ = mixture too lean, or $\lambda < 1$ = mixture too rich) by modifying the control variables, generating a control factor for use as a multiplication factor when modifying the injection duration.

Injection duration is adjusted (lengthened or reduced) and the control factor reacts by settling into a state of constant oscillation (Figure 3).

Continuing oscillation in a range with λ = 1 as its focal point is the only way to achieve optimal lambda control with a dual-state system. The precision of the closed-loop control process depends upon the speed with which the control system can adjust the control factor to counteract shifts in the excess-air factor. While waiting fuel is constantly being discharged into the combustion chamber, the O_2 sensor is located elsewhere, further back in the exhaust system. The resulting gas-transit times translate into response lag within the control circuit, with the actual delay depending on the

engine's load factor and speed. The ultimate reaction to mixture adjustments can only be measured once the lag period has elapsed. This leads to a minimum phase duration (periodicity) for the cyclical revisions in the control factor. Processing times and the sensor's response delay increase lag even further. The duration of the oscillation periods is determined by the transit times of the gas, while the ramp climb maintains largely constant amplitudes throughout the engine's load and speed range, despite variations in gas-transit times.
Radical steps in control factor during mixture adjustments (sensor jump) accelerate the reaction process, making it possible to shorten the oscillation period.

Lambda shift

Because the sensor's response pattern varies depending upon the direction of the monitored mixture transition (viz., rich to lean, or lean to rich), a symmetrical control arrangement would produce the slightly lean exhaust mixture portrayed in Figure 3b. Because catalytic converter efficiency is optimal in the $\lambda = 0.99...1.0$ range, the control system must be able to counteract this tendency. An asymmetrical controller oscillation pattern can shift the mixture into the optimal conversion range.
The required asymmetry is obtained either by delaying the switch-over of the control factor after the voltage jump (from lean to rich) at the oxygen sensor, or with an asymmetrical step function. Maxima are limited to maintain the controller's dynamic response.

Fig. 3

Dynamic two-state lambda control with typical oscillation patterns

a Signal from two-state sensor (simplified),
b Oscillation pattern of control factor,
c Switching lag with limited maximum.

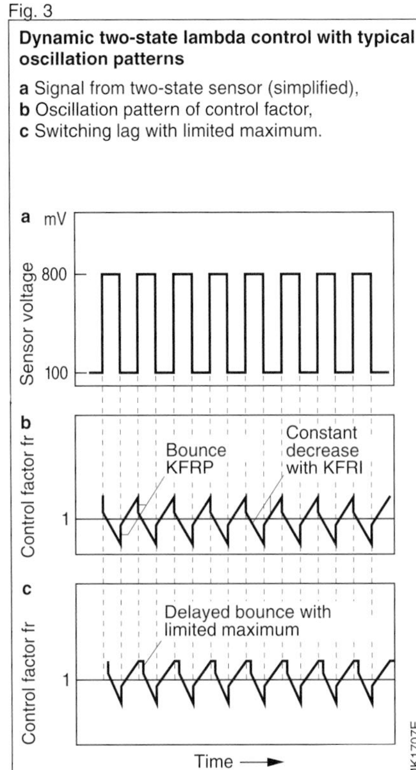

Adapting the Lambda pilot control to the Lambda closed-loop control

The lambda closed-loop control system corrects each consecutive injection event in the sequence based on previous monitoring data from the O_2 sensor.
As a result, a certain time shift arising from gas-transit times is unavoidable, and the approach to new operating points defined with maladjusted pilot control is characterized by deviations from $\lambda = 1$. This condition continues until the system's cyclical control can reestablish equilibrium.
Thus a special default (or reference) control mechanism is needed to maintain compliance with emissions limits. The pilot control is programmed when the system is adapted to the engine, and a corresponding lambda control map is stored in a ROM (program memory). However, subsequent revisions in the default control may be needed to compensate for the effects of drift factors during the vehicle's service life, including variations in the density and quality of the fuel.
If the lambda controller starts to consistently implement a single set of corrections during operation in a particular engine speed and load range, the pilot control's adaptation function will register this fact and respond by programming corresponding corrections into the system's non-volatile memory

(RAM or EEPROM with constant current supply). The corrected pilot control is then ready for immediate implementation at the next start, assuming duty until the lambda closed-loop control becomes active.
Interruptions in the power supply to the non-volatile memory are also registered; adaptation then recommences using neutral pilot-control values as a starting point.

Dual-sensor lambda closed-loop control

Installing the oxygen sensor at the back end of the catalytic converter ("cat-back" position) helps guard it against contaminants in the exhaust gas while also reducing the thermal stresses imposed on it. This type of auxiliary sensor can generate a second, overlapping control signal to augment the one from the main, ("cat-forward") sensor on the engine side and ensure stable air-fuel mixture composition over an extended period.
The superimposed control system modifies the asymmetry of the constant oscillation pattern that characterizes control mechanisms based solely on a cat-forward oxygen sensor; thus compensating for the lambda shift.
A lambda control strategy based exclusively on a sensor mounted behind the converter ("cat-back") would be handicapped by excessive control lag produced by extended gas-transit times.
While helping maintain the lambda control system's long-term operational stability, the second, "cat-back" sensor also can be employed as a tool for assessing the catalytic converter's effectiveness.

Continuous lambda closed-loop control

While the two-state sensor can only indicate two states – rich and lean – with a corresponding voltage jump, the wide-band sensor monitors deviations from $\lambda = 1$ by transmitting a continuous signal. In other words, this wide-band sensor makes it possible to implement lambda control strategies based on continuous instead of dual-state information.
The advantages are:

- A substantial improvement in dynamic response, with quantified data on deviations from the specified gas composition, and
- The option of adjusting to any values, i.e., also air factors other than $\lambda = 1$.

The second option is especially significant for strategies seeking to exploit the fuel-savings potential of lean operation (lean-burn concepts).

Evaporative-emissions control system

Source of fuel vapors

The fuel in the tank is warmed by:

- Heat radiated from external sources, and
- Excess fuel from the system return line, which is heated during its passage through the engine compartment.

This results in HC emissions which primarily emerge from the fuel tank in the form of vapor.

Limiting HC emissions

Evaporative emissions are limited by legal mandate. Limitation is by means of evaporative-emissions control systems equipped with an activated-charcoal filter (carbon canister) installed at the end of the tank's vent line. The activated charcoal in the canister binds the fuel vapors, allowing only air to escape into the atmosphere, while simultaneously providing the pressure-relief function. To support ongoing regeneration in the charcoal filter, an additional line leads from the canister to the intake manifold.
Vacuum, produced in the intake manifold whenever the engine is running, draws in a current of atmospheric air; this air flows through the charcoal on its way to the manifold. The air stream absorbs the fuel vapors stored in the activated charcoal and

Evaporative-emissions control system

1 Line from fuel tank to activated-charcoal canister, **2** Activated-charcoal canister, **3** Fresh air, **4** Canister-purge valve, **5** Line to intake manifold, **6** Throttle assembly with throttle valve.
Δp Difference between manifold pressure p_s and ambient barometric pressure p_u.

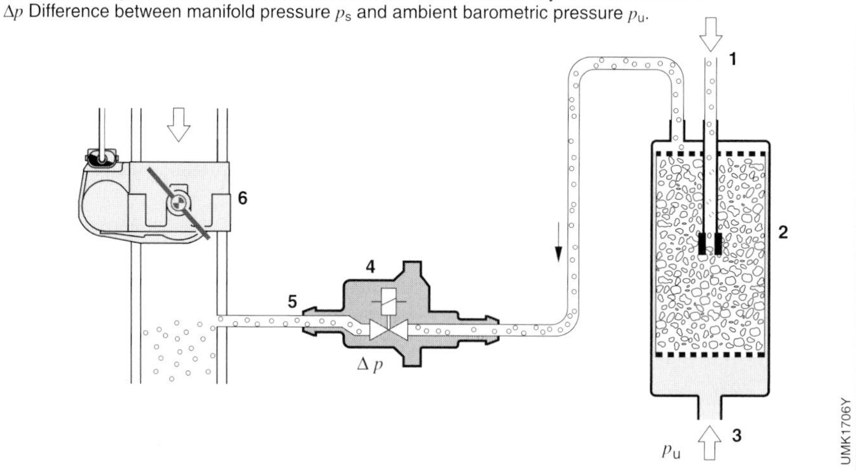

1
2
3
4
5
6
Δp
p_u
UMK1706Y

Fig. 1

takes them to the engine for combustion. A purge-valve installed in the line to the manifold meters this regenerative or "cleansing" flow (Figure 1).

Regeneration flow

The regenerative stream is an air-fuel mixture of necessarily unknown composition, as it can contain fresh air as well as substantial concentrations of gasoline extracted from the activated charcoal.

The regenerative flow thus represents a major interference factor for the lambda closed-loop control system. A regenerative flow representing 1 % of the intake stream and consisting solely of fresh air will lean out the overall intake mixture by 1 %. A flow with a substantial gasoline component on the other hand can enrichen the mixture by something in the order of 30 %, owing to the effects on the A/F ratio λ of fuel vapor with a stoichiometric factor of 14.7. In addition, the specific density of fuel vapor is twice that of air.

Canister-purge valve

The canister-purge valve ensures adequate ventilation of the carbon canister while holding lambda deviations to a minimum (Figure 2).

ECU control functions

The canister-purge valve closes at regular intervals in order to allow the mixture adaptation process to proceed without interference from the tank's ventilation flow.

During active regeneration the system selects the optimal purge quantity for

Fig. 2

Canister-purge valve

1 Hose fitting, **2** Seal flange, **3** Armature, **4** Spring, **5** Solenoid winding, **6** Solenoid core with flow channel, **7** Flow path.

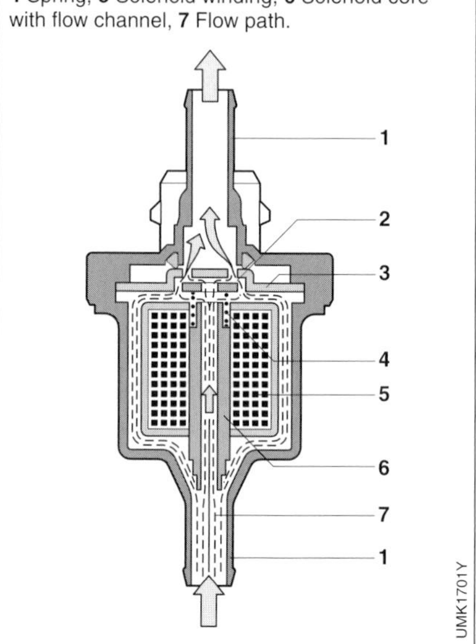

instantaneous operating conditions, with the ECU generating signals to open the valve in a ramp pattern. The purge stream absorbs a specified gaseous “fuel load,” defined using data from the previous regeneration cycle. At the same time, the system reduces injection durations to compensate for the anticipated fuel content in the purge stream. Because the mixture-adaptation function is a separate process, the system now interprets any deviations from lambda as changes in “fuel load” and responds with corrective adjustment to the initial specification.
For the “load-sensitive” control of this purge flow, ME-Motronic uses parameters familiar from the intake-manifold model, which define such factors as the manifold’s internal pressure and temperature. This facilitates precise calculation of the purge flow. The system is designed to operate with up to 40 % of the total fuel coming from the regeneration current.
With the lambda control system inactive, only a minimal regenerative current is allowed into the induction system, as until the lambda comes online there is no control mechanism capable of compensating for the mixture deviations that regeneration produces. In order to prevent unburned fuel vapors from entering the catalytic converter, the purge valve closes immediately in response to the interruption of fuel supply which occurs when the throttle is released (overrun fuel cutoff).

Knock control

Electronic control of ignition timing allows extremely precise adjustment of advance angles based on engine rpm, temperature and load factor.
Despite this precision, conventional systems must still operate with a substantial safety margin to avoid approaching the knock threshold. This margin is necessary to ensure that no cylinder will reach or go beyond the preignition limit, even when susceptibility is increased by risk factors such as engine tolerances, aging, environmental conditions and fuel quality. The engine design which results when these factors are taken into consideration features a lower compression ratio with retarded ignition which lead to sacrifices in fuel consumption and torque.
These disadvantages can be avoided by using a knock-control system. Experience confirms that such a system allows higher compression ratios, with considerable improvements in both fuel economy and torque. With this system, it is no longer necessary to specify pilot ignition-timing angles defined to reflect worst-case scenaria. Instead, ideal conditions (engine compression at tolerance threshold, maximum fuel quality, cylinder least prone to preignition) can serve as the basis for specifying ignition timing. This makes it possible for each cylinder to be operated at the preignition limit, which coincides with optimal efficiency, in virtually all ranges, and throughout the life of the engine.
The essential prerequisite for this kind of knock-control system is reliable detection of any and all preignition exceeding a specified intensity. This must embrace

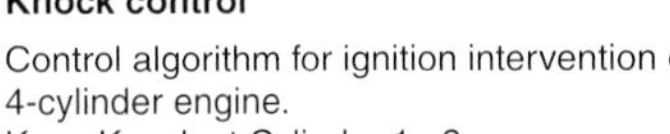

Knock control
Control algorithm for ignition intervention on a 4-cylinder engine.
$K_{1...3}$ Knock at Cylinder 1...3 (no knock on Cyl. no. 4).
a Adjusted for retarded ignition timing,
b Increment for advanced ignition timing,
c Advanced ignition timing.

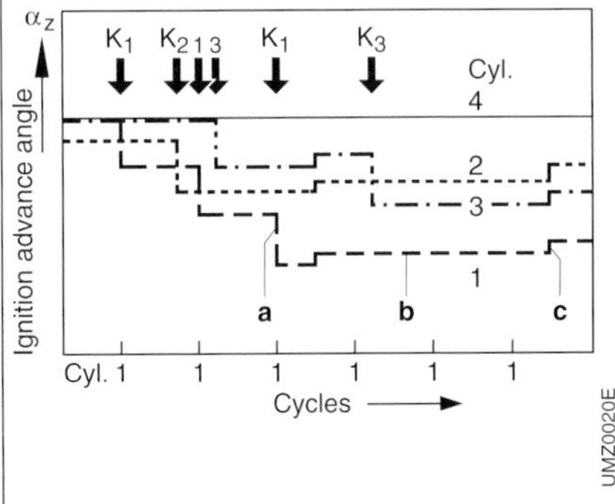

Fig. 1

every cylinder and extend throughout the engine's entire operating range.
Preignition is detected by sensors designed to register solid-borne sonic waves. Installed at one or several suitable points on the engine, these knock sensors detect the characteristic oscillation patterns produced by knock and transform them into electrical signals suitable for transmission to the Motronic ECU for subsequent processing (refer to the section on ignition for additional information). The ECU employs a special processing algorithm to detect incipient preignition in every combustion cycle and in every cylinder. Detection of knock triggers a specified, programmed reduction in ignition advance. When the knock danger subsides, the ignition for the affected cylinder is then gradually advanced back toward the pilot ignition-timing angle.
The knock-recognition and knock-control algorithms are designed to prevent the kind of preignition that results in audible knock and engine damage (Figure 1).

Adaptation

Real-world engine operation is characterized by different knock limits in different cylinders, and ignition timing must be adjusted accordingly. In order to adapt the pilot-ignition timing to reflect the individual knock limits under varying operating conditions, individual ignition-retard increments are stored for each cylinder.
These data for specific engine speeds and load factors are stored in non-volatile program maps in permanently-powered RAMs. This strategy permits the engine to be operated at maximum efficiency under all conditions without any danger of audible combustion knock, even during abrupt changes in load and rpm.
The engine can even be approved to run on low-octane fuels. Standard practice is to adapt the engine to run on premium fuel.
Operation with regular-grade gasoline can also be approved.

Boost-pressure control

Boost-pressure control mechanisms that rely on pneumatically-triggered mechanical layouts use actuators (wastegates) that are directly exposed to the pressure in the the impeller outlet. This concept allows only very limited definition of torque response as a function of engine speed. Load control is limited to the full-load bypass. There is no provision for compensation of the full-load boost

Fig. 1

Actuator for electronic boost-pressure control

1 Cycle valve.
p_2 Boost pressure,
p_D Pressure on diaphragm,
TVM Cycle valve triggering signal from ECU,
V_T Flow volume through turbine,
V_{WG} Flow volume through wastegate.

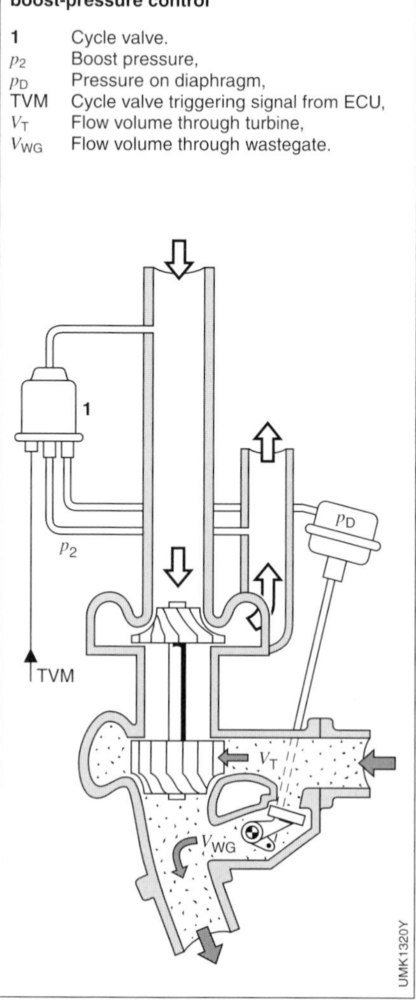

tolerances, and at part load, the closed wastegate impairs operating efficiency. Acceleration from low rpm can be marked by a delay in turbocharger response (a very pronounced "turbo lag"). These problems can be avoided with electronic boost-pressure control (Figure 1). This system can provide reductions in specific fuel consumption under some part-throttle operating conditions, controlling the wastegate's opening pattern to obtain the following results:

- The engine's back-pressure losses and the impeller's output both drop,
- Pressure and temperature at the impeller's discharge orifice fall, and
- The pressure gradient at the throttle valve is reduced.

The exhaust-gas turbocharger and its boost-control device must be precisely matched to the engine as the primary requirements for achieving these improvements.

The affected components in the boost-control device are:

- The electropneumatic cycle valve,
- The effective diaphragm surface, stroke and spring in the aneroid capsule, and
- The cross section of the valve head or valve flap in the wastegate.

ME-Motronic employs electronic boost control to regulate induction pressure to the specified value. This specified boost pressure is converted into a specification for the desired maximum cylinder charge. The torque-based control function converts this specification into a setpoint for throttle-valve aperture and a pulse-duty factor for the wastegate. The signal modifies the wastegate's control pressure and stroke to regulate the bypass opening.

Control-circuit elements compensate for the difference between the setpoint defined by current operating conditions (program map) and the actual, monitored boost pressure. The calculated value at the controller output is then included in the process used to define the maximum cylinder charge.

On turbocharged engines, the temperature of the exhaust gas between engine and turbine should not exceed certain limits. This is why Motronic's boost control operates exclusively in conjunction with knock control, as the latter represents the only means for operating the engine with maximum ignition advance throughout its service life. A result of using optimal ignition timing at all operating coordinates is extremely low exhaust-gas temperatures. Further reductions in exhaust temperature are available through intervention in cylinder charge, meaning boost pressure in this case, and/or air-fuel mixture.

Protective functions

Limiting vehicle and engine speed

Extremely high engine speeds can lead to powerplant demolition (valve train, pistons). The rpm limiting function prevents the maximum approved engine speed (redline) from being exceeded.

Incorporation of a vehicle-speed limiter may be necessary in response to specific equipment specifications as defined for vehicles in certain markets (i.e., tires, suspension). In addition, several German manufacturers have made a voluntary commitment to limit the maximum speeds of their vehicles to 250 km/h.

The functions for restricting vehicle and engine speeds operate according to the same principles. A control agorithm reduces the permitted engine output once a specified threshold is crossed. This output limit is included in ME-Motronic's torque-based control function.

Torque and power limits

It is sometimes necessary to restrict torque generation in order to reduce the loading on certain drivetrain components (such as the transmission). ME-Motronic's torque-based control function provides for the definition of such a limit. It is also possible to restrict ultimate output by governing engine speed and torque.

Limiting exhaust-gas temperatures

High exhaust-gas temperatures can damage exhaust-system components. Therefore, a model incorporated within the ECU is employed to simulate these temperatures. Extreme requirements for monitoring precision can be satisfied by installing a temperature sensor. Temperatures beyond a defined threshold trigger mixture enrichment, which cools the exhaust by extracting heat energy to vaporize the fuel. Limiting charge density and torque are additional options.

Vehicle immobilizer

To prevent unauthorized vehicle use, the Motronic ECU incorporates a feature that prevents the engine from being started until the ECU itself has been released via a special control line. The actual release mechanism is an encoded signal prepared by an external control unit. This second control unit verifies user authorization by analyzing the signal from a transmitter in the ignition key or a keypad entry code, etc.

Fig. 1

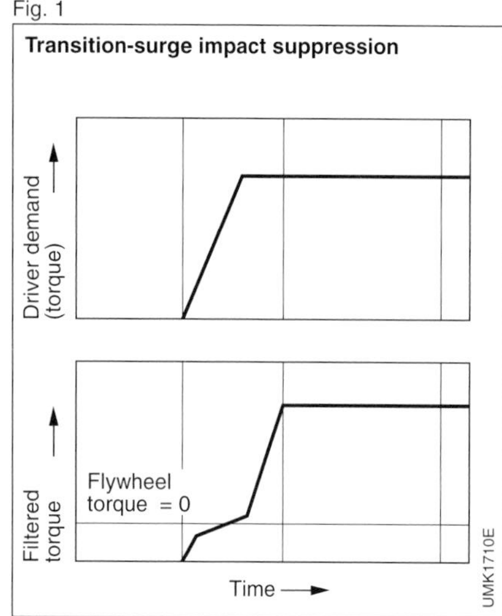

Improved drivability

Transition surge-impact suppression

Positive and negative load shifts – initiated by abruptly depressing or releasing the accelerator pedal – can produce jolts in the driveline. This effect is especially pronounced when the torque reversal transfers forces to mounting bushings or the transmission. An example is the engine, which shifts from one engine mount to the other during transitions from power-on to power-off.
This force transfer can be prevented, or at least reduced in intensity, by controlling the rates of torque rise and reduction in order to achieve gentler transitions. In order to adjust flywheel torque (Fig. 1) this strategy relies on manipulation of ignition timing and cylinder charge.

Surge-damping function

The fact that the engine and drivetrain represent a spring-mass system means that during operation this system can start to oscillate. The surge-damping function detects these oscillations and suppresses them by intervening in engine output torque in the respective phase. Oscillation recognition is based on a comparison between a reference rpm derived from driver demand and the current rpm. Intervention is through adjustments to ignition timing. Effective suppression of driveline oscillations entails implementing the torque intervention at opposed phases to the torque oscillation.

Cruise Control

The cruise control's function is to compensate for changes in rolling/aerodynamic resistance and maintain a constant vehicle speed without the driver

having to press the accelerator pedal. In addition to maintaining current vehicular velocity, these systems provide a range of supplementary functions. Cruise control thus enhances driving convenience on long trips while also facilitating compliance with posted speed limits.
Because ME-Motronic's throttle-valve actuator is already integrated in the ETC function, the supplementary effort required to integrate cruise control within the system is minimal.

Functions:

The driver uses a stalk/switch to control the following functions:
- Adopt and then maintain the current road-speed (set),
- Accelerate and then maintain predefined road-speed,
- Decelerate and then maintain predefined road-speed,
- Accelerate to a stored target speed (resume),
- Graduated incremental increase in defined road-speed (tip up),
- Graduated incremental decrease in defined road-speed (tip down),
- Cruise control deactivation at main switch and/or tip-off switch.

Control elements

Improved drivability

The driver can operate the cruise-control system with a single control element which incorporates switches for the functions:
- Set,
- Resume,
- Accelerate, and
- Decelerate.

Depending upon its particular configuration, an individual switch may govern more than one function. Thus one button may be used for set/decelerate and one for resume/accelerate.
In such a case, the function to be implemented by the system when the switch is activated will depend upon the system's current status and how long pressure is applied to the switch. The tip-up and tip-down functions are triggered by brief pressure on the switch for acceleration and deceleration. In addition to the switch components for specific functions, the control element may also include an optional main switch and a cruise-control cancelation switch. Where present, the main switch must be switched on before entries at the function switch will be registered. When the main switch is deactivated, any speeds stored previously will be lost. Active cruise control can also be interrupted by applying pressure to the brake or clutch pedal.

Fig. 1
Cruise-control operating stalk

Integrated diagnosis

Diagnostic procedure

An "On-Board Diagnosis" (OBD) system is standard equipment with Motronic. This integral diagnostic unit monitors ECU commands and system responses while also checking sensor signals for plausibility. This test program proceeds continually during normal vehicle operation.
The ECU stores recognized errors together with the operating conditions under which they arose. When the vehicle is serviced, a tester can then be used to read out and display the stored error data through a standardized interface. This information facilitates fault diagnosis procedures for service personnel.
Diagnosis processes extending far beyond those in earlier systems have been developed to comply with mandates issued by the California Air Resources Board. All components whose failure could cause a substantial increase in harmful emissions must be monitored, and detected faults must trigger a diagnosis lamp in the instrument panel. This expanded diagnosis is designated OBD II.
An OBD system adapted for European conditions is designated EOBD.

Diagnosed areas

Air-mass meter

The process for monitoring operation of the air-mass meter is an example of Motronic's self-diagnosis function. While cylinder charge is being determined based on the mass of the induction air, a supplementary reference is calculated at the same time from throttle-valve angle and engine speed. If the ECU detects excessive variation between the two, its initial response is to record the error. As vehicle operation continues, plausibility checks determine which of the sensors is defective. The ECU does not store the designated error code until it has unequivocally determined which sensor is at fault.

ETC throttle-valve actuator

Because engine output is manipulated by adapting cylinder charge, the throttle-valve actuator must satisfy stringent demands for reliability and diagnosis. The actuator monitors current throttle-valve position with two, mutually counterrotational potentiometers. The signals these produce are then compared. If a deviation occurs, the system falls back on the intake-manifold model as a basis for extrapolating throttle-valve position and restoring signal plausibility.

Combustion miss

Combustion miss, resulting from such factors as worn spark plugs, allows unburned mixture to enter the catalytic converter. This mixture can destroy the catalyst, and is also detrimental to the environment. Because even an isolated combustion miss produces higher emissions, the system must be able to detect it.

Fig. 1

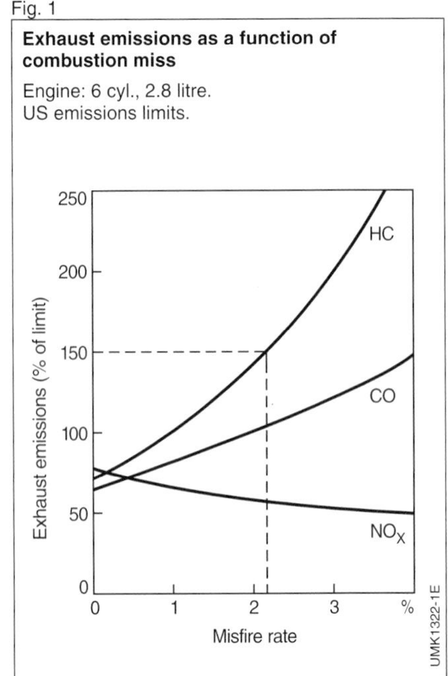

Figure 1 shows the effects of combustion miss on emissions of hydrocarbons (HC), carbon monoxide (CO) and oxides of nitrogen (NO_X).

In the search for a method to monitor combustion miss, the monitoring of inconsistencies in the crankshaft's rotational speed has emerged as the best compromise between complexity and practicability. Combustion miss is accompanied by a shortfall in torque equal to the increment that would normally have been produced in the cycle. The result is a reduction in rotation speed. At high speeds and low load factors the intervals separating firing points (periodicity) will be extended by only 0.2 %. This means that rotation must be monitored with extreme precision, while extensive computations are also required to distinguish combustion miss from other interference factors.

Catalytic converter

Yet another diagnostic function monitors the efficiency of the catalytic converter. This process relies on a second oxygen sensor mounted downstream of the converter as a supplement to the first, cat-forward unit. A healthy catalytic converter will store oxygen, thus attenuating the lambda control oscillations. As the catalyst ages, this effect deteriorates until finally the signal patterns from the two sensors start to converge. Comparison of the signals from the two O_2 sensors thus serves as the basis for assessing the catalytic converter's condition. A warning lamp alerts the driver when a defect is detected.

Lambda-Sonde

Two-state sensor (Nernst probe)

A precisely stoichiometric air-fuel mixture is vital for optimal operation of the catalytic converter.

The lambda closed-loop control system uses the signals from the oxygen sensors to maintain this mixture.

Monitoring dynamic response of lambda oxygen sensors

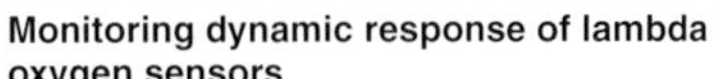

a New sensor,
b Aged Type II sensor,
c Aged Type III sensor,
T Phase duration.

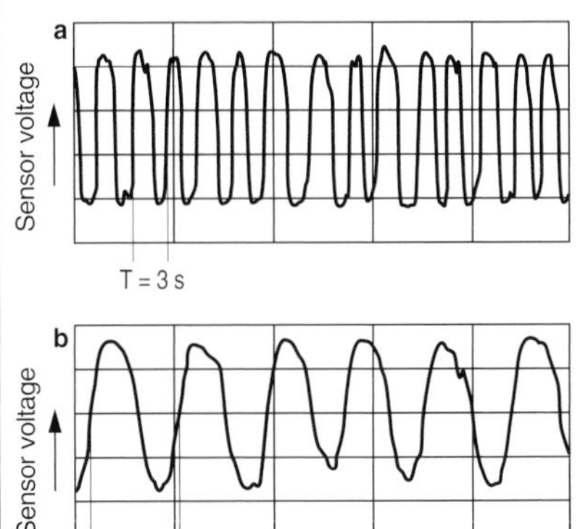

Fig. 2

The oxygen sensors are diagnosed for:

– Electrical plausibility: The system continuously assesses the sensor signal's plausibility. It reacts to implausible signals by deactivating other lambda-related control functions while simultaneously entering the corresponding error code in the fault memory.

– Sensor dynamic response (phase duration, Figure 2): An oxygen sensor exposed to excessively high temperatures for an extended period of time may start to react more slowly to changes in the air-fuel mixture. This leads to extended phase durations (periodicity) in the two-state control pattern. A diagnosis function monitors this control frequency and triggers a diagnosis lamp to alert the driver to excessive response lag in the sensor.

– Control range: The presence of two oxygen sensors in each exhaust line

makes it possible to use the post-catalyst ("cat-back") sensor to check the engine-side ("cat-forward") sensor for drift in its effective response range.
– Heater: The system checks current and voltage to the oxygen sensor's heater resistor. For this check the Motronic ECU must rely on a direct link to the heater resistor, instead of controlling it through a relay.

LSU wide-band sensor
With the introduction of the LSU wide-band oxygen sensor it is now possible to monitor specified mixtures other than $\lambda = 1$. The ongoing lambda control process uses a "cat-forward" control circuit with an LSU, complimented by a superimposed "cat-back" circuit featuring a two-state sensor. This strategy makes it possible to assess the LSU's operation with the aid of the two-state sensor. The diagnosis process includes the following elements:

– Electrical plausibility: One feature that distinguishes the LSU sensor from its two-state counterpart is that the potential spectrum of plausible signals extends throughout the entire voltage range. In addition to checking upper and lower limit values, the system also compares its signal with that being transmitted from the sensor mounted downstream of the catalytic converter.
– Sensor dynamic response: Diagnosis based on assessment of a mandatory, superimposed amplitude.
– Control range: The second, "cat-back" sensor is used to verify compliance with a delta lambda threshold.

Fig. 3

Reference leak test for detecting fuel-system leakage

1 Throttle valve, **2** Engine, **3** ECU, **4** Canister-purge valve, **5** Activated-charcoal canister, **6** Diagnosis module, **7** Reference leak, **8** Circuit-control valve, **9** Electric air pump, **10** Filter, **11** Fresh air, **12** Fuel tank, 13 Leak.

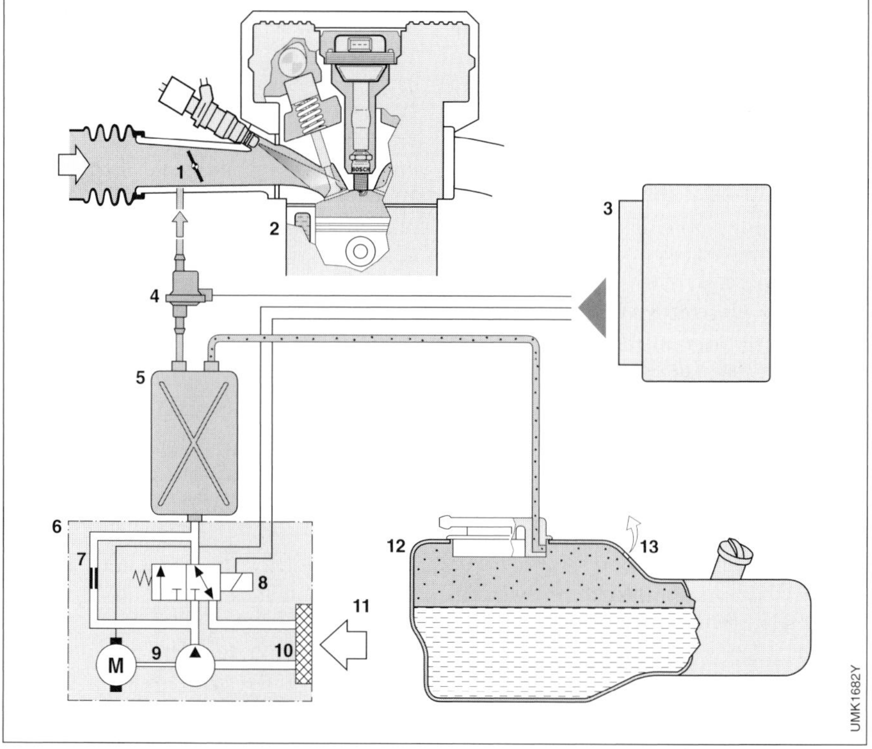

– Heater: This process is the same one used with the two-state sensor, while ongoing variations in the lambda signal are also evaluated.

Fuel supply
When the air-fuel ratio deviates from stoichiometric for extended periods of time, the system recognizes this state, with the mixture adaptation serving as one reference. If the deviations exceed specific programmed limits, this indicates that a fuel-system or fuel-metering component has shifted outside its specified tolerance range. An example would be a faulty pressure regulator or cylinder-charge sensor (for instance, the hot-film air-mass meter), while other potential error sources include leaks at the intake manifold or in the exhaust system.

Tank system
Emissions emanating from the exhaust system are not the sole element of ecological concern; vapors from the fuel tank are also a problem.
While current requirements for the European market are limited to a relatively simple check to verify correct operation of the purge valve, US mandates already demand a means of detecting leakage at any and all points within the evaporative-emissions control system.

Vacuum test
The diagnosis process relies on analysis under vacuum. A shut-off valve is used to block the fresh-air supply to the carbon canister and seal off the vapor retention system. Then, preferably with the engine running at idle, the purge valve is opened and vacuum from the intake manifold propagates throughout the system. A tank-mounted pressure sensor monitors the subsequent pressure curve to determine whether leakage is present.

Reference leak procedure
Yet another process for diagnosing fuel tank leakage (Figure 3) relies on an electric air pump (9) to pressurize the tank (12). Instead of monitoring pressure with a pressure sensor, this test uses the pump's current draw as its test parameter. The first step is calibration, based on simulating a reference leak (7) with a defined flow at the purge valve. Then a circuit-control valve (8) is used to link the pump with the activated-charcoal canister (5). The resulting current pattern will point to any leakage present in the fuel system (Figure 4).

Secondary-air injection
The secondary-air injection activated following cold starts must also be monitored, as its failure would also effect emissions. This can be done using the signals from the lambda oxygen sensor while the secondary-air injection is in operation, or the injection can be activated and observed at idle using a special lambda test function.

Exhaust-gas recirculation
Because exhaust-gas recirculation (EGR) permits reductions in the concentrations of nitrous oxides in the exhaust gas, the operational integrity of this system must also be monitored.
Opening the EGR valve conducts part of the exhaust-gas flow back into the intake manifold. As this supplementary flow of residual gases enters the manifold and

Fig. 4
Schematic portrayal of pump current curve during pressurization test on fuel system

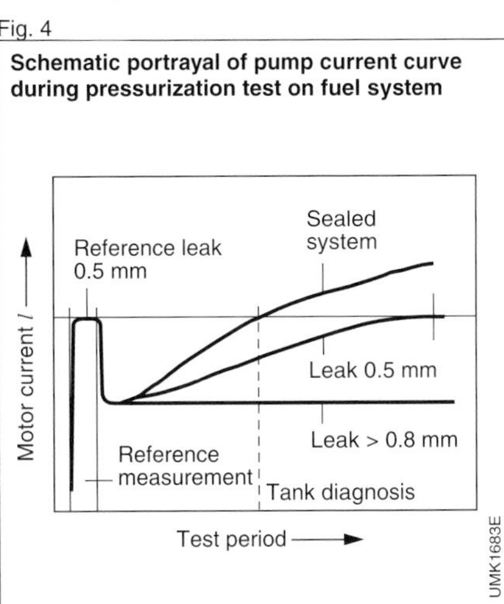

then the cylinders, it initially affects manifold vacuum, and then, subsequently, the combustion process. This characteristic opens two options for diagnosing the EGR system:

Diagnosis based on manifold vacuum
The EGR valve is briefly closed under part-throttle operation. If ETC is used to hold the flow of induction air passing through the hot-film air-mass meter at a constant level, the valve will shift the intake-manifold vacuum. This change is monitored with the manifold pressure sensor, and its magnitude is an index of the EGR system's condition.

Diagnosis based on idling stability
This method is employed on systems without hot-film air-mass meter or supplementary intake-manifold pressure sensor. At idle, the EGR valve is opened slightly. The increased mass of residual gas leads to rougher engine operation. This, in turn, is detected by the system's smooth-running monitoring system which applies the rough running to diagnose the EGR system.

Other monitoring functions
While the new emissions statutes focus on the engine-management system, other systems (such as the electronic transmission-shift control) are also monitored. These relay data on any faults to the engine-management ECU, which then assumes responsibility for triggering the diagnosis lamp.
Greater system complexity and increasingly stringent environmental legislation are making diagnosis increasingly important.

Diagnosis sequence scheduling
OBD II demands completion of all diagnosis functions at least once in the course of the emissions test cycle. The former scheduling concept triggered the individual diagnosis functions according to an invariable scheme; the pattern was dictated by the operating conditions that characterized the individual stages of the emissions test cycle. Under normal operating conditions, this strategy could lead to extended delays before the right operating conditions for initiating a specific diagnosis process were encountered. In practice this means that some diagnostic procedures might be performed only rarely, as individual driving habits could render it impossible for the system to adhere to the scheduled diagnosis sequence.
Diagnosis System Management can respond to operating conditions and initiate dynamic adjustments in the diagnostic sequence to allow optimal diagnosis processing in daily operation.

Fault memory
Detection of emissions-relevant malfunctions leads to an entry in the non-volatile fault memory. While this entry reflects the officially mandated error codes, it also includes a "freeze frame" consisting of supplementary information on the operating conditions under which the error occurs (i.e., engine speed and temperature). Depending on the system project, malfunctions of significance for vehicle servicing are also stored, although not included in the OBD II catalog.
Error codes can be read out by connecting the client's own service tester or a Bosch Motortester (Figure 5) to the ECU interface. The same equipment can be used to record test data (such as engine speed). OBD II prescribes standardized protocols for storing malfunction data as defined by the SAE (Society of Automotive Engineers). This makes it possible to access the fault memory using standard, commercially-available "scan tools."

Emergency (limp-home) mode

In the interval elapsing between initial occurrence of a fault and the subsequent vehicle workshop visit, ignition timing and air-fuel mixtures can be processed based on default values and emergency-running functions. This allows continued vehicle operation, albeit with sacrifices in comfort and convenience. The ECU responds to recognized errors in an input line by substituting data from a simulation model or by falling back on a redundant signal.

Failure in units on the output side initiate implementation of specific backup measures corresponding to the individual problem. As an example, the system can react to a defect in the ignition circuit by suppressing fuel injection at the affected cylinder to prevent damage to the catalytic converter.

The ETC throttle-valve actuator has a spring-loaded default position. This maintains engine operation at low rpm, allowing ME systems to continue in restricted operation in the event of failure in the vital ETC circuit.

Actuator diagnosis

During normal driving, many Motronic functions (e.g., EGR) are operational only under specific conditions. This renders it impossible to activate all actuators (such as the EGR valve) for operational checks while on the road.

Actuator diagnosis represents a unique case within the range of diagnostic processes. It proceeds with the engine off, and never during normal operation. This test mode relies on the engine tester, which serves as the triggering device for operational checks in the service workshop. The actual test process entails activation of each actuator in sequence. Operation can then be verified using acoustic or other means.

In this mode the injectors should only be triggered with extremely brief pulses (< 1 ms). Although this is not enough time for the injector to open completely, and no fuel is injected into the manifold, the sound can be clearly heard.

Fig. 5
Bosch KTS 500 Motortester in use

ECU

Purpose

This ECU (Figure 1) serves as the "processing and control center" for the engine-management system. It employs stored functions and algorithms (processing programs) to process the incoming signals from the sensors. These signals serve as the basis for calculating the control signals to the actuators (such as ignition coil and injectors) which it controls directly through its driver stages (3).

Physical design

The ECU has a metal housing containing a printed circuit board (2) with electronic componentry. Compact units which feature hybrid technology are available for installation directly on the engine. They are specially designed to withstand higher thermal stresses.
A multi-terminal plug connects the ECU to its sensors and actuators, as well as to its power supply. The number of individual terminals within this interface varies according to the number of functions covered by the ECU. More than 100 terminals must usually be included in the ME-Motronic connector plugs.
The PCB has a metal base underneath the output amplifier circuitry, where through-hole contacts provide good thermal transfer to its lower side. From here the heat generated by the output amplifier circuits is transferred through heat bridges to the housing.

Environmental conditions

The ECU must withstand extreme temperatures, humidity and physical stresses. Resistance to incoming electromagnetic interference and the ability to suppress outward-bound radiation of high-frequency static must also be of a high order.
Under normal operating conditions the ECU must be capable of errorless signal processing at environmental temperatures ranging from −30 °C to +60 °C, at with battery voltages that extend from 6 V (during starting) to 15 V.

Power supply

A voltage regulator (10) provides the ECU with the constant 5 V operating power needed for the digital circuitry.

Signal entry

The sensor signals enter the ECU via protective circuits, and through signal converters and amplifiers where indicated:
– Analog-digital (A/D) converters integrated in the microprocesors (4, 7) transform analog input signals (e.g., information on accelerator-pedal position, induction air mass, engine and intake-air temperature, battery voltage, air-fuel mixture ratio, etc.) into digitalized data.
– Digital input signals (such as switching signals from the air conditioner or transmission selector lever, but also including digital signals such as rpm pulses from Hall sensors) are suitable for direct processing in the microprocessor.
– Incoming signals in pulse form transmitted by inductive sensors furnish information on vehicle speed as well as on crankshaft speed and angle. These are processed in a special circuit (10) and converted into square-wave signals. Depending upon the system's integration level, initial signal processing may be partially or completely performed in the sensor itself. Input information arriving through the data bus (CAN) also arrives without need for preliminary processing.

Signal processing

Input signals are handled by the microprocessor within the ECU. In order to function, this microprocessor must be equipped with a signal-processing program stored in a non-volatile memory (ROM or EPROM, 5). This memory also contains the specific individual performance curves and program maps (data) used to govern the engine-management system.
Owing to the large number of engine and vehicle variations with their extensive

individual ranges of data requirements, the ECUs are not programmed until they reach the end of the production line. There is no need to open up the ECU for this procedure. This reduces the number of ECU configurations required by any single vehicle manufacturer.
A read/write memory (RAM) is needed for storing calculated values and adaptation factors along with any system errors that may be detected (diagnosis). RAMs rely on an uninterrupted power supply for continued operation, and lose all data if the vehicle's battery is disconnected. The ECU must then recalculate the adaptation factors after the battery is reconnected. To get around this problem, some units store required variables in an EEPROM (6, non-volatile memory) instead of a RAM.

Signal output
The microprocessor-controlled output amplifier (driver) circuits supply sufficient power for direct operation of the actuators. These driver circuits are protected against shorts to ground, fluctuations in battery voltage and the electrical overloads that could destroy them. The OBD diagnosis function is able to recognizes errors in the driver stages, and reacts by deactivating the respective output (where necessary). The error entry is stored in the RAM. The service technician can then access the error information by connecting a tester to the serial interface.
Another protective circuit operates independently of the ECU to deactivate the electric fuel pump when engine speed falls below a specified level. When the Terminal 15 power supply is interrupted at the ignition switch ("ignition off"), some ECUs rely on a holding circuit to keep the main relay open until program processing can be completed.

Fig. 1
ME7 Electronic control unit

1 Multi-terminal plug connection,
2 Printed circuit board,
3 Driver stages,
4 Microprocessor with ROM (function processor),
5 Flash EPROM (supplementary memory with program for specific vehicle),
6 EEPROM,
7 Microprocessor with ROM (expansion processor),
8 Flash EPROM (program memory for expansion processor),
9 Barometric-pressure sensor,
10 CJ910 peripheral chip (integrated 5V voltage supply and inductive sensor evaluation circuit).

RAM is on the underside of the PCB and thus invisible in this figure.

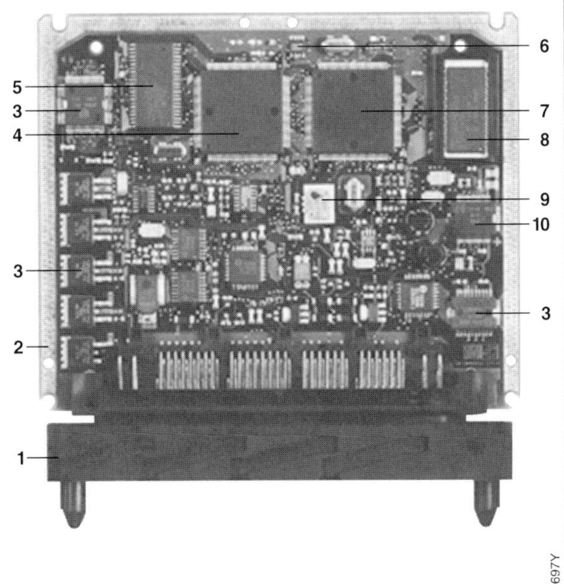

Interfaces to other systems

System overview

Increasingly widespread application of electronic control systems for automotive functions such as
- Electronic engine management (Motronic),
- Electronic transmission-shift control,
- Electronic vehicle immobilizers,
- Antilock braking systems (ABS),
- Traction control system (TCS), and
- On-board computers

has made it vital to interconnect the individual control circuits by means of networks. Data transfer between the various control systems reduces the number of sensors while also promoting exploitation of the performance potential in the individual systems.

The interfaces can be divided into two categories:
- Conventional interfaces, with binary signals (switch inputs), pulse-duty factors (pulse-width modulated signals), and
- Serial data transmission, e.g., Controller Area Network (CAN).

Conventional interfaces

In conventional automotive data-communications systems each signal is assigned to a single line. Binary signals can only be transmitted as one of two conditions: "1" or "0" (binary code). An example would be the a/c compressor, which can be "on" or "off."
Pulse-duty factors can be employed to relay more detailed data, such as throttle-valve aperture.
Increasing data traffic between various on-board electronic components means that conventional interfaces are no longer capable of providing satisfactory performance. The complexity of current wiring harnesses is already difficult to manage, and the requirements for data communications between ECUs are on the rise (Figure 1).

Serial data transmission (CAN)

These problems can be solved with a CAN, that is, a bus system (bus bar) specially designed for automotive applications.
Provided that the ECU's are equipped with a serial CAN interface, CAN can be used to relay the signals from the sources listed above.

There are three basic applications for CAN in motor vehicles:
- To link ECU's,
- Body-related and convenience electronics (multiplex), and
- Mobile communications.

The following is limited to a description of communications between ECUs.

ECU networking

This strategy links electronic systems such as Motronic, electronic transmission-shift control, etc. Typical transmission rates lie between approximately 125 kBit/s and 1 MBit/s, and must be high enough to maintain the required real-time response. One of the advantages that distinguishes serial data transfer from conventional interfaces (pulse-duty factors, switching and analog signals,

Fig. 1

Conventional data transmission

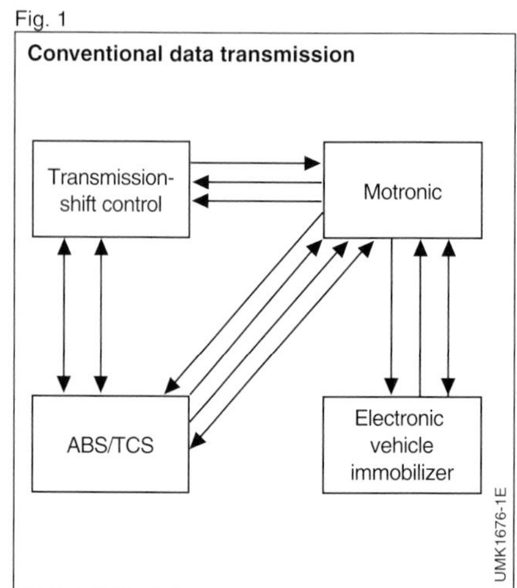

etc.) is the high speeds achieved without placing major burdens on the central processing units (CPU's).

Bus configuration

CAN works on the "multiple master" principle. This concept combines several ECU's with equal priority ratings in a linear bus structure (Figure 2).
The advantage of this structure is that failure of one subscriber will not affect access for the others. The probability of total failure is thus substantially lower than with other logical configurations (such as loop or star structures).
With loop or star architecture, failure in one of the subscribers or the central ECU will provoke total system failure.

Content-keyed addressing

The CAN bus system addresses data according to content. Each message is assigned a permanent eleven-bit identifier tag indicating the contents of the message (e.g., engine speed). Each station processes only the data for which identifiers are stored in its acceptance list (acceptance check). This means that CAN does not need station addresses to transmit data, and the interfaces do not need to administer system configuration.

Fig. 2

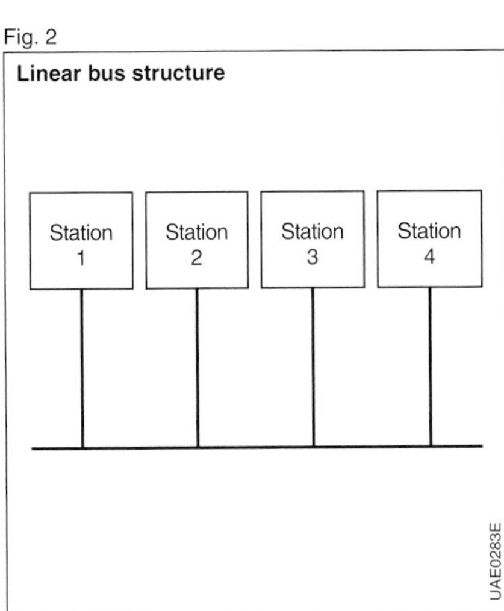

Bus arbitration

Each station can begin transmitting its highest priority message as soon as the bus is unoccupied.
If several stations initiate transmission simultaneously, the resulting bus-access conflict is resolved using a "wired-and" arbitration arrangement. This concept grants first access to the message with the highest priority rating, with no loss of either time or data bits.
When a station loses the arbitration, it automatically reverts to standby status and repeats the transmission attempt as soon as the bus indicates that it is free.

Message format

A data frame of less than 130 bits in length is created for transmissions to the bus. This ensures that the queue time until the next – possibly extremely urgent – data transmission is held to a minimum. The data frames consist of seven consecutive fields.

Standardization

The International Organization for Standardisation (ISO) has recognized a CAN standard for use in automotive applications with data rates of over 125 kBit/s, and along with two other protocols for data rates of up to 125 kBit/s.

MED-Motronic engine management

Overview

On gasoline (spark ignition) engines, gasoline direct injection not only reduces fuel consumption by as much as 20 % compared to conventional manifold injection, but it also makes a lasting contribution to reducing the CO_2 emissions.
In order for direct injection to function at all, it is necessary that a precisely defined change-over takes place between the so-called stratified charge at part load and operation with a homogeneous A/F mixture at full load (WOT).
Up to now, the following problems were among those instrumental in preventing the implementation of the above form of injection:
- The limitation of engine power during stratified-charge operation, and
- The lack of any possibility of applying catalytic NO_X aftertreatment during lean-burn operation.

Technical advances in engine management and catalytic-converter engineering helped to overcome these problems. This means that gasoline direct injection has a good chance of finding widespread use in the spark-ignition (SI) engines of the future.

Overall system Motronic MED7

Assignment

Due to its pronounced flexibility, the MED7 Motronic engine-management system permits precise control of modern SI engines equipped with gasoline direct injection. The wide range of different actuating variables involved in this system makes very high demands on the application engineering.
Essentially, the demands made upon the engine-management system are as follows:
- Precise metering of the required injected fuel quantity,
- Generation of the required injection pressure,
- Definition of the correct injection point,
- Precise, direct injection of the gasoline into the engine's combustion chambers.

The engine-management system must also coordinate the wide variety of different torque demands made on the engine so that it can then implement the necessary control actions at the engine.

The engine's indicated[1]) torque is an important system interface. The torque-control structure is subdivided into three functional areas (Fig. 1):
- Torque demand,
- Torque coordination, and
- Torque implementation.

The most important demand for torque is the one that comes from the driver's input at the accelerator pedal. The engine management interprets the accelerator pedal's setting as the driver's wish for a given torque to be developed by the engine.
Further demands for torque though can originate from the transmission-shift control, the traction control system (TCS), and the electronic stability program (ESP).

[1]) The indicated torque is the torque that the SI engine can actually deliver.

Torque coordination is implemented centrally in the engine management. This method has the following advantages:

- It is no longer necessary for an exchange of information to take place between functions which each require a torque which is suitable for their momentary status,
- It is impossible for the individual functions to interfere with each other at the manipulated-variable level,
- The individual functions have a clearly defined interface,
- The functional structure can be expanded without difficulty,
- The adaptation of functions to the engine is simplified by the lack of cross-coupling between the individual functions.

The Bosch MED7 Motronic is based on the ME7 Motronic manifold injection system. This Motronic system's innovative torque architecture will play a major part in the success of gasoline direct injection.

Design and construction

Figure 2 shows the complete direct-injection system with the most important components of the MED7 Motronic. The high-pressure injection system is designed as an accumulator injection system. In other words, the fuel can be injected directly into the cylinders at any time by means of the electromagnetically controlled high-pressure injectors.
Compared to the basic ECU for the ME7 Motronic, an additional driver stage has been included with the MED7 ECU for triggering the pressure-control valve.
The inducted air quantity can be freely adjusted by the electronically-controlled throttle valve (ETC electronic throttle control). A hot-wire air-mass meter is used for precision measurement of the intake air. The correctness of the A/F mixture is monitored by the universal LSU and LSF Lambda oxygen sensors installed in the exhaust-gas flow upstream and downstream of the catalytic converter. These serve for the closed-loop control of $\lambda = 1$ operation, as well as for the control of lean-burn operation and for catalyst regeneration. It is extremely important, particularly during dynamic operation, that the

Fig. 1

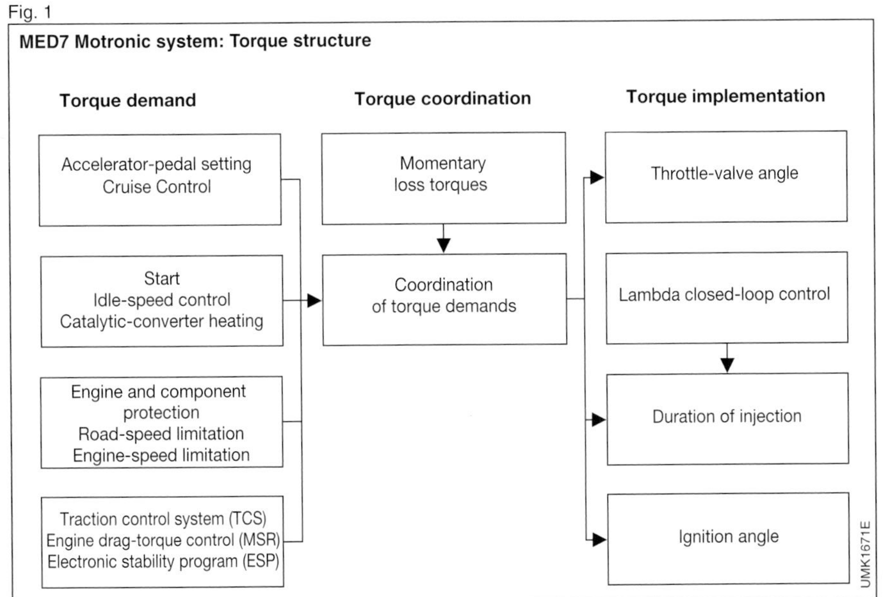

exhaust-gas recirculation (EGR) rate is precisely adjusted. This necessitates installation of a pressure sensor to measure intake-manifold pressure.

Method of operation

Fuel supply and fuel injection

The demands made on the fuel injection system are best complied with using an accumulator-type fuel-injection system. These demands include:

- Free selection of the injection point,
- Variable primary pressure.

With this system, at any desired moment in time, it is possible to inject the fuel stored under pressure in the fuel rail directly into the combustion chamber by means of an electromagnetic injector.

Low-pressure circuit

The low-pressure circuit is located in the fuel-tank side of the system. It comprises the electric fuel pump together with the parallel-connected pressure regulator and generates a pressure of 0.35 Mpa (3.5 bar). This is fed to the engine-driven high-pressure pump.

High-pressure circuit

High-pressure pump. This pump has the following assignments:

- Increase the fuel pressure from 0.35 Mpa (3.5 bar) to 12 Mpa (120 bar),
- Keep the pressure fluctuations in the fuel rail to a minimum, and
- Prevent the mixing of the fuel with the engine lube oil.

Accumulator/Fuel rail

On the one hand, the accumulator/fuel rail must have enough elasticity to handle the pressure pulsations resulting from the periodic withdrawals of fuel coupled with the delivery-flow pulsations from the high-pressure pump.

On the other, it must be so rigid that the rail pressure can react quickly enough to the engine's fuel requirements. The rail pressure is measured by a pressure sensor.

For the most part, the fuel rail's elasticity is selected as a function of the fuel compressibility and fuel-rail volume. The fuel-rail is tubular-shaped and man-

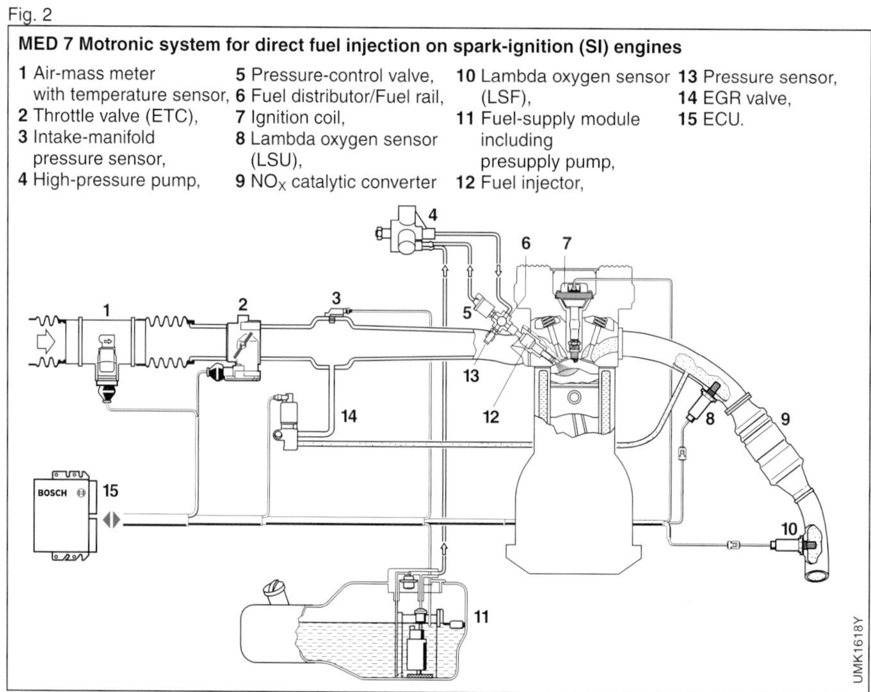

Fig. 2

MED 7 Motronic system for direct fuel injection on spark-ignition (SI) engines

1 Air-mass meter with temperature sensor, **2** Throttle valve (ETC), **3** Intake-manifold pressure sensor, **4** High-pressure pump, **5** Pressure-control valve, **6** Fuel distributor/Fuel rail, **7** Ignition coil, **8** Lambda oxygen sensor (LSU), **9** NO_X catalytic converter **10** Lambda oxygen sensor (LSF), **11** Fuel-supply module including presupply pump, **12** Fuel injector, **13** Pressure sensor, **14** EGR valve, **15** ECU.

ufactured from aluminum. It has connections for the fuel injectors, the pressure-control valve, the high-pressure pump, and for the relevant sensor technology.

Pressure sensor
This pressure sensor registers the pressure in the rail. A welded-in stainless-steel diaphragm, with the measuring resistors attached using thin-film techniques, is used as the sensor element.

Pressure-control valve
It is the pressure-control valve's job to adjust the primary pressure throughout the engine's complete operating range in accordance with the map stipulations. Here, the primary pressure is independent of the injected fuel quantity and the pump delivery quantity.
The excess fuel downstream of the pressure-control valve is a function of the load state. It is not returned to the fuel tank, but instead to the intake of the high-pressure pump. This avoids the fuel in the tank heating up and overloading the tank's canister-purge system.

Fuel injector
The fuel injector is responsible for shaping the injection jet, and must satisfy severe demands regarding installation conditions, short injection periods, and high linearity.
The injectors of the gasoline direct-injection system are connected directly to the fuel rail. Start of injection and injected-fuel quantity are defined by the injection-valve triggering signals.

A/F mixture formation and ignition
Highly complex engine management is required in order to take full advantage of the possibilities inherent in the gasoline direct-injection principle for high fuel economy and high engine power. Here, a difference is made between two basic operating conditions:

Lower load range
In the lower load range, in order to keep fuel consumption to a minimum, the engine is operated with highly stratified cylinder charge and a high level of excess air. By delaying the injection of fuel until just before the ignition point, an ideal status is aimed at in which the combustion chamber is divided into two zones: The 1st zone is a highly combustible cloud of A/F mixture at the spark plug, which is embedded in the 2nd zone, an insulating layer of air and residual exhaust gas. This enables pumping losses to be avoided, and the practically unthrottled operation of the engine.
Furthermore, since heat losses at the combustion-chamber walls are avoided, this leads to an increase in thermodynamic efficiency.
During statified-charge operation, air charge and ignition angle have practically no effect upon indicated engine torque which is practically proportional to the injected fuel quantity.
During everyday operation of the vehicle therefore, all these points add up to expected fuel savings of approx. 20% compared to manifold injection. Furthermore, a high EGR rate is aimed at in order to reduce the NO_X emissions.

Upper load range
Along with increasing engine load, the cloud of stratified-charge A/F mixture becomes increasingly richer, a fact which would lead to deterioration of the exhaust-gas figures especially with regard to soot emissions. In the upper load ranges, therefore, the engine is operated with a homogeneous cylinder charge. To a great extent, homogeneous operation can be taken over from the ME7 manifold injection. The Lambda coordination is responsible for Lambda control between $\lambda = 1$ and lean-burn operation.
To promote the efficient mixing of air and fuel, the fuel is injected already during the induction process. Similar to present-day manifold injection, the inducted air quantity is adjusted by the throttle valve in accordance with the driver's torque

requirements. The injected fuel quantity is calculated from the air mass, and corrected by the Lambda closed-loop control.

Load-range transitions

There are two quite distinct stipulations which must be fulfilled by the engine-management system if the two types of engine operation described above are to be successful (Fig. 1):

- In accordance with the operating point, the injection point must be adjustable between late (retarded) during the compression phase, and early (advanced) during the induction phase,
- In order to permit unthrottled engine operation in the lower load range, and throttle control in the upper load range, the adjustment of the intake air mass must be decoupled from the accelerator-pedal position.

During transitions between homogeneous and stratified-charge operation, it is decisive that injected fuel quantity, air charge, and ignition angle are controlled so that the engine torque input to the gearbox remains constant. The torque structure means that here too the essential functions of the electronic throttle control can be taken over directly from the ME7.

The throttle valve must be closed (Fig. 3) before the actual transition from stratified-charge to homogeneous operation.

When the intake-manifold pressure drops so does the Lambda value. During transition, two Lambda limits are decisive:

- In stratified-charge operation, in order to avoid soot Lambda has a lower limit of approx $\lambda = 1.5$,
- In homogeneous operation, due to the engine's limited lean-burn capabilities, an upper Lambda limit of approx. $\lambda = 1.3$ applies.

At the transition point, therefore, this necessitates "jumping" across the forbidden λ range $1.3 < \lambda < 1.5$ by temporarily increasing the injected fuel quantity. At the same time, to prevent an attendant torque jump, torque is reduced by briefly retarding the ignition angle.

The sequence for the transition from homogeneous to stratified-charge operation takes place in the reverse order.

Fig. 3

Transition stratified-charge/homogeneous operation

a Stratified-charge operation, **b** Homogeneous operation

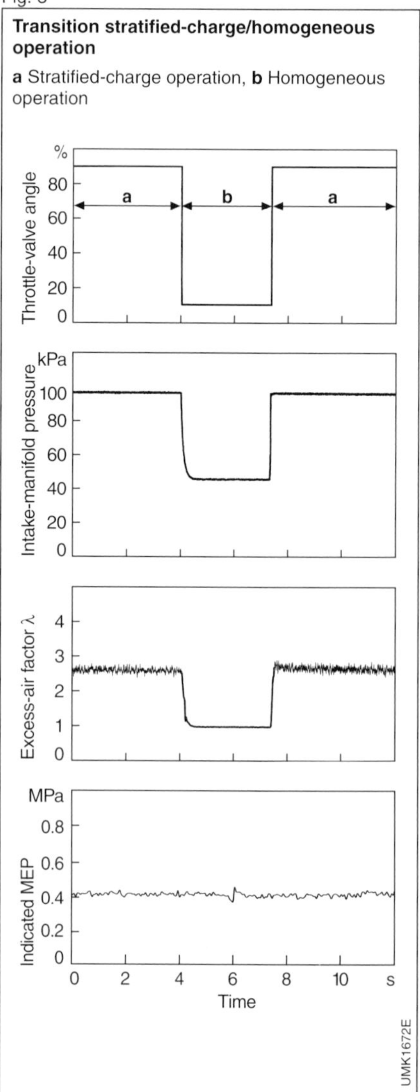

Exhaust-gas aftertreatment Catalytic-converter control

Legislation

In future, even more stringent exhaust-gas limits will apply in Europe. The engine management system is at present designed to comply with the European legislation Stage III, and will incorporate the scheduled European On-Board Diagnosis (OBD). Ongoing developments will lead to compliance with Stage IV.

Fig. 4

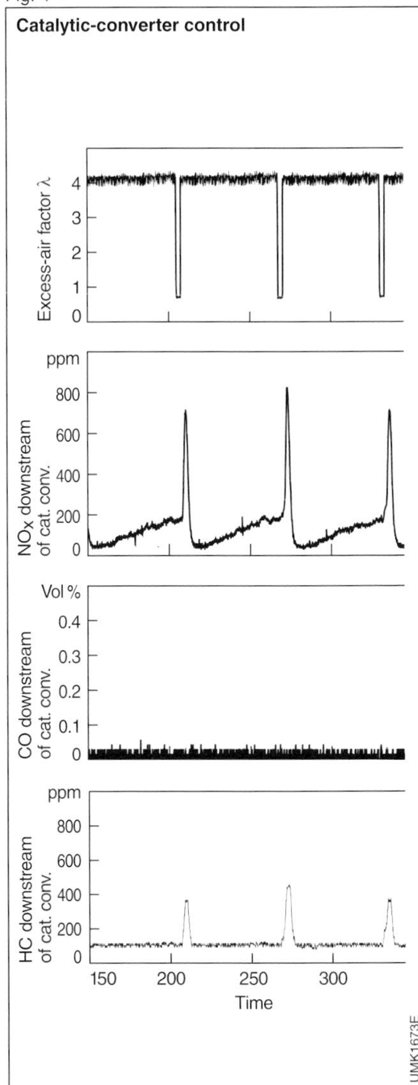

The key to series-production application of the direct-injection lean-burn engine lies in the development of catalytic-converter engineering for NO_X treatment in the lean-burn area. Here though, it will be necessary to considerably lower the sulphur content of gasoline before such catalytic converters can come into use.

Exhaust-gas recirculation

One very important aspect of gasoline direct injection is the fact that during stratified-charge operation the NO_X components in the very lean exhaust gas cannot be reduced by a three-way catalytic converter (TWC). An approx. 70% reduction of the exhaust-gas NO_X is achieved by a high rate of exhaust-gas recirculation (EGR). Compliance with the emissions-control legislation is impossible without some form of NO_X aftertreatment.

NO_X accumulator-type catalytic converter

The most promising potential for reducing the NO_X content in the exhaust gas is shown by the NO_X accumulator-type catalytic converter. Making use of the oxygen in the lean exhaust gas, it is able to store the Nitrogen oxides on its surface in the form of nitrates. As soon as the storage capacity is exhausted though, the catalytic converter must be regenerated. This is done by briefly switching over to rich homogeneous operation whereby, in combination particularly with the CO, the nitrates are reduced to nitrogen (Fig. 4). A catalytic-converter model which defines the converter's absorption and desorption characteristics is used in the control of the accumulation and regeneration phases. The exhaust-gas figures are monitored by Lambda oxygen sensors upstream and downstream of the catalytic converter.

During the cyclic switch over to rich, homogeneous operation, which only lasts a few seconds, it is particularly important that the switch over takes place without affecting vehicle handling, in other words without torque-output jumps from the engine.

Index

A

B

C

D

E

F

G

H

I

J

K

L

M

N

O

P

R

S

T

U

V

W

Comprehensive information made easy

Bosch Technical Books

Automotive electrics and electronics

Vehicle electrical systems, Symbols and circuit diagrams, EMC/interference-suppression, Batteries, Alternators, Starting systems, Lighting technology, Comfort and convenience systems, Diesel-engine management systems, Gasoline-engine management systems.

Hard cover,
Format: 17 x 24 cm,
3rd updated edition,
314 pages, bound,
with numerous illustrations.
ISBN 0-7680-0508-6

Gasoline-engine management

Combustion in the gasoline (SI) engine, Exhaust-gas control, Gasoline-engine management, Gasoline fuel-injection system (Jetronic), Ignition, Spark plugs, Engine-management systems (Motronic).

Hard cover,
Format: 17 x 24 cm,
1st edition,
370 pages, bound,
with numerous illustrations.
ISBN 0-7680-0510-8

Diesel-engine management

Combustion in the diesel engine, Mixture formation, Exhaust-gas control, In-line fuel-injection pumps, Axial piston and radial-piston distributor pumps, Distributor injection pumps, Common Rail (CR) accumulator injection systems, Single-plunger fuel-injection pumps, Start-assist systems.

Hard cover,
Format: 17 x 24 cm,
2nd updated and expanded edition,
306 pages, bound,
with numerous illustrations.
ISBN 0-7680-0509-4

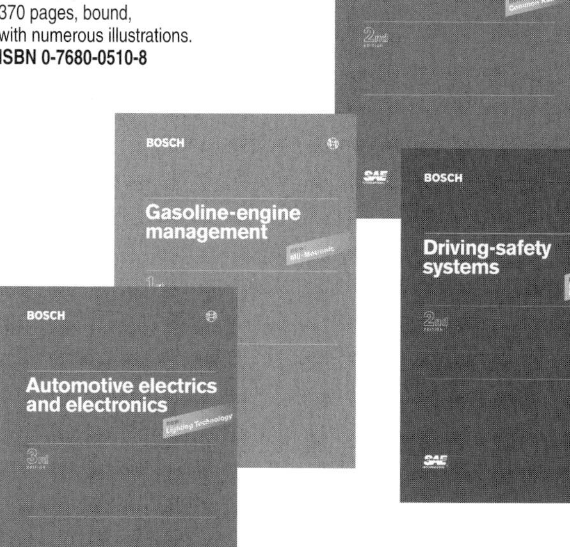

Driving-safety systems

Driving safety in the vehicle, Basics of driving physics, Braking-system basics, Braking systems for passenger cars, ABS and TCS for passenger cars, Commercial vehicles – basic concepts, systems and schematic diagrams, Compressed-air equipment for commercial vehicles, ABS, TCS, EBS for commercial vehicles, Brake testing, Electronic stability program (ESP).

Hard cover,
Format: 17 x 24 cm,
2nd updated and expanded edition,
248 pages, bound,
with numerous illustrations.
ISBN 0-7680-0511-6

Automotive terminology

4,700 technical terms from automotive technology, in German, English and French, assembled from the above Bosch Technical Books: "Automotive Electrics and Electronics", "Diesel-engine management"; "Gasoline-engine management" and "Driving-safety systems".

Hard cover,
Format: 17 x 24 cm,
1st edition
378 pages, bound.
ISBN 0-7680-0338-5

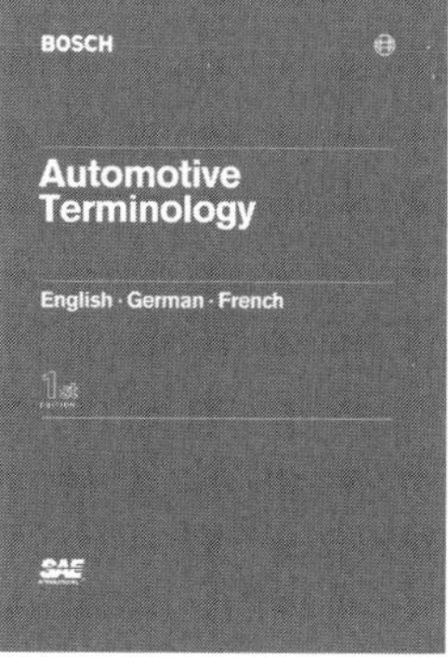